D1640994

Symmetry Theory in Molecular Physics with Mathematica

William Martin McClain

Symmetry Theory in Molecular Physics with Mathematica

A new kind of tutorial book

William Martin McClain
Department of Chemistry
Wayne State University
5101 Cass Avenue
Detroit MI 48202
USA
wmm@chem.wayne.edu

ISBN 978-0-387-73469-9 e-ISBN 978-0-387-73470-5
DOI 10.1007/b13137
Springer Dordrecht Heidelberg London New York

Library of Congress Control Number: 2009933284

Printed on acid-free paper

Springer is part of Springer Science+Business Media (www.springer.com)

Preface

Different people have different attitudes toward *Mathematica*. Some very gifted people find its step-by-step pace to be an impediment to scientific thought. If you are one of those talented people, this book is not for you. It is written by a person who understands theory only when it leads to calculation, with every logical step under full public scrutiny. That is what *Mathematica* excels at.

But *Mathematica* also permits a playful attitude toward theory, making possible little experiments and explorations that are usually left in a scientist's private notebook. These curiosity-driven excursions are an essential part of the creative process, which perhaps now, with *Mathematica* and with private websites, and in books like this one, can become a part of the public record of science.

This book was originally intended as a concise compendium of group theoretic data and algorithms. The concise statements are indeed there, in the two *Mathematica* packages that are loaded at the top of every chapter. A true theoretician, versed in the *Mathematica* language, would need nothing else. But humans are not computers, and the development and exposition of the package materials takes time and space, and one thing led to another. I have tried to break the materials into lectures of fifty minutes length (if multiple examples are omitted).

I remember with great loss those from whom I learned group theory. First, Prof. Andreas C. Albrecht taught it to his spectroscopic research students at Cornell. Afterwards, I was a post-doc with Prof. Cristopher Longuett-Higgins, Cambridge University, who pioneered the use of permutation groups in flexible molecule spectroscopy. And especially I remember Prof. Leo Falicov, Physics Department, University of California, Berkeley, who helped me with my first by-hand application of projection operators many years ago.

And of course, this book owes its very existence to the monumental achievement of Stephen Wolfram in creating and developing *Mathematica* over the last twenty years. He created a genuinely "new kind of science", even before applying *Mathematica* to complexity theory.

The Wolfram customer support group was helpful on many occasions; I would especially like to thank developers Adam Strzebonski, Andre Kuzniarek, and Buddie Richie, who pulled me out of several deep holes.

I am grateful to several colleagues for help with mathematics and chemical theory: Prof. Kay Magaard, Math Department, Wayne State; H. Bernard Schlegel and Vladimir Chernyak, Chemistry Department, Wayne State; and Robert A. Harris, Chemistry Department, University of California, Berkeley.

I thank Tom von Foerster, formerly of Springer Verlag, for agreeing to this wild new way of publishing a book directly from *Mathematica* notebooks. The style sheet vonFoerster is named after him. Many special thanks to Jeanine Burke for making Tom's agreement come true, in spite of vicissitudes too numerous to mention.

My wife, Carol Bluestone McClain, was my enabler through years of computer addiction. She is now helping me to dry out in the February summer of South America.

I thank the students of Chem 8490, Winter 2007, Wayne State University, who helped with such intellegent attention, suggestions, and corrections: Michael Cato, Armando Estillore, Hao Li, Dr. Barbara Munk, Brian Psciuk, Sushant Sahu, Fadel Shalhout, Jason Sonk, Dr. Jason Sonnenberg, Huali Wang, and Jia Zhou. The book consists of all the materials that could be presented to this class in one semester.

Wm. Martin McClain
Professor, Chemistry, Wayne State University, Detroit, Michigan
wmm@chem.wayne.edu
Mendoza, Argentina, Feb. 18, 2009

Contents

1. Introduction

1.1 What is symmetry theory?

In this book, Symmetry Theory means group theory, as applied to groups of symmetry transformations. The deepest roots of mathematical group theory deal with algebra problems and permutation problems, but students of chemistry and physics usually encounter it first in the context of geometrical symmetry problems in crystals and molecules.

All students of physical chemistry and molecular physics must learn a little group theory, if only to avoid utter mystification by phrases like "the $\mathbf{B_{2\,u}}$ states of a $\mathbf{D_{6\,h}}$ molecule" or "the diamond crystal belongs to **m3m**". There are now many undergraduate physical chemistry textbooks that treat group theory in a single chapter, but in order to do so they must leave out all the logic, presenting only lists of things to be memorized.

You should read such a chapter, if you have never seen one. It will take you quickly over the most common uses of group theory in molecular physics, but if you are at all resistant to the memorization of authority, it will leave you hungry for more. And of course, on the basis of such a chapter, you would never be able to think of anything new on your own. In this book, nothing will be presented from authority; the logic will be foremost, and you will be preparing your mind to carry symmetry theory forward into new areas.

1.2 Outline of the book

Part I
Point groups and their construction

Chapter 2 and Chapter 3 introduce you to *Mathematica*, an electronic logic engine that automates routine calculations, symbolic as well as numeric.

Chapter 4 through Chapter 8 develop the basic relationship between symmetry and mathematical group theory.

Chapter 9 introduces the *Mathematica* molecule, and its automated graphics.

W.M. McClain, *Symmetry Theory in Molecular Physics with Mathematica*,
DOI 10.1007/b13137_1, © Springer Science+Business Media, LLC 2009

Chapter 10 through Chapter 13 develop some matrix operators necessary for constructing symmetry transformation matrices, and for recognizing such matrices when they appear as a construction.

Chapter 14 introduces the character table, and the concepts of Class and Species. Nothing is derived or proved here, but the questions that will occupy the next several chapters are explicitly laid out.

Chapter 15 through Chapter 19 show how molecular point groups are made, organized, named and visualized.

Part II
Representation Theory

Chapter 20 through Chapter 22 describe subgroups and classes

Chapter 23 explains how group theory relates to quantum mechanics, and why the names that appear in the character table are used as the names of spectroscopic states.

Chapter 24 through Chapter 28 introduce representations based on symmetric objects in the most general sense, and the idea of reducible and irreducible representations.

Chapter 29 through Chapter 33 present the basic theorems of abstract group theory, culminating in the Great Orthogonality Theorem. Schur's Lemmas for the Great Orthogonality are introduced in a quasi-experimental way that may imitate the way that Schur discovered them. This brings the topic to life again, presenting questions first and answers second.

In Chapter 34 to Chapter 36 we draw out some consequences of the Great Orthogonality which clear up the final mysteries about the point groups. The character orthogonalities underlie the properties that all character tables have in common, and are of great practical consequence. They permit the automatic reduction of any reducible representation, and permit the automated construction of all possible irreducible representations for any given group.

Chapter 37 presents the concept of projection operators, which can take a mathematical object of no symmetry whatsoever and divide it systematically into parts that have simple symmetries. These operators have long been known theoretically, but their application without automated mathematical help is tedious and uncertain. Now, for the first time, the operators and tabulations in this book make them easy to use. It is quite fair to say that this book was written to make this chapter possible.

Part III
Applications, and automated construction of character tables

Chapter 38 to Chapter 46 deal with applications of group theory in chemistry. We make extensive use of all the automated operators developed in Parts I and II. Suddenly the fog lifts, and you see how true it is that if you know the structure of group theory, the applications of it in molecular theory follow a unified theme, and are easy and transparent.

In Chapter 47 and Chapter 48 you learn that if you have the multiplication table that describes a group, you can apply machinery that will turn it into the character table for that group. This is a little advanced, but it is potentially of one of the most useful algorithms in the book. New permutation groups appear all the time in the spectroscopy of flexible molecules, and their character tables are quite urgently needed.

Appendices and Bibliography

The book ends with three appendices and a bibliography. The appendices deal with calculational or mathematical issues that arise in the book, but which do not lie on the main path. They may be read only by interactive readers.

1.3 This is an interactive book

If you are reading this on paper, you may not know very much yet about *Mathematica*. It is a computer program that carries out the details of algebra, calculus, and numerical computation. It is changing the way that scientists learn and use mathematics, because it frees the mind to concentrate on the issues of logical strategy, rather than on details of symbol manipulation. Its speed and unerring accuracy gives one the energy and confidence to forge ahead with new and unexpected results, raising the creative process to new levels.

There is a paper version of this book and a CD (compact disk) version. The paper version is a reference work, providing most of the material on the CD. But you can put the CD into a computer and read the unabridged version from the screen, letting *Mathematica* recalculate all the results in the book as you go. If you do this, the *Mathematica* program will always be prepared to carry out experiments that are suggested in the text, or (even better) that you may think of on your own. This interactive mode may make this book one of the first that people will prefer to study from a computer rather than from paper. (Of course, for quick reference purposes, nothing beats reaching for a book on a shelf. That is why there is a hard copy.)

Your computer must be able to run *Mathematica* before you can use the book interactively. *Mathematica* is now widely available through institutional site licenses that are very convenient to use, or as stand-alone copies. The student price is less than the author once paid for a slide rule made of bamboo and ivory.

If you work with this book interactively, you will learn *Mathematica* and symmetry theory together, but it is the logic of symmetry theory, not the logic of *Mathematica*, that drives the book. We introduce *Mathematica* operators only when we need them for symmetry theory, so a useful application is always at hand for any programming skills you learn. When you finish, your knowledge of symmetry theory will be quite solid, as far as this book takes you. You will feel that it is easy and natural to carry out certain group theoretic calculations which in the past were always tedious and prone to human error, the domain only of specialists.

The interactive mode of this book lets you learn group theory in a way that partly imitates the way it was discovered by research mathematicians. The book provides axioms, definitions, and leading questions. The questions are then explored before definitive answers are given, and you are encouraged to use *Mathematica* yourself to do numerical experiments; to see if you can arrive at solutions before they are presented. Thus this book teaches the general analytical skills required for all research, as well as teaching *Mathematica* and symmetry theory.

1.4 Learning *Mathematica*

People often say that *Mathematica* has a very steep learning curve, but in over five years of introducing undergraduates to *Mathematica*, the author has not found this to be true. After a few hour's instruction, as given in Chapter 2 and Chapter 3 of this book, along with model examples of particular problems, students can begin to operate with *Mathematica*.

It is more accurate to say that *Mathematica,* like mathematics itself, has a very long learning curve. Stephen Wolfram, the chief architect of *Mathematica*, frequently says that he is always learning to be a better user of *Mathematica*. The program was designed to grow, and in more than a decade of development by a large staff of professionals, it has become a vast but strictly orderly program. It now encompasses much of the knowledge contained in the mathematical tables and handbooks that line the reference shelves of university libraries, and in a form much easier to use.

The only way to learn *Mathematica* is to use it. After you run through the basic tutorials of Chapter 2 and Chapter 3 and work with the examples in the first few sections of the book, you will be begin to be able to use *Mathematica* on your own. If you finish the book, you will have learned enough tricks to make you a quite competent user of *Mathematica*. Also, you will have learned how to use *Mathematica*'s own on-line documentation to teach yourself any new area you might need, and you will feel that you can use *Mathematica* to tackle problems that arise in your own research.

The primary printed reference on *Mathematica* is Stephen Wolfram's The Mathematica Book. It describes *Mathematica* up to the introduction of Version 6. The new Version 6 operators are missing, but very little in it is obsolete. Currently the primary reference is the online **Documentation Center**, accessible as the first item in the Help menu of *Mathematica* itself. It provides a document for every operator, as well as many concise tutorials on specific areas. It sets a new gold standard for computer program documentation.

Click here for a very selective list of Mathematica books, all more tutorial than Wolfram's book, but all less complete.

1.5 The human's view of *Mathematica*

There is a version of *Mathematica* for every major kind of computer. From a human's point of view they all work the same way, and they all produce the same kind of Text files, called notebooks. Humans can exchange notebooks electronically without concern about differences among computers.

When a person first begins to use a computer, it is difficult to get used to the machine's combination of brilliance and stupidity. The simplest, most obvious variations you make in syntax are just not understood (like using the letter **x** to indicate multiplication). The iron discipline of sticking to a precise syntax is for beginners the hardest thing to learn. But this is true of all computers programs, not just *Mathematica* .

When you first launch *Mathematica*, only the "front end" is activated. A notebook will appear, either blank or with material in it, depending on how you launch. With the front end alone, you may read the notebook and edit its prose. But you cannot calculate until you start the *Mathematica* "kernel". The kernel starts up automatically the first time you ask it to evaluate something. (Therefore, the first answer may be slow to come up.)

Mathematica provides typeset mathematical notation. This includes Greek and other alphabets, specialized mathematical characters, and standard mathematical two-dimensional notation for integrals, powers, and fractions, as well as matrices and vectors, and other objects. Here, for instance, is a nicely typeset *Mathematica* input-output pair :

$$\int \mathbf{x}^{\mathbf{p}} \,\mathbf{sin}(\mathbf{x})\, d\mathbf{x}$$

$$-\frac{1}{2} x^p \left(x^2\right)^{-p} (\Gamma(p+1, i\, x)(-i\, x)^p + (i\, x)^p\, \Gamma(p+1, -i\, x))$$

It might take you some time to find that in a table of integrals. But *Mathematica* did not look it up in a table; it uses a general symbolic algorithm, unlike anything normally taught to humans, and it handles variants as easily as simple cases :

$$\int \mathbf{x}^{\mathbf{p}} \,\mathbf{sin}(\mathbf{a\, x + b})\, d\mathbf{x}$$

$$-\frac{1}{2\, a} x^p \left(a^2\, x^2\right)^{-p} (\Gamma(p+1, i\, a\, x)(\cos(b) - i\sin(b))(-i\, a\, x)^p + (i\, a\, x)^p\, \Gamma(p+1, -i\, a\, x)(\cos(b) + i\sin(b)))$$

That might not be in any table, even the largest. You could of course deduce it from the result above, but it would take some time, and you could easily make mistakes in the process. Welcome to the world of computer-assisted mathematics!

1.6 Does *Mathematica* make errors ?

There is no way to make an absolute guarantee against error in a computer program. Confidence that a program is error-free comes from years of testing under all kinds of unforeseen circumstances. The public use of a program puts it through more unforeseeable twists and turns than any testing program ever could. The parts of *Mathematica* we will use in this book have withstood public use for over ten years, and we have great confidence that any errors in the basic central code have long since been found and corrected.

Nevertheless, the wise user of *Mathematica* sets up frequent checks for self-consistency. This is one of the things we will teach as we go. We have never found a *Mathematica* error with these checks, but we have found plenty of human errors. Consistency checking is the way that rigorous minded scientists deal with all the black boxes they must use in this electronic age.

2. A tutorial on notebooks

- If you want to read this book "live" (as intended) you will need to read this chapter on screen and with *Mathematica* running, and do the things it says to do. You won't get much out of it by just reading the hard copy, but here it is for quick reference :

2.1. What are *Mathematica* notebooks?

Every document produced by the *Mathematica* front end is a "notebook", and every chapter of the CD version of this book is a "notebook". The notebook allows a document to present itself in outline form, with each topic and subtopic opening up for reading on command, or closing up again on command. Notebooks require a little explanation, but they soon seem very natural.

Notebooks are divided into cells, which are rather like paragraphs. The individual cells are usually grouped by enclosing brackets of increasing length. When you read this book on-screen, you will see nested blue brackets on the right side of the screen. In the printed book, this chapter is the only one that displays the nested brackets. Look ⟶

This paragraph-like cell is grouped by a secondary Section bracket that encloses all of Section 2.1. All Section brackets are in turn enclosed by an outer bracket that begins at the top of the notebook and runs all the way to the bottom, enclosing all of Chapter 2. If you click once on the outer bracket, it will turn dark and you will have "selected" the whole notebook, as if for Copying or (God forbid!) Deleting. Well, of course, you can delete the loaded copy from RAM, but not from the CD.

If you click *twice* on the outermost bracket, the notebook will fold up and you will be left with only the Chapter title showing. To reopen, click twice on the outer folded bracket (with a small dark triangle, like a down-pointing half arrow), and the notebook will reappear.

Go ahead. Do it. The whole reason for putting this book on a computer is to let you try things as we go.

To open the Section heading below ("2.2. A basic notebook tutorial"), click twice on the middle bracket to its right, the one with the little dark

W.M. McClain, *Symmetry Theory in Molecular Physics with Mathematica*,
DOI 10.1007/b13137_2, © Springer Science+Business Media, LLC 2009

triangle at the bottom It will unfold, showing the subsection headers. They open similarly.

To fold it up again, click twice on the same (but now expanded) bracket. It will shrink and the little dark triangle (or downward half-arrow) will reappear, indicating that the bracket is expandable. All multiply bracketed cells can be unfolded and folded up again in this way.

2.2. A basic notebook tutorial

2.2.1. Magnification

Every notebook window has a magnification option on its bottom edge, just left of the horizontal scroll bar. By default, the chapters of this book are set for 200%. Click the little dark triangle and other magnification choices will drop down. Try some other sizes.

2.2.2. Make a new cell with **`1+1`** in it

Move the cursor around until, between two existing cells, it turns to a horizontal bar with split ends. Click once, and it turns to a long horizontal "insertion line". This is where the new cell will appear, if you start typing. For instance, you can do it just below, typing "1+1" (without the quotes) :

The "1+1" you just typed is in an **`Input`** cell, distinguished its **`Bold Courier`** type face. This is the kind of cell you want when preparing input for calculations. (Go ahead, do it! Nothing terrible can happen!)

Now we discuss how to get *Mathematica* to compute **`1+1`** for you.

2.2.3 The initial computation

Put your cursor anywhere in the **`Input`** cell you just created, and hit SHIFT-ENTER or SHIFT-RET. One of these should cause something to happen; remember which one does the job. We will call it the **`Process`** command.

In this book, the Process command will be represented by ↵.

On a Macintosh, you may use ENTER by itself, or the SHIFT-ENTER combination, or the SHIFT-RET combination. On PCs, the SHIFT is required.

If you are starting up a new notebook, you may now experience a little distraction. A question will appear in a box :

Do you want to evaluate all the initialization cells in this notebook?

For notebooks from this book, you must always click YES. Otherwise, the notebook may not work properly.

In this book, the initialization cells are mostly in the **`Preliminaries`** section at the top of every chapter. (You can open them and see them, but they are hard to understand; don't try until you know more.)

After the question disappears, an **`Output`** cell appears, bracketed with its **`Input`** cell. These cell types can be distinguished at a glance:

> **`Input`** cells have **`Bold Courier`** typeface;
> `Output` cells have `Plain Courier` typeface.

After processing, notice the appearance of the blue In[ ] and Out[ ] cell labels. These will prove very useful. Unprocessed **`Input`** cells (a major cause of error for beginners) will not have the blue label.

Try a few other simple sums until you are very familiar with the **`Process`** command. On a Macintosh note that both ENTER and SHIFT-ENTER can act as a Process command. Then clean up after yourself. Select *the innermost bracket* of each new cell and hit the **delete** key. Or, leave one example standing. It's up to you.

2.2.4 Text cells

Notebooks have **`Text`** cells as well as **`Input`** and **`Output`** cells. **`Text`** cells provide a place for commentary on your calculation (or for most of the text of this book). This cell is a **`Text`** cell. All **`Text`** cells have a short horizontal line near the top of the innermost cell bracket.

To make a new **Text** cell, make an insertion line and start typing. A new cell will appear, and it will be, by default, an **Input** cell. After a few characters, stop and issue **Option-Command-7** (that's OPTION-CMD-**7**, or OPTION-⌘-**7**). The cell will turn to a **Text** cell, distinguished by its 10-point Times type face.

2.2.5 Heading cells

Notebooks also have heading cells of several different types. Make an insertion line and start typing a heading. After a few characters stop and issue **Option-Command-4** (OPTION-CMD-**4**, or OPTION-⌘-**4**). It will become a cell of type **Section**, in **14** point type.

Other headings are made by commands of the form **Option-Command-digit** (OPTION-CMD-**digit** or OPTION-⌘-**digit**), where **digit** is any digit from **1** to **7** . In the *vonFoerster* style sheet used by this book, digits **1-6** are headings of decreasing rank, and **7** is **Text**, **8** is a magnification toggle, and **9** is **Input**. You may switch a cell freely among any of these types by putting the cursor anywhere inside it, holding down **Option-Command** (OPTION-CMD, or OPTION-⌘), and typing different digits. Try it.

In this book every chapter uses the hierarchy
ChapterLine (autonumbered, OPTION-⌘-**1**)
  Section (autonumbered, OPTION-⌘-**4**)
    Subsection (autonumbered, OPTION-⌘-**5**)
      Subsubsection (not numbered, OPTION-⌘-**6**)

2.2.6 Processing cell groups

Select any grouping bracket, open or not, and issue a **Process** command. This will processing all the selected cells. If you select the outermost bracket that encloses the whole notebook, you will **Process** the whole notebook with one hit.

2.2.7 The active way to read a notebook

With the cursor right here in this cell, do a SHIFT-ENTER . The **1+1** below will be selected and another SHIFT-ENTER sends it to the processor. A third selects the next evaluatable cell, and a fourth evaluates that. Try it :

```
1 + 1
```

```
2 + 2
```

You can click down the whole notebook by keeping the **Shift** key depressed and repeatedly typing **Enter**. This is a good way to evaluate a pre-written notebook at a human pace, following the logic as you click down. You can stop at any point and carry out experiments of your own.

If you want to do further work in a newly opened notebook, select everything above your starting point and ↵. It will run down to your starting point, making all necessary the definitions as it goes. Then you can start work.

2.2.8 Fold and unfold groups of cells

Select a grouping bracket and do a SHIFT-⌘-} (shift-command-close-Bracket). The selected bracket (and all sub-brackets within it) will close. If you select the whole notebook, you can close the whole notebook with one stroke, leaving only the title showing. Then click twice on the outermost bracket, and you get a nice compact index of Sections for the whole notebook.

Similarly, SHIFT-⌘-{ (shift-command-openBracket) opens a selected section, and all its subsections.

This is really all you need to know to get started using notebooks.

3. A basic *Mathematica* tutorial

- If you want to read this book "live" (as intended) you will need to read this chapter on screen and with *Mathematica* running, and do the things it says to do. You won't get much out of it by just reading the hard copy, but here it is for quick reference :

Preliminaries

Section 3.1 This tutorial

This tutorial is for Version 6 of *Mathematica.* It is a framework for running quickly over the basic terms and concepts. It is probably a good way to get started, but you will not really learn these things except by practice, bringing them up in your consciousness from your own memory. Like riding a bicycle, it becomes very automatic, but only with practice.

If you have not yet read Chapter 2, the NotebooksTutorial, it will help a lot if you do so now.

> In this tutorial, nothing on a colored background will evaluate or copy. You should retype it in a cell of your own between the colored cells, and then ⏎ . Typing forces you to notice important details that escape you if you merely read. Type in everything exactly as shown (particularly punctuation, and upper and lower cases). After you see the given example work correctly, make variations of your own.

As you work the examples, pass nothing over with a shrug; let nothing mystify you; everything that happens, happens on your orders. If you see something strange, examine your last input for typos before anything else. A small accidental space in the middle of a variable name, or some other trivial typo, can make all the difference. Explanations, when needed, follow the examples.

W.M. McClain, *Symmetry Theory in Molecular Physics with Mathematica*,
DOI 10.1007/b13137_3,

Section 3.2 On-line help

3.2.1 Quick, basic help: the question mark

No one can remember all the details of all the *Mathematica* operators, so there is an excellent on-line help facility. Just prepend a question mark to the name of the operator:

```
?Solve ↵
```

You will see the syntax of **Solve**, and a basic description of its action. Here is an example :

```
Solve[x-2y==0,x]
```

The output is a little surprising, but you will learn to read it. For humans, the little arrow "→" basically means "=" . For the computer it has a meaning you will learn below.

Sometimes the exact name is what you have forgotten. Type **?** and then some part of the name that you remember, surrounded by wildcard asterisks:

```
?*Plot*
```

This brings up every operator that contains **Plot** anywhere in its name. Click on the one you want, and help will appear. For a good relevant example, click on **Plot** itself and try to read what comes up.

3.2.2 More detailed help

No computer program has more detailed help than *Mathematica*. Click on the >> symbol at the end of the thumbnail sketch above, and a whole notebook devoted to the **Plot** operator will appear. Do it. The notebook is too long to make it a part of this document, so when you have finished looking at it, just click it closed and come back here.

3.2.3 The Documentation Center

The first item under the Help menu is the Documentation Center (DC). Click on it and you will see a clickable outline of everything in the DC. But the best way to use it is to type a term into the long thin box at the top, and ↵. *Mathematica* will search the DC for all occurrences of your term, and present you with a clickable list of DC destinations. It is very similar to a Google search.

Section 3.3 Basic operations

3.3.1 Simple arithmetic and algebra

(a) Add, subtract, multiply, and divide

2+3 ↵

2-3 ↵

6/3 ↵

2*2 ↵ The * means multiply, as in most languages

2[SPACE]**2** ↵ The [SPACE] also means multiply (only in *Mathematica*)

Make your own cell and try these things out. Click above or below this cell to get the insertion line, and then start typing the material from the boxed cell above.

(b) Powers

The old-fashioned (all-on-the-baseline) way to write a power is

2^3 ↵ The carat ^ means "to the power". It is the uppercase 6 key.

Here is the way to get the power up off the baseline (where it belongs!) :

2[CTRL]**^** Makes a superscript template on the **2**, like $2^{\square}$. Fill it :

2^3 ↵

Come back to the baseline with [CTRL][SPACE] .

3.3.2 Three useful palettes

The BasicInput palette

Follow the click chain

`PalettesMenu` ▷ `BasicMathinput`.

A palette will appear, containing templates for a number of mathematical notations. Drag it to the side by its top bar. Make an insertion line and click on a palette item. The item will appear in a new Input cell. Inside an existing cell, the new item will appear where ever you leave the blinking cursor.

The SpecialCharacters palette

Follow the click chain

`PalettesMenu` ▷ `SpecialCharacters`.

You may click to insert any character that *Mathematica* has. Under Letters his there are five alphabet styles; under Symbols there are seven whole palettes of math symbols. Many of these were in the past available only to professional typesetters. Look at them all.

The AlgebraicManipulation palette

Follow the click chain

`PalettesMenu` ▷ `AlgebraicManipulation`. In an input cell containing a square of some fomula, select the formula and click on Expand. Then selectthe result and click on Factor. If you are in an Input cell, it will happen right in-place. If not, a new Input cell will open, containing the expression, as changed. Then try some of the other operations.

3.3.3 The most basic operators

Set (the " = " sign)

The mathematical equal sign (**=**) is left-right symmetric for a very good reason: The expressions that stand on either side of it must evaluate to the same numerical value. But this is NOT how the " = " sign is used in computer languages. The earliest languages used it in commands like

```
aMemoryLocation = aNumber
```

This sent the given number into storage at the given location. This is not mathematical equality, but this usage cannot be stopped. So the name of "=" is **`Set`**. In *Mathematica*, memory locations are referred to by symbolic names, like **`q`**. Try

`q` ↵ If the location **`q`** is free, the symbol **`q`** will be returned. If not, see **`Clear`** (below).

`q = 7` ↵ Location **`q`** is **`Set`** to **`7`**, and **`7`** is returned.

`q` ↵ This time, the number **`7`** is returned.

Until the content of location **`q`** is changed, *Mathematica* will always automatically turn symbol **`q`** into **`7`**.

Thus **`7`** has become the "meaning" of **`q`**.

(e) Clear

`Clear[q]` ↵ This undoes the **`Set`** operation, freeing up the given memory location.

(f) Equal (the double "==" sign)

Since the "=" is already taken by **Set**, mathematical equality needs another symbol. It is two equal signs typed together with no space :

1 == 1↵

1 == 2↵

a == b↵ *Mathematica* returns an answer when it can, and repeats the question when it can't.

Unlike primitive numerical languages, *Mathematica* can put symbolic expressions into memory locations. You can give a name to an equation :

eqName = 2 x == 3 y ↵

eqName↵

(e) Use a decimal or not? It matters.

Sin[2.] ↵ Note the decimal point.

Sin[2] ↵ Note the absence of the decimal point

This requires explanation. The decimal point tells *Mma* that you are willing to accept machine accuracy; the absence of the decimal means you want an exact result. That is why it refuses to do anything with **Sin[2]**, which is an irrational number that has no exact representation.

(f) Symbols for irrational numbers

c = $\sqrt{2}$ ↵ The $\sqrt{\square}$ template is typed as CTRL 2 .

c^2 ↵ The **c** behaves as it should when squared.

(g) Rule and ReplaceAll

Clear[x] ↵

Rules are similar to equalities, except that they have a right-pointing arrow "→" in them instead of an equal sign. We write a rule that we name as **xRule** .
Use ESC - > ESC to make the nice little arrow. That's *escape-minus-greaterthan-escape*):

```
xRule = x→2 ↵
```

Rules would be useless without the **ReplaceAll** operator. It is written as a two-stroke combination **/.** between an expression and a **Rule**. It causes the **Rule** to be applied to the expression; in this case it causes every **x** in the expression to be replaced by **2**.

```
x^2 /. xRule ↵
```

You don't have to give names to **Rules**. You can just say

```
x^3 /. x→3 ↵
```

The great thing about **Rule** and **ReplaceAll** is that it does nothing permanent to the symbol on the left side of the **Rule** (the **x**, in this case). Ask what *Mathematica* has now for the value of **x** :

```
x ↵
```

Nothing! It is still an unset symbol, though we used it above as a temporary container for two different numerical values. This will be more useful than you can possibly imagine right now.

Section 3.4 The FrontEnd

Mathematica consists of two executable programs, the **FrontEnd** and the **Kernel**. This section describes the **FrontEnd**; the next section describes the **Kernel**.

The **FrontEnd** constructs all the displays on your screen and interprets all input from the keyboard and the mouse. The **Kernel** does all symbolic and numeric calculation. When you click twice on a notebook icon to open it, only the **FrontEnd** starts up. The **Kernel** starts automatically the first time you send anything to the processor. For instance, **1**↵ will start the **Kernel**. It takes a perceptible time to start. The next time you open a new notebook, look for this.

3.4.1 A few FrontEnd tricks

Greek letters, and other strange characters

Greek letters may be clicked from a palette, or may be input right from the keyboard.

Greek letters, and other strange characters

The **`FileMenu ▷ Palettes ▷ CompleteCharacters`** palette contains many specialized characters available in the past only to professional typesetters. It provides an alternative way to get Greek letters, among many others, by clicking. Check it out.

There are keyboard shortcuts for everything on all palettes. The Greek letters are particularly easy; just type the Latin equivalent, surrounded by escapes :

ESC	stands for the escape key, *esc*. Try
ESC**`a`**ESC	Greek alpha
ESC**`b`**ESC	Greek beta
ESC**`p`**ESC	If you want your **`pi`** in Greek, type this.
ESC**`ph`**ESC	ϕ Some need two strokes.
ESC**`ps`**ESC	ψ
ESC**`S`**ESC	Σ Capitals, too
`Sin[`ESC**`pi`**ESC **`/ 3]`** ↵	π has a preassigned meaning, the same as "**`Pi`**".
`Sin[`ESC**`Pi`**ESC **`/ 3]`** ↵	but Π is not π.

Subscripts and superscripts

We have already seen how to make a superscript template with CTRL^ (**`control-carat`**) to raise an expression to a power. The subscript is quite similar, though there is a distinction to be made. But first, some painfully detailed instructions on how to type a subscript.

Type a symbol, then type CTRL- (**`control-minus`**). This produces a selected subscript template below and to the right of the symbol. Fill the template with something simple. Now come up to the main line with CTRL-SPACE (**control-space**). Now you have a subscripted object that acts for most purposes like a symbol. Later you will learn to **Symbolize** it when needed so that it becomes a true symbol.

```
c₁ = 2.0;
{c₁, c₂}
```

You can use it as input to functions :

```
{Log[c₁], c₁²}
```

The high-low toggle (CTRL 5)

Here is a handy trick. Suppose you want to write c_1 squared. If you write the subscript first and then the superscript, it will look like ${c_1}^2$. The is not terrible, but sometimes you would rather have the superscript directly over the subscript. For this, write c_1, and then with the cursor blinking behind the **sub-1**, do a CTRL 5 (**control-5**). This will create an insertion box directly above the subscript. Actually, you may also do it in the opposite order- the **control-5** operation is a toggle between matching high and low insertion boxes. It also toggles between **overscript** and **underscript** boxes.

Section 3.5 The Kernel

3.5.1 Introduction

The **Kernel** knows nothing about what the **FrontEnd** has on any line until you give that line an *Enter* or *Shift-Return* (↵). This is particularly true of a freshly opened notebook that has lots of In-Out pairs in it. The **FrontEnd** displays them, but the **Kernel** know nothing about them. The very commonest mistake in *Mathematica* is to forget this. Everybody does it; all you can do is learn to recognize the symptoms and scroll back up to run the lines that were skipped.

If you want the Kernel to be aware of all the work that you did in an old notebook, you must rerun it FROM THE TOP. This is easy. Just select the outermost bracket for the whole notebook, and hit *Enter* or *Shift-Return* (↵).

To do this at a more human pace, hold down **Shift** while repeatedly hitting

`Enter`. This jumps you from one executable cell to the next (and then executes it), so you can follow what is going on, fix errors as they are encountered, etc.

3.5.2 Five kinds of brackets

1. Square brackets [...] (Operators)

The processor works **only** on square brackets. Indeed, square brackets must never be used for anything other than an **`Operator[operand]`** construction. In particular, they must never be used to group algebraic expressions.

Many expressions contain nested square brackets. The processor looks for the innermost pair and processes it first. Then the expression is simply recycled back into the processor afresh, and again the innermost pair is located. When all the square brackets are gone, the fully evaluated expression is displayed. Ordinary functions are just operators that operate on numbers, so we can make a simple example of this by nesting several functions around a single number :

> First evaluate **`ArcTan[Log[Sin[1.23]]]`**.
>
> Then do them one at a time: Evaluate **`Sin[1.23]`**, then paste the numerical value into the square brackets of **`Log[]`**, and then the **`Log`** into the **`ArcTan`**, to verify the "innermost first" rule.

2. Parentheses (...) (Algebraic grouping)

Parentheses are used exclusively for algebraic grouping:

```
(u+v)/(f (b^2 + c^2))
```

This means you will have to give up your old favorite, **`f(x)`**. In *Mathematica*, **`f(x)`** means "**`f`** times **`x`**" and never anything else. The function "**`f`** of **`x`**" must be written with square brackets as **`f[x]`**, or as one of its aliases, like **`x//f`**. Do not ever try to group symbols using curly braces **`a{b+c}`** or square brackets **`a[b+c]`**. The processor understands these expressions in a quite different way.

```
2 Cos[φ] Sin[φ] (Cos[φ]^2 - Sin[φ]^2)
```

After typing the above and looking at its output, copy it and paste it into **`TrigReduce[]`** to see what *Mathematica* knows about trig :

```
TrigReduce[2 Cos[ϕ] Sin[ϕ] (Cos[ϕ]^2 - Sin[ϕ]^2)]
```

3. Curly braces { ... } (Lists)

Curly braces are used exclusively to enclose lists of things. The things listed are separated by commas. The list is the one and only data structure used by *Mathematica*. It is a theorem that any data structure is equivalent to a nested list structure, so there is no need for anything else. Below, note that **abcd** is a single symbol, whereas **{a,b,c,d}** is a list of four symbols.

```
abcd = {a,b,c,d};
ABCD = {A,B,C,D};
```

Many clever algebraic operations can be carried out on lists.

```
abcd . ABCD  (* the Dot product *)
abcd * ABCD  (* the Times product *)
```

Many functions have the **Listable** property. The **Log** function is an example :

```
Log[{1., 2., 3., 4.}] ↵
```

$$\frac{1}{\{1., 2., 3., 4.\}}$$ ↵

Here is a nested list structure (a matrix), and the basic use for it with a vector :

```
matrix = {{a,b},{c,d}};
vector = {e,f};
matrix.vector
vector.matrix
```

4. Comments (* ... *) (ignore)

Sometimes you want to put a short note beside some input, without going to the trouble of making a Text cell for it. This is done by enclosing your comment in the two-stroke combination (* as the opener, and another two-stroke combination *) as the closer. Everything inside will be ignored by the processor.

```
Factor[a² - b²]     (*example of factoring*)
```

5. Double brackets ⟦ ...⟧ (Part)

Double square brackets, or double-struck square brackets, are the human-friendly guise of the **Part** operator. In the list **{a,b,c}**, the "**b**" is Part 2. You bring it out on its own with

```
{a,b,c}[[2]]  ↵
```

The better-looking form below is typed as ESC **[[** ESC and as ESC **]]** ESC.

```
{a,b,c}⟦ 3⟧  ↵
```

Many expression have parts that have parts. Here is an example of a List of Lists.

```
{{a,b,c},{d,e,f}}⟦ 2,3⟧  ↵
```

Part 2 of the main expression is **{d,e,f}** . Part 3 of that is **f**.

Section 3.6 *Mathematica* graphics

3.6.1. Two dimensional graphics

Two- dimensional graphics are made by the **Graphics** operator working on a graphic "primitive", like **Circle** :

```
circ = Graphics[Circle[], ImageSize → 72]
```

The **Head** of the output is also called **Graphics**; thus, the output is said to be a "Graphics object".

```
Head[circ]
```

```
Graphics
```

There are only 13 **Graphics** primitives. Some of the most useful are **Circle**, **Line**, **Arrow**, **Text**, **Rectangle**, and **Polygon**. Look up **Graphics** in the Doc Center and open **MoreInfo** to see a list of them all.

Usually, the **Graphics** operator is applied to a **List** of *primitives* interspersed with various *directives*. First, make such a list:

```
grList = {PointSize[1 / 10], Red,
   Point[{0, 0}], Blue, Point[{2, 0}], Green,
   Point[{4, 0}], White, Point[{6, 0}], Black,
   Point[{8, 0}], Yellow, Point[{10, 0}]};
```

Read the **grList** carefully. The only *primitive* that appears is **Point[x,y]**, which says to draw a point with center at **x,y**. Everything else is a *directive*. The list starts with the directive **PointSize[1/10]**, which says that every following point is to have a diameter **1/10** the total width of the figure. The colors names are also directives, each establishing a color that will apply to all following primitives (until another color directive is given). Now feed **grList** to the **Graphics** operator, along with options to control the **PlotRange**, **ImageSize** and the **Background** color :

```
bigDots =
 Graphics[grList, PlotRange → {{-2, 12}, {-1, 1}},
  ImageSize → 200, Background → GrayLevel[0.6]]
```

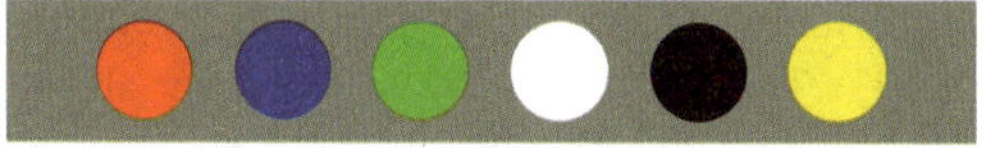

The output is the graphic itself, a graphics object named **bigDots**.

Look at the **Head** of **bigDots**. Then change their **x,y** coordinates and restructure the list to make interesting pattern of **Red** and **Black** dots with different sizes.

Make a thick line located on the same background:

```
thickLine = Graphics[
  {Thickness[0.02`], Line[{{-1, 0}, {11, 0}}]},
  ImageSize → 200, PlotRange → {{-2, 12}, {-1, 1}},
  Background → GrayLevel[0.6`]]
```

Thickness[.02] directs all following lines to have a thickness 2% of the width of the figure.

Graphics objects can be shown again, or combined, by the **Show** operator.

```
lineInFront = Show[{bigDots, thickLine}]
```

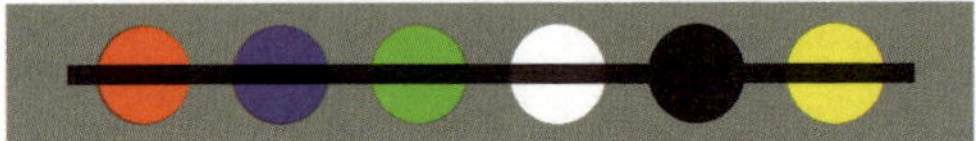

```
dotsInFront = Show[{thickLine, bigDots}]
```

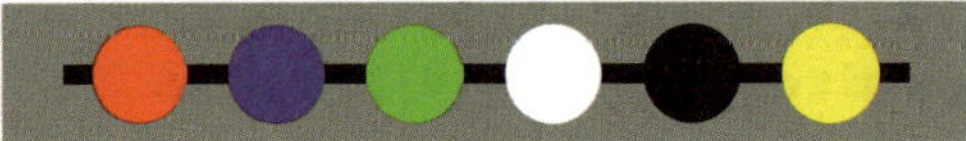

In the **Show** list, the first object is drawn first; then later objects overlay each other in list order.

3.6.2. Three dimensional graphics

Three dimensional graphics work like two dimensional graphics, except that the operator is **Graphics3D**, and the primitives are specialized for three dimensions. For instance, **Point** now requires three coordinates, not two. Others, like **Cuboid**, **Cylinder**, and **Sphere** are intrinsically **3D** only. Also, the directives **Opacity**, **Specularity**, **EdgeForm**, and **FaceForm** address issues that do not arise in 2D.

We show a single example, making use of a handy operator **Graphics⋱ Complex[ptList,grList]**. The idea behind it is to specify all the 3D points up front in **ptList**, and then refer to these points in the **grList** simply as integers **1,2,...** . For a diatomic, we need only two points:

```
diatomicMolecule = GraphicsComplex[
   (*ptList*){{-1, 0, 0}, {+1, 0, 0}},
   (*grLIST*){
    Brown, Cylinder[{1, 2}, .1],
    Red, Sphere[1, 0.5],
    Blue, Sphere[2, 0.4]}];
```

Make the molecule visible by wrapping it in **Graphics3D**, with an option that controls the **ImageSize** :

```
dM = Graphics3D[diatomicMolecule, ImageSize → 80]
```

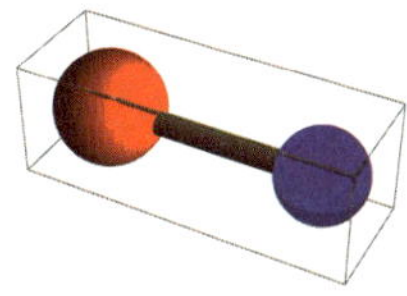

Book readers will see nothing very remarkable about this. But interactive readers can use the mouse to drag this image around into any desired orientation:

```
{dM, dM, dM}
```

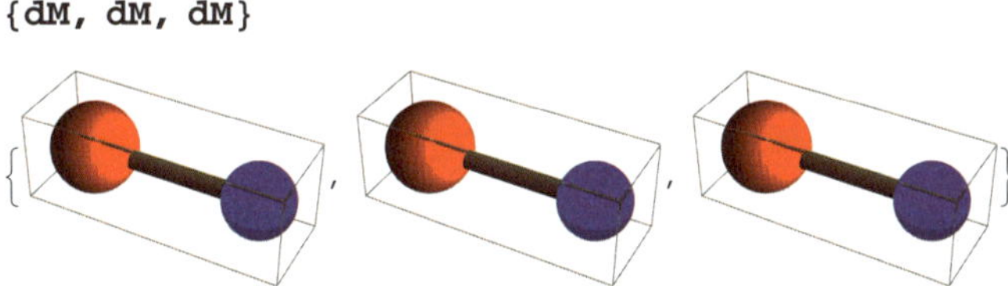

This is one of the very best things about *Mathematica*'s 3D graphics. If the structure is complicated you can maneuver it around until you understand it exactly. Biochemists have elaborate programs for doing this with macromolecule structures, but here it is just part of the core of *Mathematica*.

4. The meaning of symmetry

Preliminaries

4.1. Symmetry and its undefined terms

Proportion, *harmony*, *balance*, and *beauty of form* are terms invoked in dictionaries to explain the meaning of the word *symmetry*. But the formal study of symmetry must begin with a much simpler, more precise definition.

> **Symmetry**
> We will say that a set of objects has a symmetry if we know of a transformation rule that leaves the set unchanged.

This is a formal definition of the word **symmetry**. One of the great discoveries of modern logic is that every fundamental definition must contain undefined terms. The words *set*, *object*, and *transformation rule* are the main undefined terms in this definition.

For instance, you could define a ***set*** as a ***collection*** **without regard to *order*.** This is a good informal explanation of the word *set*, but it is not a formal definition, because what is the formal definition of *collection*, or of *order* ? We are worse off than before, because we have traded one undefined term for two others. It is an endless cycle, and it is best to recognize it as such, and agree upon the terms that will be accepted as undefined.

The undefined terms of a logical system are not a limitation or a flaw; they are the source its power. Because they are undefined, you are free to interpret them in any way that you can make sense of them.

Below, in Section 4.2 (Geometric Symmetry) and Section 4.3 (Algebraic Symmetry) we will give two different interpretations of the words *object* and *transformation rule*.

The most creative thing that a person can do with a formal logical system is to find a new interpretation of its undefined terms; this has happened many times in the study of symmetry, and it will happen again. If you can find new interpretations for *object* and *transformation rule*, all the many rigorous results of symmetry theory apply to the new object, immediately illuminating and clarify-

W.M. McClain, *Symmetry Theory in Molecular Physics with Mathematica*,
DOI 10.1007/b13137_4,

ing its properties. An outstanding example of this was the interpretation of spin as a new symmetric object, after it was discovered experimentally.

Although it is useless to give formal definitions of "object" or "transformation rule", one can still discuss them informally. We will see two kinds of "objects" in this chapter, geometric objects and algebraic objects. A "transformation rule" must be something that would effect a change in most of the objects to which it applies, so that the few left unchanged by it are indeed very special ones that deserve the special adjective *symmetric*.

So already we have a subtle insight: According to our formal definition, a symmetric set of objects is never a thing entirely on its own; it is symmetric only with respect to the transformation rules that humans formulate and lay upon it. When an object is "intuitively" symmetric, the rule is a subconscious one. When this process becomes conscious, with rules formulated according to an intent, we extend and refine the idea of symmetry in elegant and useful ways. We will see examples of this. But first, in this chapter, the basics.

4.2. Geometric symmetry

4.2.1. Links to background brush-up materials

If you have done the Tutorial and you still feel a little rusty on vectors, matrices, and the **Dot** product, click here to go to a more detailed review at the end of the book. It is long and comprehensive, but the first parts of it are what you need.

4.2.2. Geometric meanings for the undefined terms

We now take specific meanings for the undefined terms in the definition of *symmetry*. We let ***set of objects == a set of points in 3-dimensional space***, and (to get started) we let ***transform == rotation or reflection of all the points***.

The center of gravity of the object is taken as the origin, and the rotation axes and the reflection planes must all contain the origin. Later this list of kinds of geometrical transforms will be extended, but these two simple kinds are enough to get us started. Repeating the definition of symmetry with these explicit meanings:

"We will say that a set of geometric points has a symmetry if we know of a rotation or a reflection that leaves the point set unchanged."

As described above, geometric symmetry is a matter of visualizing things in one's mind. But people vary a lot in the their visualizations, and visions are hard to describe convincingly to other people. To do mathematics with this definition, we must make use of the great discovery of René Descartes, that every mathematical point in 3-dimensions may be described as a triple of numbers **`{a,b,c}`**, called a *vector*.

As we will see below, both rotations and reflections can be described by *matrices* of size 3×3 that multiply the vectors, using the row-by-column rule, or the **`Dot`** product.

But visualization cannot and should not be supplanted, and a large part of *Mathematica* is devoted to graphics operators that turn lists of vectors into line drawings. So we have the best of both worlds: automated matrix multiplication for speed and absolute accuracy in calculation; and automated, exact perspective line drawings of the results, á la Leonardo, for visual understanding.

4.2.3. Reflection matrices

Definition of "reflection"

The "reflection" used in symmetry theory is not an optical reflection. However, it gets its name from the optical diagram of a light ray reflection, below. A ray from a street light bounces from the surface of a puddle and into an eye (or actually, into two stereoscopic eyes). The brain then uses the straight dashed extension of the eye's line of sight to interpret the position of the light as a point below the surface of the ground, at a depth equal to the height of the light above the ground.

In Symmetry Theory, we forget about the eye and the true path of the ray, and take the relation between Source, Mirror, and Image as the "reflection".

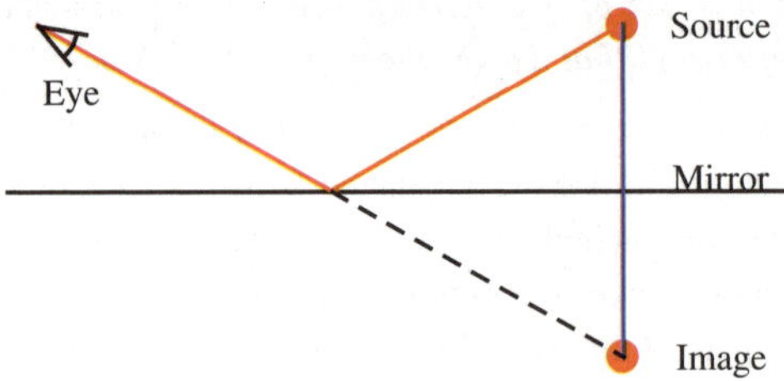

Fig. 4.1 Optical reflection diagram

The U. S. Capitol building has reflection symmetry

Consider an object that seems intuitively symmetric; for instance, the United States capitol building. Of course, it is not absolutely symmetric, but we idealize. If our formal definition cannot be applied to this idealized object, then it is nonsense; so let's begin here.
What is a transformation rule that leaves the building unchanged? Think of a vertical plane passing through the center of the rotunda and dividing the whole building into equal halves, East Wing and West Wing. This is a *mirror plane* for the building, in the sense that the West Wing looks just like the reflection of the East Wing in a giant mirror placed in this plane.

To sharpen this up, we establish two axes that lie in the mirror plane: vertical axis **z** running up through the center of the rotunda, and horizontal axis **y** running out the front door. Then **x** is an axis that runs perpendicular to the mirror plane, say, an East-West axis that runs in along the central hallway of one wing, through the rotunda and mirror plane, and out through a similar hall in the other wing.

Consider a set of vectors that point to all the major architectural elements of the Capitol. One of its members, **{x,y,z}**, points from the origin to, say, the center of a window in the East Wing. Mirror symmetry means that the vector **{-x,y,z}** points to the center of a similar window in the West wing. The transformation rule that leaves the object unchanged is

$$\{x,y,z\} \to \{-x,y,z\}$$

In words, the rule is: *Using an origin that lies in the mirror, in each architectural vector negate the component that is perpendicular to the mirror plane.* This transform changes every East-wing vector into a West-wing vector, and vice-versa. If the two wings have true mirror symmetry, the vector set is the same before and after the transform (ignoring the order of the vectors, as always in a *set*). If there is no picture below, select and enter the little closed cell below.

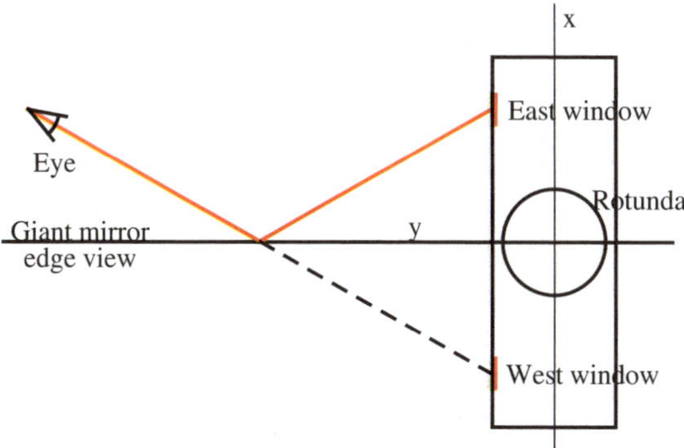

Fig. 4.2 Architectural symmetry

Ammonia has reflection symmetry

There is an obvious analogy with molecules. Below we call up from the Molecules` package our specification of the ammonia molecule, in a format that we adopt as standard for all molecules. Each row specifies one atom by its element name, its Cartesian position in the molecule (in Angstroms), and its atom tag.

```
MoleculeToList["ammonia"];
% // ColumnForm
```

$$\{N, \{0, 0, \frac{903}{25000}\}, \{1, C3v\}\}$$
$$\{H, \{\frac{2469}{2500}, 0, -\frac{1673}{10000}\}, \{1\}\}$$
$$\{H, \{-\frac{2469}{5000}, \frac{2469\sqrt{3}}{5000}, -\frac{1673}{10000}\}, \{2\}\}$$
$$\{H, \{-\frac{2469}{5000}, -\frac{2469\sqrt{3}}{5000}, -\frac{1673}{10000}\}, \{3\}\}$$

Note that we use no decimal approximations in the specification of molecules. The above is based on a four-digit experimental result for the ammonia molecule, but we have rationalized the Cartesian coordinates and used the exact irrational $\sqrt{3}$ to make the threefold rotational symmetry exact.

The operation of reflection in the **x, z** plane is performed by the **Rule**

{x,y,z}→{x,-y,z}.

This may be called either a reflection parallel to **y**, or a reflection in the **xz** plane. We generalize this definition :

> **Reflection plane**
> Given any plane, a Cartesian coordinate system can be defined such that two axes lie in the plane and the third is perpendicular. Reflection is the negation of the perpendicular coordinate.

Is negation of the **y** axis a symmetry operation for the ammonia molecule, as given above? By inspection, one can see that it is. If the sign of each **y**-component is changed, the first two atoms are unchanged, while the **H**-atoms with tags **2** and **3** (the last two atoms in the list) merely exchange places.

This changes the **List**, of course, because order is an important property of a **List**. But we regard the molecule as a **set** of atoms, not as a **List** of atoms, and order means nothing in a **set**. This idea is enforced by all our functions that operate on molecule objects. We have graphics functions, a molecular weight function, a center-of-mass function, an inertia tensor function, and others. But in all of them the output is independent of the order of the atoms in the molecule object.

To see if two molecule objects represent the same molecule, apply **Sort** to each molecule and then apply the **Equal** question. If they are truly identical except for atom order, the answer will be **True**.

Matrix formula for reflection in the y direction

In a computer, an easy way to change vector **{x,y,z}** into vector **{x,-y,z}** is to multiply it by a simple matrix. If the vector is written as a column, it must be written on the right side of the matrix. The usual row-by-column multiplication rules are embodied by the **Dot** operator in *Mathematica*, written quite literally as a normal punctuation **period** between the matrix and the column vector. We apply **MatrixForm** to the result to have it written as a column

$$\begin{pmatrix} 1 & 0 & 0 \\ 0 & -1 & 0 \\ 0 & 0 & 1 \end{pmatrix} . \begin{pmatrix} x \\ y \\ z \end{pmatrix} \text{ // MatrixForm}$$

$$\begin{pmatrix} x \\ -y \\ z \end{pmatrix}$$

In a molecule with four atoms, you can do all the atoms at once if the atom position vectors are written as the columns of a 3-by-4 matrix.

$$\begin{pmatrix} 1 & 0 & 0 \\ 0 & -1 & 0 \\ 0 & 0 & 1 \end{pmatrix} . \begin{pmatrix} x_1 & x_2 & x_3 & x_4 \\ y_1 & y_2 & y_3 & y_4 \\ z_1 & z_2 & z_3 & z_4 \end{pmatrix} \text{ // MatrixForm}$$

$$\begin{pmatrix} x_1 & x_2 & x_3 & x_4 \\ -y_1 & -y_2 & -y_3 & -y_4 \\ z_1 & z_2 & z_3 & z_4 \end{pmatrix}$$

Matrix multiplication is an extremely fast operation, so even pictures with thousands of points can be transformed this way very quickly. Many of the computer graphics animations that you see in movies and television advertising are based on frame-by-frame multiplication of a **3×3** transform matrix (on the left) dotted into a **3 × N** point matrix (on the right). This operation is so fast it can be done in real time in some movies that you see in computer games. Later, we will see a similar matrix scheme, of size **(4×4).(4×N)**, for translating and rotating at the same time.

It is traditional to give all reflection matrices a name that starts with σ. In the **Symmetry`** package we defined matrices **σx**, **σy**, and **σz** , and in the preliminaries we made them equivalent to the subscripted forms σ_x, σ_y, and σ_z:

```
Map[MatrixForm, {σx, σy, σz}]
```

$$\left\{\begin{pmatrix} -1 & 0 & 0 \\ 0 & 1 & 0 \\ 0 & 0 & 1 \end{pmatrix}, \begin{pmatrix} 1 & 0 & 0 \\ 0 & -1 & 0 \\ 0 & 0 & 1 \end{pmatrix}, \begin{pmatrix} 1 & 0 & 0 \\ 0 & 1 & 0 \\ 0 & 0 & -1 \end{pmatrix}\right\}$$

Numerical example of molecule reflection

We look at a real example. The **{x,y,z}** coordinates of the ammonia molecule can be brought up by the command

```
RestPoints["ammonia"] // MatrixForm
```

$$\begin{pmatrix} 0 & 0 & \frac{903}{25000} \\ \frac{2469}{2500} & 0 & -\frac{1673}{10000} \\ -\frac{2469}{5000} & \frac{2469\sqrt{3}}{5000} & -\frac{1673}{10000} \\ -\frac{2469}{5000} & -\frac{2469\sqrt{3}}{5000} & -\frac{1673}{10000} \end{pmatrix}$$

But above, the coordinates of each atom are written as a row. We need them as columns, which is easily done by the **Transpose** operator :

```
ammoniaCoordinates =
  Transpose[RestPoints["ammonia"]];
% // MatrixForm
```

$$\begin{pmatrix} 0 & \frac{2469}{2500} & -\frac{2469}{5000} & -\frac{2469}{5000} \\ 0 & 0 & \frac{2469\sqrt{3}}{5000} & -\frac{2469\sqrt{3}}{5000} \\ \frac{903}{25000} & -\frac{1673}{10000} & -\frac{1673}{10000} & -\frac{1673}{10000} \end{pmatrix}$$

Now we can multiply from the left by the reflection matrix σ_y :

```
reflectedAmmonia = σy.ammoniaCoordinates ;
% // MatrixForm
```

$$\begin{pmatrix} 0 & \frac{2469}{2500} & -\frac{2469}{5000} & -\frac{2469}{5000} \\ 0 & 0 & -\frac{2469\sqrt{3}}{5000} & \frac{2469\sqrt{3}}{5000} \\ \frac{903}{25000} & -\frac{1673}{10000} & -\frac{1673}{10000} & -\frac{1673}{10000} \end{pmatrix}$$

Columns **3** and **4** are both **H** atoms, and have traded places. But column order is meaningless (as long as both columns refer to the same kind of atom) so the molecule is unchanged. The technical language for this is that the atom set is *invariant* under the given transform.

Pictures of symmetry and non-symmetry reflections

We have defined an operator called **ShowOperation** which draws a **before** picture of any given molecule, extracts the rest points of its atoms as a **3×N** matrix, transforms them by the given symmetry matrix, and then draws an **after** picture of the molecule after transformation. Below, we ask it to use matrix σ_y on the molecule **"ammonia"**. We also include the option **ViewPoint→{4,2,3}**, which puts the viewing eye at this position in 3D space. The eye looks directly toward the origin, and the view that it sees is rendered in mathematically perfect perspective.

We first make a **Graphics3D** object that shows what we want, but with quite a bit of unwanted white space around it. The 3D object cannot be cropped. So we **Rasterize** it to make it into a two-dimensional **Graphics** object, and crop it using **PlotRange** inside the **Show** operator.

```
fig3 = ShowOperation[σy, "ammonia", ViewPoint → {4, 2, 3},
   Style → "TB", ImageSize → {300, 300}, Ticks → None];
fig3R = Rasterize[fig3, ImageResolution → 300];
Show[fig3R, PlotRange → {{40, 270}, {100, 230}},
 ImageSize → {200, 100}]
```

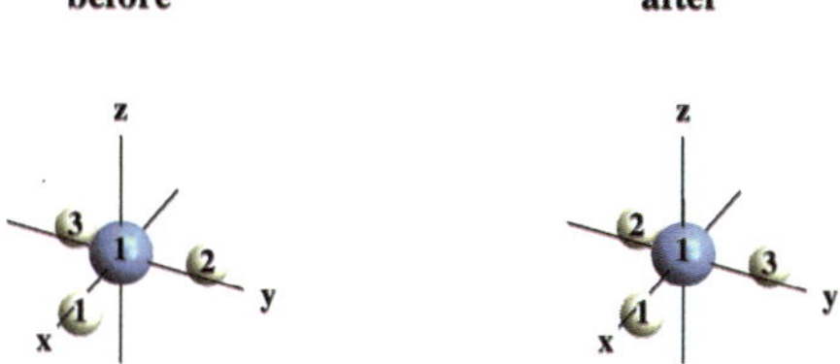

Fig. 4.3 Ammonia with numbered atoms, before and after reflection in the **z,x** plane. H atoms 2 and 3 switch places. But they are identical, so the molecule is unchanged. This is symmetry.

The molecule looks the same **before** and **after** (except for atom labels, which don't count), so according to our definition σ_y is a symmetry operation for the ammonia molecule- it is a transform that leaves the object unchanged. But to really understand this, you must also see an operation on a molecule that is NOT a symmetry operation:

```
fig4 = ShowOperation[σx, "ammonia", ViewPoint → {4, 2, 3},
   Style → "TB", ImageSize → {300, 300}, Ticks → None];
fig4R = Rasterize[fig4, ImageResolution → 300];
fig4S = Show[fig4R, PlotRange → {{40, 270}, {100, 230}},
  ImageSize → {200, 100}]
```

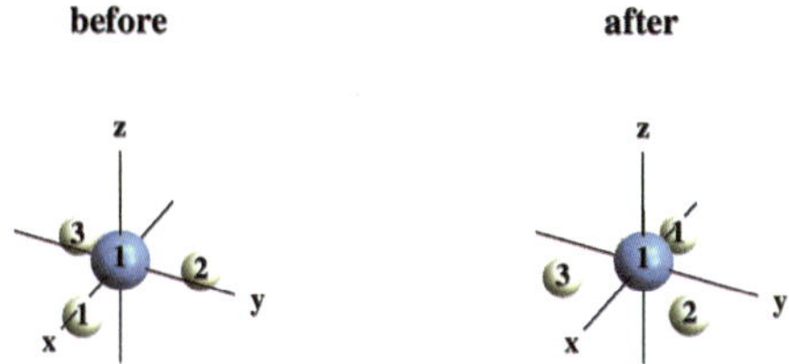

Fig. 4.4 Ammonia, before and after reflection in the **y,z** plane. The atoms moved, so **y,z** reflection is not a symmetry of ammonia.

Each atom "reflects through" the **z,y** plane (or stays put, if it lies in the plane). The molecule **after** looks quite different than **before**, so σ_x is NOT a symmetry operation for this molecule.

4.2.4. Rotation matrices

Rotational symmetry

Rotational symmetry is another common interpretation of the word "symmetry"; it is embodied by the rotunda and dome of the capitol, considered apart from the rest of the building. It has 24 identical segments, each with a little window and other features, so we say that it has "24-fold" rotational symmetry about a **z** axis that rises vertically through its exact center.

Physically, we mean that if the dome were jacked up and turned by $2\pi/24$, the Capitol would be exactly the same before as after. As a public works project, this would certainly bring home the meaning of rotational symmetry to the entire nation.

Our model ammonia molecule has a similar rotation symmetry. It is a little umbrella, where the umbrella handle is the z axis, and the blue **N** atom is at the top of the umbrella. But in this case the symmetry rotation is only a third of a turn.

```
amm = Show[{AtomGraphics["ammonia", AtomLabels -> False],
    BondGraphics["ammonia", 1.2, .015]}, Boxed -> False];
AxialViews[amm, ImageSize → {200, 200}]
```

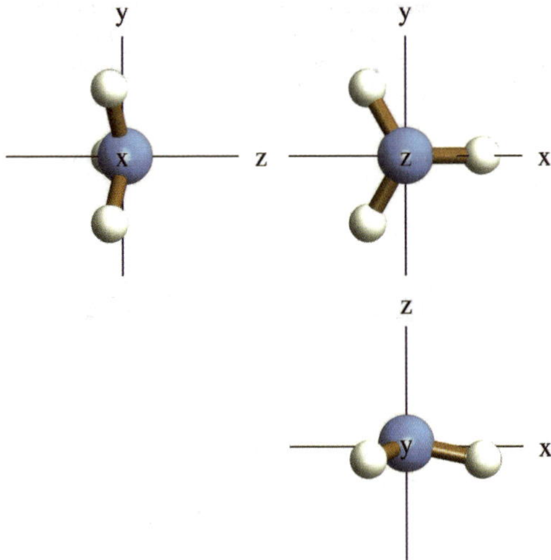

Fig. 4.5 Complete information on the structure of ammonia.

A matrix formula for rotation about the z axis

Before we rotate whole objects, let us think about how to rotate a single mathematical point around the **z**-axis. We begin by writing a general Cartesian point **{x,y,z}** in cylindrical coordinates **{r,α,z}**, taking the cylinder axis as the **z** axis. The general point before transformation is

```
ptBefore = {x, y, z} /. {x → r Cos[α], y → r Sin[α]}
```

```
{r Cos[α], r Sin[α], z}
```

Before transformation, the point could be anywhere, as specified by the values of **r**, **α**, and **z**. The final point, however, must be related to the initial point by a simple rotation through some angle **β** about the **z** axis. This means that **r** and **z** must stay constant and the angle must change from **α** to **α + β**. Thus we write

```
ptAfter = ptBefore /. α → α + β
```

```
{r Cos[α + β], r Sin[α + β], z}
```

TrigExpand rewrites trigonometric functions of sums, difference, or multiples of angles in terms of the individual angles.

```
ptAfter2 = TrigExpand[ptAfter]
```

```
{r Cos[α] Cos[β] - r Sin[α] Sin[β],
 r Cos[β] Sin[α] + r Cos[α] Sin[β], z}
```

In this form we can reverse the original substitution, trading off **α** and **r** in favor of the original **x** and **y** :

```
ptAfter3 = ptAfter2 /. {r Cos[α] → x, r Sin[α] → y}
```

```
{x Cos[β] - y Sin[β], y Cos[β] + x Sin[β], z}
```

This is the position of the point after rotation, in terms of the original position **{x,y,z}** and the rotation angle **β**. Each of the three vector components is linear in **x**, **y**, and **z**. Therefore, we can rewrite this expression as a *rotation matrix* times the original vector **{x,y,z}**.

In fact, the matrix may be extracted by a handy operator called **MatrixOfCoefficients**. It takes two operands; first, a list of expressions, and second, a list of variables, in which the expressions must be linear. First, we try it on a purely mathematical example. Consider two expressions linear in **x**, **y**, and **z**.

```
twoExpressions = {
    c11 x + c12 y + c13 z,
    c21 x + c22 y };
```

The matrix of coefficients extracted by the operation

```
MatrixOfCoefficients[twoExpressions, {x, y, z}] //
 MatrixForm
```

$$\begin{pmatrix} c_{11} & c_{12} & c_{13} \\ c_{21} & c_{22} & 0 \end{pmatrix}$$

Note that any missing variable is automatically assigned a zero coefficient, so the matrix is always rectangular. Now we are ready to extract the square matrix of coefficients for rotation about the **z** axis by angle β :

```
rotMat = MatrixOfCoefficients[ptAfter3, {x, y, z}];
MatrixForm[rotMat]
```

$$\begin{pmatrix} \text{Cos}[\beta] & -\text{Sin}[\beta] & 0 \\ \text{Sin}[\beta] & \text{Cos}[\beta] & 0 \\ 0 & 0 & 1 \end{pmatrix}$$

This is a beautiful result. It says that in order to rotate any point **{x,y,z}** around the **z** axis by angle β, we simply left-multiply the point by this matrix. Let's check it, using the starting expression **ptAfter3** above for the right side :

$$\begin{pmatrix} \text{Cos}[\beta] & -\text{Sin}[\beta] & 0 \\ \text{Sin}[\beta] & \text{Cos}[\beta] & 0 \\ 0 & 0 & 1 \end{pmatrix} . \begin{pmatrix} x \\ y \\ z \end{pmatrix} == \begin{pmatrix} x\,\text{Cos}[\beta] - y\,\text{Sin}[\beta] \\ y\,\text{Cos}[\beta] + x\,\text{Sin}[\beta] \\ z \end{pmatrix}$$

```
True
```

Even better, the format **matrix.oldVector == newVector** is exactly the same as for reflection. If we keep at it, we can develop matrices representing all possible reflections and rotations.

Perhaps it's obvious, but this format also means that we can rotate many vectors at once, just as we reflected many vectors at once, using formulas with column vector points, like

$$\begin{pmatrix} \text{Cos}[\beta] & -\text{Sin}[\beta] & 0 \\ \text{Sin}[\beta] & \text{Cos}[\beta] & 0 \\ 0 & 0 & 1 \end{pmatrix} . \begin{pmatrix} x_1 & x_2 & x_3 & x_4 \\ y_1 & y_2 & y_3 & y_4 \\ z_1 & z_2 & z_3 & z_4 \end{pmatrix} \text{ // GridForm}$$

Cos[β] x_1 - Sin[β] y_1	Cos[β] x_2 - Sin[β] y_2	Cos[β] x_3 - Sin[β] y_3	Cos[β] x_4 - Sin[β] y_4
Sin[β] x_1 + Cos[β] y_1	Sin[β] x_2 + Cos[β] y_2	Sin[β] x_3 + Cos[β] y_3	Sin[β] x_4 + Cos[β] y_4
z_1	z_2	z_3	z_4

Two-dimensional rotations (about an axis implicitly perpendicular to the page) must be

```
(Cos[β]  -Sin[β])   (x1  x2)
(Sin[β]   Cos[β]) . (y1  y2)  // GridForm
```

Cos[β] x_1 - Sin[β] y_1	Cos[β] x_2 - Sin[β] y_2
Sin[β] x_1 + Cos[β] y_1	Sin[β] x_2 + Cos[β] y_2

Standard notation for rotations

C_n denotes a rotation by **2π/n** radians about an unspecified axis.

$C_{n,x}$ denotes a rotation by **2π/n** radians about the **x** axis.

C$_x$[ϕ] is a matrix for rotation by **ϕ** radians about the **x** axis, and similarly for **y** and **z**.

Does the **Symmetry`** package know this notation ? Subscripts are forbidden in packages, but in the **Preliminaries** we set some notational equalities that permit us to write

```
Cx[ϕ] // GridForm
```

1	0	0
0	Cos[ϕ]	-Sin[ϕ]
0	Sin[ϕ]	Cos[ϕ]

right handed rotation

Position your right hand with the thumb extended and pointing in the positive direction along the axis of rotation, and curl your fingers. The fingers then point in the sense of a right handed rotation.

Another way of saying it:

Right handed rotations are counterclockwise, as viewed from a point on the positive side of the axis of rotation, looking toward the origin.

Right handed is meaningless if the axis of rotation does not have plus and minus ends. Thus, if an axis of rotation in a drawing is just a simple line with no indication of the positive end (such as a plus sign, or an arrowhead, or an axis name), then there is no way to know which sense of rotation is right handed.

The undefined terms of the handedness definition are *right hand* and *counter-clockwise*. This has given rise to some deep thoughts. What would you understand here if you knew nothing about hands or clocks? The definition rests only on human conventions, yet the world is full of handedness asymmetries; most notably in the realm of biomolecules. The electromagnetic force, that governs the structure of molecules, has no intrinsic handedness, so right- and left-handed versions of handed molecules have exactly the same energy in all standard quantum chemical calculations. The energies split in a magnetic field, but the effect is very tiny for fields the size of the earth's.

At a more fundamental level, the "weak" interaction, which does apply to electrons, is handed. But it is so small that it cannot influence chemistry. Yet the amino acids extracted from living materials are all 100% handed, and the handedness is always the same throughout all of biology. This asymmetry remains a mystery, one of the few that 20th century science failed to illuminate. Most scientists by now believe that the source of this asymmetry is not to be sought in physics, but in the extreme early stages of Darwinian evolution, when barely living polymer systems were first emerging from a primordial chemical soup. If mixed handedness ever existed, then some early evolutionary advantage developed in only one handedness, in a way that could not be adopted by the other. Perhaps it was a small polymer that could bind left-X (but not right-X) and convert it to something useful. So the organisms that had only right-X fell into a disadvantage, and dwindled to extinction.

Rotations about the **x**-axis and the **y**-axis may be derived just like the **z**-rotations, above. (This is an excellent exercise; try it!) The results are already in the **Symmetry`** package :

```
Cy[β] // MatrixForm
```

$$\begin{pmatrix} \text{Cos}[\beta] & 0 & \text{Sin}[\beta] \\ 0 & 1 & 0 \\ -\text{Sin}[\beta] & 0 & \text{Cos}[\beta] \end{pmatrix}$$

```
Cz[β] // MatrixForm
```

$$\begin{pmatrix} \text{Cos}[\beta] & -\text{Sin}[\beta] & 0 \\ \text{Sin}[\beta] & \text{Cos}[\beta] & 0 \\ 0 & 0 & 1 \end{pmatrix}$$

Numerical example of molecule rotation

You need to see a numerical example to get the full meaning here. Above, we defined

```
ammoniaCoordinates =
  Transpose[RestPoints["ammonia"]];
ammoniaCoordinates // MatrixForm
```

$$\begin{pmatrix} 0 & \frac{2469}{2500} & -\frac{2469}{5000} & -\frac{2469}{5000} \\ 0 & 0 & \frac{2469\sqrt{3}}{5000} & -\frac{2469\sqrt{3}}{5000} \\ \frac{903}{25000} & -\frac{1673}{10000} & -\frac{1673}{10000} & -\frac{1673}{10000} \end{pmatrix}$$

The first column represents an atom is on the **z** axis, which must be the nitrogen atom. The other three columns are the **H** atoms. Now we rotate all the atom positions at once around the **z** axis by the matrix formula

```
rotatedAmmonia =
 Cz[2 π / 3].ammoniaCoordinates  // MatrixForm
```

$$\begin{pmatrix} 0 & -\frac{2469}{5000} & -\frac{2469}{5000} & \frac{2469}{2500} \\ 0 & \frac{2469\sqrt{3}}{5000} & -\frac{2469\sqrt{3}}{5000} & 0 \\ \frac{903}{25000} & -\frac{1673}{10000} & -\frac{1673}{10000} & -\frac{1673}{10000} \end{pmatrix}$$

The nitrogen atom (first column) was unmoved because it lies on the axis of rotation. The rotation has switched the three **H**-atom columns around, but column order (atom order) is meaningless, so the molecule is unchanged. As expected, a three-fold rotation about **z** is a symmetry operation for ammonia.

Pictures of symmetry and non-symmetry rotations

```
ShowOperation[Cz[2 π / 3], "ammonia",
 ViewPoint -> {0, 0, 4}, ImageSize → 300]
```

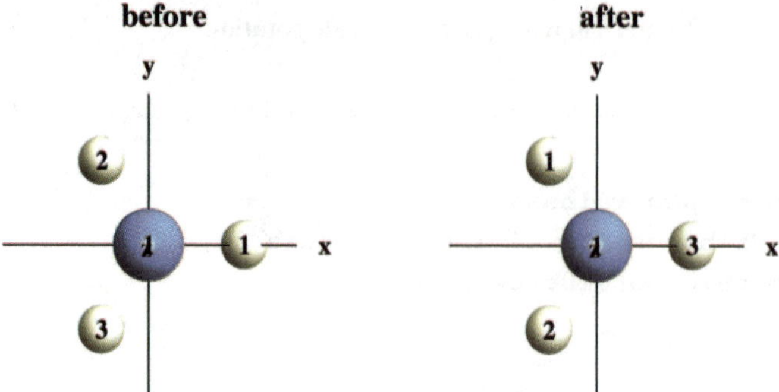

Fig. 4.6 Rotation of ammonia by a third of a turn IS a symmetry operation.

Without the labels, one cannot tell above that anything happened. This is a symmetry transform. Look at rotation by half a turn:

```
ShowOperation[Cz[π], "ammonia",
 ViewPoint -> {0, 0, 4}, ImageSize → 300]
```

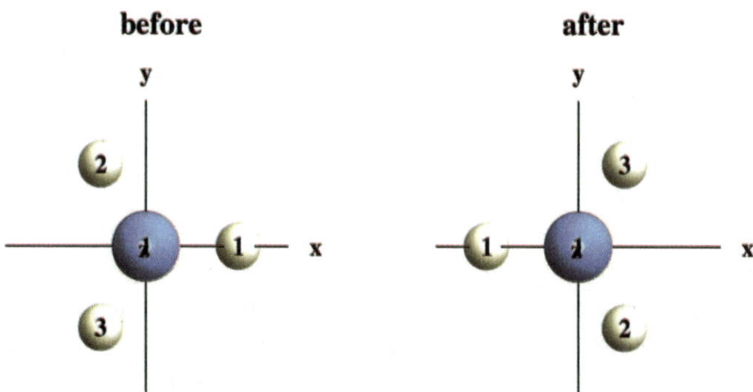

Fig. 4.7 Rotation of ammonia by a half turn is NOT a symmetry transform.

Here you do not need labels to see that something happened. This is NOT a symmetry transform.

4.2.5. New symmetries from old

If you know of two transforms that leave an object unchanged, then it must be true that applying first one and then the other also leaves the object unchanged. This simple idea has a very important consequence. Look at an example.

```
ammoniaCoordinates // MatrixForm
```

$$\begin{pmatrix} 0 & \frac{2469}{2500} & -\frac{2469}{5000} & -\frac{2469}{5000} \\ 0 & 0 & \frac{2469\sqrt{3}}{5000} & -\frac{2469\sqrt{3}}{5000} \\ \frac{903}{25000} & -\frac{1673}{10000} & -\frac{1673}{10000} & -\frac{1673}{10000} \end{pmatrix}$$

If we apply first the reflection and then the rotation about **z** by a third of a turn, the calculation is

```
Cz[2 π / 3].(σy.ammoniaCoordinates) // MatrixForm
```

$$\begin{pmatrix} 0 & -\frac{2469}{5000} & \frac{2469}{2500} & -\frac{2469}{5000} \\ 0 & \frac{2469\sqrt{3}}{5000} & 0 & -\frac{2469\sqrt{3}}{5000} \\ \frac{903}{25000} & -\frac{1673}{10000} & -\frac{1673}{10000} & -\frac{1673}{10000} \end{pmatrix}$$

The result is the same molecule, as it must be, though we have not seen this particular order of the **H** atoms before. Matrix multiplication is associative, so this is the same as

```
(Cz[2 π / 3].σy).ammoniaCoordinates // MatrixForm
```

$$\begin{pmatrix} 0 & -\frac{2469}{5000} & \frac{2469}{2500} & -\frac{2469}{5000} \\ 0 & \frac{2469\sqrt{3}}{5000} & 0 & -\frac{2469\sqrt{3}}{5000} \\ \frac{903}{25000} & -\frac{1673}{10000} & -\frac{1673}{10000} & -\frac{1673}{10000} \end{pmatrix}$$

This shows that the single new matrix, given by

```
Cz[2 π / 3].σy // MatrixForm
```

$$\begin{pmatrix} -\frac{1}{2} & \frac{\sqrt{3}}{2} & 0 \\ \frac{\sqrt{3}}{2} & \frac{1}{2} & 0 \\ 0 & 0 & 1 \end{pmatrix}$$

is also a symmetry transform of the ammonia molecule. Later we will see that this matrix represents reflection in a line rotated by 1/3 turn from the y-axis. It is different from either of the other two that we have been considering, which we repeat just below for comparison:

```
{Cz[2 π / 3] // MatrixForm, σx // MatrixForm}
```

$$\left\{\begin{pmatrix} -\frac{1}{2} & -\frac{\sqrt{3}}{2} & 0 \\ \frac{\sqrt{3}}{2} & -\frac{1}{2} & 0 \\ 0 & 0 & 1 \end{pmatrix}, \begin{pmatrix} -1 & 0 & 0 \\ 0 & 1 & 0 \\ 0 & 0 & 1 \end{pmatrix}\right\}$$

This illustrates an important principle: If you multiply any two matrices that are symmetry transform matrices of the same object, you get another symmetry transform matrix of that object. Another way of saying it: Two symmetries imply the existence of a third symmetry, which may be different from the first two, and which you may or may not have noticed directly.

4.3. Algebraic symmetry

4.3.1. Algebraic meanings for the undefined terms

We now take specific meanings for the undefined terms in the definition of *symmetry*. We let ***object* == algebraic *expression***, and we let ***transform* == *replacement of variables*** that belong to the expression. Repeating the definition of symmetry with these explicit meanings:

"We will say that a set of algebraic expressions has a symmetry if we know of a replacement of variables that leaves the set unchanged."

Replacement of variables is an operation that lies at the heart of *Mathematica,* but usually, it is NOT a symmetry transform. For instance, in the expression x^3 we can replace **x** by **-x** if we write

```
x^3 /. x → -x
```

$-x^3$

The output is algebraically different from the input and there is no symmetry. When we say *unchanged* we mean it quite literally. Look at the following, perhaps the most famous of examples :

```
x^2 /. x → -x
```

x^2

Now there, that is what we mean by symmetry in an algebraic expression. The output expression is EXACTLY the same as the input expression Or, as we will see, is algebraically equivalent to input expression.

4.3.2. Examples of algebraic symmetry

Sine and cosine under x, y exchange

Take a set of two expressions, the Cartesian expressions for sine and cosine, and then swap **x** and **y** :

$$\left\{\frac{y}{\sqrt{x^2+y^2}}, \frac{x}{\sqrt{x^2+y^2}}\right\} /. \{x \to y, y \to x\}$$

$$\left\{\frac{x}{\sqrt{x^2+y^2}}, \frac{y}{\sqrt{x^2+y^2}}\right\}$$

The second becomes the first and first become the second. The **List** is different, because the order has changed. But a **set** is a collection without regard to order, and the **set** is the same.

From matrix to rule

Every geometric symmetry transform can be described as either a matrix or as a replacement rule. Many transform matrices are tabulated in our **Symmetry`** package, and there is a simple way to turn them into rules. Here it is:

```
Thread[{x, y, z} → mat.{x, y, z} ]
```

```
{x → mat.{x, y, z}, y → mat.{x, y, z}, z → mat.{x, y, z}}
```

To see what **Thread** does, take it away and rerun. (It turns a **3D** vector replacement into three **1D** scalar replacements.)

Matrix **σz** is defined in the **Symmetry`** package :

```
σz // MatrixForm
```

$$\begin{pmatrix} 1 & 0 & 0 \\ 0 & 1 & 0 \\ 0 & 0 & -1 \end{pmatrix}$$

The corresponding rule is

```
Thread[{x, y, z} → σz.{x, y, z} ]
```

```
{x → x, y → y, z → -z}
```

Later, when you are more familiar with the names of groups and elements, you can use a package operator to call them down by the name **SymmetryRule**. Call up its syntax statement (using the ?) and read it.

Distance from the origin, under rotation

This one illustrates a little different twist. We will show that the distance of a point from the origin is invariant under rotation about the origin. The original position is **{x,y,z}** and its distance from the origin is $\sqrt{x^2 + y^2 + z^2}$. We rotate around the **z** axis by angle α :

```
Cz[α].{x, y, z}
```

```
{x Cos[α] - y Sin[α], y Cos[α] + x Sin[α], z}
```

The corresponding rotational transform rules are

```
rotRule = Thread[{x, y, z} → Cz[α].{x, y, z} ]
```

```
{x → x Cos[α] - y Sin[α], y → y Cos[α] + x Sin[α], z → z}
```

Now apply these rules to the radius formula :

$$\text{transformedRadius} = \sqrt{x^2 + y^2 + z^2} \;\; /.\;\; \text{rotRule}$$

$$\sqrt{z^2 + (y\,\text{Cos}[\alpha] + x\,\text{Sin}[\alpha])^2 + (x\,\text{Cos}[\alpha] - y\,\text{Sin}[\alpha])^2}$$

This certainly does not look the same before and after replacement. The point of this example is, initial appearance is not the criterion. The criterion is whether, after replacement, one can use valid algebraic replacement rules to return it to the original form :

```
Map[TrigExpand, transformedRadius]
```

$$\sqrt{x^2 + y^2 + z^2}$$

There it is. As we knew, the radius expression is indeed invariant under rotation.

Coulomb field of one H nucleus

The algebraic symmetry of Hamiltonian operators is of central importance in molecular physics. If the Hamiltonian $\mathcal{H}_{x,y,z}$ is symmetric under a given algebraic transform, then the ground state of the molecule ψ_0**[x, y, z]** (but not necessarily higher states!) will be symmetric under the same transform. We again take the ammonia molecule as our example.

We look first at the potential energy part of the Hamiltonian. The potential energy of an electron at **{x,y,z}** due to a proton at the origin is given (in atomic units) by $-1/\sqrt{x^2 + y^2 + z^2}$. If the proton is not at the origin, but at **{a,b,c}**, this energy becomes $1/\sqrt{(x-a)^2 + (y-b)^2 + (z-c)^2}$. We can make this into a function of the vector **{a,b,c}** :

```
Clear[CoulombEnergy];
CoulombEnergy[{a_, b_, c_}] :=
```

$$\frac{-1}{\sqrt{(x-a)^2 + (y-b)^2 + (z-c)^2}}$$

We test the function out:

```
CoulombEnergy[{x1, y1, z1}]
```

$$-\frac{1}{\sqrt{(x-x_1)^2 + (y-y_1)^2 + (z-z_1)^2}}$$

Coulomb field of all the H nuclei in ammonia

In our ammonia molecule the three **H** atoms are at

```
ptH1 = RestPoints["ammonia"][[2]];
ptH2 = RestPoints["ammonia"][[3]];
ptH3 = RestPoints["ammonia"][[4]];
{ptH1, ptH2, ptH3} // ColumnForm
```

$$\left\{\frac{2469}{2500}, 0, -\frac{1673}{10000}\right\}$$

$$\left\{-\frac{2469}{5000}, \frac{2469\sqrt{3}}{5000}, -\frac{1673}{10000}\right\}$$

$$\left\{-\frac{2469}{5000}, -\frac{2469\sqrt{3}}{5000}, -\frac{1673}{10000}\right\}$$

The three Coulomb energy expressions may be written in a list by the command

```
threeEnergies =
 {CoulombEnergy[ptH1], CoulombEnergy[ptH2],
   CoulombEnergy[ptH3]} // ExpandAll
```

$$\left\{-\frac{1}{\sqrt{\frac{20066861}{20000000}-\frac{2469\,x}{1250}+x^2+y^2+\frac{1673\,z}{5000}+z^2}},\right.$$

$$-\frac{1}{\sqrt{\frac{20066861}{20000000}+\frac{2469\,x}{2500}+x^2-\frac{2469\sqrt{3}\,y}{2500}+y^2+\frac{1673\,z}{5000}+z^2}},$$

$$\left.-\frac{1}{\sqrt{\frac{20066861}{20000000}+\frac{2469\,x}{2500}+x^2+\frac{2469\sqrt{3}\,y}{2500}+y^2+\frac{1673\,z}{5000}+z^2}}\right\}$$

Rotational symmetry of the Coulomb field

Now we want to see what happens to these expressions when we rotate the coordinate **{x,y,z}** by one-third of a turn. We already have this transform above in terms of the general rotation angle α, so we just replace α appropriately.

```
oneThirdTurn = (rotRule /. α → 2 π / 3)
```

$$\left\{x \to -\frac{x}{2}-\frac{\sqrt{3}\,y}{2},\ y \to \frac{\sqrt{3}\,x}{2}-\frac{y}{2},\ z \to z\right\}$$

Perform the replacements :

```
rotatedEnergies =
 threeEnergies /. oneThirdTurn // ExpandAll
```

$$\left\{-\frac{1}{\sqrt{\frac{20066861}{20000000}+\frac{2469\,x}{2500}+x^2+\frac{2469\sqrt{3}\,y}{2500}+y^2+\frac{1673\,z}{5000}+z^2}},\right.$$

$$-\frac{1}{\sqrt{\frac{20066861}{20000000}-\frac{2469\,x}{1250}+x^2+y^2+\frac{1673\,z}{5000}+z^2}},$$

$$\left.-\frac{1}{\sqrt{\frac{20066861}{20000000}+\frac{2469\,x}{2500}+x^2-\frac{2469\sqrt{3}\,y}{2500}+y^2+\frac{1673\,z}{5000}+z^2}}\right\}$$

A very interesting thing has happened: each Coulomb expression has been transformed into one of the other two.

```
{threeEnergies⟦1⟧ == rotatedEnergies⟦2⟧,
 threeEnergies⟦2⟧ == rotatedEnergies⟦3⟧,
 threeEnergies⟦3⟧ == rotatedEnergies⟦1⟧}
```

```
{True, True, True}
```

Clearly, the sum of the three is the same before and after this transform, so the sum is an algebraic expression that is symmetric under rotation by one third of a turn.

Reflection symmetry of the Coulomb field

The Coulomb energy should also be symmetric under the reflection transform **y → -y**. You can see it by inspection, but we make it explicit:

```
threeEnergiesReflected = threeEnergies /. y → -y //
  ExpandAll
```

$$\left\{-\frac{1}{\sqrt{\frac{20066861}{20000000}-\frac{2469\,x}{1250}+x^2+y^2+\frac{1673\,z}{5000}+z^2}},\right.$$

$$-\frac{1}{\sqrt{\frac{20066861}{20000000}+\frac{2469\,x}{2500}+x^2+\frac{2469\sqrt{3}\,y}{2500}+y^2+\frac{1673\,z}{5000}+z^2}},$$

$$\left.-\frac{1}{\sqrt{\frac{20066861}{20000000}+\frac{2469\,x}{2500}+x^2-\frac{2469\sqrt{3}\,y}{2500}+y^2+\frac{1673\,z}{5000}+z^2}}\right\}$$

Comparing the reflected expression to the original, we see that the last two terms have been transformed into each other, while the first remains invariant. Again, the sum will be invariant, so **y**-reflection is a symmetry transform for it.

4.3.3. Visualization of algebraic symmetry

If all this is true, the contour plot of this function should show a threefold rotational symmetry and a y-reflection symmetry. This is easy to verify. We add up the three expressions, to give the total potential energy of the electron due to the three H nuclei, and (preparing to make a 2-D graphic) we specialize to the plane **z == 0** :

```
energyH3z0 = Apply[Plus, threeEnergies] /. z → 0
```

$$-\frac{1}{\sqrt{\frac{20066861}{20000000}-\frac{2469\,x}{1250}+x^2+y^2}}-$$

$$\frac{1}{\sqrt{\frac{20066861}{20000000}+\frac{2469\,x}{2500}+x^2-\frac{2469\sqrt{3}\,y}{2500}+y^2}}-$$

$$\frac{1}{\sqrt{\frac{20066861}{20000000}+\frac{2469\,x}{2500}+x^2+\frac{2469\sqrt{3}\,y}{2500}+y^2}}$$

The **Apply** operator, as used above, is worth noting in detail. We take a simple example and write it out in **FullForm**, to show that **Apply** is an operator that changes the head of an expression :

```
Apply[Plus, List[A, B, C]]
```

A + B + C

and can therefore be used to add up the items in a **List**. Returning to **energyH3z0** function, we make a contour plot for it:

```
ContourPlot[energyH3z0, {x, -2, 2}, {y, -2, 2},
 PlotPoints -> 40, BaseStyle → "TR", Axes -> True,
 AxesLabel -> {"x", "y"}, Epilog -> {Thickness[.005], Red,
   Line[{{-2, 0}, {2, 0}}], Blue, Line[{{1, -√3}, {-1, √3}}],
   Line[{{1, √3}, {-1, -√3}}]}, ImageSize → 200]
```

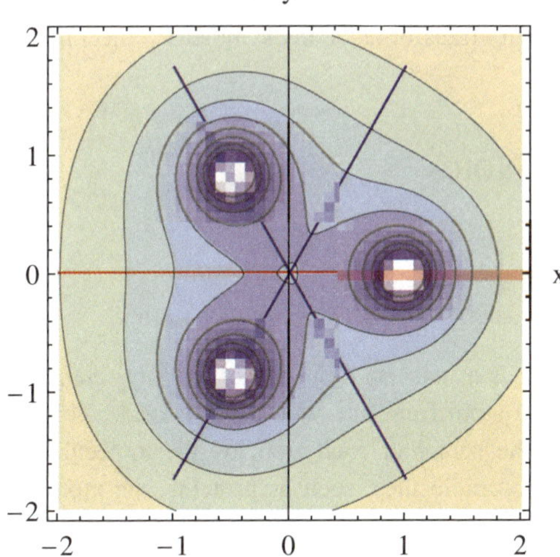

Fig. 4.8 Cross section of the potential of an electron in the three protons of the ammonia molecule, taken through the center of gravity, perpendicular to the threefold axis.

The potential does indeed have the expected symmetries: a threefold rotational symmetry, as well as a reflection symmetry in the red horizontal line.

This plot also shows two other symmetries, reflections in the two blue lines rotated **±1/3** turn from the red horizontal line. You can see rather intuitively that if any object has a reflection plane that contains a threefold rotation axis, it has to have the two other similar reflection planes. Finding all the symmetries implied by a small number of symmetries is an important part of group theory that we will study in detail in Chapter 15 (Make Matrix Group).

4.4. Summary and preview

We have considered two basic kinds of symmetry, reflection and rotation, focussing on the symmetry transforms themselves, in two forms.

1. Geometric symmetry transforms were carried out by matrix multiplications.
2. Algebraic symmetry transforms were carried out by substitutions of variables.

In the next chapter (Axioms) we will introduce the concept of a mathematical group; in the chapter after that we will see many examples of groups; then in Chapter 7 we will prove the fundamental theorem of symmetry theory; namely, that all the symmetry transforms of any symmetric object must form a group.

4.5. End Notes

Atom tags

The full atom tag is a list. Its first (and often only) element is an integer that distinguishes each atom from others of its own kind. This integer is automatically printed in the center of each atom by the molecular graphics functions. But in complicated molecules, such as proteins, the atom tag list can include other information, such as to which amino acid the atom belongs, or to which chain of a protein complex. Specialized molecular graphics functions can extract this information to color whole amino acids, or whole chains.

5. Axioms of group theory

Preliminaries

5.1. Undefined terms in the axioms

In this book, we will be concerned almost exclusively with groups of symmetry transforms. But the "group" is one of the most general concepts in mathematics, and we must begin with the axioms of group theory, stated in a form that applies to all kinds of groups.

We consider a set of **elements** `{A,B,C,... }` and an **operator** written as ⊗ which takes two members of the set as its arguments, in expressions like **A⊗B**. The elements of the set form a "group under ⊗" if the conditions of Axioms 1 through 4 are met.

Comments:

(1) The terms **elements** and **operator** are undefined. They can take on any meanings for which they make sense.

(2) The set of elements may be finite or infinite in number.

(3) When **A⊗B** goes to the *Mathematica* processor, it is automatically translated into a square bracket form like `f[A,B]`, where the symbolic ⊗ has become an explicit `f`, like **Times**, **Dot**, **Plus**, or **Permute**. In all these cases, **A** and **B** will be simple, definite things that the operator can work on.

(4) The number of elements in a group is called its *group order*. This has absolutely nothing to do with sorting order. The group elements may be named or listed in any sorting order.

(5) There is some confusion whether the elements `{A,B,C,... }` must all be different, or whether repeats are allowed. This confusion is tolerated because it affects nothing but the language with which some of the theorems are stated. Usually we will mean the elements to be distinct, but when they are not we will call the group a *redundant group*.

Symmetry groups are associated with groups of matrices called representations, and representations are often redundant groups.

W.M. McClain, *Symmetry Theory in Molecular Physics with Mathematica*,
DOI 10.1007/b13137_5, © Springer Science+Business Media, LLC 2009

5.2. The four axioms

5.2.1. Closure Axiom

> **Axiom 1.**
> If **M** and **N** are members of a group under ⊗, the operation **M⊗N** yields a unique value, which is also a member of the group.

Comments:

(1) No matter what the operator ⊗ actually is, it is conventionally called *multiplication*. For instance, we will often specify that the elements of the group are *matrices* and that the operation is *matrix multiplication,* In this case **M ⊗N** will be carried out as **M.N** , which in *Mathematica* is **Dot[M,N]**. Or **M** and **N** might be ordered lists, and **M⊗N** could mean a reordering of list **M** according to instructions in list **N**. This is permutation, carried out in *Mathematica* as **Part[M,N]**.

(2) "Multiplication" may be commutative or non-commutative. That is, it may or may not be true that **M⊗N** is always identical to **N⊗M**. For numbers under numerical multiplication, it is true. But when ⊗ is matrix multiplication or permutation of a list, the two products are generally different. Commutative groups are often called **Abelian** groups, after the Norwegian mathematician Niels Hendrik Abel (1802- 1829), forever young.

5.2.2. Unit Element Axiom

> **Axiom 2.**
> Every group contains a unit element **E**, such that for every member **X** of the group,
> **X⊗E == E⊗X == X**.

Comments:

(1) The letter **E** comes from the German word *Einheit*, meaning *unity*. This notation is so firmly embedded in group theory that it absolutely cannot be changed. However, in *Mathematica* the CapitalE is defined as the exponential *e*. Rather than disrupt this system, we will use a string "E" for the E of group theory. In Input cells it makes a big difference :

```
{E, "E"} // N
{2.71828, E}
```

(2) For groups of numbers under numerical multiplication, **E** is indeed the number **1**. For groups of matrices, **E** is the unit matrix (with **1**'s down the diagonal and zeroes everywhere else).

(3) For a more perverse example, consider the set of all numbers with ⊗ as numerical addition. Here the number zero plays the role of the unit element **E**, because **0 + n = n + 0 = n**.

(4) When listing the members of a group, the unique element **E** is conventionally given first.

5.2.3. Inverse Axiom

> **Axiom 3.**
> For every element **X** of the group, there is an element **Y** in the group such that
>
> **X⊗Y == Y⊗X == E.**

Comments:

(1) Elements **X** and **Y** are said to be *mutually inverse*, and you may write **X** as Y^{-1}, or **Y** as X^{-1}. It is allowed that **X** and **Y** be the same. For instance, setting **X → E** in **Axiom 3**, above, we see an instant proof that **Y** has to be **E**, so that **E** is its own inverse.

(2) The algorithm for finding the inverse depends on what the elements are. For numbers under **Times**, the inverse of **x** is **1/x**, as carried out by long division. For matrices under **Dot**, it is the matrix inverse operation, as carried out by the *Mathematica* operator **Inverse[mat]**.

(3) This axiom excludes **0** from every group of numbers under multiplication, just as it excludes singular matrices (matrices that have no matrix inverse) from all groups of matrices under **Dot**.

5.2.4. Associative Axiom

> **Axiom 4.**
> For any three elements **A**, **B**, and **C** of the group, it must be **True** that
>
> **A⊗(B⊗C)) == (A⊗B)⊗C.**

Comments:

(1) This property of the operator ⊗ is called **associativity**.

(2) This statement must be read in the light of the universal convention that at each step in an evaluation, only the innermost groupings are evaluated. Thus, in **A ⊗ (B ⊗ C)** , the inner **B ⊗ C** evaluates as some **D**, then **A ⊗ D** is evaluated. But in **(A ⊗ B) ⊗ C** , the inner **A ⊗ B** evaluates first as some **F**, then **F ⊗ C** is evaluated.

(3) Every operator must be tested for associativity according to its own properties. Associativity is a well known property of numerical multiplication, and of matrix multiplication.

(4) Because of associativity, the triple product **A ⊗ B ⊗ C** is uniquely defined, and does not really need any grouping.

(5) This axiom excludes numerical division as a possible group operation, because, for example, **8/(4/2)** is **4**, whereas **(8/4)/2** is **1**.

5.2.5. Coda

These axioms are surprisingly rich in content. All of the many subtle theorems of abstract group theory trace back to these four statements and to nothing else. The terms "class", "subgroup", "irreducible representation", and "character" do not occur in the axioms. Yet all these concepts are defined only in terms of language in the axioms, and all their properties are logical consequences of the axioms.

There are also surprisingly many different kinds of mathematical objects that form groups. In the next Chapter, we will look in detail at several of them.

6. Several kinds of groups

Preliminaries

6.1 Numbers under Times

6.1.1 Groups and subgroups of numbers

All complex numbers (excluding zero, which has no inverse) form a group under `Times`. We have already mentioned the relevant properties in the Comments following each axiom.

Further, certain finite subsets of the complex numbers also form groups.

> **Subgroup**
> If $\mathcal{G}$ is a group and $\mathcal{H}$ is a subset of $\mathcal{G}$ that is also a group, we say that $\mathcal{H}$ is a **subgroup** of $\mathcal{G}$.

Now you should use these concepts on your own. This will carve them into the stone of your mind.

6.1.2 Some practice questions

(1) Which of the following is false? (The operator is `Times` in all cases.)

- **(a)** The real numbers are a subgroup of the complex numbers.
- **(b)** The rationals are a subgroup of the reals.
- **(c)** The integers are a subgroup of the rationals.
- **(d)** There is one finite subgroup within the integers.

If you can't see which statement is false, then after really, really trying, click here.

(2) The numbers `1` and `-1` form a group under `Times`. Prove that if another real number is joined in, it is no longer a group. When you have it, click here to see a standard proof.

(3) What is the unit element for the group of all complex numbers under `Times`? What is a simple expression for the inverse of a complex number, in terms of the number itself? Don't click here; this is too easy.

(4) Many finite sets of complex numbers form groups. But they are all somewhat similar. The idea is a straightforward extension of the group `{1,-1}` for

W.M. McClain, *Symmetry Theory in Molecular Physics with Mathematica*,
DOI 10.1007/b13137_6, © Springer Science+Business Media, LLC 2009

the reals. It will come to you if you start by thinking about whether successive powers of a given complex number migrate toward the origin, or away from it.. If, after serious effort, you still need help, click here.

6.2 Matrices under Dot

6.2.1 All nonsingular 2×2 matrices

All square matrices of the same size (excluding singular matrices, which have no inverse) form a group under **Dot**, the matrix multiplication operator. We make a detailed examination of all the axioms, for 2×2 matrices :

Axiom 1. Closure

$$\begin{pmatrix} a & b \\ c & d \end{pmatrix} . \begin{pmatrix} e & f \\ g & h \end{pmatrix} \text{ // GridForm}$$

ae+bg	af+bh
ce+dg	cf+dh

The product is another 2-by-2 matrix. This by itself is not quite enough to show closure for nonsingular matrices. We must also show that the product of two nonsingular matrices is nonsingular. This has to be true, because of two facts:

(1) `Det[A]==0` if and only if matrix **A** is singular (has no inverse), and

(2) `Det[A.B] == Det[A]*Det[B]`. This is easily verified for the 2×2 case :

$$\text{mat1} = \begin{pmatrix} a & b \\ c & d \end{pmatrix}; \text{ mat2} = \begin{pmatrix} e & f \\ g & h \end{pmatrix};$$

```
Det[mat1.mat2] == Expand[Det[mat1] Det[mat2]]
```

```
True
```

Since we used no numerical values in the matrices, this is a general proof for the 2×2 case : If neither `Det[A]` nor `Det[B]` vanishes, then `Det[A.B]` cannot vanish either. Therefore non-singularity comes down in the bloodline of 2×2 matrices, and Axiom 1 is satisfied.

Axiom 2. Unit element

We test the matrix $\begin{pmatrix} 1 & 0 \\ 0 & 1 \end{pmatrix}$ to see if it has the properties required of a unit element :

$$\begin{pmatrix} 1 & 0 \\ 0 & 1 \end{pmatrix} . \begin{pmatrix} a & b \\ c & d \end{pmatrix} == \begin{pmatrix} a & b \\ c & d \end{pmatrix} . \begin{pmatrix} 1 & 0 \\ 0 & 1 \end{pmatrix} == \begin{pmatrix} a & b \\ c & d \end{pmatrix}$$

```
True
```

That True proves it.

Axiom 3. Inverse

Take the inverse of a general 2-by-2 :

$$\text{Inverse}\left[\begin{pmatrix} a & b \\ c & d \end{pmatrix}\right]$$

$$\left\{\left\{\frac{d}{-b\,c+a\,d}, -\frac{b}{-b\,c+a\,d}\right\}, \left\{-\frac{c}{-b\,c+a\,d}, \frac{a}{-b\,c+a\,d}\right\}\right\}$$

This fails only if the denominator **a d - b c = 0**. But this is the determinant,

$$\text{Det}\left[\begin{pmatrix} a & b \\ c & d \end{pmatrix}\right]$$

$$-b\,c+a\,d$$

and matrices with zero determinant are excluded by definition.

Axiom 4. Associativity of the Dot operator

Try it, using three perfectly general 2×2 matrices :

$$\text{Expand}\left[\begin{pmatrix} a & b \\ c & d \end{pmatrix} . \left(\begin{pmatrix} e & f \\ g & h \end{pmatrix} . \begin{pmatrix} i & j \\ k & m \end{pmatrix}\right)\right] ==$$
$$\text{Expand}\left[\left(\begin{pmatrix} a & b \\ c & d \end{pmatrix} . \begin{pmatrix} e & f \\ g & h \end{pmatrix}\right) . \begin{pmatrix} i & j \\ k & m \end{pmatrix}\right]$$

```
True
```

All the axioms are satisfied; therefore the set of all nonsingular 2×2 matrices form a group, q.e.d.

6.2.2 All unitary 2×2 matrices

It will turn out that all symmetry transform matrices have determinant **+1** or **-1**. These are called *unitary* matrices.

If **Det[A] = ±1** and **Det[B] = ±1**, then **Det[A.B] = Det[A] Det[B] = ±1**. This proves that unitarity comes down in the bloodline. Therefore the subset of *unitary* 2×2 matrices form a group within the group of *all nonsingular* 2×2 matrices. This is your first example of the general concept of subgroups. The unitary matrices are a subgroup of the nonsingular matrices.

6.2.3 All nonsingular n×n matrices

The results above can be extended to square matrices of any size by an inductive proof : Assume **Det[A.B] = Det[A] Det[B]** for matrices of size **n**× **n**. Use this to prove it for matrices of size **(n+1)**×**(n+1)**. Don't use any special case, such as **n=2**. It is important to prove it for the general **n**.

Remember expansion by minors? Use it on the expression **Det[A] Det[B]** for matrices of size **(n+1)**×**(n+1)**. Expand by minors along the top row of each matrix, giving everything in terms of **n**×**n** matrices. Then work with just one term from each expansion. The relation holds term-by-term, so you will have shown that IF it is true for **n**×**n** matrices THEN it must be true for **(n+1)** ×**(n+1)** matrices.

When you have it for the general **n**, reason as follows: We know by direct calculation that non-singularity comes down in the bloodline for **n=2**. Therefore it is true for **n=3** . Therefore, it is true for **n=4**. And so on ... .

q.e.d
This finishes the proof that all nonsingular matrices of size **n**×**n** form a group, whatever **n** may be.

6.2.4 All unitary n×n matrices

The inductive proof in 6.2.3 (All nonsingular **n**×**n** matrices) can be applied without change to 6.2.2, showing that all unitary **n**×**n** matrices form a group.

6.3 Axial rotation groups

In Chapter 4 we showed that in three dimensions, rotation about the **z** axis is performed by the matrix

```
Cz[ϕ] // MatrixForm
```

$$\begin{pmatrix} \text{Cos}[\phi] & -\text{Sin}[\phi] & 0 \\ \text{Sin}[\phi] & \text{Cos}[\phi] & 0 \\ 0 & 0 & 1 \end{pmatrix}$$

Cutting back to two dimensions, we define

$$\textbf{R}[\phi\_] := \begin{pmatrix} \textbf{Cos}[\phi] & -\textbf{Sin}[\phi] \\ \textbf{Sin}[\phi] & \textbf{Cos}[\phi] \end{pmatrix}$$

It is easy to show that all the 2×2 **R** matrices produced as ϕ runs around the whole circle form a group under **Dot**, called the "axial rotation group".

```
(* 1, Closure *)  R[a].R[b] == (R[a + b] // TrigExpand)
```

```
True
```

```
(* 2, Identity *)  R[0] == {{1, 0}, {0, 1}}
```

```
True
```

```
(* 3, Inverse *)  (R[a].R[-a] // TrigExpand ) == R[0]
```

```
True
```

(* 4, Associative *) Always `True` for **Dot**

The name "axial rotation group" is symbolized by physicists and chemists as C_∞, or by mathematicians as **SO(2)**, meaning **SpecialOrthogonal(2D)**. These matrices are **Orthogonal** because their transposes are their inverses

```
R[φ].Transpose[R[φ]] // TrigExpand
```

```
{{1, 0}, {0, 1}}
```

and they are **Special** because any closed curve in the plane has the same area before and after any rotation.

6.4 Permutations under Permute

6.4.1 About permutations

Exchange of identical particles, both electrons or nuclei, is an important symmetry operation of the molecular Hamiltonian. The labels that distinguish the particles are just the first **n** integers, and the order of this list changes as particles are exchanged. But this is exactly what mathematicians call a *permutation*.

> **Permutation**
>
> A permutation of length **n** is a **List** containing the first **n** integers, in any order.

Mathematica has a fast and simple way of using a permutation as an instruction for rearranging a list of **n** objects. It is the **Part** operator. Its simplest use is just to pull one item out of a List :

```
Part[{a, b, c}, 2]
```

```
b
```

More often this is written as

```
{a, b, c}[[3]]
```

```
c
```

But **Part** also responds to a list of addresses:

```
{a, b, c}[[{2, 3, 1}]]
```

```
{b, c, a}
```

In fact it has permuted the list **{a,b,c}** according to the instructions in **{2,3,1}**. Going straight down the list of integers, it first took item **#2**; then it took item **#3**, then it took item **#1**.

This gets more interesting when the first list is also a permutation of the same integers :

```
{3, 1, 2}[[{2, 3, 1}]]
```

```
{1, 2, 3}
```

Now we have two objects combining to produce a third of the same kind. This sounds like it could be a group operator. We call the group operator **Permute**, and define it by the statement

```
Permute[permutee, permuter] := permutee[[permuter]]
```

Try it out :

```
Permute[{3, 1, 2}, {3, 2, 1}]
```

```
{2, 1, 3}
```

Stare at this until you see exactly what happened.

6.4.2 `Permute` satisfies the axioms

We will now show that **Permute** can be the operator for a group.

Closure
There are only **n!** permutations of the first **n** integers, so the **permutee**, the **permuter** and their product are all members of the same finite set. There is no way the permutation operation can wander off to infinity. It only hops around among the **n!** possibilities, or some subset thereof. Closure is assured.

Unit element

The unit element is the list of the first **n** integers in natural order. It must be tested in both positions. First we try it as the **permuter**, then as **permutee** :

```
unit = {1, 2, 3, 4, 5};
perm = {3, 2, 5, 1, 4};
{Permute[unit, perm] == Permute[perm, unit] == perm}
```

```
{True}
```

It works as a unit element should: **E⊗P = P⊗E = P**.

Inverse

There is a package operator that finds the inverse of any permutation. First we demonstrate it, then we discuss it. Again let

```
perm = {3, 5, 1, 2, 4};
```

and ask for its inverse:

```
permInv = InversePermutation[perm]
```

```
{3, 4, 1, 5, 2}
```

Does **permInv** have the fundamental properties of an inverse?

```
{Permute[perm, permInv], Permute[permInv, perm]}
```

```
{{1, 2, 3, 4, 5}, {1, 2, 3, 4, 5}}
```

Yes! It works as it should: $\mathbf{P}\otimes\mathbf{P}^{-1} = \mathbf{P}^{-1}\otimes\mathbf{P} = \mathbf{E}$. If it walks like an inverse and talks like an inverse, it is an inverse.

How does the **InversePermutation** operator work? It does exactly what a human would do. (To avoid complicated general notation, we work on an example.) If you need the inverse of **{3,5,1,2,4}** you think to yourself: I want the permutation that puts this in natural order. So I take the third one (the **1**) first, the fourth one (the **2**) second, etc. Working it out in your head, you should come up with **{3,4,1,5,2}**, exactly as calculated above.

But there is a **Position** operator that can automate this thought process. Try it on **perm = {3,5,1,2,4}** :

```
Position[perm, 3]
```

```
{{1}}
```

Yes, the **3** is in position **1**. Try some others on your own. We can define a function named **permPosition** that does this for element **n** of the example **perm** :

```
permPosition[n_] := Position[perm, n];
```

Try it out:

```
permPosition[4]
```

```
{{5}}
```

It says that **4** is in the **5**th position of **perm**. This is true. So we just **Map** our new operator onto the integers in natural order:

```
permInverseRaw = Map[permPosition, {1, 2, 3, 4, 5}]
```

```
{{{3}}, {{4}}, {{1}}, {{5}}, {{2}}}
```

Remember, **perm** was **{3,5,1,2,4}**. So the result above says: the **1** is in **3**rd place, the **2** is in **4**th place, etc. That is what we want, except for all those curly brackets. Fortunately, there is an operator called **Flatten** that takes any multiply bracketed object and removes all interior brackets, turning it into a simple **List** :

```
automatedInv = permInverseRaw // Flatten
```

```
{3, 4, 1, 5, 2}
```

Is this the inverse that we found at the top by thinking it out, or by using **InversePermutation** ?

```
automatedInv == permInv
```

```
True
```

So it worked. But remember, this is not a formal proof, only an example.

> **On your own**
> Think about how you could turn this example into a general proof.

Associativity

Numerical examples are not proofs, but they can be suggestive. Let

```
aPerm = RandomPermutation[16]
```

```
{14, 9, 13, 8, 6, 12, 1, 2, 15, 7, 4, 10, 11, 5, 3, 16}
```

```
bPerm = RandomPermutation[16]
```

```
{7, 13, 3, 6, 5, 12, 10, 11, 15, 16, 9, 4, 1, 8, 2, 14}
```

```
cPerm = RandomPermutation[16]
```

```
{10, 14, 8, 13, 6, 4, 1, 7, 9, 16, 12, 5, 3, 15, 2, 11}
```

Below, the red pairs are calculated first. If **Permute** is associative, the triple products will be the same:

```
Permute[aPerm, Permute[bPerm, cPerm]] ==
Permute[Permute[aPerm, bPerm], cPerm]
```

```
True
```

If you ever find an **aPerm**, **bPerm,** and **cPerm** that produce different results above, please send them, with full details, to the American Mathematical Society, 201 Charles St., Providence, RI, USA Phone (401) 455-400, Fax (401) 331-3842. They get a lot of crackpot stuff and they will know what to do.

The associative relation

Permute[a,Permute[b,c]]==Permute[Permute[a,b],c]

is a famous relation, and a true one, but no quick and easy proof is known. Bogus proofs of it abound, some published under very big names. If you are offered a quick and easy proof, try this: Substitute **Divide** for **Permute**, and see if the proof still seems to hold.

```
{Divide[a, Divide[b, c]], Divide[Divide[a, b], c]}
```

$$\left\{\frac{a\,c}{b},\ \frac{a}{b\,c}\right\}$$

Now there is a rigorous and complete proof that **Divide** is not associative. If you have a "proof" that it is associative, think again.

Probably the clearest valid proof that **Permute** is associative follows this sketch: First prove that there is a one-to-one correspondence between permutations and permutation matrices (**aPerm↔aMat**, **bPerm↔bMat**) (click here), and then that **Permute[aPerm,bPerm]** corresponds to **Dot[aMat,bMat]**. Since we know that matrices are associative under **Dot**, it follows that permutations are associative under **Permute**.

6.5 Fruit flies under reproduction (a non-group)

A bottle of fruit flies does not form a group under the operation of biological reproduction. Biological reproduction gives meaning to the closure expression **flyA⊗flyB = flyC**, but many things are missing. Thinking of them all in detail, axiom by axiom, is an excellent exercise. Try it, then click here.

6.6 End Notes

6.6.1 The false claim

The false claim is (**c**). The integers do not form a group under **Times** because the inverse of an integer is not an integer (except in two special cases; namely, 1 and -1).

6.6.2 Real numbers

Consider the set **{1,-1,a}**, where **a** is any other real number. The closure axiom says the set must contain $\mathbf{a}^2$ if it is to be a group under **Times**. But it doesn't, so it isn't. Depending on the value of **a**, absolute values of further powers grow steadily larger or steadily smaller, so closure is never possible.

6.6.3 Unit and inverse of complex numbers

The unit element of the complex numbers under **Times** is **1+ⅈ*0** . In other words, it is just the usual **1** itself. The inverse of **a+ⅈ*b** is $(\mathbf{a} - i * \mathbf{b}) \big/ \left(\mathbf{a}^2 + \mathbf{b}^2\right)$. If you don't believe it, multiply it out. You will see.

6.6.4 Finite groups of complex numbers

The "**n**th roots of unity" are a set of **n** complex numbers that form a group under multiplication Starting at **{1,0}** in the complex plane, they are regularly spaced around the unit circle by **2π/n** radians. Powers of any number inside the unit circle move steadily closer to the origin; outside, they move steadily away from the origin. But powers of numbers on the unit circle stay on the unit circle. We generate the five fifth roots of unity :

```
soln = Solve[u^5 == 1, u] // N;
roots5 = soln /. (u → r_) → r // Flatten
```

```
{1., -0.809017 - 0.587785 ⅈ, 0.309017 + 0.951057 ⅈ,
 0.309017 - 0.951057 ⅈ, -0.809017 + 0.587785 ⅈ}
```

We used a little tricks above. The output of **Solve** is a list of **Rules**, not a list of numbers, but we converted it to a list of numbers by a pattern-matching trick **/.(u→r_)→r**. Take a look at the Argand diagram (a plot in the complex plane) for these numbers :

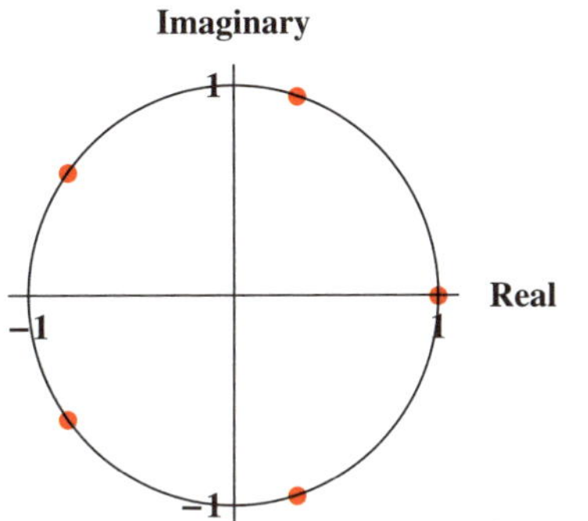

Fig. 6.1 The five red dots are the fifth roots of unity.

Take the fifth power of each one, just to make sure:

```
roots5^5 // ChopInteger
```

```
{1, 1, 1, 1, 1}
```

Look at this step by step to see the function of all its parts. First look at **roots5**, then at **roots5**^{5}, then with **ChopInteger**.

6.6.5 Fruit flies

1. The Closure Axiom is not really fulfilled, because flies are either male or female, and the expression is **True** only for

(male fly)⊗(female fly) = fly.

2. The Unit Axiom might in an incomplete sense be fulfilled by a peculiar **flyE** in which every gene is recessive. Then as far as the phenotype is concerned, **flyA⊗flyE = flyA** . But although the offspring might look the same as **flyA**, it would not really be the same; it would have a lot more recessive genes.

3. Given **flyA**, is there a **flyAinverse** such that **flyA⊗flyAinverse** always yields **flyE**, with all-recessive genes? This might happen once in a zillion tries, (or it could happen more quickly under artificial selection) but Mendelian inheritance is probabilistic, and it could not happen repeatably for any given **flyA**.

4. The associative axiom definitely does not hold, because **(flyA⊗flyB)⊗flyC** has half its genes from **flyC**, whereas **flyA⊗(flyB⊗flyC)** has only a quarter of its genes from **flyC**.

Too bad. Group theory just cannot help with Mendelian genetics.

6.6.6 Permutation matrices

We must show that there is a one-to-one correspondence between permutations and permutation matrices. Begin by constructing a permutation matrix **Pmat**, an **n**×**n** matrix that multiplies leftward into an ordered **List** of **n** objects to produce the required permutation of those objects. Totally general notation for this gets confusing, so we simply take an example. What is the permutation matrix that changes **{a,b,c,d}** into **{b,c,d,a}** ? We define it as the matrix **Pmat** that obeys

```
{a, b, c, d}.Pmat == {b, c, d, a}          (6.1)
```

You might prefer that the left side be **Pmat.{a,b,c,d}**, but we have a reason to put **Pmat** on the right, and we stick firmly with it. We can solve this for **Pmat** using the **MatrixOfCoefficients** operator :

- **MatrixOfCoefficients[exprs,vars]** returns matrix **MC**, where **MC.vars = exprs**. It may fail if the **vars** are not simple symbols, and it will fail if the **exprs** are not linear in the **vars**.

Actually this is not exactly what we want. We would prefer an **MC** such that **vars.MC == exprs**, but we can get the same thing if we simply apply a **Transpose** to the result of this operator.

```
Pmat = MatrixOfCoefficients[
    {b, c, d, a}, {a, b, c, d}] // Transpose;
% // MatrixForm
```

$$\begin{pmatrix} 0 & 0 & 0 & 1 \\ 1 & 0 & 0 & 0 \\ 0 & 1 & 0 & 0 \\ 0 & 0 & 1 & 0 \end{pmatrix}$$

Permutation matrices are easy to recognize at a glance: they must be all zeroes except for a single 1 in each row and column. Make sure it really works:

$$\{a, b, c, d\}.\begin{pmatrix} 0 & 0 & 0 & 1 \\ 1 & 0 & 0 & 0 \\ 0 & 1 & 0 & 0 \\ 0 & 0 & 1 & 0 \end{pmatrix} == \{b, c, d, a\}$$

```
True
```

We have an operator that converts a single permutation into a single matrix, so the correspondence **permutation→matrix** is always one-to-one, What about the reverse correspondence, **matrix→ permutation**? It is sitting right up above. Given any permutation matrix, **Dot** it leftward into the unit permutation, and the answer is the permutation that corresponds to the matrix. So the correspondence **permutation ↔ matrix** is one-to-one both ways.

For the closure axiom to hold we need to show that the product of any two permutation matrices is a permutation matrix. In other words, the property of being a permutation matrix must come down in the bloodline of matrices. The question is : If **pMatA** and **pMatB** are permutation matrices, and if **pMatA.p**
MatB is **pMatAB**, can we be sure that **pMatAB** is a permutation matrix? We can settle this easily using the correspondence just established.

Start with two permutations **A** and **B**, and find their product. Then convert all three permutations to permutation matrices using **permToMat**. This guarantees that **pMatAB** is a permutation matrix. Finally, test the dot product relation to see if it is true. We diagram this process below :

```
permA ⊗ permB   ⃗==   permAB
  ↓       ↓             ↓
pMatA . pMatB   =?=   pMatAB
```

Do a concrete example : Let

```
permA = {2, 5, 3, 1, 4};
permB = {3, 2, 4, 5, 1};
```

The operation **permA ⊗ permB** is implemented as

```
permAB = Permute[permA, permB]
```

```
{3, 5, 1, 4, 2}
```

Now convert all three permutations to matrices :

```
{pMatA, pMatB, pMatAB} =
  Map[permToMat, {permA, permB, permAB}];
% // MatrixList // Size[6]
```

$$\left\{\begin{pmatrix}0&0&0&1&0\\1&0&0&0&0\\0&0&1&0&0\\0&0&0&0&1\\0&1&0&0&0\end{pmatrix},\begin{pmatrix}0&0&0&0&1\\0&1&0&0&0\\1&0&0&0&0\\0&0&1&0&0\\0&0&0&1&0\end{pmatrix},\begin{pmatrix}0&0&1&0&0\\0&0&0&0&1\\1&0&0&0&0\\0&0&0&1&0\\0&1&0&0&0\end{pmatrix}\right\}$$

Now **pMatAB** is known to be a permutation matrix because it was created from a permutation by **permToMat**. So carry out the test

```
pMatA.pMatB == pMatAB
```

```
True
```

This is only an example, not a proof. But there is nothing special about the **permA** and **permB** we chose, and in fact the relation is always true.

7. The fundamental theorem

Preliminaries

7.1. Statement and commentary

> **The fundamental theorem of symmetry theory**
> The symmetry transforms of an object form a group under the operation of sequential application.

This Chapter puts together the definition of *symmetry* given in Chapter 4, and the definition of *group* given in Chapter 5.

You might think the simplest way to prove this theorem would be just to go down the axioms and show that the symmetry transforms of an object fulfill each axiom. This almost works, but it runs into trouble at the Inverse Axiom. For instance, consider all the symmetry rotations of the cube; there are only twelve of them. As will be shown in detail in the next chapter, they can be constructed by closure from only two rotations, so the Closure Axiom is satisfied. There is a no-rotation element, so the Unit Axiom is satisfied. All matrix multiplications are associative, so the Associative Axiom is satisfied. But how can we be sure (without detailed examination, which we want to avoid in this general theorem) that within this little closed set, every element has an inverse? Strangely, no one has been able to create a simple, direct line of reasoning that assures it.

Fortunately, a clever trick has been found that does the job without too much distraction. The trick begins by remembering that the twelve matrices in question are just twelve chosen from the universe of all unitary 3-by-3 matrices, a set which is known to form a group. Then there are two things to show:
(1) Every member of the closed set is a symmetry transform of the object, and
(2) The closed set inherits, from its infinite universe group, all the properties that make it a finite group. The trick makes this quite easy.

W.M. McClain, *Symmetry Theory in Molecular Physics with Mathematica*,
DOI 10.1007/b13137_7, © Springer Science+Business Media, LLC 2009

7.2. Proof of the fundamental theorem

7.2.1. The universe of transforms

We will write this proof specifically for 3-by-3 matrix transforms. It could be made more general, but that would involve a lot of blather that does not really clarify the essentials. That is available already in plenty of math books.

In Chapter 6 we showed that all unitary n-by-n matrices form a group under **Dot**. In particular, this applies to 3-by-3 matrices. This infinite group is the universe-group from which our symmetry transforms are chosen.

In this proof, we will sometimes **Dot** the matrices into themselves, as in Chapter 6, and sometimes we will **Dot** them into an array of **n** point columns on their right, as in Chapter 4. Thus the application operator ⊗ is well defined for both kinds of operands.

7.2.2. The symmetric object

Here we review, in more general language, some things that were shown by example in Chapter 4.

A complex geometric object is represented by a **set** of points, which we symbolize as $\langle p_1, p_2, ..\rangle$. The pointed brackets enclosing the **set** mean that the contents are like a list, except that the order in which they are written is meaningless. When a transform **T** is applied to such an object,

$$T \otimes \langle p_1, p_2, ..\rangle \text{ means } \langle T \otimes p_1, T \otimes p_2, ..\rangle.$$

Generally, **T** will change each point into another point, and the transformed set will be different from the original set. But it can happen that **T** either leaves a point unchanged, or else just changes it to another point already in the set. If no old points are lost, and no new ones created, the set is symmetric under **T**, according to our formal definition in 4.1 .

Let $sym = \langle p_1, p_2, ..\rangle$ be symmetric under two transforms,

$$T_1 \otimes sym = sym \quad \text{and} \quad T_2 \otimes sym = sym$$

Then the associative axiom requires

$$T_1 \otimes (T_2 \otimes sym) = (T_1 \otimes T_2) \otimes sym = sym$$

and we see that the product $T_1 \otimes T_2$ is also a symmetry operator for **sym**. We give it a name like those of the other symmetry operators $T_3 = T_1 \otimes T_2$. For a concrete example, click back to Chapter 4, "New symmetries from old".

7.2.3. Construct a closed set of transforms

The process described below is so useful that we will make a *Mathematica* operator **MakeGroup** that carries it out automatically. But first, we describe the process formally.

Say we start with a set of two symmetry transforms, $\langle \mathbf{T_1}, \mathbf{T_2} \rangle$. We make no effort to include the unit element, or to make them mutually inverse. We then compute all pairwise products $\mathbf{T_1} \otimes \mathbf{T_1}$, $\mathbf{T_1} \otimes \mathbf{T_2}$, $\mathbf{T_2} \otimes \mathbf{T_1}$, $\mathbf{T_2} \otimes \mathbf{T_2}$. This might create nothing new, or it might create as many as four new transforms. We append everything new to the original set of two, creating $\langle \mathbf{T_1}, \mathbf{T_2}, \mathbf{T_3}, \ldots \rangle$. Then we make a new table of pairwise products, including products like $\mathbf{T_1} \otimes \mathbf{T_3}$, which we did not have before. We again append any new products to the original set, and recycle. And so on, and on. We will see some groups where you have to start with more than just two transforms. But the closure process is entirely similar.

We *assume* now that after a certain finite number of iterations, this process ceases to make new products; or in other words, we assume that the set closes. (This always happens if the object has a simple geometric symmetry. But since "simple" has been given no formal meaning, this statement is not part of the proof; it is just explanatory. But you know what we mean.)

7.2.4. Finite subgroup theorem

Theorem
If $\mathcal{G}$ is a group, and $\mathcal{H}$ is a finite subset of $\mathcal{G}$ that is closed under multiplication, **then** $\mathcal{H}$ is a group.

Comment: $\mathcal{G}$ may be finite or infinite, but as we work, think of it as the infinite group of all unitary **3**-by-**3** matrices under **Dot**.

Lemma 1
Every element of $\mathcal{H}$ is the basis of a cyclic power sequence within $\mathcal{H}$

Let **A** be any member of $\mathcal{H}$ and consider the infinite sequence of powers of **A**. All members of the sequence must be in $\mathcal{H}$ because $\mathcal{H}$ is closed under multiplication. Further, because $\mathcal{H}$ is finite, there must be among the powers of **A** a recurrent element **B** from which whole sequence repeats; otherwise the number of elements in $\mathcal{H}$ would be infinite. Suppose that $\mathbf{B}$ is $\mathbf{A}^{m+1}$, and that it repeats with period $\mathbf{n}$. (Lemma 2 will show that $\mathbf{A}^{n+1}$is $\mathbf{E}$.) Then the sequence must read

$$\mathbf{A},\ \mathbf{A}^2,\ \ldots,\ \mathbf{A}^m,\ \mathbf{B},\ \mathbf{B}\otimes\mathbf{A},\ \ldots,$$
$$\mathbf{B}\otimes\mathbf{A}^n,\ \mathbf{B},\ \mathbf{B}\otimes\mathbf{A},\ \ldots,\ \mathbf{B}\otimes\mathbf{A}^n,\ \mathbf{B},\ \ldots$$

But $\mathbf{B}$ is $\mathbf{A}^{m+1}$, so we can rewrite without using $\mathbf{B}$ at all :

$$\mathbf{A},\ \mathbf{A}^2,\ \ldots,\ \mathbf{A}^m,\ \mathbf{A}^{m+1},\ \mathbf{A}^{m+2},\ \ldots,$$
$$\mathbf{A}^{m+n},\ \mathbf{A}^{m+1},\ \ldots,\ \mathbf{A}^{m+n},\ \mathbf{A}^{m+1},\ \ldots$$

Note that after the power $\mathbf{m+n}$, the power drops back to $\mathbf{m+1}$. Now $\mathbf{A}$ was any element, not originally claimed as a recurring element. But looking at the first part of the sequence, we see that where $\mathbf{A}^{m+1}$occurs, we can count back $\mathbf{m}$ places and we will be at $\mathbf{A}$ again. So in this sequence, every element is a recurring element with period $\mathbf{m}$. (We originally said the period of $\mathbf{B}$ was $\mathbf{n}$, but now we see that every element has the same period, so $\mathbf{n} = \mathbf{m}$.) So the sequence is really

$$\mathbf{A},\ \mathbf{A}^2,\ \ldots,\ \mathbf{A}^m,$$
$$\mathbf{A},\ \mathbf{A}^2,\ \ldots,\ \mathbf{A}^m,$$
$$\mathbf{A},\ \mathbf{A}^2,\ \ldots,\ \mathbf{A}^m,\ \ldots$$

So the set of the all powers of $\mathbf{A}$ that occur in its own power sequence is just a finite little set, $\mathcal{P}\ [\mathbf{A}] = \langle \mathbf{A},\ \mathbf{A}^2,\ \ldots,\ \mathbf{A}^m \rangle$. In other words, every element of $\mathcal{H}$ is the basis of a cyclic power sequence, which was to be proved.

Lemma 2
The power sequence for every element $\mathbf{A}$ in $\mathcal{H}$ includes the identity.

Since $\mathbf{A}^m$ is followed by $\mathbf{A}$, it must be true that $\mathbf{A}^m \otimes \mathbf{A} = \mathbf{A}$ and regrouping, also that $\mathbf{A} \otimes \mathbf{A}^m = \mathbf{A}$. But by definition, only the unit element $\mathbf{E}$ behaves like $\mathbf{A}^m$ in these equations, so $\mathbf{A}^m = \mathbf{E}$.

Lemma 3
The power sequence for every element $\mathbf{A}$ in $\mathcal{H}$ includes the inverse $\mathbf{A}^{-1}$.

Since $\mathbf{E}$ is preceded by $\mathbf{A}^{m-1}$, we have $\mathbf{A}^{m-1} \otimes \mathbf{A} = \mathbf{E}$ and regrouping, $\mathbf{A} \otimes \mathbf{A}^{m-1} = \mathbf{E}$. But the element that behaves like $\mathbf{A}^{m-1}$ is by definition $\mathbf{A}^{-1}$, the inverse of $\mathbf{A}$. Since we started by saying "Let $\mathbf{A}$ be any member of $\mathcal{H}$...", the inverse of every element in the sequence is also present somewhere in the sequence.

Proof of the theorem

Nothing in Lemma 1 says that the power sequence based on **A**, $\mathcal{P}$[**A**], contains all the elements of $\mathcal{H}$ Except in pure cyclic groups, it does not. So if **C** is an element of $\mathcal{H}$ not contained in $\mathcal{P}$[**A**], we start again with **C** and construct its power sequence $\mathcal{P}$[**C**]. We continue to construct power sequences until all the elements of $\mathcal{H}$ appear in some power sequence. Then we combine all the power sequences, discarding duplicate elements. (In *Mathematica*, this is done by the **Union** operator.) The **Union** of all power sequences of $\mathcal{H}$ is the same as $\mathcal{H}$ itself, because:

(a) The power sequences do not omit anything in $\mathcal{H}$ because every element of $\mathcal{H}$ either occurred in a power sequence, or was used to start a new power sequence.

(b) The power sequences cannot contain anything new, because we began by saying "let $\mathcal{H}$ be a finite subset ... closed under multiplication".

Now we can successfully examine the group axioms, one by one.

1. Closure The set $\mathcal{H}$ is closed under multiplication by hypothesis.
2. Unit By Lemma 2, $\mathcal{H}$ includes the unit element.
3. Inverse By Lemma 3, $\mathcal{H}$ includes an inverse partner for each element.
4. Associative The multiplication operator for $\mathcal{H}$ is the same as for $\mathcal{G}$. Since $\mathcal{G}$ is a group, the operator is associative.

All four axiomatic requirements are met. Therefore, if $\mathcal{H}$ is closed under multiplication, then $\mathcal{H}$ is a group, as was to be shown.

7.2.5. Example of power sequences in a group

The central claim of the proof is that all finite groups can be split up into non-overlapping power sequences. Just to make this absolutely clear, we carry it out explicitly for a group named **"D3h"** , which is tabulated in the **Symmetry`** package. The standard names of its elements are

```
allGroupNames = ElementNames["D3h"]
```

```
{E, C3a, C3b, C2a, C2b, C2c, σh, S3a, S3b, σva, σvb, σvc}
```

We convert one of the names to the matrix itself :

```
matS3a = "S3a" /. NamesToRepMats["D3h"]
```

$$\left\{\left\{-\frac{1}{2}, -\frac{\sqrt{3}}{2}, 0\right\}, \left\{\frac{\sqrt{3}}{2}, -\frac{1}{2}, 0\right\}, \{0, 0, -1\}\right\}$$

We use **NestList** (look it up) to construct the power sequences based on an element named **"S3a"** . Since the group has only 12 elements, we make a sequence of 12 elements. This has to be long enough.

```
𝒫[S3a] = NestList[#.matS3a &, matS3a, 12] /.
  RepMatsToNames["D3h"]
```

```
{S3a, C3b, σh, C3a, S3b, E, S3a, C3b, σh, C3a, S3b, E, S3a}
```

This power sequence has a repeat length of 6. We eliminate the repeats :

```
𝒰𝒫[S3a] = 𝒫[S3a] // Union
```

```
{C3a, C3b, E, S3a, S3b, σh}
```

Which elements of group D3h do not appear in this sequence?

```
Complement[allGroupNames, 𝒰𝒫[S3a]]
```

```
{C2a, C2b, C2c, σva, σvb, σvc}
```

Six elements did not appear. The next step is to make the power sequence based on the first such element :

```
matC2a = "C2a" /. NamesToRepMats["D3h"]
```

```
{{1, 0, 0}, {0, -1, 0}, {0, 0, -1}}
```

```
𝒫[C2a] = NestList[#.matC2a &, matC2a, 12] /.
  RepMatsToNames["D3h"]
```

```
{C2a, E, C2a, E, C2a, E, C2a, E, C2a, E, C2a, E, C2a}
```

```
𝒰𝒫[C2a] = 𝒫[C2a] // Union
```

```
{C2a, E}
```

The repeat length is only 2. This says that element **C2a** is its own inverse. This is true; **C2a** is a twofold rotation. The same is true for any element with a **C2_** name.

Similarly, all **σ_** names are reflections, and also give power sequences with only two members.

Therefore, the whole group is the **Union** of the power sequences based on seven of its elements: **S3a**, **C2a**, **C2b**, **C2c**, **σva**, **σvb**, and **σvc**. This is a very typical situation.

7.3. How this theorem helps

Stephen Jay Gould, late lamented natural history essayist and Agassiz Professor of Biology at Harvard, told this story about his eponymous predecessor, Louis Agassiz.

Prof. Agassiz assigned a new student to draw a fish pickled in formaldehyde. At the end of the day, Agassiz glanced at the drawings and said, "No! You have not seen one of the most essential features of this animal!". This went on for several days. The student's drawings became more and more detailed in fin and scale and marking, but Agassiz always rejected them, never explaining why. Finally, on the sixth day, fainting from formaldehyde fumes, the student saw it: His error was not in the details, but in the picture as a whole. The fish had bilateral reflection symmetry.

We repeat this story to remind ourselves that humans are not necessarily very good at seeing symmetry transforms "instinctively". The fundamental theorem, however, gives us a way of finding the ones we fail to see. But wait, you may ask, absolutely all of them, without fail? Well, yes, but with one little reservation. We discuss the reservation later; first, we describe the basic procedure:

To construct the group of an object, begin by examining the object. Put a well chosen Cartesian coordinate system on it, and write down all of its symmetry transforms that you can see, in the form of 3-by-3 Cartesian transform matrices. (You will see many examples of these matrices just ahead.) You will probably miss some of the object's symmetry transforms, but don't worry; this will not matter.

When you have a list of as many symmetry matrices as you can see, make a multiplication table from them. Any new matrix that appears in the table is a symmetry matrix that you missed by direct observation. Look at the object again, and you will see it this time. Now join the new matrices onto the old list, and calculate the new, larger table. Continue this until you get a table with nothing new in it. At this point you will have a **group** of symmetry transforms that characterizes the object.

Now comes the unanswerable question: Is this the set of ALL symmetry transforms of the given object? You can never be mathematically certain that it is, because it is possible that in your examination, you failed to notice a symmetry element that is there, but completely independent from the ones that you did see. If, at a later time, this new symmetry element comes to your attention, you may add it to your table and expand the table until it closes again. Your description of the symmetries of the object will be improved; but then of course, the question comes again: Do we have ALL the symmetries this time? The answer, alas, must be the same.

8. The multiplication table

Preliminaries

8.1. The generalized "multiplication" table

Consider a group with unspecified operator ⊗ and a list of **n** elements, named **{E, A, B, ..., Z}**. The Closure Axiom says that a "product" is defined for every ordered pair of these elements, giving $\mathbf{n}^2$ products in all. It seems natural to display them in a square "multiplication" table. We have written a display operator for such tables, called **BoxUp**. Here is what it produces if we take the elements to be **{E, A, B, C}**, and the entries to be blanks :

```
blankMat = Table["", {i, 4}, {j, 4}];
BoxUp[blankMat,
 {" E ", " A ", " B ", " C "}, "", 1.1]
```

	E	A	B	C
E				
A				
B				
C				

We must decide once and for all whether the entries in such a table will be "row ⊗ column" or "column ⊗ row". The nearly universal convention for naming the elements of matrices is **{row,column}**, so we pick the form closest to this; namely, **row ⊗ column**. Thus the simple matrix of table entries will be

```
els = {"E", "A", "B", "C"};
productMat =
  Table[els[[i]]⊗els[[j]], {i, 4}, {j, 4}];
% // GridForm
```

E⊗E	E⊗A	E⊗B	E⊗C
A⊗E	A⊗A	A⊗B	A⊗C
B⊗E	B⊗A	B⊗B	B⊗C
C⊗E	C⊗A	C⊗B	C⊗C

To put row and column headings on it, use **BoxUp** , with a blank for the upper left element :

W.M. McClain, *Symmetry Theory in Molecular Physics with Mathematica*,
DOI 10.1007/b13137_8, © Springer Science+Business Media, LLC 2009

```
BoxUp[productMat, {"E", "A", "B", "C"}, "", 1.1]
```

	E	A	B	C
E	E⊗E	E⊗A	E⊗B	E⊗C
A	A⊗E	A⊗A	A⊗B	A⊗C
B	B⊗E	B⊗A	B⊗B	B⊗C
C	C⊗E	C⊗A	C⊗B	C⊗C

If the unit element **E** comes first in the element list, then the first row and firs column simplify by the Unit Axiom. Now we take the default "**Left⊗**$^{\textbf{Right}}$" in the upper left to remind ourselves that the row label goes on the left of the ⊗ and the column label goes on the right, in operations of the form **Left⊗Right** (or in square bracket notation, **op[Left,Right]**). Here it is:

```
simplMat =
  productMat /. "E"⊗X_ -> X /. X_⊗"E" -> X;
BoxUp[simplMat, {E, A, B, C}]
```

Left⊗^{Right}	E	A	B	C
E	E	A	B	C
A	A	A⊗A	A⊗B	A⊗C
B	B	B⊗A	B⊗B	B⊗C
C	C	C⊗A	C⊗B	C⊗C

The redundant row and column labels (top row and left column, above) are often omitted.

The Closure Axiom says each product must evaluate as **E**, **A**, **B**, or **C**. But the other axioms put significant restrictions on how the table may fill out. The argument is simple and quick, and is given below.

8.2. Latin square theorem

8.2.1. Statement

We start with a definition :

> **Latin square**
> If, in a square table, every element appears exactly once in each column and once in each row, the table is a Latin square.

This definition makes it easy to state the theorem :

> **Latin square theorem**
> **If** an operation and a list of elements form a non-redundant group, **then** the multiplication table of the group is a Latin square.

8.2.2. Proof

First, we show that no element may appear twice in any row or column of this table. Consider the **A** row, and assume that some element **X** appears twice in this row. This means that there are two columns, headed by different elements **U** and **V**, in which

$$\mathbf{A \otimes U == A \otimes V == X}.$$

But the Inverse Axiom says that **A** has an inverse called $\mathbf{A^{-1}}$, and multiplying both members of the first equality by $\mathbf{A^{-1}}$ from the left, we have

$$\mathbf{A^{-1} \otimes (A \otimes U) == A^{-1} \otimes (A \otimes V)}$$

The Associative Axiom allows us to rewrite this as

$$\mathbf{(A^{-1} \otimes A) \otimes U == (A^{-1} \otimes A) \otimes V},$$

and the Inverse Axiom then allows simplification to

$$\mathbf{E \otimes U == E \otimes V}$$

The Unit Axiom then says

$$\mathbf{U == V}$$

contradicting our assumption that the group is non-redundant. The assumption that some **A** appears twice in some row is therefore inconsistent with the axioms, and with the assumption of non-redundancy. The same argument, applied to the transpose of the multiplication table, shows that no **A** can appear twice in any column. So the multiplication table must be a Latin square, which was to be proved.

8.2.3. A corollary: the rearrangement theorem

A corollary is a quick and easy theorem that follows from another theorem. It is often just an alternative statement of the theorem, or a little different way of looking at the result. The Latin Square theorem has the following corollary :

> **Corollary : The Rearrangement Theorem**
> **If** all the elements of a group are multiplied by some element of the group,
> **then** the result is a rearrangement of the group.

Let the unit element come first in the first row of a group multiplication table, and take the first row as the standard order of the group. Every other row is just a left-multiplication of this row by some element of the group, and by the Latin Square theorem, each row is just a rearrangement of the same group, which was to be proved.

8.3. The Latin square converse: True or False?

8.3.1. Converses in general

The converse of a theorem interchanges the premise (the **If** part) and the conclusion (the **then** part). Sometimes this makes a valid theorem and sometimes not. In the present case, the converse statement would be:

> **Converse of the Latin square theorem :**
>
> **If** the multiplication table of a list of elements is a Latin square,
> **then** the list of elements is a group.
>
> Is this converse **`True`** or **`False`** ?

Keep this question in mind as we proceed. Remember, a single counter-example is enough to disprove a proposed theorem. If you just have to peek ahead, click here.

8.3.2. Exploring an example

Here is the multiplication table of a certain operator and a certain element list, disguised by being transformed into letter-names. We hide the calculation of the table to avoid giving the answer away too soon.

LtnSq =

	A	B	C	D
A	C	D	A	B
B	D	A	B	C
C	A	B	C	D
D	B	C	D	A

The multiplication table is a Latin square, as predicted for groups. So the Latin square theorem does not exclude the possibility that it is a group, but it does not prove it either. The converse of Theorem 1 says it is a group, but we do not know if the converse is true or false.

We do not know what the elements are or what the operation is. To determine whether we have a group we must show, from the structure of the multiplication table alone, that each of the four axioms holds. We reason carefully, axiom by axiom, only on the basis of what we see in the multiplication table :

1. By inspection of the table, the Closure Axiom holds.

2. The third row and third column are identical to the original element list, as seen in the top and left headings. The label of this row and column is **C**; thus **C** is the unit element, and the Unit Axiom holds.

3. Every row and every column contain one unit element **C**, so every element has an inverse. Furthermore, wherever there is a **C** at position **⟦i,j⟧**, it also appears at position **⟦j,i⟧**. Therefore, the Inverse Axiom holds.

4. It remains now only to verify the Associative Axiom. Remember, we want to draw our conclusion solely from the content of the table. This cannot be done by a simple inspection; the Associative Axiom deals with triple products, whereas the table gives direct information only about double products. The only way to test the Associative Axiom is to use the table to form all possible triple products and test them for associativity. In a 4-by-4 table, there will be $4^3 = 64$ cases.

The test is easily automated, using the integer name trick. Instead of using the actual element names, we replace each element by an integer, 1 through 4. When we "multiply" these number-names we use the given multiplication table itself to look up the value of the product. In the hidden calculation above we defined

```
LtnSq
```

```
{{C, D, A, B}, {D, A, B, C}, {A, B, C, D}, {B, C, D, A}}
```

Check by inspection that this is the content of the boxed table above. We begin by translating **LtnSq** into number-names:

```
LtnSq
```

```
{{C, D, A, B}, {D, A, B, C}, {A, B, C, D}, {B, C, D, A}}
```

```
LtnSqN = LtnSq /. {"A" → 1, "B" → 2, "C" → 3, "D" → 4};
MatrixForm[LtnSq] → MatrixForm[LtnSqN]
```

$$\begin{pmatrix} C & D & A & B \\ D & A & B & C \\ A & B & C & D \\ B & C & D & A \end{pmatrix} \rightarrow \begin{pmatrix} 3 & 4 & 1 & 2 \\ 4 & 1 & 2 & 3 \\ 1 & 2 & 3 & 4 \\ 2 & 3 & 4 & 1 \end{pmatrix}$$

Now on the right, the product **1⊗1** is **3**, for instance. (This is OK; they are names, not integers.) In this way of naming things, the value of product **i⊗j** is given by the content of the table at position **i,j**. This makes it easy to construct an exhaustive association test. We simply write the association test for the triple product **i⊗j⊗k** (namely, **i⊗(j⊗k)==(i⊗j)⊗k**) and then run **i**, **j**, and **k** over all possible values. Remember, the ⊗ is carried out by table look-up. Each test yields either **True** or **False**. The numerical Latin square on the right above is called **LtnSqN**,

The exhaustive test is

```
Table[LtnSqN[[i, LtnSqN[[j, k]] ]] ==
    LtnSqN[[ LtnSqN[[i, j]], k]], {i, 4},
   {j, 4}, {k, 4}] // Flatten // Union
```

```
{True}
```

To see the full output, take away the **Union** operator and rerun. If even one case had produced a **False**, the return from **Union** would have reported **{True,False}**. But it did not, so the Associative Axiom holds. Since all four axioms have been verified, we have a group, as determined solely by the properties of the group multiplication table.

If you are curious, the operator was **Times** and the element list was **{-1,ⅈ,1,-ⅈ}**. We violated the usual convention of writing the unit element first, just to show that it can still be identified by its fundamental properties.

8.3.3. A different Latin Square to test

Below you see the multiplication table for an operator and an element list that do NOT form a group. What's wrong, why not? Just repeat all the tests made above, and you will see.

```
LtnSq2 =
  {{C, B, A, D}, {D, C, B, A}, {A, D, C, B}, {B, A, D, C}};
GridForm[LtnSq2]
```

C	B	A	D
D	C	B	A
A	D	C	B
B	A	D	C

After you verify that this Latin square is not the multiplication table of a group, you will have a counter-example which proves that the converse of the Latin Square theorem is false.

If for some reason you can't get it to work, click here.

8.4. Automated multiplication tables

8.4.1. Demonstration of `MultiplicationTable`

There is a **`MultiplicationTable`** operator that comes with the **`Symmetry`**` package of this book, loaded automatically during the preliminaries of each chapter. If you give it the group "multiplication" operator, a list of the group elements in concrete form, and a list of names for the group elements, it will carry out each possible multiplication, replace each product by its name, and display the result as a square table. Since it is a package operator, you may call up a thumbnail description of it. Do so.

We give the group of four numbers a name :

```
groupN4 = {1, ⅈ, -1, -ⅈ};
```

Make its **`MultiplicationTable`**, using **`Times`** as the group operator:

```
tableN4 = MultiplicationTable[
   Times, groupN4, {"E", "A", "B", "C"}];
% // GridForm
```

E	A	B	C
A	B	C	E
B	C	E	A
C	E	A	B

Section 8.5, below, shows in detail how this operator works. But first, in 8.4.2 just below, we continue this example to illustrate a basic issue of group theory.

8.4.2. Abstract groups and representations of groups

Here is a four-element group of matrices that perform rotations about **z** by **0** turn, **1/4** turn, **2/4** turn and **3/4** turn, respectively:

```
groupC4 = {
   {{1, 0, 0}, {0, 1, 0}, {0, 0, 1}},
   {{0, -1, 0}, {1, 0, 0}, {0, 0, 1}},
   {{-1, 0, 0}, {0, -1, 0}, {0, 0, 1}},
   {{0, 1, 0}, {-1, 0, 0}, {0, 0, 1}}};
```

Make its **MultiplicationTable**, using **Dot** as the group operator :

```
tableC4 = MultiplicationTable[
   Dot, groupC4 , {"E", "A", "B", "C"}];
% // MatrixForm
```

$$\begin{pmatrix} E & A & B & C \\ A & B & C & E \\ B & C & E & A \\ C & E & A & B \end{pmatrix}$$

```
tableC4 == tableN4
```

```
True
```

It is the same multiplication table again! Now we will produce it again with yet another group:

Here are the cyclic permutations of **{1,2,3,4}** :

```
groupP4 = {{1, 2, 3, 4},
   {2, 3, 4, 1}, {3, 4, 1, 2}, {4, 1, 2, 3}};
```

Make its **MultiplicationTable**, using **Permute** as the operator :

```
tableP4 = MultiplicationTable[
```

```
   Permute, groupP4 , {"E", "A", "B", "C"}];
% // MatrixForm
```

$$\begin{pmatrix} E & A & B & C \\ A & B & C & E \\ B & C & E & A \\ C & E & A & B \end{pmatrix}$$

Review what we have done :

```
tableN4 == tableC4 == tableP4
```

```
True
```

The multiplication tables are exactly the same. These groups have a deep relationship, and we need words that express the similarity as well as the differences. Here they are:

> An **abstract group** is defined completely by its symbolic multiplication table. A group of concrete mathematical objects (matrices, permutations, etc.) that obeys that table (and does not obey any smaller table) is a **faithful representation** of the abstract group.

How would you express the deep relationship in words, in the example above? These are all cyclic groups of four; it's just the cycling object that is different. In the first example, the cycling object is the number i ; the group is given in order by $\{i^0, i^1, i^2, i^3\}$; then i^4takes you back to i^0.

In the second example the cycling object is the matrix

```
C4 = {{0, -1, 0}, {1, 0, 0}, {0, 0, 1}};
```

taken to **MatrixPower**s **0** through **3**. Matrix power **4** brings you back to power **0**, the unit matrix.

```
MatrixPower[C4, 4]
```

```
{{1, 0, 0}, {0, 1, 0}, {0, 0, 1}}
```

In the third example the cycling object is the list **{1,2,3,4}**, rotated left by one place, zero through three times. The fourth leftward rotation then returns you to the original.

```
RotateLeft[{1, 2, 3, 4}]
```

```
{2, 3, 4, 1}
```

> **On your own**
> **(a)** Dot matrix $\mathbf{C_4}$ with itself **1** time, **2** times, **3** times, and **4** times.
> **(b)** Apply **RotateLeft** similarly to the list **{1,2,3,4}**.

If abstract groups had systematic names, we would have to call this one something like **FourCycle**. We would then say that **groupN4**, **groupC4**, and **groupP4** are **faithful representations** of the abstract group **FourCycle**.

In the next few chapters, we will continue to work primarily with 3×3 Cartesian transform matrices. Just keep in mind that they are only one kind of **representation** of the underlying **abstract** groups.

Far ahead in Chapter 25 we will learn an algorithm for constructing matrix representations that imitate the symmetries of any set of symmetric objects. Click to it now for a preview of the powerful ideas that lie ahead. Don't worry if you do not understand it completely right now. This book builds up systematically, and when you get to Chapter 25 you will understand it.

8.5. How MultiplicationTable works

8.5.1. The basic core operator, Outer

In this Section we show in detail how the **MultiplicationTable** operator was constructed, and people who are willing to take it on faith can skip on to the next chapter. However, there are some interesting non-obvious issues that arise, and to those at all interested in symbolic computation itself, this section should be worth reading.

Mathematica has a core operator that constructs exactly the kind of multiplication table implied by Axiom1. It is the **Outer** operator, which constructs every possible pairwise product from any two given lists, using any desired operator. As an initial exhibit of its powers, we use the undefined symbolic operator **CircleTimes**:

```
Outer[CircleTimes, {a, b}, {a, b}] // GridForm
```

a⊗a	a⊗b
b⊗a	b⊗b

Here is a concrete numerical example where it works well:

```
Outer[Times, {1, ⅈ}, {1, ⅈ}] // GridForm
```

1	i
i	-1

8.5.2. A problem with matrices, and its solution

Our main concern will be groups of matrices, and a problem arises when we try to make a multiplication table for matrices using this method. We take two matrices that are easy to multiply together mentally

$$\textbf{aMat} = \begin{pmatrix} 1 & 0 \\ 0 & 1 \end{pmatrix};$$

$$\textbf{bMat} = \begin{pmatrix} 3 & 0 \\ 0 & 3 \end{pmatrix};$$

Now we attempt to calculate a multiplication table for the list **{aMat,bMat}** :

```
Outer[Dot, {aMat, bMat}, {aMat, bMat}]
```

```
{{{{{{1.1, 1.0}, {1.0, 1.1}}, {{1.3, 1.0}, {1.0, 1.3}}},
   {{{0.1, 0.0}, {0.0, 0.1}}, {{0.3, 0.0}, {0.0, 0.3}}}},
  {{{{0.1, 0.0}, {0.0, 0.1}}, {{0.3, 0.0}, {0.0, 0.3}}},
   {{{1.1, 1.0}, {1.0, 1.1}}, {{1.3, 1.0}, {1.0, 1.3}}}}},
 {{{{{3.1, 3.0}, {3.0, 3.1}}, {{3.3, 3.0}, {3.0, 3.3}}},
   {{{0.1, 0.0}, {0.0, 0.1}}, {{0.3, 0.0}, {0.0, 0.3}}}},
  {{{{0.1, 0.0}, {0.0, 0.1}}, {{0.3, 0.0}, {0.0, 0.3}}},
   {{{3.1, 3.0}, {3.0, 3.1}}, {{3.3, 3.0}, {3.0, 3.3}}}}}}
```

How bizarre and ugly! The **Outer** operator failed because **Dot** acted at too low a level, trying to **Dot** integers with integers instead of matrices with matrices.

There is a very simple way to get what we want: just append a trailing **1** to the list of parameters that we feed to **Outer**. This tells the **Dot** operator to act at "level **1**" of the given lists. In **{aMat,bMat}**, the matrices are at "level **1**", the matrix rows are at "level **2**", and the elements themselves are at "level **3**". If you omit the level instruction, the operator acts by default at the lowest level. In this case, it tries to **Dot** integers with integers, yielding a mess. Now we redo it with **Dot** acting at the correct level, level **1** :

```
matTable =
  Outer[Dot, {aMat, bMat}, {aMat, bMat}, 1];
% // MatrixForm
```

$$\begin{pmatrix} \begin{pmatrix} 1 & 0 \\ 0 & 1 \end{pmatrix} & \begin{pmatrix} 3 & 0 \\ 0 & 3 \end{pmatrix} \\ \begin{pmatrix} 3 & 0 \\ 0 & 3 \end{pmatrix} & \begin{pmatrix} 9 & 0 \\ 0 & 9 \end{pmatrix} \end{pmatrix}$$

This is the desired answer. What a difference that trailing **1** makes!

8.5.3. Symbolic output

The output above, a matrix of matrices, suggests another issue. In a group table, each element is a member of the original list, and it has a name given in the name list. So we ought to be able to write the final output in terms of names again. This will be a great advantage when doing groups of large matrices, or other objects that are hard to identify at a glance.

$$\texttt{elements} = \left\{\begin{pmatrix} 1 & 0 \\ 0 & 1 \end{pmatrix}, \begin{pmatrix} -1 & 0 \\ 0 & -1 \end{pmatrix}\right\};$$

```
names = {"E", "M"};
```

We make a recognition **Rule**:

```
matRule = Thread[elements → names]
```

```
{{{1, 0}, {0, 1}} → E, {{-1, 0}, {0, -1}} → M}
```

Make the table, and use the rule :

```
Outer[Dot, elements, elements, 1] /. matRule //
 GridForm
```

E	M
M	E

Now you can see it at a glance.

On your own

Consider

E	M
M	E

. Is this the multiplication table of a group?

Give an iron-clad answer, based on a theorem you have seen.

8.5.4. The precision problem

It is easy to identify and replace objects made of symbols or exact integers or other exact quantities. It is a little more difficult to identify objects that contain roundoff error, because two versions of an object that differ insignificantly in the final digit may not be identified as the same. Here are two numbers that differ in the only fourteenth digit:

```
a = 0.12345678901234;
b = 0.12345678901237;
{a, b}
```

```
{0.123457, 0.123457}
```

Does that 14th digit make any difference to **Equal**?

```
a == b
```

```
False
```

Mathematica never lies. Often two different routes to the same answer may yield differences in the 15th digit, and this can be a problem. But once recognized, it can be handled. The **SetPrecision** operator effectively rounds a real number at a given number of decimal digits. We round the numbers **a** and **b** above to **12** digits, and ask again if they are "equal" :

```
a12 = SetPrecision[a, 12];
b12 = SetPrecision[b, 12];
{a12, b12}
```

```
{0.123456789012, 0.123456789012}
```

```
a12 == b12
```

```
True
```

The only remaining problem is the roundoff zero problem. Even when pared back to 12 significant digits, small numbers are not identified as a true zero.

```
tiny = SetPrecision[0.1234567890123456 * 10^-23, 12]
```

$1.23456789012 \times 10^{-24}$

```
tiny == 0
```

```
False
```

The **Chop** operator does what we want:

```
Chop[{1. * 10^-9, 1. * 10^-10, 1. * 10^-11, 1. * 10^-12}]
```

$\{1. \times 10^{-9},\ 1. \times 10^{-10},\ 0,\ 0\}$

The results above may differ a little bit on different computers, but the output should be a list that contains some small numbers and some zeroes where even smaller numbers have been set to an exact zero. Using **Chop** when we test **tiny**, it is now identified as effectively zero:

```
Chop[tiny] == 0
```

```
True
```

Therefore, we set up **MultiplicationTable** so that it will truncate all numbers to **prc** significant figures and **Chop** all very small numbers to an exact zero, before trying to make identifications. The significant figures parameter **prc** is by default **12**, but it can take other values.

On your own
Pull down the syntax statement for **MultiplicationTable** and read it carefully.

8.5.5. What to do with unidentifiable products

What happens when recognition fails? Suppose we apply our old **matRule** to the following table :

```
badTable = {{{{1, 0}, {0, 1}}, {{-1, 0}, {0, -1}}},
   {{{-1, 0}, {0, -1}}, {{2, 0}, {0, 2}}}};
% // GridForm
```

{{1, 0}, {0, 1}}	{{-1, 0}, {0, -1}}
{{-1, 0}, {0, -1}}	{{2, 0}, {0, 2}}

Note that **badTable** contains $\begin{pmatrix} 2 & 0 \\ 0 & 2 \end{pmatrix}$, which will not be recognized.

```
badTable2 = badTable /. matRule;
% // GridForm
```

E	M
M	{{2, 0}, {0, 2}}

When recognition fails, the table will be neater if it reports "?" instead of the literal unrecognized matrix.

```
useQM[x_] := If[MemberQ[names, x], x, "?", "?"];
Map[useQM, badTable2, {2}] ;
% // GridForm
```

E	M
M	?

8.6. Test `MultiplicationTable`

We will test **`MultiplicationTable`** using calls to the package operator **`CartRep[gpNm]`**. This is short for **`CartesianRepresentation`**, and **`gpNm`** is the string name of any group in the **`GroupCatalog`**. This operator goes into the package and returns the Cartesian transformation matrices that we have been calling " the group". But technically speaking an abstract group is defined by its multiplication table. The **`CartRep`** is a list of 3×3 Cartesian transform matrices that obey the same table as the abstract group.

The package **`Symmetry``** contains a number of predefined matrix groups. We used **`"C3v"`** above; now we test the package operator using another one called **`"D2h"`**.

```
matsD2h = CartRep["D2h"];
% // GridList
```

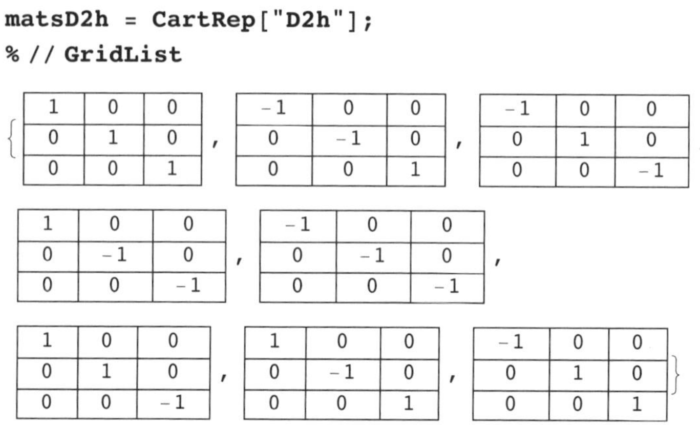

There are also predefined element names :

```
namesD2h = ElementNames["D2h"];
% // FullForm
```

```
List["E", "C2z", "C2x", "C2y", "inv", "σz", "σy", "σx"]
```

All the names are strings. We ask for the group multiplication table (under the operator for matrix multiplication, which is **`Dot`**) :

```
mtD2h =
  MultiplicationTable[Dot, matsD2h, namesD2h];
% // GridForm
```

E	C2z	C2x	C2y	inv	σz	σy	σx
C2z	E	C2y	C2x	σz	inv	σx	σy
C2x	C2y	E	C2z	σy	σx	inv	σz
C2y	C2x	C2z	E	σx	σy	σz	inv
inv	σz	σy	σx	E	C2z	C2x	C2y
σz	inv	σx	σy	C2z	E	C2y	C2x
σy	σx	inv	σz	C2x	C2y	E	C2z
σx	σy	σz	inv	C2y	C2x	C2z	E

On your own

There is a short form of this operator that works on tabulated groups :

```
MultiplicationTable["D2h"]
```

Try it.

So far, so good. Now we try it on a non-group, the list **{1,ⅈ}**under **Times**. Numerically the multiplication table will be $\begin{pmatrix} 1 & ⅈ \\ ⅈ & -1 \end{pmatrix}$. If we name **1** as **A** and **ⅈ** as **B**, we should get a question mark in the lower right, since **ⅈ*ⅈ** will not be recognized.

```
MultiplicationTable[Times, {1, ⅈ}, {A, B}] //
 GridForm
```

A	B
B	?

Have we found all the bugs in the **MultiplicationTable** routine? You can never say yes to this question. In the interest of full disclosure, open UrSymmetryV6.nb and Find the exact text of the operator. For a cautionary tale on program bugs, click here.

8.7. Redundant groups

Redundant groups will arise naturally when we get into representation theory, but here we make a simple redundant group by simply joining together two copies of a non-redundant group. We take a small one, $\mathbf{C_3}$:

```
redundantGroup =
```

```
Join[CartRep["C3"], CartRep["C3"]];
% // MatrixList // Size[5]
```

$$\left\{\begin{pmatrix}1&0&0\\0&1&0\\0&0&1\end{pmatrix}, \begin{pmatrix}-\frac{1}{2}&-\frac{\sqrt{3}}{2}&0\\\frac{\sqrt{3}}{2}&-\frac{1}{2}&0\\0&0&1\end{pmatrix}, \begin{pmatrix}-\frac{1}{2}&\frac{\sqrt{3}}{2}&0\\-\frac{\sqrt{3}}{2}&-\frac{1}{2}&0\\0&0&1\end{pmatrix}, \begin{pmatrix}1&0&0\\0&1&0\\0&0&1\end{pmatrix}, \begin{pmatrix}-\frac{1}{2}&-\frac{\sqrt{3}}{2}&0\\\frac{\sqrt{3}}{2}&-\frac{1}{2}&0\\0&0&1\end{pmatrix}, \begin{pmatrix}-\frac{1}{2}&\frac{\sqrt{3}}{2}&0\\-\frac{\sqrt{3}}{2}&-\frac{1}{2}&0\\0&0&1\end{pmatrix}\right\}$$

We name the matrices **E**, **A**, **B**, **U**, **V**, and **W**, respectively:

```
MultiplicationTable[Dot, redundantGroup,
  {"E", "A", "B", "U", "V", "W"}];
% // MatrixForm
```

$$\begin{pmatrix}E&A&B&E&A&B\\A&B&E&A&B&E\\B&E&A&B&E&A\\E&A&B&E&A&B\\A&B&E&A&B&E\\B&E&A&B&E&A\end{pmatrix}$$

The names **U,V,** and **W** do not appear at all; other names appear multiple times in each row and column. That is because the name replacement rule is used as soon as the first match is seen. Matrices **A** and **V** are identical, and the **A** rule is always tried and used before **V** rule. Thus redundant groups really look redundant and can be identified at a glance from their multiplication tables.

8.8. End Notes

8.8.1. Not a group table

The following is a Latin Square, but is it the multiplication table of a group?

```
Clear[LtnSq2, A, B];
```

LtnSq2 =

C	B	A	D
D	C	B	A
A	D	C	B
B	A	D	C

[[1]]

```
{{C, B, A, D}, {D, C, B, A}, {A, D, C, B}, {B, A, D, C}}
```

Note: **Part** 1 of the **GridForm** of a matrix is the matrix itself, as you see above.

Test whether the multiplication operation is associative :

```
LtnSq2N = LtnSq2 /. {A → 1, B → 2, C → 3, D → 4};
MatrixForm[LtnSq2] → MatrixForm[LtnSq2N]
```

$$\begin{pmatrix} C & B & A & D \\ D & C & B & A \\ A & D & C & B \\ B & A & D & C \end{pmatrix} \to \begin{pmatrix} 3 & 2 & 1 & 4 \\ 4 & 3 & 2 & 1 \\ 1 & 4 & 3 & 2 \\ 2 & 1 & 4 & 3 \end{pmatrix}$$

```
Table[LtnSq2N[[i, LtnSq2N[[j, k]] ]] ==
    LtnSq2N[[ LtnSq2N[[i, j]], k]], {i, 4},
   {j, 4}, {k, 4}] // Flatten // Union
```

```
{False, True}
```

There was at least one **False** among the 64 cases, so the operator is not associative. (Take away the **Flatten** and **Union** operators to see all 64 cases.) Therefore it is not a group table, in spite of being a Latin Square. This one counter-example proves that the converse of the Latin Square theorem is false.

8.8.2. The great aerial warfare program

The Israeli Air Force wrote an immense aerial warfare program that used comprehensive radar coverage of the Middle East to keep track of everything flying. This information was fed to an artificial intelligence routine that told Israeli fighter planes how to attack, or how to evade. It was used in the 1982 air war in the Bekaa Valley, and it gave Israel a decisive advantage. In fact, the Israelis shot down about 100 Syrian MiGs without losing a single fighter of their own. (Click to the Wikipedia) So you have to say this was very successful software, tested repeatedly in the very fire of battle.

After the war, an Israeli plane crossed the equator under the control of this program. Immediately, the plane flipped upside down, and the pilot had scramble to save himself. The bug was traced to the function **Tan[θ]**, which has a singularity at $\pi/2$, the latitude of the equator:

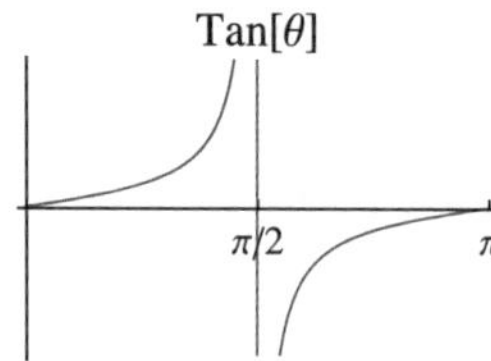

The Bekaa Valley is well north of the equator, and the program never had to deal with negative values of **Tan[θ]** before. This can happen whenever an algorithm is taken into new territory. I hate to think what may happen with some of mine ...

9. Molecules

Preliminaries

9.1 Molecule definitions in *Mathematica*

Our *Mathematica* molecules are written and maintained by a simple form of object-oriented programming, which insures that errors are caught at the time of molecule entry, rather than during their use.

One of the predefined molecules in the **Molecules`** package is "**methane**". Look at the data contained in it :

```
MoleculeToList["methane"] // GridForm
```

H	$\left\{\frac{6287}{10\,000}, \frac{6287}{10\,000}, \frac{6287}{10\,000}\right\}$	{1, Td}
H	$\left\{-\frac{6287}{10\,000}, -\frac{6287}{10\,000}, \frac{6287}{10\,000}\right\}$	{2}
H	$\left\{-\frac{6287}{10\,000}, \frac{6287}{10\,000}, -\frac{6287}{10\,000}\right\}$	{3}
H	$\left\{\frac{6287}{10\,000}, -\frac{6287}{10\,000}, -\frac{6287}{10\,000}\right\}$	{4}
C	{0, 0, 0}	{1}

On your own
Also look at this without the **GridForm** display wrapper.

Each row of this matrix (i.e., each sublist of this list) contains the data for one atom, in the form of an "atom list". Each atom list contains three items :

(1) The element symbol for the atom

(2) The Cartesian coordinates for the atom, in Angstrom units $\left(1\,\text{Å} = 10^{-10}\,\text{Meter}\right)$· Although they are experimental values good (at most) to 4 or 5 sig figs, they are written above as infinite precision numbers (e.g., 0.6287 Å → $\frac{6287}{10\,000}$) . As you will see below, this allows us make the symmetry exact even when the known bond distances are only approximate.

(3) A list of tags that distinguish it from other atoms of the same kind. The tag list of each atom must be a **List** of **Strings**, and it must contain at least one item. The first item in the tag list is displayed in the center of the atom by graphics operators. If you want no tag, put a null string ("") in first position. But you cannot leave it blank.

W.M. McClain, *Symmetry Theory in Molecular Physics with Mathematica*,
DOI 10.1007/b13137_9, © Springer Science+Business Media, LLC 2009

(4) Every molecule has an extra element in the tag list of its first atom; namely, the point group to which the molecule belongs. It must always be the last item in the first tag list. In this position it will not interfere with any programming that makes use of the preceding tags. If you do not know the group, enter **"?"** as the group tag.

Now we look at the first atom of the molecule in **FullForm** :

```
MoleculeToList["methane"][[1]] // FullForm
```

```
List["H",
 List[Rational[6287, 10000], Rational[6287, 10000],
  Rational[6287, 10000]], List["1", "Td"]]
```

Now you can see that the names of the elements and the tags are all **Strings** so that they cannot be confused with algebraic or symbolic quantities. When writing a new molecule data list from scratch, remember that the quotes are required. Note also that quotes are required around the names of molecules.

What sort of thing, besides an atom number, might go into the tag list of an atom? In a protein, each amino acid could carry a common tag on all its atoms. This tag would make it possible for a special display operator to color, say, all the lysines green. Another tag could record a calculated local electrical potential at the atom, or the atom's magnetic shielding constant. It is up to the user to define the tag list, and to keep track of its meaning. You can bring out a list of all the tags with

```
AtomTagList["methane"]
```

```
{{1, Td}, {2}, {3}, {4}, {1}}
```

Note that group $\mathbf{T_d}$ is the last tag of the first atom.The first tags, used by **Atom⋱ Graphics**, are brought out by

```
FirstTagList["methane"]
```

```
{1, 2, 3, 4, 1}
```

The list of elements can also be called up:

```
ElementList["methane"]
```

```
{H, H, H, H, C}
```

```
AtomNameList["methane"]
```

$\{H_1, H_2, H_3, H_4, C_1\}$

The atom number subscripts are in a distinctive type face ("Copperplate") to diminish confusion between, e.g., the second H atom H_2 and the molecule H_2.

On your own
Enter **MoleculeCatalog** to see the names of all molecules currently defined. Look at the data lists of a few of them.

9.2 Molecule "objects"

To construct a molecule, begin by assembling a list of data for it, in the format given above. Say you give it the name **molList**. However, none of our molecule operators can work on **molList** as it stands. To make it useful you must call the object-making operator

```
ListToMolecule[molList,"molName"]
```

which checks the format of your **molList**, and then (and only if the format is correct) sets it equal to **molecule["molName"]**. It has fairly explicit error messages, designed to help you get you molecules correct. Here is a molecule list with a typo in it :

$$\text{badMethaneList} = \Big\{\Big\{\text{"H"}, \Big\{\frac{6287}{10\,000}, \frac{6287}{10\,000}, \frac{6287}{10\,000}\Big\}, \{\text{"1"}, \text{"Td"}\}\Big\},$$
$$\Big\{\text{"H"}, \Big\{-\frac{6287}{10\,000}, -\frac{6287}{10\,000}, \frac{6287}{10\,000}\Big\}, \{\text{"2"}\}\Big\},$$
$$\Big\{\text{"H"}, \Big\{-\frac{6287}{10\,000}, \frac{6287}{10\,000}, -\frac{6287}{10\,000}\Big\}, \{\text{"3"}\}\Big\},$$
$$\Big\{\text{H}, \Big\{\frac{6287}{10\,000}, -\frac{6287}{10\,000}, -\frac{6287}{10\,000}\Big\}, \{\text{"4"}\}\Big\},$$
$$\{\text{"C"}, \{0, 0, 0\}, \{\text{"1"}\}\}\Big\};$$

```
ListToMolecule[badMethaneList, "badMethane"]
```

```
non-String element name
```

Yes, one **H** is not a string (did not have quotes around it). This "object oriented" format check keeps our molecule formats correct. All molecule objects are stored together in a place not directly accessible to you. The only way to put them there is to use the **ListToMolecule** operator. If you need to see the list again, you must go through the molecule reading operator

```
MoleculeToList["molName"]
```

For example, the ammonia molecule is in the computer somewhere as an object having **molecule** as its **Head**, but you do not have direct access to it. Try to call it up:

```
molecule["ammonia"]
```

```
molecule[ammonia]
```

It returns unevaluated. This is just a slight protection to remind you that the format of the ammonia molecule has been checked and certified to be correct, and you should not mess around with it. This slight protection is easy to defeat, but the best advice is, don't try. The molecule objects are guaranteed to have the correct format, and any direct meddling might introduce an error.

> **On your own**
> Look at ammonia using **MoleculeToList["ammonia"]**.

9.3 Molecule object operators

We have a number of operators that work on molecule objects. Here is one that goes into the current **molecule** object collection, and pulls out all the molecule names :

```
originalMC = MoleculeCatalog
```

```
{1,3,5-trichlorobenzene, ammonia, benzene,
 brick, C2H2Br2Cl2 staggered, carbonate anion,
 CopperIIhexanitriteAnion4, cyclobutadiene,
 cyclopropenyl, cyclopropenylCation, eclipsedEthane,
 equilateral ozone, ethylene, formaldehyde,
 haloTet, hexachlorobenzene, icosahedron, methane,
 methaneOLD, naphthaleneCatoms, octaComplex, PCl5,
 PtCl4dianion, RNAdimer, staggeredEthane, terrylene,
 transHB3H2D2, twoTriangles, UF9, water, XeF4}
```

If you define new molecules, they will appear in the catalog after they are put away by **ListToMolecule**. A few more simple operators :

```
MolecularWeight["ammonia"]
```

```
17.0305
```

```
CenterOfMass["methane"]
```

```
{0.×10^-7, 0.×10^-7, 0.×10^-7}
```

```
EnclosingRadius["ammonia"]
```

```
1.00167
```

Use the **?** to see the precise definition of **EnclosingRadius**. The next one may be a little less familiar,

```
InertiaTensor["ammonia"]
```

```
{{1.57756, 0.×10^-7, 0.×10^-7},
 {0.×10^-7, 1.57756, 0.×10^-7}, {0.×10^-7, 0.×10^-7, 2.94929}}
```

This is used to find the symmetry axes of molecules that are entered in an asymmetric orientation.

9.4 How to make a molecule From NIST to *Mathematica*

You may write a new molecule from scratch. Note, for instance that there is no "chloroform" molecule in the MoleculeCatalog. Here is how to add it:

1. Click on http://webbook.nist.gov/chemistry .
2. Under SearchOptions, you want **Name**. Click on it.
3. Type in *chloroform*. This will take you to the data on chloroform.
4. Click on "Computational Chemistry Comparison and Benchmark Database".
5. Click on the blue underlined x at the intersection of the *Experiment* column and the *Internal Coordinates* row.
6. Scroll down to *Geometric Data*, and find the table.
7. **Copy** the content of the *Cartesians* table and **Paste** directly into a *Mathematica* notebook. The data will appear as an unpunctuated **Input** cell. Do a **Format ▷ ClearFormatting** on it and provide it with whatever it needs to become a *Mathematica* molecule list :

```
chloroformList = {
{"C", {0.0000, 0.0000, 0.5231}, {"1", "C3v"}},
{"H", {0.0000, 0.0000, 1.5961}, {"1"}},
{"Cl", {1.6562, 0.0000, -0.0928}, {"1"}},
{"Cl", {-0.8281, 1.4343, -0.0928}, {"2"}},
{"Cl", {-0.8281, -1.4343, -0.0928}, {"3"}}};
```

Above, the **Cl** atoms nearly, but not exactly, a third of a turn apart. If this is a problem, click to an End Note to see how to make the rotational symmetry exact. Make it into a molecule object:

```
ListToMolecule[chloroformList, "chloroform"]
```

Has chloroform has been added to the catalog ? We pull down the current catalog and use **Complement** to compare to the old catalog, pulled down and named above. Is there anything new :

```
newMC = MoleculeCatalog;
Complement[newMC, originalMC]
```

```
{chloroform}
```

There is. We will use it below to demonstrate the **AtomGraphics** operator.

9.5 Molecule graphics

9.5.1. AtomGraphics and BondGraphics

The **Molecules`** package defines several operators for making molecules into graphics objects. The wo basic ones are **AtomGraphics** and **BondGraphics**. If you want to see the atoms of the molecule "chloroform" (created above, in Section 9.4), the command is

```
Options[AtomGraphics] // Column
```

```
AxisLength → Automatic
AtomDiameterFactor → 1
AtomLabels → True
ViewPoint → {-3, -24, 20}
BaseStyle → Text
Boxed → False
ImageSize → {250, 250}
Lighting → Neutral
```

```
AtomGraphics["chloroform",
 ImageSize → 150, ViewPoint → {3, -24, 10}]
```

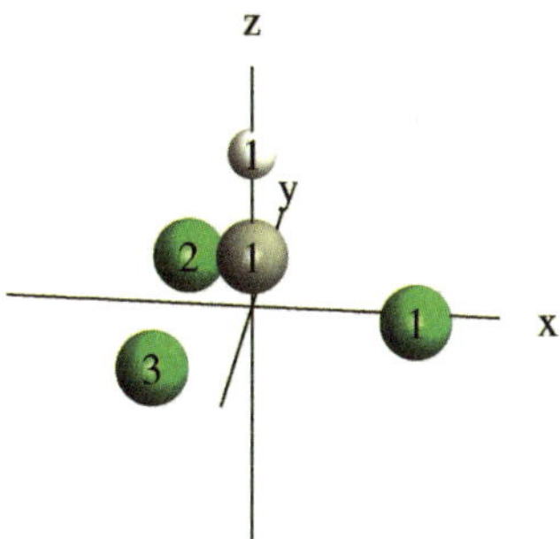

Each element has its own color code. Above, **C** is gray, **Cl** is green, and **H** is white. Below, the default colors for the common elements :

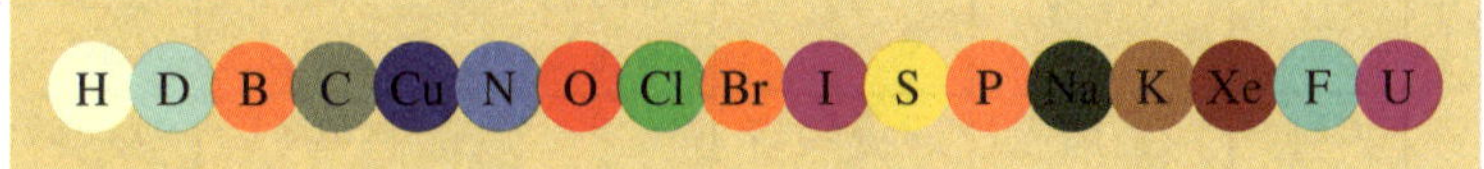

Usually chemists want to see the bonds as well as atoms. For this we use **BondGraphics**:

```
CHCl3pic = Show[{
BondGraphics["chloroform"],
AtomGraphics["chloroform"]},
ImageSize → 150, ViewPoint → {3, -24, 10}]
```

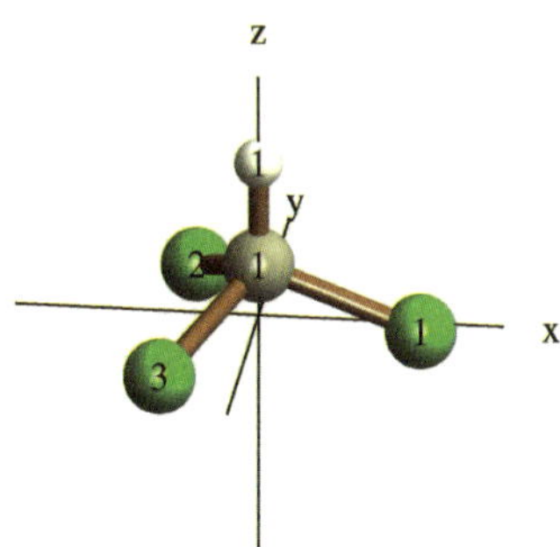

On your own
Grab this image with the mouse and turn it to a new orientation.
Show the bonds all by themselves with no atoms.
Show the molecule with Cl-Cl bonds as well as C-Cl bonds.

There is nothing in the molecule object that says which atoms are bonded to which. Instead, we draw bonds based strictly on inter-atom distances; by default, a bond will be drawn between any two atoms closer than **1.8 Å**. But if the default does not do what you want, look at the **DistanceMatrix** of the molecule :

```
DistanceMatrix["chloroform"] //
 GridForm
```

mol	C_1	H_1	Cl_1	Cl_2	Cl_3
C_1	0.	1.073	1.767\ 01	1.767	1.767
H_1	1.073	0.	2.365\ 46	2.365\ 45	2.365\ 45
Cl_1	1.767\ 01	2.365\ 46	0.	2.868\ 62	2.868\ 62
Cl_2	1.767	2.365\ 45	2.868\ 62	0.	2.8686
Cl_3	1.767	2.365\ 45	2.868\ 62	2.8686	0.

From this you can find a bond length parameter of **BondGraphics** that will draw the bonds you want. As you see, **1.8** Å is a good guess for chloroform: **C-H** and **C-Cl** bonds were drawn, but no others. 1.8 is nearly always a good first guess.

No single 3D projection, even if perfectly drawn, can convey all the three dimensional information about a molecule. Orientational mouse drags will tell the full story for interactive readers, but for documents we need another strategy. Below in Section 9.6 and Section 9.7 you will see two quite different ways of presenting static multiple views that do convey all the three dimensional information.

9.5.2. The default molecule

When you are working a lot with one particular molecule, it is nice to set it up as a default so you don't have to keep retyping its name so much. Taking a new example, we set

```
$DefaultMolecule = "haloTet";
```

Now every operator that uses a molecule name will use this one by default. But note that you cannot use defaults and options together in the same operator. If you want display options, you have to use the **Show** wrapper, and make the display options into options of **Show**:

```
Show[{BondGraphics[], AtomGraphics[]},
  ImageSize → 150, ViewPoint → {3, -24, 10}]
```

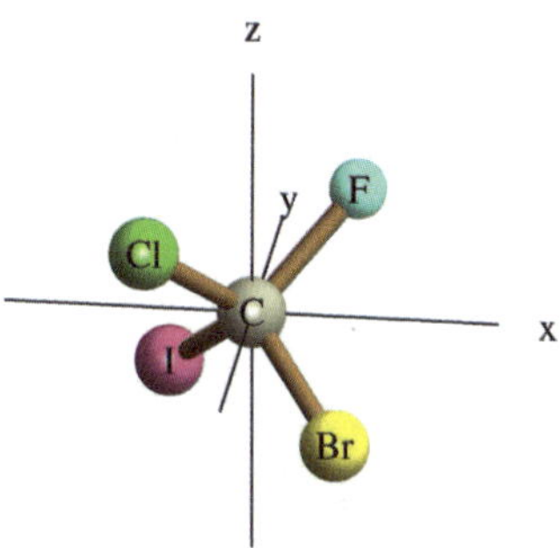

This is "**haloTet**". It is like methane, but with four different halogens instead of hydrogens. It is a particularly simple handed molecule.

> **On your own**
> A bond is drawn by default between every pair of atoms closer than **1.8** Å . Of course, this is not always appropriate. Take a look at the **DistanceMatrix** for this molecule to see that it is appropriate in this case.
>
> To practice using explicit bonding distances in **BondGraphics**, change the default molecule to "ethylene", and show a picture of it with C-C bonds and C-H bonds, but no H-H bonds.
>
> Then just for fun, add H-H "bonds" to the picture. Chemists, forebear. This is just a mathematical exercise.

9.5.3. Options for AtomGraphics

We want to make the "haloTet" picture easier to read. Redraw it without coordinate axes, with bigger atoms, and with better-looking atom labels. Call it **ag**:

```
T8I =
  {FontFamily → "Times", 10, FontSlant → "Italic"};
ag = AtomGraphics["haloTet",
  (*Option1*) AxisLength → 0,
  (*Option2*) AtomDiameterFactor → 2,
  (*Graphics options*) BaseStyle → T8I,
  ImageSize → 100, ViewPoint → {3, -24, 10}]
```

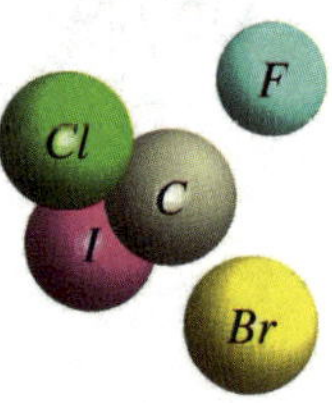

Now it is easy to put the bonds with the atoms :

```
bg = BondGraphics[];
haloTetObj3D = Show[{ag, bg}, ImageSize → 100]
```

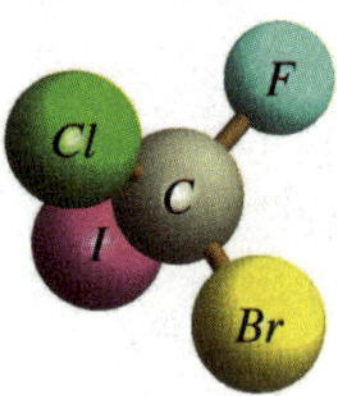

9.6 StereoView

Let's look at a stereo graphic of **haloTet**. The left figure is for the left eye; the right, for the right. Can you make the image pop up in 3D?

```
svhT = StereoView[haloTetObj3D, ImageSize → 200]
```

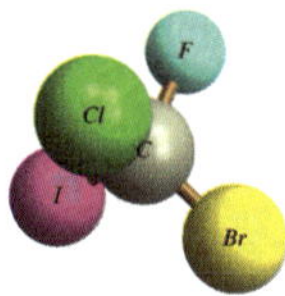

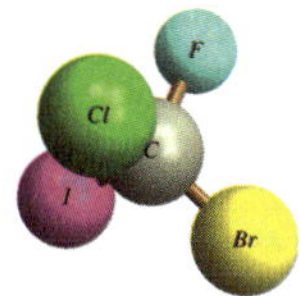

Nearly everyone can see these figures in 3D with a little practice. If you have trouble, click once in the middle of the figure and drag the corner handles make the image smaller. Relax and let your eyes wander into double vision. Don't get scared; it is OK to see double. This may be the hardest part. Just think far-off thoughts, and try to enjoy seeing four figures floating and wandering. Tilt your head left or right until all four **C** atoms are at the same height. Hold there, and gradually let the middle two float together. They will fuse suddenly into a single 3D image, flanked by two rather unfocused peripheral images which you just ignore. You will feel completely relaxed and unstrained when you achieve this. You will see clearly that Cl, C, F defines one plane and I, C, Br defines a perpendicular plane.

If you are nearsighted, it may help to take off your glasses and lean forward. Also, the pairs look much better when printed than they do on your computer screen. Print the image pair above, and try it with the hard copy.

Programming the stereo pair in Mathematica was very easy, doing a little calculation to get a slightly different **ViewPoint** for each eye, and then using the **GraphicsRow** command to place the two pictures side by side. Open up the **Molecules`** package and see how it was done, if you like.

9.7 AxialViews (a 3D shop drawing)

No single view of a three-dimensional object can tell everything about it, no matter how well-drawn it may be. If the single views provided by **AtomGraphics** are ambiguous, try **AxialViews**, which gives three views down the three coordinate axes:

AxialViews[obj3D(,opts)] returns three panels arranged in a way that is standard for machine shop drawings. Its main input must be something that is already a **Graphics3D** object, so we use the chloroform molecule that we evaluated above :

```
AxialViews[CHCl3pic, ImageSize → 220]
```

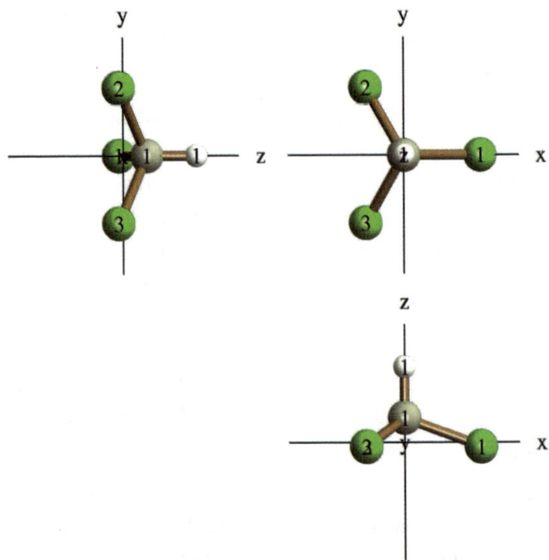

On the upper right you see a "top" view, on the upper left is a "side" view, and on the lower right a "front" view. If you cut this figure out and folded down the front and side views, you would have three sides of a cube containing the object, with its projection drawn on each of the three faces. Theoretically, the object is described completely in these drawings. With a little practice and a little thought, you can grasp its structure so completely that you could make the object in a machine shop.

On the compute screen, you can grab and rotate any of the three axial views using the mouse. But this ruins it, or course, for pasting into a document.

9.8 End Notes

9.8.1. Infinite precision coordinates

The coordinates of the chloroform molecule, as pulled down from NIST, were good to the fourth decimal:

```
chloroformList = {
{"C", {0.0000, 0.0000, 0.5231}, {"1", "C3v"}},
{"H", {0.0000, 0.0000, 1.5961}, {"1"}},
{"Cl", {1.6562, 0.0000, -0.0928}, {"1"}},
{"Cl", {-0.8281, 1.4343, -0.0928}, {"2"}},
{"Cl", {-0.8281, -1.4343, -0.0928}, {"3"}}};
```

But its symmetry is not exact. We show here how to make it exact. The first step is to turn all decimal numbers into ratios of exact integers. This is done by the core **Rationalize** operator, with optional second parameter **0** :

```
chloroformList2 =
 Map[Rationalize[#, 0] &, chloroformList] // Column
```

$\left\{C, \left\{0, 0, \frac{5231}{10000}\right\}, \{1, C3v\}\right\}$

$\left\{H, \left\{0, 0, \frac{15961}{10000}\right\}, \{1\}\right\}$

$\left\{Cl, \left\{\frac{8281}{5000}, 0, -\frac{58}{625}\right\}, \{1\}\right\}$

$\left\{Cl, \left\{-\frac{8281}{10000}, \frac{14343}{10000}, -\frac{58}{625}\right\}, \{2\}\right\}$

$\left\{Cl, \left\{-\frac{8281}{10000}, -\frac{14343}{10000}, -\frac{58}{625}\right\}, \{3\}\right\}$

Chlorine atom **1** is in the **x,z** plane, as required by convention. So we leave it alone. To put chlorine atom 2 exactly in its place, use

$$\texttt{RotationMatrix3Dz}[-2\,\pi\,/\,3].\left\{\frac{8281}{5000}, 0, -\frac{58}{625}\right\}$$

$$\left\{-\frac{8281}{10000}, -\frac{8281\sqrt{3}}{10000}, -\frac{58}{625}\right\}$$

Note the infinite precision radical that has appeared. Now chlorine atom 3 must be at

$$\texttt{RotationMatrix3Dz}[-4\,\pi\,/\,3].\left\{\frac{8281}{5000}, 0, -\frac{58}{625}\right\}$$

$$\left\{-\frac{8281}{10\,000},\ \frac{8281\sqrt{3}}{10\,000},\ -\frac{58}{625}\right\}$$

Copy and Paste these positions into to list below, and Enter it :

```
chloroformList3 = {
   {"C", {0, 0, 5231/10000}, {"1", "C3v"}},
   {"H", {0, 0, 15961/10000}, {"1"}},
   {"Cl", {0, 8281/5000, -58/625}, {"1"}},
   {"Cl", {8281 Sqrt[3]/10000, -8281/10000, -58/625}, {"2"}},
   {"Cl", {-8281 Sqrt[3]/10000, -8281/10000, -58/625}, {"3"}}};
```

Finally, put the molecule into the package :

```
ListToMolecule[chloroformList3 ,"chloroformExact"]
```

10. The point groups

Preliminaries

10.1. Introduction

So far we have seen two kinds of symmetry matrices that apply to molecules: rotation and reflection. But the fundamental theorem makes it clear that any new kind of matrix that can be generated from these by multiplication must also be accepted as a symmetry matrix on its own. In this chapter we will do exactly that, producing three new kinds of symmetry matrices that apply to molecules. That will make five kinds in all.

We will then reclassify them as just two kinds, and show that nothing fundamentally new can come from multiplying them together.

The groups made from matrices of these kinds are called *point groups*.

10.2. The unit matrix

The first example is rather trivial, but it points the way to two other developments that are not so trivial. Suppose you examine a flat, but otherwise unsymmetrical molecule. The plane of the molecule is a reflection plane in which every atom remains unmoved, so it is a symmetry element for the molecule. Let the **x,y** axes lie in the plane of the molecule, with **z** perpendicular to it. The reflection matrix σ_z is predefined in the **Symmetry`** package, as is the unit matrix **Emat** :

```
{σz, Emat} // MatrixList
```

$$\left\{\begin{pmatrix} 1 & 0 & 0 \\ 0 & 1 & 0 \\ 0 & 0 & -1 \end{pmatrix}, \begin{pmatrix} 1 & 0 & 0 \\ 0 & 1 & 0 \\ 0 & 0 & 1 \end{pmatrix}\right\}$$

Multiply σ_z by itself :

```
σz.σz == Emat
```


DOI 10.1007/b13137_10, © Springer Science+Business Media, LLC 2009

```
True
```

Further multiplications yield nothing new, so the Fundamental Theorem says that **{σ_z, Emat}** must be a group, and **Emat** must be accepted as a kind of symmetry matrix for the molecule. So now we have three distinct kinds of symmetry matrix: σ matrices (reflections), **C** matrices (rotations), and **Emat** (the unit, or identity, matrix).

10.3. The inversion matrix

The symmetry group of a brick is one of the easiest to visualize. Simple as it is, the brick generates a surprise gift; a new kind of symmetry transform, not a rotation nor a reflection nor the identity, but another kind of symmetry matrix that is required to exist by the fundamental theorem.

Imagine before yourself a perfect brick, centered squarely in an **x,y,z** coordinate system. (If you do not see one, choose the closed cell just below and process it, and such a brick will appear.)

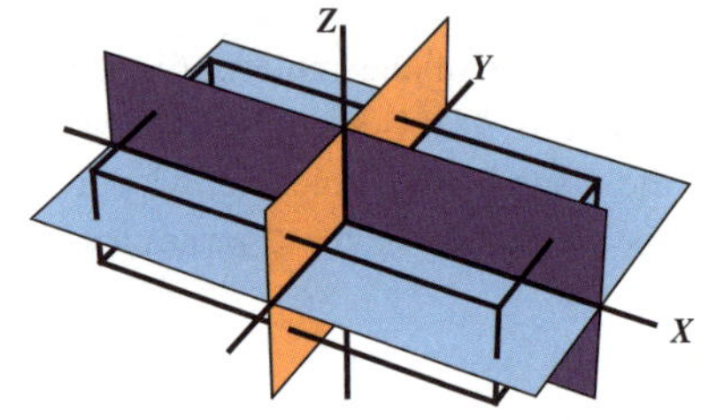

Fig. 10.1 A brick, centered squarely on the axes, with reflecting planes

We notice that each of the coordinate planes (**x = 0**, **y = 0**, and **z = 0**) is a reflection plane for the brick. We put them together, along with **Emat**, as a list of matrices

```
mats1 = {Emat, σx, σy, σz};
Map[MatrixForm, mats1]
```

$$\left\{\begin{pmatrix}1&0&0\\0&1&0\\0&0&1\end{pmatrix},\begin{pmatrix}-1&0&0\\0&1&0\\0&0&1\end{pmatrix},\begin{pmatrix}1&0&0\\0&-1&0\\0&0&1\end{pmatrix},\begin{pmatrix}1&0&0\\0&1&0\\0&0&-1\end{pmatrix}\right\}$$

Give them string symbolic names :

```
names1 = {"E", "σx", "σy", "σz"};
```

Now using these names, we test the list of matrices for closure, to see if they form a group:

```
MultiplicationTable[Dot, mats1, names1] // MatrixForm
```

$$\begin{pmatrix}E&\sigma_x&\sigma_y&\sigma_z\\\sigma_x&E&?&?\\\sigma_y&?&E&?\\\sigma_z&?&?&E\end{pmatrix}$$

No they do not; all the $\sigma.\sigma$ products are missing. For example, $\sigma_x.\sigma_y$ is missing.

```
σx.σy // MatrixForm
```

$$\begin{pmatrix}-1&0&0\\0&-1&0\\0&0&1\end{pmatrix}$$

A little thought shows that this ought to be the same as a rotation by half a turn about **z**, but we check to make sure :

```
Cz[π] == σx.σy
```

```
True
```

Yes, of course, two-fold rotations about the coordinate axes are indeed symmetry operations for the brick. They are just as obvious as the reflections, once they are pointed out. This little mistake emphasizes nicely that the rotation axes are a logical consequence of the reflection axes, and not some completely new kind of symmetry. We extend our list of matrices and names to include the rotations :

```
C2x = Cx[π];
C2y = Cy[π];
C2z = Cz[π];
C2mats = {C2x, C2y, C2z};
C2names = {"C2x", "C2y", "C2z"};
mats2 = Join[mats1, C2mats];
names2 = Join[names1, C2names];
Map[MatrixForm, mats2]
```

$$\left\{\begin{pmatrix}1&0&0\\0&1&0\\0&0&1\end{pmatrix}, \begin{pmatrix}-1&0&0\\0&1&0\\0&0&1\end{pmatrix}, \begin{pmatrix}1&0&0\\0&-1&0\\0&0&1\end{pmatrix}, \begin{pmatrix}1&0&0\\0&1&0\\0&0&-1\end{pmatrix}, \begin{pmatrix}1&0&0\\0&-1&0\\0&0&-1\end{pmatrix}, \begin{pmatrix}-1&0&0\\0&1&0\\0&0&-1\end{pmatrix}, \begin{pmatrix}-1&0&0\\0&-1&0\\0&0&1\end{pmatrix}\right\}$$

All these matrices are very simple; they are all based on various negations of the **1**'s on the diagonal of the unit matrix . Now, with the three rotations added, does the table close?

```
MultiplicationTable[Dot, mats2, names2] // MatrixForm
```

$$\begin{pmatrix} E & \sigma_x & \sigma_y & \sigma_z & C_{2x} & C_{2y} & C_{2z} \\ \sigma_x & E & C_{2z} & C_{2y} & ? & \sigma_z & \sigma_y \\ \sigma_y & C_{2z} & E & C_{2x} & \sigma_z & ? & \sigma_x \\ \sigma_z & C_{2y} & C_{2x} & E & \sigma_y & \sigma_x & ? \\ C_{2x} & ? & \sigma_z & \sigma_y & E & C_{2z} & C_{2y} \\ C_{2y} & \sigma_z & ? & \sigma_x & C_{2z} & E & C_{2x} \\ C_{2z} & \sigma_y & \sigma_x & ? & C_{2y} & C_{2x} & E \end{pmatrix}$$

Not quite. For instance, $\sigma_x . C_{2x}$ must still be included. We look at it:

```
Imat = σx.C2x;   MatrixForm[Imat]
```

$$\begin{pmatrix}-1&0&0\\0&-1&0\\0&0&-1\end{pmatrix}.$$

This matrix, which we call **Imat**, has the only pattern of diagonal negation that was missing from the list **mats2**. **Imat** is a new kind of symmetry operation that we have not seen before; it is not the identity, nor a rotation about any axis, nor a reflection in any plane. It is called the *inversion*, and is perhaps best described as a three simultaneous reflections in the three coordinate planes.

Inversion
The inversion is the simultaneous negation of all coordinates;
i.e., it is a reflection through the origin.

With the inclusion of the inversion matrix **Imat** , does our list of matrices close?

```
mats3 =  Join[mats2, {Imat}];
names3 = Join[names2, {"I"}];
mt3 = MultiplicationTable[Dot, mats3, names3];
MatrixForm[mt3]
```

$$\begin{pmatrix} E & \sigma_x & \sigma_y & \sigma_z & C_{2x} & C_{2y} & C_{2z} & I \\ \sigma_x & E & C_{2z} & C_{2y} & I & \sigma_z & \sigma_y & C_{2x} \\ \sigma_y & C_{2z} & E & C_{2x} & \sigma_z & I & \sigma_x & C_{2y} \\ \sigma_z & C_{2y} & C_{2x} & E & \sigma_y & \sigma_x & I & C_{2z} \\ C_{2x} & I & \sigma_z & \sigma_y & E & C_{2z} & C_{2y} & \sigma_x \\ C_{2y} & \sigma_z & I & \sigma_x & C_{2z} & E & C_{2x} & \sigma_y \\ C_{2z} & \sigma_y & \sigma_x & I & C_{2y} & C_{2x} & E & \sigma_z \\ I & C_{2x} & C_{2y} & C_{2z} & \sigma_x & \sigma_y & \sigma_z & E \end{pmatrix}$$

It does close, so now we have four kinds of symmetry elements: **C** matrices, σ matrices, **Emat**, and **Imat**. (This group will later be named systematically as $\mathbf{D_{2\,h}}$.)

It may seem a little strange to give a name and an independent existence to a symmetry operation that is easily visualized as the product of three simple reflections. Remember, we do this in order to have a complete group of symmetry operations, to which we may apply all the many subtle theorems of abstract group theory. A little strangeness here in the beginning yields a big payoff down the road.

Molecules that contain an inversion center play an important role in molecular spectroscopy, because for very fundamental reasons (which we will analyze in due course) the spectroscopy of such molecules is cleaner and simpler than that of similar molecules that lack the inversion center. So inversion plays a larger role in the theory of spectroscopy than one might at first imagine.

Once we admit the inversion as a symmetry operation on its own, it does not have to occur as a consequence of other symmetries. For instance, **Emat** and **Imat** make a two-element group all by themselves. (This group is called C_i.)

```
MultiplicationTable[Dot, {Emat, Imat}, {"E", "I"}] //
 MatrixForm
```

$$\begin{pmatrix} E & I \\ I & E \end{pmatrix}$$

10.4. Roto-reflection matrices

10.4.1. A group based on "eclipsedEthane" (an S_3 axis)

With one more example we will complete our discoveries of new types of symmetry operators. Consider the ethylene molecule in its eclipsed configuration. This molecule will lead us to a whole new class of symmetry transforms. Here is the molecule, in a view that shows one of the H-C-C-H planes edge-on. We let this plane be the **y, z** plane, or the **x=0** plane.

```
eclEth = Show[AtomGraphics["eclipsedEthane",
    AtomLabels → False, AxisLength → 2],
  BondGraphics["eclipsedEthane", 1.5`],
  ViewPoint → {0, 3, 1}, ImageSize → 120, Boxed → False]
```

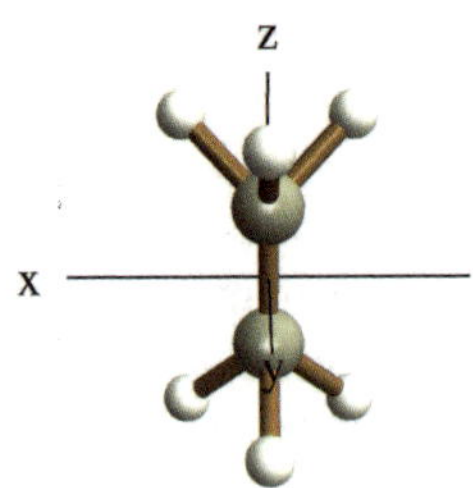

Fig. 10.2 Ethane, in eclipsed configuration

One view is never enough to see for sure the structure of a three-dimensional molecule. The **AxialViews** operator shows everything :

```
AxialViews[eclEth, ImageSize -> 200]
```

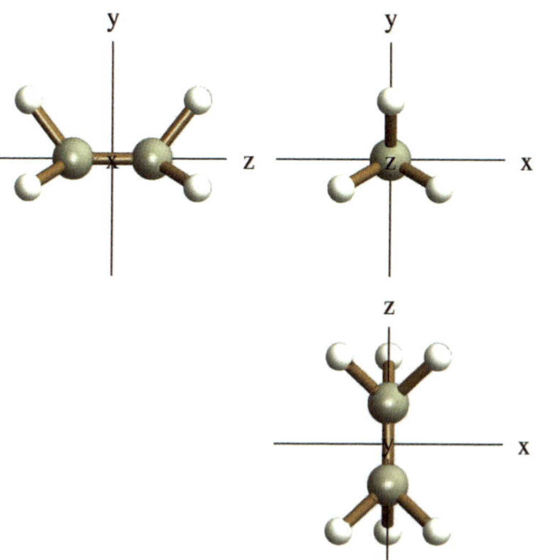

Fig. 10.3 A shop drawing of eclipsedEthane

The view down the **z** axis shows only three **H** atoms because the top three are "eclipsing" the bottom three. The threefold rotation about **z** and the reflection in the **xy** plane (parallel to the **z** axis) are easily noticed.

We have already defined matrices **Emat** and σ_z ; now we need a rotation C_{3z} and its square, C_{3z}^2.

```
C3z = RotationMatrix3Dz[2 π / 3]; C3z^2 = C3z.C3z;
{C3z, C3z^2} // MatrixList
```

$$\left\{\begin{pmatrix} -\frac{1}{2} & -\frac{\sqrt{3}}{2} & 0 \\ \frac{\sqrt{3}}{2} & -\frac{1}{2} & 0 \\ 0 & 0 & 1 \end{pmatrix}, \begin{pmatrix} -\frac{1}{2} & \frac{\sqrt{3}}{2} & 0 \\ -\frac{\sqrt{3}}{2} & -\frac{1}{2} & 0 \\ 0 & 0 & 1 \end{pmatrix}\right\}$$

We put the four matrices together and test for closure:

```
MultiplicationTable[Dot, {Emat, C3z, C3z^2, σz},
  {"E", "C3z", "C3z^2", "σz"}] // MatrixForm
```

$$\begin{pmatrix} E & C_{3\,z} & C_{3\,z}^2 & \sigma_z \\ C_{3\,z} & C_{3\,z}^2 & E & ? \\ C_{3\,z}^2 & E & C_{3\,z} & ? \\ \sigma_z & ? & ? & E \end{pmatrix}$$

There are four question marks. Two of them are due to

```
{ C3z.σz, σz. C3z} // Map[MatrixForm, #] &
```

$$\left\{\begin{pmatrix} -\frac{1}{2} & -\frac{\sqrt{3}}{2} & 0 \\ \frac{\sqrt{3}}{2} & -\frac{1}{2} & 0 \\ 0 & 0 & -1 \end{pmatrix}, \begin{pmatrix} -\frac{1}{2} & -\frac{\sqrt{3}}{2} & 0 \\ \frac{\sqrt{3}}{2} & -\frac{1}{2} & 0 \\ 0 & 0 & -1 \end{pmatrix}\right\}$$

These are identical. The other two question marks are due to

```
{ C3z^2.σz, σz. C3z^2} // Map[MatrixForm, #] &
```

$$\left\{\begin{pmatrix} \frac{1}{4} & \frac{3}{4} & 0 \\ \frac{3}{4} & \frac{1}{4} & 0 \\ 0 & 0 & -1 \end{pmatrix}, \begin{pmatrix} \frac{1}{4} & \frac{3}{4} & 0 \\ \frac{3}{4} & \frac{1}{4} & 0 \\ 0 & 0 & -1 \end{pmatrix}\right\}$$

and these are identical also. So the group needs at least two more matrices which we name as $\mathbf{S}_{3\,z}$ and $\mathbf{S}_{3\,z}^2$.

```
S3z = C3z.σz;
S3z^2 = C3z^2.σz;
Map[MatrixForm, {S3z, S3z^2}]
```

$$\left\{\begin{pmatrix} -\frac{1}{2} & -\frac{\sqrt{3}}{2} & 0 \\ \frac{\sqrt{3}}{2} & -\frac{1}{2} & 0 \\ 0 & 0 & -1 \end{pmatrix}, \begin{pmatrix} -\frac{1}{2} & \frac{\sqrt{3}}{2} & 0 \\ -\frac{\sqrt{3}}{2} & -\frac{1}{2} & 0 \\ 0 & 0 & -1 \end{pmatrix}\right\}$$

We rearrange the group in an order that may seem peculiar, but which we know from experience will be instructive. Then we test again for closure :

```
MultiplicationTable[
  Dot, {Emat, S3 z, C3 z^2, σz, C3 z, S3 z^2},
    {"E", "S3 z", "C3 z^2", "σz", "C3 z", "S3 z^2"}] //
 MatrixForm
```

$$\begin{pmatrix} E & S_{3z} & C_{3z}^2 & \sigma_z & C_{3z} & S_{3z}^2 \\ S_{3z} & C_{3z}^2 & \sigma_z & C_{3z} & S_{3z}^2 & E \\ C_{3z}^2 & \sigma_z & C_{3z} & S_{3z}^2 & E & S_{3z} \\ \sigma_z & C_{3z} & S_{3z}^2 & E & S_{3z} & C_{3z}^2 \\ C_{3z} & S_{3z}^2 & E & S_{3z} & C_{3z}^2 & \sigma_z \\ S_{3z}^2 & E & S_{3z} & C_{3z}^2 & \sigma_z & C_{3z} \end{pmatrix}$$

Finally it closes, so this is indeed a group of matrices under **Dot**. Also, we notice immediately (by the purity of the counter-diagonal) that it is a *cyclic* group. But what is it that is cycling? It is our newly constructed matrix $\mathbf{S_{3\,z}}$, which suddenly comes into focus as the fundamental generator of this group. In other words, all the other members of the group are just matrix powers of $\mathbf{S_{3\,z}}$ (including the unit element, which is its zeroth power and its sixth power).

```
MatrixPower[S3 z, 6]
```

```
{{1, 0, 0}, {0, 1, 0}, {0, 0, 1}}
```

This is a group, but it is not the whole group of the eclipsed ethane molecule because we ignored the **yz** reflection, and two similar reflections. This is exactly how humans can easily err in assigning a molecule to a group.

10.4.2 The new symmetry element : the $S_{3\,z}$ matrix

Compare the matrix $\mathbf{S_{3\,z}}$ with the threefold rotation matrix $\mathbf{C_{3\,z}}$:

```
Map[MatrixForm, {C3 z, S3 z}]
```

$$\left\{\begin{pmatrix} -\frac{1}{2} & -\frac{\sqrt{3}}{2} & 0 \\ \frac{\sqrt{3}}{2} & -\frac{1}{2} & 0 \\ 0 & 0 & 1 \end{pmatrix}, \begin{pmatrix} -\frac{1}{2} & -\frac{\sqrt{3}}{2} & 0 \\ \frac{\sqrt{3}}{2} & -\frac{1}{2} & 0 \\ 0 & 0 & -1 \end{pmatrix}\right\}$$

They are the same, except for the negation of the **z,z** element, caused by multiplying $\mathbf{C_{3\,z}}$ with $\sigma_{\mathbf{z}}$, in either order. Matrix $\mathbf{S_{3\,z}}$ is not a rotation about any axis nor a reflection in any plane, not the inversion, nor the identity. It is a fifth kind of symmetry matrix called a *roto-reflection*, a combination of a rotation about an axis followed (or preceded) by reflection in a plane perpendicular to that axis. We also see that it is a required new independent element in any group that contains a rotation and a reflection parallel to that rotation's axis.

Just as rotation axes are always denoted by **C**, rotoreflection axes are always denoted by **S**.

Again, you may find it strange to give a name and an independent existence to a combination that is so easily visualized in terms of two more intuitive operations. And again we can only say that this is an essential part of creating groups of symmetry operations, which has such a big payoff later on.

Look at the six **H** atom lists of the "eclipsedEthane" molecule :

```
sixHlistEclipsed =
  Take[MoleculeToList["eclipsedEthane"], 6];
Column[sixHlistEclipsed]
```

$$\left\{H, \left\{0, 1, \frac{3}{2}\right\}, \{1, S3\}\right\}$$

$$\left\{H, \left\{-\frac{\sqrt{3}}{2}, -\frac{1}{2}, -\frac{3}{2}\right\}, \{2\}\right\}$$

$$\left\{H, \left\{\frac{\sqrt{3}}{2}, -\frac{1}{2}, \frac{3}{2}\right\}, \{3\}\right\}$$

$\{\text{H}, \{0, 1, -\frac{3}{2}\}, \{4\}\}$

$\{\text{H}, \{-\frac{\sqrt{3}}{2}, -\frac{1}{2}, \frac{3}{2}\}, \{5\}\}$

$\{\text{H}, \{\frac{\sqrt{3}}{2}, -\frac{1}{2}, -\frac{3}{2}\}, \{6\}\}$

Make a "molecule" that omits the two **C** atoms, for visual clarity :

```
ListToMolecule[sixHlistEclipsed, "sixHeclipsed"];
```

Starting with the coordinates of **H** atom **1**, we can generate the positions of all the other H atoms by repeated application of $S_{3\,z}$. The operator **NestList** does exactly this. Use the **?** to look at its syntax.

```
NestList[Sin, α, 3]
```

```
{α, Sin[α], Sin[Sin[α]], Sin[Sin[Sin[α]]]}
```

As advertised, it applied the **Sin** operator to α, **0** times through **3** times. We want to take a particular position in space (the location of **H** atom **1** in **eclipsedEthane**) and multiply is from the left by matrix $S_{3\,z}$, **0** times through **6** times. To avoid having to name this multiplication operator, we write in the anonymous form **$S_{3\,z}$.# &** .

Note: If you want to understand this completely right now, go the Documentation Center and read the tutorial on "Pure Functions". *Pure* functions have no function name. So it might be better to call them "anonymous functions".

```
Hpos1 = {0, 1, 3 / 2};
lineEclipsedH6 = NestList[S3z.# &, Hpos1, 6]
```

$$\left\{\left\{0, 1, \frac{3}{2}\right\}, \left\{-\frac{\sqrt{3}}{2}, -\frac{1}{2}, -\frac{3}{2}\right\}, \left\{\frac{\sqrt{3}}{2}, -\frac{1}{2}, \frac{3}{2}\right\}, \left\{0, 1, -\frac{3}{2}\right\}, \left\{-\frac{\sqrt{3}}{2}, -\frac{1}{2}, \frac{3}{2}\right\}, \left\{\frac{\sqrt{3}}{2}, -\frac{1}{2}, -\frac{3}{2}\right\}, \left\{0, 1, \frac{3}{2}\right\}\right\}$$

```
Options[AtomGraphics]
```

```
{AxisLength → Automatic, AtomDiameterFactor → 1,
 AtomLabels → True, ViewPoint → {-3, -24, 20},
```

```
    BaseStyle → Text, Boxed → False,
    ImageSize → {250, 250}, Lighting → Neutral}
```

Note that these points are identical to the **H** atoms positions in **sixHlist`Eclipsed**. The sixth one is identical to the first one, so it is a cycle. Make a red track that helps the eye to follow the atoms in order :

```
zigzagEcl = Graphics3D[{Red,
    Thickness[0.01`], Line[lineEclipsedH6]}];
Show[{AtomGraphics["sixHeclipsed",
   AtomDiameterFactor → 1.6, AxisLength → 0],
  zigzagEcl}, ViewPoint → {0, 4, 2.2},
 ImageSize → {200, 200}]
```

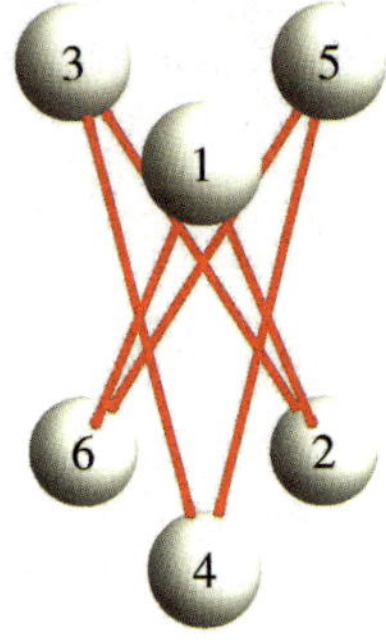

Fig. 10.4 Atom positions generated by repeated application of $S_{3\,z}$ to position 1. Grab it and rotate it to see it from other view points.

Letting your eye follow the numbers, you see what $S_{3\,z}$ does to each **H** atom: It advances by **1/3** turn, and also flips between bottom and top. Note particularly that it has to go around the circle twice to hit all six atoms.

10.4.3. Generalization to S_n

This is such a useful idea that we immediately generalize it :

> **S_n operation**
> Operation S_n is a right handed rotation by $2\pi/n$ radians, plus a reflection in a plane perpendicular to the rotation axis, and passing through the origin .

> **On your own**
> Show that it does not matter whether the rotation or the reflection is done first.

Carrying out the definition above as **RotationMatrix3D[β].σ_z** , we find that roto-reflection by angle β about the z axis is given by

```
Sz[β_] := ( Cos[β]  -Sin[β]   0 )
          ( Sin[β]   Cos[β]   0 )
          (   0        0     -1 )
```

We evaluate the case of roto-reflection by **1/6** of a turn :

```
S6z = Sz[2 π / 6]; MatrixForm[S6z]
```

$$\begin{pmatrix} \frac{1}{2} & -\frac{\sqrt{3}}{2} & 0 \\ \frac{\sqrt{3}}{2} & \frac{1}{2} & 0 \\ 0 & 0 & -1 \end{pmatrix}$$

Mathematica has a handy core operator for taking matrix powers. We use it to take the **5**th matrix power of S_{6z}; i.e., the result of matrix multiplying S_{6z} with itself **5** times :

```
S6z^5 = MatrixPower[S6z, 5]; MatrixForm[S6z^5]
```

$$\begin{pmatrix} \frac{1}{2} & \frac{\sqrt{3}}{2} & 0 \\ -\frac{\sqrt{3}}{2} & \frac{1}{2} & 0 \\ 0 & 0 & -1 \end{pmatrix}$$

Just as a check, we also do it the primitive way :

$$S_{6\,z}^{5} == S_{6\,z}.S_{6\,z}.S_{6\,z}.S_{6\,z}.S_{6\,z}$$

```
True
```

This is very different from the numerical **5**th power of $\mathbf{S_{6\,z}}$, given by

$$(S_{6\,z})^5$$

$$\left\{\left\{\frac{1}{32},\ -\frac{9\sqrt{3}}{32},\ 0\right\},\ \left\{\frac{9\sqrt{3}}{32},\ \frac{1}{32},\ 0\right\},\ \{0,\ 0,\ -1\}\right\}$$

Do you remember what the numerical power of a matrix is? If you are puzzled, take the numerical fifth power of a symbolic matrix **{{a,b},{c,d}}**.
Hint: This arises because **Power** is **Listable** , which means "automatically mapped". To see this for yourself, look at **Attributes[Power]**.

Because rotations about **z** commute with reflections parallel to **z**, it is easy to show that all $\mathbf{S_n^m}$ matrices about the **z** axis are of the form

$$S_{n\_}^{m\_} := \begin{pmatrix} \mathrm{Cos}\left[\frac{2\pi m}{n}\right] & -\mathrm{Sin}\left[\frac{2\pi m}{n}\right] & 0 \\ \mathrm{Sin}\left[\frac{2\pi m}{n}\right] & \mathrm{Cos}\left[\frac{2\pi m}{n}\right] & 0 \\ 0 & 0 & (-1)^m \end{pmatrix}$$

We will not use this form very often, but here it helps to make an important point. Below, for the case **n = 6**, we tabulate the rotation angle **2π*m/6** together with the associated matrix $\mathbf{S_6^m}$:

```
Table[{2 π m / 6, S6^m}, {m, 0, 6}] //
  Map[MatrixForm, #, {2}] & // Size[7]
```

$$\left\{\left\{0,\ S_6^0\right\},\ \left\{\frac{\pi}{3},\ S_6^1\right\},\ \left\{\frac{2\pi}{3},\ S_6^2\right\},\ \left\{\pi,\ S_6^3\right\},\ \left\{\frac{4\pi}{3},\ S_6^4\right\},\ \left\{\frac{5\pi}{3},\ S_6^5\right\},\ \left\{2\pi,\ S_6^6\right\}\right\}$$

Above we see that the matrices of an $\mathbf{S_6}$ axis start out as a unit matrix when the angle is **0**, and come back to the unit matrix after a rotation of **2π**, giving a

group of **6** matrices. All even-fold axes do a qualitatively similar thing.
But look at the behavior of an odd axis. We take $\mathbf{S_3}$ as the example :

```
Table[{2 π m / 3, S3^m}, {m, 0, 6}] //
  Map[MatrixForm, #, {2}] & // Size[7]
```

$\{\{0, S_3^0\}, \{\frac{2\pi}{3}, S_3^1\}, \{\frac{4\pi}{3}, S_3^2\}, \{2\pi, S_3^3\}, \{\frac{8\pi}{3}, S_3^4\}, \{\frac{10\pi}{3}, S_3^5\}, \{4\pi, S_3^6\}\}$

After a rotation of 2π, the $\mathbf{S_3}$ axis gives a reflection matrix. It does not return to a unit matrix until it has completed a rotation of 4π, so the cycle has **6** matrices in it, just as did the $\mathbf{S_6}$ axis. The general rule is not hard to see:

> When **n** is even, the $\mathbf{S_n}$ axis yields **n** matrices, and has a cycle of 2π radians. But when **n** is odd, the $\mathbf{S_n}$ axis yields **2n** matrices and has a cycle of 4π radians.

10.4.4. The "staggeredEthane" molecule; an S_6 axis

```
{σz, Emat}
```

```
{{{1, 0, 0}, {0, 1, 0}, {0, 0, -1}},
 {{1, 0, 0}, {0, 1, 0}, {0, 0, 1}}}
```

This whole section is almost a mechanical copy of parts of Subsection 10.4.1, on "eclipsedEthane". We have changed it to a discussion of "staggeredEthane", another predefined molecule in the package **Molecules`**, by making the minimum necessary changes.

```
stgEth = Show[AtomGraphics["staggeredEthane",
    AtomLabels → False, AxisLength → 0],
   BondGraphics["staggeredEthane", 1.5`],
   ViewPoint → {0, 30, 6}, ImageSize → 90, Boxed → False]
```

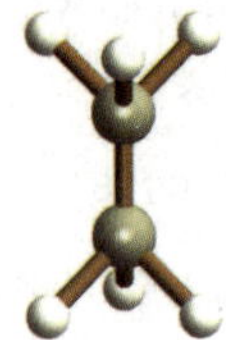

Fig. 10.5 Ethane in its "staggered" configuration.

The difference is that the bottom CH_3 group has been rotated by half a turn about the `C-C` axis.

```
AxialViews[stgEth,
 AtomLabels -> False, ImageSize -> 200]
```

Fig. 10.6 Shop drawing of staggered ethane.

Now the view down the C-C axis shows all six H atoms; formerly, the top three "eclipsed" the bottom three.

```
sixHlistStaggered =
  Take[MoleculeToList["staggeredEthane"], 6];
Column[sixHlistStaggered]
```

$\{H, \{0, 1, \frac{3}{2}\}, \{1, S6\}\}$

$\{H, \{-\frac{\sqrt{3}}{2}, \frac{1}{2}, -\frac{3}{2}\}, \{2\}\}$

$\{H, \{-\frac{\sqrt{3}}{2}, -\frac{1}{2}, \frac{3}{2}\}, \{3\}\}$

$\{H, \{0, -1, -\frac{3}{2}\}, \{4\}\}$

$\{H, \{\frac{\sqrt{3}}{2}, -\frac{1}{2}, \frac{3}{2}\}, \{5\}\}$

$\{H, \{\frac{\sqrt{3}}{2}, \frac{1}{2}, -\frac{3}{2}\}, \{6\}\}$

We make this atom list into an official molecule (omitting the **C** atoms for visual clarity) :

```
ListToMolecule[sixHlistStaggered, "sixHstaggered"];
```

Make a list of all the staggered H atom positions, in sequence :

```
lineStaggeredH6 = NestList[S6z.# &, {0, 1, 3 / 2}, 6]
```

$$\{\{0, 1, \frac{3}{2}\}, \{-\frac{\sqrt{3}}{2}, \frac{1}{2}, -\frac{3}{2}\}, \{-\frac{\sqrt{3}}{2}, -\frac{1}{2}, \frac{3}{2}\}, \{0, -1, -\frac{3}{2}\}, \{\frac{\sqrt{3}}{2}, -\frac{1}{2}, \frac{3}{2}\}, \{\frac{\sqrt{3}}{2}, \frac{1}{2}, -\frac{3}{2}\}, \{0, 1, \frac{3}{2}\}\}$$

Put the picture together :

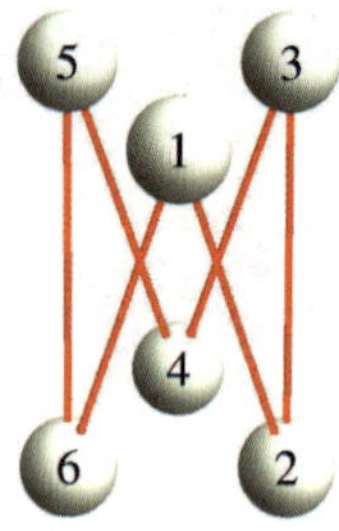

Fig. 10.7 Repeated applications of $S_{6\,z}$ to position 1.

Letting your eye follow the numbers, you see what $S_{6\,z}$ does to each **H** atom: It advances by **1/6** turn, and also flips between bottom and top. It hits each atom during one full rotation, because 6 is even. This differs from the eclipsed situation, where we used $S_{3\,z}$. Because 3 is odd, it took two full rotations to hit each of the six atoms once.

10.4.5. The nature of S_2 : it's the same as the inversion

Visualize a roto-reflection $S_{2\,z}$ about the **z** axis by half a turn. For any point **{x,y,z}**, the reflection part changes the sign of the **z**, while the rotation by half a turn about **z** changes the sign of both **x** and **y**. So it is exactly the same as an inversion.
We construct such a twofold rotoreflection matrix :

```
S2 z = RotationMatrix3Dz[π] . σz ;
MatrixForm[S2 z]
```

$$\begin{pmatrix} -1 & 0 & 0 \\ 0 & -1 & 0 \\ 0 & 0 & -1 \end{pmatrix}$$

This is the inversion matrix **Imat**, confirming the answer by visualization.
Now we show that the same is true, regardless of the axis you use. In Chapter 12 (LieRotations), (click for a preview) we will develop a general operator for rotoreflection about an arbitrary axis by arbitrary angle, and an operator for reflection in a mirror with arbitrary normal. We jump ahead a little and use

them to make the matrix of an $\mathbf{S_2}$ operator with an arbitrarily oriented axis. First we make the axis :

```
arbitraryAxis = RandomReal[{0, 1}, 3]
```

```
{0.409513, 0.954485, 0.166154}
```

Reflection parallel to this axis is

```
σarb = AxialReflection[arbitraryAxis];
% // MatrixForm
```

$$\begin{pmatrix} 0.696839 & -0.706601 & -0.123003 \\ -0.706601 & -0.646933 & -0.286694 \\ -0.123003 & -0.286694 & 0.950093 \end{pmatrix}$$

Rotation by half a turn about it is

```
Carb[π] = AxialRotation[arbitraryAxis, π];
% // MatrixForm
```

$$\begin{pmatrix} -0.696839 & 0.706601 & 0.123003 \\ 0.706601 & 0.646933 & 0.286694 \\ 0.123003 & 0.286694 & -0.950093 \end{pmatrix}$$

So rotoreflection by half a turn about the arbitrary axis is

```
(σarb.Carb[π]) // ChopInteger // MatrixForm
```

$$\begin{pmatrix} -1 & 0 & 0 \\ 0 & -1 & 0 \\ 0 & 0 & -1 \end{pmatrix}$$

It's still the inversion matrix. Therefore, $\sigma_{arb}.C_{arb}[\pi]$ (or $C_{arb}[\pi].\sigma_{arb}$, as you may try for yourself) is the inversion matrix, regardless of the axis direction.

10.5. Only two kinds of matrices

Now that we have the rotoreflection axis as a concept, we begin to see our situation more clearly. Every symmetry transform we have discussed is either *proper* or *improper* .

> The *proper* transforms are all rotations plus the identity.
>
> The *improper* transforms are all retroflections, all reflections, and the inversion.

We will show below that the generalized multiplication table for **Proper** and **Improper** transform matrices is

Dot	Proper	Improper
Proper	Proper	Improper
Improper	Improper	Proper

Fig. 10.8 Multiplication for proper and improper unitary matrices.

10.5.1. Proper transforms

For instance, what is rotation by zero angle? It's nothing; or in more formal words, it is the identity operation, carried out by **Emat**. Look at the general and special cases together :

```
{Cz[ϕ], Cz[0]} // MatrixList
```

$$\left\{\begin{pmatrix} \text{Cos}[\phi] & -\text{Sin}[\phi] & 0 \\ \text{Sin}[\phi] & \text{Cos}[\phi] & 0 \\ 0 & 0 & 1 \end{pmatrix}, \begin{pmatrix} 1 & 0 & 0 \\ 0 & 1 & 0 \\ 0 & 0 & 1 \end{pmatrix}\right\}$$

The determinant of all proper rotations is **1** :

```
Det[Cz[ϕ]] // TrigReduce
```

1

10.5.2. Improper transforms

Similarly, what is rotoreflection by zero angle? The rotation does nothing, but the reflection still operates, so this is what we have so far called "reflection". For example,

```
{Sz[ϕ], Sz[0], σz} // MatrixList
```

$$\left\{\begin{pmatrix} \text{Cos}[\phi] & -\text{Sin}[\phi] & 0 \\ \text{Sin}[\phi] & \text{Cos}[\phi] & 0 \\ 0 & 0 & -1 \end{pmatrix}, \begin{pmatrix} 1 & 0 & 0 \\ 0 & 1 & 0 \\ 0 & 0 & -1 \end{pmatrix}, \begin{pmatrix} 1 & 0 & 0 \\ 0 & 1 & 0 \\ 0 & 0 & -1 \end{pmatrix}\right\}$$

So **{rotations,identity}** are proper and **{rotoreflections, reflections}** are improper. This leaves only the inversion matrix as a possibly different kind of operation. But think about rotoreflection by half a turn. In the plane perpendicular to the rotation axis, each of the in-plane axes has its sign reversed by the half-turn rotation, and then the sign of the rotation axis is changed by the reflection. Together, these makes an inversion; so the inversion, too, is just a special case of rotoreflection. For instance,

```
{Sz[ϕ], Sz[π], Imat} // MatrixList
```

$$\left\{\begin{pmatrix} \text{Cos}[\phi] & -\text{Sin}[\phi] & 0 \\ \text{Sin}[\phi] & \text{Cos}[\phi] & 0 \\ 0 & 0 & -1 \end{pmatrix}, \begin{pmatrix} -1 & 0 & 0 \\ 0 & -1 & 0 \\ 0 & 0 & -1 \end{pmatrix}, \begin{pmatrix} -1 & 0 & 0 \\ 0 & -1 & 0 \\ 0 & 0 & -1 \end{pmatrix}\right\}$$

The determinant of all improper rotations is **-1** :

```
Det[Sz[ϕ]] // TrigReduce
```

```
-1
```

10.5.3. Nothing left to discover

We state a pair of theorems that tie this chapter up quite neatly, but which we are not yet ready to prove. The proof will appear later in the book .

A pair of theorems

If the determinant of a matrix is **+1**, the matrix represent a *proper* rotation.
If the determinant of a matrix is **-1**, the matrix represents a rotoreflection (an *improper* rotation).

Now we couple this with a standard theorem from the theory of determinants. Using the *Mathematica* operator **Det** for the determinant, it is

If **Amat** and **Bmat** are square matrices of the same size, then

```
Det[Amat] Det[Bmat] == Det[Amat.Bmat]
```

On your own

Make two different 3×3 matrices filled with random real numbers, and test this out. Random real numbers, uniformly distributed from **0** to **1**, are returned by **Random[]**.

In view of the theorems above, the generalized multiplication table for Proper and Improper transform matrices must be that shown above as Fig. 10.7.

There is something in this table that many people find surprising: an **Improper** times an **Improper** is always a **Proper**. Therefore, a reflection times a reflection is a rotation. Ponder this. It does make sense.

The closure of this table means that we can never make anything fundamentally new by combining **Proper** and **Improper** transforms. It also means that all **Proper** and **Improper** matrices of the same size (think 3×3) form a group under **Dot**.

Point Group

The group of all proper rotation and improper rotation matrices is the **Point Group**. Chemists and physicists rarely use the whole point group.

Point Groups

The plural refers to finite subgroups of the whole infinite **Point Group**. These small subgroups describe molecular symmetries.

The name **Point Group** reminds us that only one point (namely, the origin) is symmetric under all of these transforms. Every molecule of finite size has a unique point, its center of mass. This is the point which must be used as the origin of all the symmetry rotations and rotoreflections of the molecule. It is returned by the package operator **CenterOfMass["molecule name"]**. Also, a correct set of symmetry axes for the molecule is generated automatically by **InertiaTensor["molecule name"]**.

10.5.4. Coming attractions

The point group is of infinite order, but any given molecule is described by a small finite subgroup of this infinite group. The description and naming of all possible finite subgroups of the infinite point group will be the subject of Chapter 17. But first, we must study how groups are constructed, because the names are actually very succinct coded instructions on how to construct them.

So in Chapter 11 and Chapter 12 we study the construction of individual rotation and rotoreflection matrices.

In Chapter 13 we make an inverse operator that works on any point group matrix, returning its axis and rotation (or roto-reflection) angle.

Then in Chapter 15 we make an operator that accepts two or three point group matrices as input, returning the whole groups implied by the input matrices.

Finally in Chapter 17 we will be ready for the naming of the Point Groups.

11. Euler rotation matrices

Preliminaries

11.1 The Euler idea

We have so far considered only rotations about the coordinate axes. But molecules often contain rotations axes at some cockeyed angle, and we now have to learn how to deal with that. We will show two ways, the Euler rotation matrix (this chapter) and the Lie rotation matrix (next chapter).

Consider a solid object, oriented in some standard position. To be concrete, think of a cylindrical drinking glass in a defined standard position: standing upright, with a decoration on one side facing, say, South. Consider also a Cartesian coordinate system, **z** axis pointing up and **x** axis pointing East. How can we use this coordinate system to specify a series of rotations that will take the drinking glass into absolutely any other orientation?

Here is the answer given by Leonard Euler (pronounced "oiler", Swiss mathematical genius, 1707-1783): Think of the object as possessing a body-fixed axis (in our case, say, the cylindrical axis of the glass). If you can turn the body axis (and the body with it) in any desired direction, and if you can then turn the glass about its body axis (to make the decoration take any desired direction around the body axis) then you have finished the job.

How many rotations will this require? Just as positions on the earth require a latitude angle and a longitude angle, the positive end of the body axis requires two coordinate angles to specify its direction. The final rotation about the body axis is a third rotation, and thus we have "the three Euler angles".

This sounds simple enough, but in a bare **x,y,z** coordinate system there are many ways to specify those latitude and longitude angles, and each way requires a different final body axis angle. The devil is in the details. We will adopt the Euler conventions of A. R. Edmonds, who is consistent with Herzberg and uniquely consistent with the Condon-Shortley convention for the spherical harmonics, now almost universally adopted.

W.M. McClain, *Symmetry Theory in Molecular Physics with Mathematica*,
DOI 10.1007/b13137_11, © Springer Science+Business Media, LLC 2009

11.2 The Euler rotations visualized

In **Fig. 1** below, we see four superposed coordinate systems with flags of different colors. After each motion one coordinate system will be left in place as a marker of where the moving system has been.

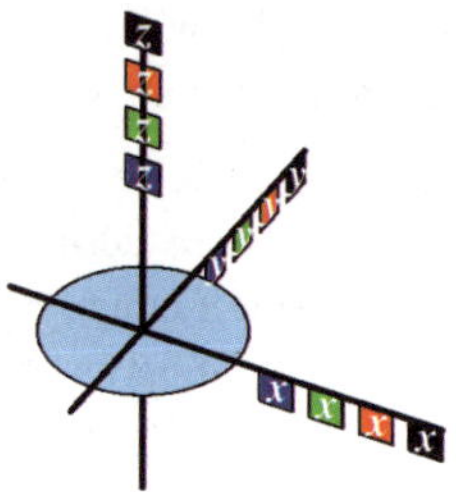

Fig. 11.1 Four flagged axis systems in original position

For **Fig. 2**, rotate by ϕ about the $\mathbf{z}$-axis. In the figure below, $\phi = 2\pi/6$. The black "space-fixed" axes are left behind. The rotation is counterclockwise as viewed from the positive $\mathbf{z}$ axis, as it should be for a positive angle.

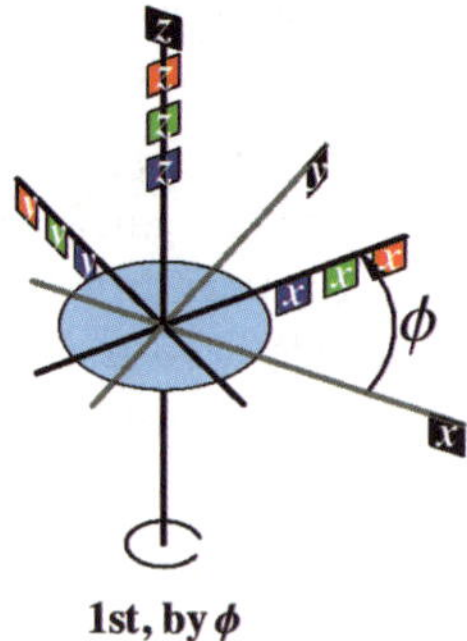

Fig. 11.2 First Euler rotation, by ϕ about z. The black space-fixed system stays behind.

For **Fig. 3**, rotate by θ around the body-fixed $\mathbf{y}$ axis (here red and green, accompanied temporarily by blue). The only axes that move are the coincident

z and **x** axes in green and blue. The matrix is $\mathbf{Rot_2} = \mathbf{Rot_1.C_y[\theta].Rot_{-1}}$, where in **Fig. 3** below, $\theta = 2\pi/15$.

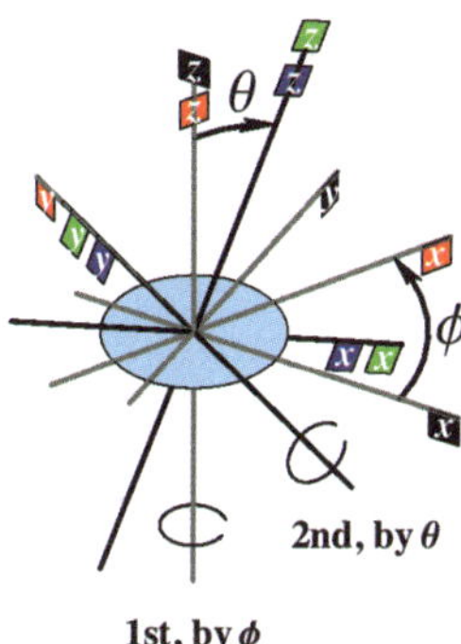

Fig. 11.3 Second Euler rotation. The red system stays behind as the blue and green systems rotate about the blue and green y-axis. The blue z axis (the body axis) is in final position.

For **Fig. 4**, rotate about the body axis by $\mathbf{Rot_3}$. For this, undo the second rotation, then undo the first rotation then rotate by ψ about **z**, then redo the first two rotations: $\mathbf{Rot_3} = \mathbf{Rot_2.Rot_1.C_z[\psi].Rot_{-1}.Rot_{-2}}$. We take $\psi = .41\,\pi$.

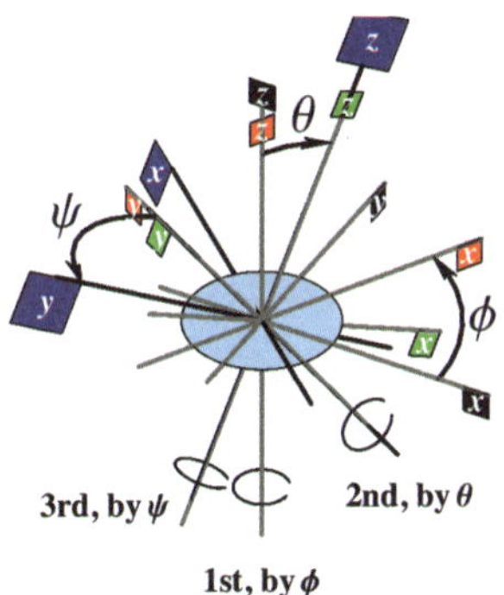

Fig. 11.4 Third Euler rotation. The green system stays behind as the blue system rotates about its own z-axis (the body axis). The large blue flags indicate the final coordinate system position.

11.3 The Euler matrix theorem

Now you are about to witness an algebraic miracle; a geometric property that is true, but essentially impossible for humans to visualize. We begin by collecting together the three Euler rotations we used in the last section, but we use capital axis indices because (for example) $\mathbf{C_z}$ is automatically evaluated, but $\mathbf{C_Z}$ is not. The first rotation and its inverse are

```
Clear[φ, θ, ψ];
Rotn1 = CZ[φ];
Rotn-1 = CZ[-φ];
```

The second rotation and its inverse are

```
Rotn2 = Rotn1.CY[θ].Rotn-1;
Rotn-2 = Rotn1.CY[-θ].Rotn-1;
```

The third rotation is

```
Rotn3 = Rotn2.Rotn1.CZ[ψ].Rotn-1.Rotn-2;
```

Now the total Euler rotation matrix is given by

```
Rotn3.Rotn2.Rotn1
```

```
CZ[φ].CY[θ].CZ[-φ].CZ[φ].CZ[ψ].CZ[-φ].
  CZ[φ].CY[-θ].CZ[-φ].CZ[φ].CY[θ].CZ[-φ].CZ[φ]
```

This result can be simplified a lot. We repeat it below, with cancelling neighbors colored red :

```
CZ[φ].CY[θ].CZ[-φ].CZ[φ].CZ[ψ].CZ[-φ].
  CZ[φ].CY[-θ].CZ[-φ].CZ[φ].CY[θ].CZ[-φ].CZ[φ]
```

Cancelling the red pairs, we create one more cancelling pair :

```
CZ[φ].CY[θ].CZ[ψ].CY[-θ].CY[θ]
```

The final miraculous result is

```
CZ[φ].CY[θ].CZ[ψ]
```

We recapitulate this result as sacred words, never to be violated for any reason :

> **Euler angles**
> First, rotate the solid body by ψ about **z**.
> Second, by θ about **y**.
> Third, by ϕ about **z** (again).
>
> Axes **x**, **y**, and **z** are fixed in space and do not rotate.

11.4 The function EulerMatrix[ϕ, θ, ψ]

Now it is very easy to calculate the Euler matrix in terms of trigonometric functions. Just replace the dummy capital letter axes by lower case letters; then C_x, C_y, and C_z automatically turn into a rotation matrix formulas:

```
EuMat = Cz[ϕ].Cy[θ].Cz[ψ];
EuGrid = Grid[EuMat, Dividers → All]
```

Cos[θ] Cos[ϕ] Cos[ψ] - Sin[ϕ] Sin[ψ]	-Cos[ψ] Sin[ϕ] - Cos[θ] Cos[ϕ] Sin[ψ]	Cos[ϕ] Sin[θ]
Cos[θ] Cos[ψ] Sin[ϕ] + Cos[ϕ] Sin[ψ]	Cos[ϕ] Cos[ψ] - Cos[θ] Sin[ϕ] Sin[ψ]	Sin[θ] Sin[ϕ]
-Cos[ψ] Sin[θ]	Sin[θ] Sin[ψ]	Cos[θ]

The whole calculation is **(3rd.2nd.1st).vec**., where "ψ about **z**" is **1st**, etc. The Euler matrix is always at hand in the **Symmetry`** package:

```
? EulerMatrix
```

```
EulerMatrix[ϕ,θ,ψ] or EulerMatrix[{ϕ,θ,ψ}] returns a
three dimensional rotation matrix for rotation by
Euler angles{ϕ,θ,ψ}, using Edmonds's convention: First
by ψ about z, then by θ about y, then by ϕ about z.
All rotations are right handed.  See Edmonds, p.8.
```

We check that the **EulerMatrix** in **Symmetry`** is the same as **EuMat** defined above :

```
EulerMatrix[ϕ, θ, ψ] == EuMat
```

```
True
```

For a further comment on Edmonds, click here, or flip to the End Notes.

11.5 Euler headaches

11.5.1 The 96 possible Euler matrices

In defining a space-fixed Euler rotation, there are 3 coordinate axes from which to pick the first rotation axis, and 2 from which to pick the second. (The third is always the same as the first.) Aside from that, you could in principle take any of the rotations to be right handed or left handed. That makes (3*2)*(2*2*2) = 48 possible Euler matrices right there, and probably all of them have been used at one time or another. And of course any of them could be defined to act to the right or to the left, doubling the confusion again to 96.

If an author does not state his Euler convention clearly, it may take many tries for a reader to hit the right combination.

11.5.2 Even worse

Even if everyone could agree on a standard Euler rotation, our troubles would not be over. Very often you need to rotate a body from one non-standard orientation to another. There are always Euler angles that will do this, but they may be fiendishly difficult to visualize.

For example, consider a unit cube and its `{1,1,1}` diagonal axis :

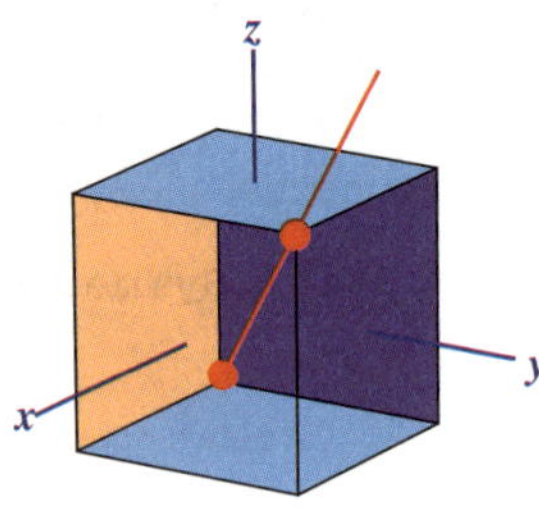

Question: What Euler angles describe a rotation of `1/3` turn about this axis? Think about this a little. This is a particularly easy and symmetric case, but if you get the answer without peeking at the end of the chapter, you may declare yourself a visual genius.

For all their importance in physics, the Euler angles are a dreadful way to do geometry. In the next chapter, we show a much better way.

11.6 The inverse function EulerAngles[matrix]

11.6.1 Basic algorithm

Given a rotation matrix, it is quite possible to find the Euler angles of the rotation it represents. We repeat the general Euler matrix :

```
EuGrid
```

Cos[θ] Cos[ϕ] Cos[ψ] - Sin[ϕ] Sin[ψ]	-Cos[ψ] Sin[ϕ] - Cos[θ] Cos[ϕ] Sin[ψ]	Cos[ϕ] Sin[θ]
Cos[θ] Cos[ψ] Sin[ϕ] + Cos[ϕ] Sin[ψ]	Cos[ϕ] Cos[ψ] - Cos[θ] Sin[ϕ] Sin[ψ]	Sin[θ] Sin[ϕ]
-Cos[ψ] Sin[θ]	Sin[θ] Sin[ψ]	Cos[θ]

We take a numerical example :

```
rotmat = EulerMatrix[1., 2., 4.];
% // Grid[#, Dividers → All, Spacings → {2, 2}] &
```

0.783796	0.379859	0.491295
-0.180012	-0.618179	0.765147
0.594356	-0.688159	-0.416147

Pretend that we do not to know the Euler angles **{1.,2.,4.}** that created this matrix. Can we find them just from the numerical matrix alone? We can certainly extract θ by taking the **ArcCos** of element **[[3,3]]**, a simple **Cos[θ]** :

```
ArcCos[rotmat[[3, 3]]]
```

```
2.
```

Yes, **2.** is exactly the angle we used as input for θ. But of course there is the domain issue. Check out *Mathematica*'s **ArcCos** function:

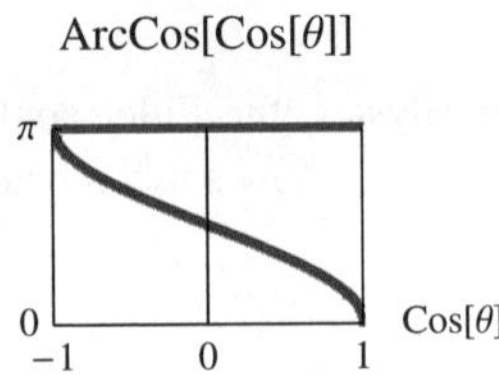

The returned angle will always be between **0** and π, which is nice, because that is the defined domain of θ.

We can calculate ψ similarly because element **[[3,2]]** divided by the negative of element **[[3,1]]** is **Tan[ψ]**. If we use the two-parameter **ArcTan** function, we will get ψ without ambiguity in the range $-\pi$ to $+\pi$. We use the same numerical example again :

```
oldψ = ArcTan[-rotmat[[3, 1]], rotmat[[3, 2]]]
```

```
-2.28319
```

Our input was **4.**, not this. But the trouble is trivial; *Mathematica*'s **ArcTan** has branch cuts at **±(odd n)**π, and it has reported our input, minus 2π .

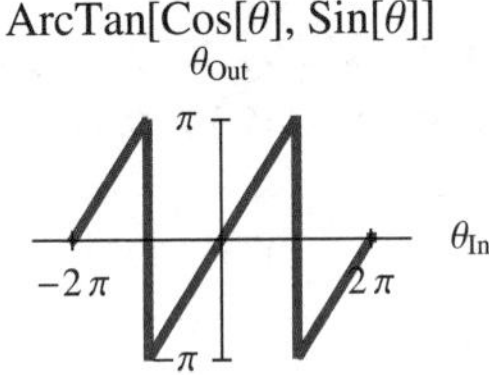

This is easily fixed. The function **Mod[x,n]** divides **x** by **n** and returns the remainder, which is always a number between **0** and **n**. So if we **Mod** by 2π the function becomes

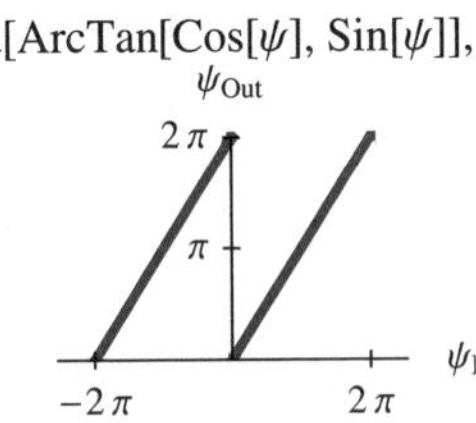

Now the output is always between **0** and 2π , the defined domain of ψ.

```
newψ = Mod[ArcTan[-rotmat⟦3, 1⟧, rotmat⟦3, 2⟧], 2 π]
```

```
4.
```

That was exactly the input. The **Mod** operator is only for the mental convenience of humans; *Mathematica* does not care about the **{0,2π}** convention :

```
{Cos[oldψ] == Cos[newψ], Sin[oldψ] == Sin[newψ]}
```

```
{True, True}
```

Now finally we find ϕ from a similar pair of rotation matrix elements in a similar way :

```
ϕ = Mod[ArcTan[rotmat⟦1, 3⟧, rotmat⟦2, 3⟧], 2 π]
```

```
1.
```

There, we have recovered all three Euler angles. This algorithm is embodied in the package operator **EulerAngles.** We try it out on our demo case :

```
EulerAngles[rotmat]
```

```
{1., 2., 4.}
```

Each Euler angle is in its conventional range, to avoid confusing humans.

11.6.2 Exceptional cases

Most rotation matrices lead to a unique set of Euler angles. But some do not, and the **EulerAngle** operator must deal with them correctly. For instance, when $\theta = 0$, the matrix represents two consecutive rotations about **z**, and clearly the angles ϕ and ψ should add.

```
eum102 = EulerMatrix[1., 0, 2.];
% // MatrixForm
```

$$\begin{pmatrix} -0.989992 & -0.14112 & 0 \\ 0.14112 & -0.989992 & 0 \\ 0 & 0 & 1 \end{pmatrix}$$

Its inverse is not well defined, but if you try it you get the best answer available :

```
EulerAngles[eum102]
```

```
Rotation about the z axis
Euler angles are not unique
General solution is {ϕ,0,ψ},
where (in the given case) ϕ+ψ = 3.
```

```
{0, 0, 3.}
```

Another exceptional case occurs when $\theta = \pi$. Here the final body axis is the negative of the initial body axis, so the first and third Euler angles operate around the same axis, but in opposite directions. Give this one to our operator:

```
eum1π2 = EulerMatrix[1., π, 2.];
eum1π2 // MatrixForm
```

$$\begin{pmatrix} -0.540302 & 0.841471 & 0 \\ 0.841471 & 0.540302 & 0 \\ 0 & 0 & -1 \end{pmatrix}$$

```
EulerAngles[eum1π2]
```

```
Rotation about an axis in the xy plane
Euler angles are not unique
General solution is {ϕ,π,ψ},
where (in the given case) ϕ-ψ = -1.
```

```
{0, 3.14159, 1.}
```

The **EulerAngles** operator again gives the best available answer. But check that it is really correct :

```
EulerMatrix[0, π, 1.] == EulerMatrix[1., π, 2.]
```

```
True
```

To see how the operator does this, open up the **Private** section of the file **UrSymmetryV6.nb** and **Find** the operator name.

11.7 End notes

11.7.1 Edmonds's two Euler matrices

We use the Euler matrix from p. 8 of A. R. Edmonds, Angular Momentum in Quantum Mechanics. But there is another one on p. 53 that is not the same. The difference is that on p.8 the reference frame is fixed and the body moves, whereas on p.53 the body is fixed and the reference frame moves. Therefore, the two Euler matrices are inverse to each other. Throughout this book, the frames are fixed and the bodies move, as on p. 8 of Edmonds.

11.7.2 Answer to the Euler puzzle

What axial Euler rotations are equivalent to a third of a turn about axis **{1,1,1}**?

Going left to right, top to bottom, the figures show

(1) original position

(2) after $\psi = \pi/2$ about **z**

(3) after $\theta = \pi/2$ about **y**

This completes the puzzle: the yellow, green and red balls have rotated a third of a turn about the diagonal axis, through the final blue ball position.
For completeness, the fourth figure shows a **Null** rotation about **z** again :

(4) after $\phi = 0$ about **z**.

Since it really needs only two axial rotations, this should be especially easy for humans. But it isn't.

Original position

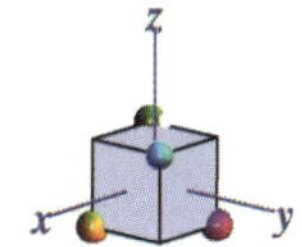

1. After π/2 about z

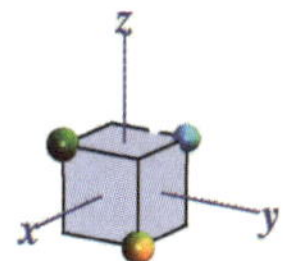

2. After π/2 about y

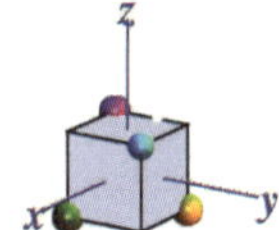

3. After 0 about z

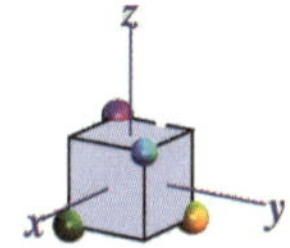

12. Lie's axis-angle rotations

Preliminaries

12.1 Introduction

This chapter presents an alternative to the Euler rotation formula, a formula that returns the rotation matrix as a function of any given axis and rotation angle. But the method used is extendable in the most surprising directions, and with the most profound consequences for symmetries in physics. The method was invented by the Norwegian mathematician Sophus Lie (1842-1899), and it lies at the basis of what are called Lie groups.

As a warm-up we will use Lie's method to construct the general 2D rotation matrix. Here we produce only the well known result, but by new thinking. Then we will extend the method to 3D, where the result is actually useful as the axis-angle formula for the rotation matrix.

12.2 Rotation matrices in 2D form a Lie group

The two-dimensional rotation matrix formula derived in Chapter 4 is

$$\mathbf{R_{2\,D}}[\alpha\_] := \begin{pmatrix} \mathbf{Cos}[\alpha] & -\mathbf{Sin}[\alpha] \\ \mathbf{Sin}[\alpha] & \mathbf{Cos}[\alpha] \end{pmatrix};$$

where α is a free parameter. This formula produces an infinite number of rotation matrices that form a group under **Dot** because

(1) When two $\mathbf{R_{2\,D}}$ matrices are dotted together, they yield a third $\mathbf{R_{2\,D}}$ matrix :

$$\begin{pmatrix} \mathbf{Cos}[\alpha] & -\mathbf{Sin}[\alpha] \\ \mathbf{Sin}[\alpha] & \mathbf{Cos}[\alpha] \end{pmatrix} . \begin{pmatrix} \mathbf{Cos}[\beta] & -\mathbf{Sin}[\beta] \\ \mathbf{Sin}[\beta] & \mathbf{Cos}[\beta] \end{pmatrix} \quad //$$

TrigReduce // MatrixForm

$$\begin{pmatrix} \mathrm{Cos}[\alpha+\beta] & -\mathrm{Sin}[\alpha+\beta] \\ \mathrm{Sin}[\alpha+\beta] & \mathrm{Cos}[\alpha+\beta] \end{pmatrix}$$

(2) The unit matrix is $\mathbf{R_{2\,D}}[0]$;

(3) The inverse of $\mathbf{R_{2\,D}}[\alpha]$ is $\mathbf{R_{2\,D}}[-\alpha]$, and

(4) Associativity is assured by the **Dot** product.

This group is of infinite order because there is a different group element for

W.M. McClain, *Symmetry Theory in Molecular Physics with Mathematica*,
DOI 10.1007/b13137_12, © Springer Science+Business Media, LLC 2009

every different value of parameter α in the interval $[-\pi, \pi]$. These matrices therefore meet all the requirements for a **Lie group** :

> **Lie group**
> If the elements of an infinite group are given by a function of one or more continuously variable parameters, and if the function is infinitely differentiable with respect to its parameters, then the group is a Lie group.

12.3 Lie group for 2D rotation

The concept of matrix exponentials lie at the heart of Lie's method. If you need a discussion of them, click here. to go into A3, the MatrixReview appendix.

Here is a little calculation that must have struck Lie like a thunderbolt: he devoted his entire career to developing its amazing generalizations. It begins with the rather innocent thought that a large rotation can be built up from many repeated small rotations. What is the rotation matrix for a very small rotation ? Expand the $\mathbf{R_{2\,D}}$ rotation matrix as a power series, correct to third order in α:

```
Series[R2D[α], {α, 0, 3}] // MatrixForm
```

$$\begin{pmatrix} 1 - \frac{\alpha^2}{2} + O[\alpha]^4 & -\alpha + \frac{\alpha^3}{6} + O[\alpha]^4 \\ \alpha - \frac{\alpha^3}{6} + O[\alpha]^4 & 1 - \frac{\alpha^2}{2} + O[\alpha]^4 \end{pmatrix}$$

The zeroth order term (containing no α) is $\begin{pmatrix} 1 & 0 \\ 0 & 1 \end{pmatrix}$ and the first order term (linear in α) is $\begin{pmatrix} 0 & -\alpha \\ \alpha & 0 \end{pmatrix}$. We make the rotation very small, dividing the first order term by some large **N**, and then we compound the small rotation by performing it **N** times. Passing immediately to the limit of infinite **N**

```
Timing[
 Limit[MatrixPower[(1 0; 0 1) + (0 -α/N; α/N 0), N], N → ∞] //
   Expand // ExpToTrig // MatrixForm]
```

$$\left\{0.702947, \begin{pmatrix} \mathrm{Cos}[\alpha] & -\mathrm{Sin}[\alpha] \\ \mathrm{Sin}[\alpha] & \mathrm{Cos}[\alpha] \end{pmatrix}\right\}$$

There is no error or approximation. The limit of **N** rotations by α/N is

EXACTLY the rotation by α. Alternatively, and more rapidly,

```
MatrixExp[(0  -α)] // MatrixForm
          (α   0)
```

```
( Cos[α]  -Sin[α] )
( Sin[α]   Cos[α] )
```

The thunderbolt is that the first order term of the matrix expansion TAKEN ALONE contains all the information in the whole matrix function. The information is extracted by Lie's trick; namely, by wrapping the first order term in **MatrixExp**. Click here for a discussion of how *Mathematica* actually computes the **MatrixExp**.

But there is something strange here. There are plenty of matrix functions whose power series begins as $\begin{pmatrix} 1 & 0 \\ 0 & 1 \end{pmatrix} + \begin{pmatrix} 0 & -\alpha \\ \alpha & 0 \end{pmatrix} + \ldots$ Why is it that only the rotation matrix $\mathbf{R}_{2\,D}$ reproduces itself when its first order term is exponentiated?

Lie produced the definitive answer to this question. It happens because there is no other matrix function that begins this way and also produces a **group** of matrices as α varies. In other words, he showed that $\mathbf{R}_{2\,D}$ performs the **MatrixExp** trick because it also possesses the property

```
(R2D[ϵ].R2D[ϵ] // TrigReduce) == R2D[2 ϵ]
```

```
True
```

On your own

There is another useful **2×2** matrix function that produces a group of matrices. It is given by $\begin{pmatrix} 1 & \Delta x \\ 0 & 1 \end{pmatrix}$, and it represents translation in one dimension in the sense that

$$\begin{pmatrix} 1 & \Delta x \\ 0 & 1 \end{pmatrix} \cdot \begin{pmatrix} x \\ 1 \end{pmatrix} = \begin{pmatrix} x + \Delta x \\ 1 \end{pmatrix}$$

It is a little strange that the second row is essential for calculation, but is discarded in the interpretation. However, mathematically, nothing prevents our using it in this way,

(a) Prove that matrices $\begin{pmatrix} 1 & \alpha \\ 0 & 1 \end{pmatrix}$ form a Lie group with parameter α.

(b) Make it perform Lie's trick: Find the first order term and take its **MatrixExp**.

(c) As a counter-example, do the same thing on $\begin{pmatrix} 1 & \text{Sin}[\alpha] \\ 0 & 1 \end{pmatrix}$. Why does the counter-example fail?

Click here if you have to peek at the answer.

12.4 Lie group for 3D rotation

12.4.1 Lie's trick for axis-angle rotation in 3D

Take $\underline{\vec{n}}$ to be a fixed rotation axis of unit length, and let plane **P** be perpendicular to $\underline{\vec{n}}$. In plane **P** let $\underline{\vec{r}}$ be a vector from the origin to any point on a unit circle. Thus $\underline{\vec{n}}$ and $\underline{\vec{r}}$ are orthonormal. Some matrix $Q[\alpha \mid \underline{\vec{n}}]$ (unknown to us at this point) rotates vector $\underline{\vec{r}}$ by angle α about axis $\underline{\vec{n}}$ according to the formula

$$Q[\alpha \mid \underline{\vec{n}}] \cdot \underline{\vec{r}} = \underline{\vec{r}} + \underline{\Delta \vec{r}}[\alpha \mid \underline{\vec{n}}]$$

The vertical bar in this symbol reminds us that α is the Lie parameter that produces different group elements as it varies, whereas $\underline{\vec{n}}$ is just an auxiliary parameter that is constant for all members of the group. The derivative

$\mathbb{Q}'\left[\alpha \mid \underset{\rightharpoondown}{\mathbf{n}}\right]$ means differentiation with respect to $\boldsymbol{\alpha}$ only. So the series expansion will be

$$\mathbb{Q}\left[\alpha \mid \underset{\rightharpoondown}{\mathbf{n}}\right] = \begin{pmatrix} 1 & 0 & 0 \\ 0 & 1 & 0 \\ 0 & 0 & 1 \end{pmatrix} + \alpha \begin{pmatrix} ? & ? & ? \\ ? & ? & ? \\ ? & ? & ? \end{pmatrix} + \alpha^2 \begin{pmatrix} ? & ? & ? \\ ? & ? & ? \\ ? & ? & ? \end{pmatrix} + \ldots$$

Of course, for Lie's trick we need only the linear term. We can figure it out with the help of the following diagram :

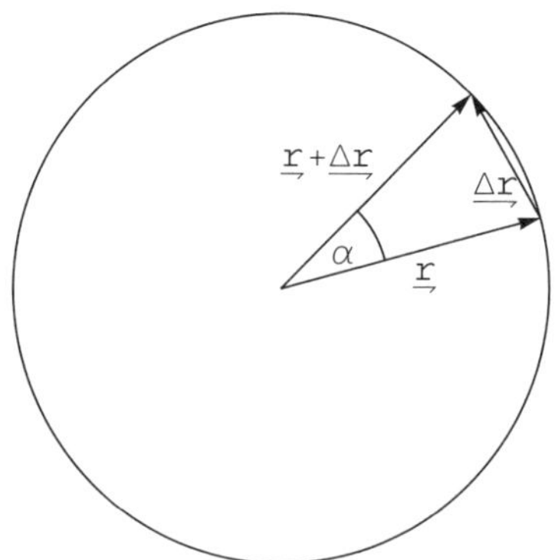

Fig. 12.1 Figure for thinking about the derivative matrix $\mathbb{Q}'\left[\alpha \mid \underset{\rightharpoondown}{\mathbf{n}}\right]$, evaluated at $\alpha = 0$. Axis $\underset{\rightharpoondown}{\mathbf{n}}$ (not shown) has unit length, and rises out of the page at the center of the circle. Vectors $\underset{\rightharpoondown}{\mathbf{r}}$ and $\underset{\rightharpoondown}{\mathbf{r}} + \Delta\underset{\rightharpoondown}{\mathbf{r}}$ also have unit length.

As $\Delta\underset{\rightharpoondown}{\mathbf{r}}$ shrinks to zero, it becomes perpendicular to $\underset{\rightharpoondown}{\mathbf{n}}$, and therefore parallel to $\underset{\rightharpoondown}{\mathbf{n}} \times \underset{\rightharpoondown}{\mathbf{r}}$. (We have the sign right: If you curl your right hand fingers to push $\underset{\rightharpoondown}{\mathbf{n}}$ toward $\underset{\rightharpoondown}{\mathbf{r}}$, your thumb points in the direction of $\Delta\underset{\rightharpoondown}{\mathbf{r}}$.) Also, the length of $\Delta\underset{\rightharpoondown}{\mathbf{r}}$ approaches the length of the little arc between $\underset{\rightharpoondown}{\mathbf{r}}$ and $\underset{\rightharpoondown}{\mathbf{r}} + \Delta\underset{\rightharpoondown}{\mathbf{r}}$. By definition, angle $\boldsymbol{\alpha}$ is that little arc length divided by $\mathbf{2}\,\pi$. Now $\underset{\rightharpoondown}{\mathbf{n}} \times \underset{\rightharpoondown}{\mathbf{r}}$ has unit length, so the limit of $\Delta\underset{\rightharpoondown}{\mathbf{r}}$ is given by $\alpha\, \underset{\rightharpoondown}{\mathbf{n}} \times \underset{\rightharpoondown}{\mathbf{r}}$. Put all this into the definition of the derivative:

$$\left(\mathbb{Q}\left[0 \mid \underset{\rightharpoondown}{\mathbf{n}}\right] . \underset{\rightharpoondown}{\mathbf{r}}\right)' = \underset{\alpha \to 0}{\mathbf{Lim}}\left[\frac{\mathbb{Q}[\alpha] . \underset{\rightharpoondown}{\mathbf{r}} - \mathbb{Q}[0] . \underset{\rightharpoondown}{\mathbf{r}}}{\alpha - 0}\right] = \underset{\alpha \to 0}{\mathbf{Lim}}\left[\frac{\alpha\, \underset{\rightharpoondown}{\mathbf{n}} \times \underset{\rightharpoondown}{\mathbf{r}}}{\alpha}\right] = \underset{\rightharpoondown}{\mathbf{n}} \times \underset{\rightharpoondown}{\mathbf{r}}$$

The prime can be moved onto the $\mathbb{Q}$ only, because $\underset{\rightharpoondown}{\mathbf{r}}$ is independent of $\boldsymbol{\alpha}$. This deserves a box:

$$\mathbb{Q}'[0 \mid \underset{\rightharpoondown}{\mathtt{n}}].\underset{\rightharpoondown}{\mathtt{r}} = \underset{\rightharpoondown}{\mathtt{n}} \times \underset{\rightharpoondown}{\mathtt{r}}$$

Cross products may always be written as the **Dot** product of an antisymmetric matrix and a vector. Taking the unit axis of rotation to be $\underset{\rightharpoondown}{\mathtt{n}} = \{n_x, n_y, n_z\}$ and $\underset{\rightharpoondown}{\mathtt{r}} = \{x, y, z\}$, the identity is

$$\begin{pmatrix} 0 & -n_z & n_y \\ n_z & 0 & -n_x \\ -n_y & n_x & 0 \end{pmatrix}.\{x, y, z\} == \{n_x, n_y, n_z\} \times \{x, y, z\}$$

True

If we rewrite Eq. 12.1 in this format,

$$\mathbb{Q}'[0 \mid \underset{\rightharpoondown}{\mathtt{n}}].\underset{\rightharpoondown}{\mathtt{r}} = \begin{pmatrix} 0 & -n_z & n_y \\ n_z & 0 & -n_x \\ -n_y & n_x & 0 \end{pmatrix}.\underset{\rightharpoondown}{\mathtt{r}}$$

the value of $\mathbb{Q}'[0]$ is suddenly staring us in the face. It is

$$\mathbb{Q}'[0 \mid \underset{\rightharpoondown}{\mathtt{n}}] = \begin{pmatrix} 0 & -n_z & n_y \\ n_z & 0 & -n_x \\ -n_y & n_x & 0 \end{pmatrix}$$

Now we can write the series expansion of $\mathbb{Q}[\alpha \mid \underset{\rightharpoondown}{\mathtt{n}}]$ about $\alpha = 0$ correct to first order in α. It is

$$\mathbb{Q}'[\alpha \mid \underset{\rightharpoondown}{\mathtt{n}}] = \begin{pmatrix} 1 & 0 & 0 \\ 0 & 1 & 0 \\ 0 & 0 & 1 \end{pmatrix} + \begin{pmatrix} 0 & -n_z & n_y \\ n_z & 0 & -n_x \\ -n_y & n_x & 0 \end{pmatrix} \alpha + ..$$

We are confident that the matrices generated by this series form a group. Therefore, the first order term must perform Lie's trick; namely,

$$\mathbb{Q}[\alpha \mid \underset{\rightharpoondown}{\mathtt{n}}] == \mathbf{MatrixExp}\left[\begin{pmatrix} 0 & -n_z & n_y \\ n_z & 0 & -n_x \\ -n_y & n_x & 0 \end{pmatrix} \alpha\right] \qquad (12.2)$$

This is evaluated in Subsection 12.4.3, below.

12.4.2 Simple cases: rotations about coordinate axes

Before charging ahead, make sure this is consistent with what we already know. When **{n_x, n_y, n_z} = {0, 0, 1}** we have the important special case of rotation about the **z** axis. Then

```
Q[α | z] = MatrixExp[{{0, -1, 0}, {1, 0, 0}, {0, 0, 0}} α];
```

```
% // MatrixForm
```

$$\begin{pmatrix} \text{Cos}[\alpha] & -\text{Sin}[\alpha] & 0 \\ \text{Sin}[\alpha] & \text{Cos}[\alpha] & 0 \\ 0 & 0 & 1 \end{pmatrix}$$

This, of course, is the usual rotation matrix for rotation about the **z** axis. Astonishing. That **MatrixExp** really works.

> **On your own**
> Do the same thing for the **x** and **y** axes.

These "generator matrices" play such an important role in group theory that we have made them callable from the **Symmetry`** package as **Ix**, **Iy**, and **Iz**. The **I** stands for **InfinitesimalRotation**. Other people call them **Rx**, **Ry**, and **Rz**; but really, they are not rotation matrices; they are *infinitesimal* rotation matrices.

```
Map[GridForm, {Ix, Iy, Iz}]
```

{

0	0	0
0	0	-1
0	1	0

,

0	0	1
0	0	0
-1	0	0

,

0	-1	0
1	0	0
0	0	0

}

12.4.3 The general axis-angle rotation matrix

Formula 12.2 (click here) evaluates symbolically. Enforcing the unit length on $\underset{\rightarrow}{n}$, it yields a formula for a 3-D rotation matrix that is not any worse than the Euler angle formula :

```
LieRotMat1 =
```

```
MatrixExp[( 0    -n_z   n_y
            n_z   0    -n_x
           -n_y   n_x   0  ) α] /. {n_x^2 + n_y^2 + n_z^2 → 1,
    Sqrt[-n_x^2 - n_y^2 - n_z^2] → ⅈ} // ComplexExpand
```

```
{{n_x^2 + Cos[α] n_y^2 + Cos[α] n_z^2, n_x n_y - Cos[α] n_x n_y - Sin[α] n_z,
  Sin[α] n_y + n_x n_z - Cos[α] n_x n_z},
 {n_x n_y - Cos[α] n_x n_y + Sin[α] n_z, Cos[α] n_x^2 + n_y^2 + Cos[α] n_z^2,
  -Sin[α] n_x + n_y n_z - Cos[α] n_y n_z},
 {-Sin[α] n_y + n_x n_z - Cos[α] n_x n_z,
  Sin[α] n_x + n_y n_z - Cos[α] n_y n_z, Cos[α] n_x^2 + Cos[α] n_y^2 + n_z^2}}
```

Abbreviating the trig functions as (**Cos[α] → C** and **Sin[α] → S**) and using **Collect** to bring together the coefficients of $\mathbf{n_x}$, $\mathbf{n_y}$, and $\mathbf{n_z}$,

```
LieRotMat2 = LieRotMat1 /. {Cos[α] → C, Sin[α] → S} //
   Collect[#, {n_x, n_y, n_z}] &;
LieRotMat2 // GridForm
```

$n_x^2 + C\,n_y^2 + C\,n_z^2$	$(1-C)\,n_x n_y - S\,n_z$	$S\,n_y + (1-C)\,n_x n_z$
$(1-C)\,n_x n_y + S\,n_z$	$C\,n_x^2 + n_y^2 + C\,n_z^2$	$-S\,n_x +$ $(1-C)\,n_y n_z$
$-S\,n_y +$ $(1-C)\,n_x n_z$	$S\,n_x + (1-C)\,n_y n_z$	$C\,n_x^2 + C\,n_y^2 + n_z^2$

A human can trim the diagonal terms a little more:

$$n_a^2 + C\,n_b^2 + C\,n_c^2 == n_a^2 + C\left(n_b^2 + n_c^2\right) ==$$
$$n_a^2 + C\left(1 - n_a^2\right) == n_a^2 + C - C\,n_a^2 == C + (1 - C)\,n_a^2$$

leaving

```
LieRotMat = {
   {C + (1 - C) n_x^2,       (1 - C) n_x n_y - n_z S, (1 - C) n_x n_z + n_y S},
   {(1 - C) n_x n_y + n_z S, C + (1 - C) n_y^2,       (1 - C) n_y n_z - n_x S},
   {(1 - C) n_x n_z - n_y S, (1 - C) n_y n_z + n_x S, C + (1 - C) n_z^2}};
```

12.4.4 Lie's matrix in tensor notation

The better to appreciate its symmetry, separate **LieRotMat** into isotropic, symmetric, and antisymmetric parts :

$$C\begin{pmatrix}1&0&0\\0&1&0\\0&0&1\end{pmatrix}+(1-C)\begin{pmatrix}n_x^2&n_x n_y&n_x n_z\\n_x n_y&n_y^2&n_y n_z\\n_x n_z&n_y n_z&n_z^2\end{pmatrix}+S\begin{pmatrix}0&-n_z&n_y\\n_z&0&-n_x\\-n_y&n_x&0\end{pmatrix}$$

Note the cyclic symmetry in the off-diagonal terms. In tensor notation this is

$$\delta\epsilon\text{formula} = C\,\delta_{i,j} + (1-C)\,n_i\,n_j - S\,\epsilon_{i,j,k}\,n_k;$$

Let's check this. *Mathematica* does not automatically understand the Einstein convention, so in the last term we must do an explicit sum over repeated index **k**. We can sum over **{1,2,3}**, but not over **{x,y,z}**; so we must convert $\{n_x, n_y, n_z\}$ to $\{n_1, n_2, n_3\}$, and then back again when the sum is done. Also, we have to teach *Mathematica* about the alternating tensor $\epsilon_{i,j,k}$ and the Kronecker delta function $\delta_{i,j}$. It actually knows them, but not in the familiar notation. Call up **?LeviCivitaEpsilon** and **?KroneckerDelta** to see. You can use subscripts as the arguments of functions, but only if you are careful not to declare that all subscripted quantities are symbols. We often make that declaration in the preliminaries, but in this chapter we did not.

```
Head[x₀]
```

```
Subscript
```

If that returned `Symbol` instead of `Subscript`, the following will not work. You must quit *Mathematica* and start this chapter in a fresh session.

$$\epsilon_{i\_,j\_,k\_} := \text{LeviCivitaEpsilon}[i,\ j,\ k]$$

Now we are ready for *Mathematica* to evaluate **δϵformula**:

$$\delta\epsilon\text{result} = \text{Table}\Big[C\,\delta_{i,j} + (1-C)\,n_i\,n_j - S\sum_{k=1}^{3}\epsilon_{i,j,k}\,n_k,\ \{i,1,3\},\ \{j,1,3\}\Big]\ /.\ \{n_1\to n_x,\ n_2\to n_y,\ n_3\to n_z\};$$

```
δϵresult == LieRotMat
```

```
True
```

It checks; the **δϵformula** is correct.

12.5 The operator AxialRotation[axis, angle]

The 3-D Lie rotation matrix is in the **Symmetry`** package as **AxialRota⧵tion**. Use the **?** to call up its syntax statement.

> **On your own**
> Check all three types of input. Make a vector with **Random[]** Cartesian components, calculate its length, set the length equal to α, and then test all three types of input.

In Chapter 13 (RecognizeMatrix) we will construct an inverse for the function **AxialRotation**. It will be named **RecognizeMatrix**, and it will find the axis and rotation angle for any given rotation matrix, among other things.

12.5.1 Check the handedness; boring but necessary

You can never be too careful about handedness. Here is a small rotation of **{1,0,0}** about the **z** axis, using a positive angle of **0.1 Radian** :

```
AxialRotation[{0, 0, 1}, 0.1].{1, 0, 0}
```

```
{0.995004, 0.0998334, 0.}
```

Is this a right-handed rotation? Look toward the origin from a **+Z** position, with **+X** rightward and **+Y** upward. When your right thumb points over your left shoulder (in the **+Z** direction), the fingers curl up and left. Therefore the tip of the **+X** vector, rotated through a small positive angle, should move up and left.

In the **x** direction the point moved from **1** to **.995** (left), and in the **y** direction it moved from **0** to **0.099** (up). So all is well.

Just permute the symbols and you have it for another case.

12.5.2 The {1,1,1} axis example

In Chapter 11 we saw how difficult it is to use Euler angles to describe a rotation by **1/3** turn about the diagonal of the unit cube. Remember, this is the rotation that permutes the axes (**x→y, y→z, z→x**). Now it is easy. The desired axis is **{1,1,1}** and the angle is **2π/3** :

```
AxialRotation[{1, 1, 1}, 2 π / 3]
```

```
{{0, 0, 1}, {1, 0, 0}, {0, 1, 0}}
```

The answer is just a permutation matrix, serving as a rotation matrix.

12.6 End notes

12.6.1 The `Collect` operator

Watch it work:

```
Collect[a x + b y - c x - d y, {x, y}]
```

```
(a - c) x + (b - d) y
```

12.6.2 Symmetric word substitution

We define a **ZaxisStatement** in the little closed cell below. It was originally defined in an **Input** cell, with a **String** in **Text** style as the right side of the **Set** statement, but when it is **Saved**, *Mathematica* converts it to a horrible unreadable format, so don't look at it. But do evaluate it; the return is perfectly legible.

Look toward the origin from a **+Z** position,with **+X** rightward and **+Y** upward. When your right thumb points over your left shoulder (in the **+Z** direction), the fingers curl up and left. Therefore the tip of the **+X** vector, when rotated through a small positive angle, should move up and left.

Take the **ZaxisStatement** and permute the axes:

```
XaxisStatement = StringReplace[
  ZaxisStatement, {"X" → "Y", "Y" → "Z", "Z" → "X"}]
```

Look toward the origin from a **+X** position,with **+Y** rightward and **+Z** upward. When your right thumb points over your left shoulder (in the **+X** direction), the fingers curl up and left. Therefore the tip of the **+Y** vector, when rotated through a small positive angle, should move up and left.

```
YaxisStatement = StringReplace[
  XaxisStatement, {"X" → "Y", "Y" → "Z", "Z" → "X"}]
```

Look toward the origin from a **+Y** position,with **+Z** rightward and **+X** upward. When your right thumb points over your left shoulder (in the **+Y** direction), the fingers curl up and left. Therefore the tip of the **+Z** vector, when rotated through a small positive angle, should move up and left.

Lots of books say "just permute the symbols and you have it for another case". This is the first book that really does it.

12.6.3 Translation group solution

(a) The group proof is just examination of the four axioms.
Closure: Two matrices of this pattern yield a third one of the same pattern under **Dot**

$$\begin{pmatrix} 1 & \alpha \\ 0 & 1 \end{pmatrix} . \begin{pmatrix} 1 & \beta \\ 0 & 1 \end{pmatrix}$$

```
{{1, α + β}, {0, 1}}
```

The unit axiom and the inverse axiom are easily satisfied:

$$\texttt{unitMatrix} == \begin{pmatrix} 1 & \alpha \\ 0 & 1 \end{pmatrix} \; /. \; \alpha \to 0$$

```
unitMatrix == {{1, 0}, {0, 1}}
```

$$\texttt{invMatrix} == \begin{pmatrix} 1 & \alpha \\ 0 & 1 \end{pmatrix} \; /. \; \alpha \to -\alpha$$

```
invMatrix == {{1, -α}, {0, 1}}
```

$$\begin{pmatrix} 1 & \alpha \\ 0 & 1 \end{pmatrix} . \begin{pmatrix} 1 & -\alpha \\ 0 & 1 \end{pmatrix}$$

```
{{1, 0}, {0, 1}}
```

(b) The series expansion happens to terminate exactly after two terms. It is

$$\begin{pmatrix} 1 & \alpha \\ 0 & 1 \end{pmatrix} = \begin{pmatrix} 1 & 0 \\ 0 & 1 \end{pmatrix} + \begin{pmatrix} 0 & \alpha \\ 0 & 0 \end{pmatrix}$$

The linear term performs Lie's trick :

$$\texttt{MatrixExp}\left[\begin{pmatrix} 0 & \alpha \\ 0 & 0 \end{pmatrix}\right] \; // \; \texttt{MatrixForm}$$

$$\begin{pmatrix} 1 & \alpha \\ 0 & 1 \end{pmatrix}$$

(c) The matrices of the counter-example do not form a group.

$$\begin{pmatrix} 1 & \texttt{Sin}[\alpha] \\ 0 & 1 \end{pmatrix} . \begin{pmatrix} 1 & \texttt{Sin}[\beta] \\ 0 & 1 \end{pmatrix} \; // \; \texttt{MatrixForm}$$

$$\begin{pmatrix} 1 & \texttt{Sin}[\alpha] + \texttt{Sin}[\beta] \\ 0 & 1 \end{pmatrix}$$

No way is **Sin[α]+Sin[β]** identical for all **α,β** to **Sin[someFn[α,β]]**.

So the matrix pattern is not preserved under **Dot**. These matrices do not close, do not form a group, and do not perform Lie's trick.

```
Series[(1  Sin[α]
        0    1   ), {α, 0, 1}] // MatrixForm
```

$$\begin{pmatrix} 1 & \alpha + O[\alpha]^2 \\ 0 & 1 \end{pmatrix}$$

If you try to make it do Lie's trick, it produces the translation group formula.

On your own

(a) To verify the commutation rubric above, put the matrix factors in different multiplication orders and recalculate.

(b) Calculate to second differential order, retaining mixed differentials of total order 2, and note that multiplication order already matters.

On your own

Prove this by direct calculation on three general 3×3 matrices.

On your own

Replace **DotComm** by **Dot** in the **Outer** operator that produces the commutation table above. You will see that **Dot** produces nothing of great interest; neither a group, nor anything else recognizable. Only the commutators produce something suggestive.

13. Recognizing matrices

Preliminaries

13.1 The problem, and a strategy

13.1.1 Rotational eigensystem theorem

Rotation matrices and rotoreflection matrices, in numerical form, usually have no distinguishing characteristic identifiable by inspection. Unless they are especially simple, they look just about like any other 3-by-3 real numerical matrix filled with values between **-1** and **+1**. In this chapter we develop numerical tests to distinguish them, and to extract the axis and the angle. These tests are collected together in a useful operator named **RecognizeMatrix**, based on the following theorem :

Rotational eigensystem theorem

If an axis is given by the unit vector **{a,b,c}** and an angle by α,

Then the matrix for the specified rotation has eigenvectors given by the rows of the matrix

$$\textbf{evecRows} = \begin{pmatrix} a & b & c \\ \frac{-b^2-c^2}{\sqrt{2}\sqrt{b^2+c^2}} & \frac{a\,b-i\,c}{\sqrt{2}\sqrt{b^2+c^2}} & \frac{i\,b+a\,c}{\sqrt{2}\sqrt{b^2+c^2}} \\ \frac{-b^2-c^2}{\sqrt{2}\sqrt{b^2+c^2}} & \frac{a\,b+i\,c}{\sqrt{2}\sqrt{b^2+c^2}} & \frac{-i\,b+a\,c}{\sqrt{2}\sqrt{b^2+c^2}} \end{pmatrix}$$

with corresponding eigenvalues $\{1,\ e^{+i\alpha},\ e^{-i\alpha}\}$.

Comment: Notice that the rotation angle α does not appear anywhere in the eigenvectors, nor does the axis appear in the eigenvalues. How clean!

13.1.2 Proof

We want the rotation matrix for a rotation axis of unit length. So here is a rule enforcing normalization:

W.M. McClain, *Symmetry Theory in Molecular Physics with Mathematica*,
DOI 10.1007/b13137_13, © Springer Science+Business Media, LLC 2009

```
unitAxis = {a² → 1 - b² - c², -a² → -1 + b² + c²,
    u_/(a² + b² + c²) → u, -u_/Sqrt[a² + b² + c²] → -u};
```

and here is the rotation matrix :

```
Rotmat = AxialRotation[{a, b, c}, α] //. unitAxis;
Rotmat // MatrixForm // Size[7]
```

$$\begin{pmatrix} 1-b^2-c^2+(b^2+c^2)\,\mathrm{Cos}[\alpha] & a\,b\,(1-\mathrm{Cos}[\alpha])-c\,\mathrm{Sin}[\alpha] & a\,c\,(1-\mathrm{Cos}[\alpha])+b\,\mathrm{Sin}[\alpha] \\ a\,b\,(1-\mathrm{Cos}[\alpha])+c\,\mathrm{Sin}[\alpha] & b^2+(1-b^2)\,\mathrm{Cos}[\alpha] & b\,c\,(1-\mathrm{Cos}[\alpha])-a\,\mathrm{Sin}[\alpha] \\ a\,c\,(1-\mathrm{Cos}[\alpha])-b\,\mathrm{Sin}[\alpha] & b\,c\,(1-\mathrm{Cos}[\alpha])+a\,\mathrm{Sin}[\alpha] & c^2+(1-c^2)\,\mathrm{Cos}[\alpha] \end{pmatrix}$$

Eigensystem can calculate the eigenvectors of this matrix symbolically, but the expressions returned are extremely long. (Try it; you will be appalled. It takes less than two seconds at 1 GHz , or 30 seconds at 60 MHz.) We could do this and then prove the theorem by dragging you through a tedious simplification process, but that would be too boring. Instead, we will prove it by eigensystem diagonalization, as discussed in the MatrixReview. The idea is that for any given matrix, there are just a few matrices which diagonalize it by the formula

evecs†.Rotmat.evecs = diagMat

Matrix **evecs** is unique up to a rearrangement of its columns, which are the eigenvectors of **Rotmat**. No other columns can possibly produce a diagonal matrix on the right side. The adjoint of **evecs** is the conjugate of **evecRows**, as given in the theorem statement :

$$\text{evecs}\dagger = \begin{pmatrix} a & b & c \\ \frac{-b^2-c^2}{\sqrt{2}\sqrt{b^2+c^2}} & \frac{a\,b-i\,c}{\sqrt{2}\sqrt{b^2+c^2}} & \frac{i\,b+a\,c}{\sqrt{2}\sqrt{b^2+c^2}} \\ \frac{-b^2-c^2}{\sqrt{2}\sqrt{b^2+c^2}} & \frac{a\,b+i\,c}{\sqrt{2}\sqrt{b^2+c^2}} & \frac{-i\,b+a\,c}{\sqrt{2}\sqrt{b^2+c^2}} \end{pmatrix};$$

so that **evecs** itself is

$$\text{evecs} = \begin{pmatrix} a & \frac{-b^2-c^2}{\sqrt{2}\sqrt{b^2+c^2}} & \frac{-b^2-c^2}{\sqrt{2}\sqrt{b^2+c^2}} \\ b & \frac{a\,b+i\,c}{\sqrt{2}\sqrt{b^2+c^2}} & \frac{a\,b-i\,c}{\sqrt{2}\sqrt{b^2+c^2}} \\ c & \frac{-i\,b+a\,c}{\sqrt{2}\sqrt{b^2+c^2}} & \frac{i\,b+a\,c}{\sqrt{2}\sqrt{b^2+c^2}} \end{pmatrix};$$

Verify orthonormality of these vectors :

```
(evecs†.evecs // Simplify) /. unitAxis
```

```
{{1, 0, 0}, {0, 1, 0}, {0, 0, 1}}
```

Now for the proof that they are the eigenvectors of **Rotmat**. We test whether **Rotmat** is diagonalized by a similarity transform between **evecs†** on the left and **evecs** on the right:

```
(evecs†.Rotmat.evecs // Simplify) /. unitAxis //
   Simplify // TrigToExp;
% // MatrixForm
```

$$\begin{pmatrix} 1 & 0 & 0 \\ 0 & e^{i\alpha} & 0 \\ 0 & 0 & e^{-i\alpha} \end{pmatrix}$$

That's it; **q.e.d.** The given vectors have to be the eigenvectors because no other vectors can perform this diagonalization. Furthermore, the diagonal elements above and the columns of **evecs** must be in corresponding order.

> **On your own**
> Redo this proof (by **Copy** and **Paste**, with minimal alteration) for the general rotoreflection matrix. You might need just one hint: The real eigenvalue will be **-1** .
>
> If you have trouble getting started, click here.

We now have everything we need to extract the axis and the angle from a numerical rotation matrix.

13.2 Recognize a numerical rotation matrix

Make a rotation matrix with input numbers that are easy to remember. It is important to this discussion that we use a negative input angle :

```
rotMatN = AxialRotation[{1., 2., 3.}, -2 π / 5];
rotMatN // MatrixForm
```

$$\begin{pmatrix} 0.358373 & 0.861253 & -0.360293 \\ -0.66383 & 0.506441 & 0.550316 \\ 0.656429 & 0.041955 & 0.75322 \end{pmatrix}$$

Pretending that we know only the matrix itself, we find its **Eigensystem** :

```
{evalsN, evecsN} =
  ChopInteger[Eigensystem[rotMatN]];
Size[7][evalsN]
Size[7][Column[evecsN]]
```

```
{1, 0.309017 + 0.951057 i, 0.309017 - 0.951057 i}

{0.267261, 0.534522, 0.801784}
{0.681385, -0.104828 + 0.588348 i, -0.157243 - 0.392232 i}
{0.681385, -0.104828 - 0.588348 i, -0.157243 + 0.392232 i}
```

Check that the eigensystem operator did its job :

```
evecsNT = Transpose[evecsN];
rotMatN.evecsNT == evecsNT. DiagonalMatrix[evalsN]
```

```
True
```

Pair up each eigenvalue with its own eigenvector :

```
epairs = Transpose[{evalsN, evecsN}];
epairs⟦1⟧ // Size[7]
epairs⟦2⟧ // Size[7]
epairs⟦3⟧ // Size[7]
```

```
{1, {0.267261, 0.534522, 0.801784}}

{0.309017 + 0.951057 i,
 {0.681385, -0.104828 + 0.588348 i, -0.157243 - 0.392232 i}}

{0.309017 - 0.951057 i,
 {0.681385, -0.104828 - 0.588348 i, -0.157243 + 0.392232 i}}
```

As predicted in the theorem statement, we have two complex eigenvalues conjugate to each other, and another eigenvalue equal to **1**. Preparing for automation of this process, we use the **Cases** operator to pull out the case in which **eval** is **1** :

```
{eval₁, evec₁} = Cases[
   Transpose[{evalsN, evecsN}], {1, {_, _, _}}]⟦1⟧
```

```
{1, {0.267261, 0.534522, 0.801784}}
```

Check the basic eigenvector property :

```
rotMatN.evec₁
```

```
{0.267261, 0.534522, 0.801784}
```

The vector did not change. The only points that stay put under a rotational transform are those that lie on the axis of rotation. We have recovered the axis of rotation, up to a multiplicative constant. In fact, it is normalized :

```
axis = evec1;
axis.axis
```

```
1.
```

Is this the axis that we used as input? It will be more familiar if we scale it so that its first element equal to **1** :

```
axis / axis[[1]]
```

```
{1., 2., 3.}
```

Indeed, there are the simple numbers we used as axis input. Now for the rotation angle. We know the eigenvalues are of the form **Exp[± ⅈ α]**, so we **Select** for the presence of a **Complex** number :

```
complexEvals =
 Select[evalsN, Not[FreeQ[#, Complex]] &]
```

```
{0.309017 + 0.951057 ⅈ, 0.309017 - 0.951057 ⅈ}
```

The theorem says these numbers should be $\{e^{+i\alpha}, e^{-i\alpha}\}$. The angles themselves are found by

```
angles = π (Arg[complexEvals] / π // Rationalize)
```

$$\left\{\frac{2\pi}{5}, -\frac{2\pi}{5}\right\}$$

Note that we divided by π before rationalizing, and then re-multiplied afterwards. So the rotation angle is **2π/5**, with an unavoidable ambiguity in sign. This is linked to an unavoidable ambiguity in the sign of the rotation axis.

> **On your own :**
> Verify that the axis-angle pair **{{a,b,c},α}** produces exactly the same rotation matrix as **{{-a,-b,-c},-α}.**

We must settle this ambiguity by adopting a convention. Here it is :

> The rotation angle is chosen as a positive acute angle (a number in **[0,π]**). We feed this angle and the rotation axis to AxialRotation. If the output agrees with the input matrix, we are done. If not, the sign of the axis is changed, and agreement is then assured.

The input angle was actually **-2 π/5** ; but we "do not know" this. Try to reconstruct the input matrix using a positive angle :

```
AxialRotation[axis, 2 π / 5] == rotMatN
```

```
False
```

The only possible error is that we chose the wrong sign for the angle (as we know we did). To compensate, we change the sign *of the axis* and recompute :

```
AxialRotation[-axis, 2 π / 5] == rotMatN
```

```
True
```

We can automate this check as part of the operator. The angle will always be positive, and the axis will always come out with the right sign.

13.3 Recognize a rotoreflection matrix

The proof for rotoreflection goes just like the proof for rotation, above, with only a few changes in details. It would be an excellent exercise for you to make a copy of that section and run it, making modifications as they are needed. If you run into trouble, click here.

13.4 Recognize a reflection matrix

Reflection matrices are just rotoreflection matrices with zero rotation angle. Use **?AxialReflection** to see the syntax statement.

```
inputAxis = RandomReal[{0, 1}, 3]
```

```
{0.414379, 0.0268549, 0.98056}
```

```
Refmat = AxialReflection[inputAxis]
```

```
{{0.697141, -0.0196275, -0.716665},
 {-0.0196275, 0.998728, -0.0464453},
 {-0.716665, -0.0464453, -0.695869}}
```

```
{evals, evecs} = Eigensystem[Refmat] // ChopInteger
```

```
{{1, -1, 1}, {{0.270397, -0.958718, -0.0880115},
  {-0.389139, -0.0252191, -0.920834},
  {0.921179, -0.0106535, -0.388993}}}
```

As you would expect by setting α to zero in the rototreflection result, the key feature of the eigensystem of a reflection matrix is the eigenvalue *set* **⟨1,1,-1⟩**. But the position of the **-1** eigenvalue is not dependable, so you must **Sort** before doing an automated comparison.

13.5 Operator RecognizeMatrix

The process of recognizing symmetry transform matrices is so simple that it really does not need a human to carry it through. So we have made an operator called **RecognizeMatrix**, which carries out the whole process quickly and accurately. Its syntax statement is

- **RecognizeMatrix[mat(,prcsn)]** is a function inverse to the three **Axial...** operators. The return is {matrixType,fold⋱ Number,{axis,angle}}. The foldNumber is 2π/angle. Returned angles are always positive, and rotations are right handed as viewed from the returned axis point. Optional parameter **prcsn** is the number of digits used for comparison, with default **12**. If **mat** is not a numerical symmetry transform matrix, the return will be "anomalous".

It begins by testing for matrix size (3×3), reality, and unitarity. Failure here leads to immediate rejection. Then it finds the eigensystem of the given matrix, and examines the eigenvalues. First it looks for the special cases of identity matrix **{1,1,1}** and inversion matrix **{-1,-1,-1}** and reports just the names if it finds them. Next it tests whether the matrix is a reflection, by looking for the eigenvalue set **{-1,1,1}**. If it finds this it reports "reflection", along with the reflection axis (the eigenvector that goes with **-1**). If not, it goes on to the cases of nonzero angle, and reports as described in the thumbnail sketch, above.

If you want to see exactly how all this is done, open up the **Symmetry`** package and have a look. It is a lengthy case analysis based on **If**, **Goto**, and **Label**, just as in any computer language.

> **On your own**
> **(a)** Try a few examples, just to see it work. You will see many more examples in the chapters ahead.
>
> **(b)** Make a rotation matrix and add 10^{-6} to one of its elements. It will be slightly non-unitary (no longer a true point group symmetry operation) and **RecognizeMatrix** ought to detect this and reject it.

13.6 The complex eigenvectors

Give the rotation axis a name, zeta :

```
ζ = {3., 2., 1.};
```

Here we will finally investigate the meaning of the complex eigenvectors.

```
Mrot = AxialRotation[ζ, 2 π / 7];
Mrot // MatrixForm
```

$$\begin{pmatrix} 0.865532 & -0.0475917 & 0.498587 \\ 0.370315 & 0.731064 & -0.573073 \\ -0.337226 & 0.680647 & 0.650383 \end{pmatrix}$$

Find its **Eigensystem** :

```
{evalsN, evecRowsN} =
  Eigensystem[Mrot] // ChopInteger;
esysN = Transpose[{evalsN, evecRowsN}];
esysN [[1]] // Size[7]
esysN [[2]] // Size[7]
esysN [[3]] // Size[7]
```

```
{0.62349 + 0.781831 i, {0.157243 + 0.392232 i, 0.104828 - 0.588348 i, -0.681385}}

{0.62349 - 0.781831 i, {0.157243 - 0.392232 i, 0.104828 + 0.588348 i, -0.681385}}

{1, {-0.801784, -0.534522, -0.267261}}
```

The two complex **{eval,evec}** pairs are conjugate to each other, and therefore contain the same information. We work with only one of them.

```
{evalComplex, evecComplex} =
 Cases[esysN, {_Complex, {_, _, _}}][[1]]
```

```
{0.62349 + 0.781831 i,
 {0.157243 + 0.392232 i, 0.104828 - 0.588348 i, -0.681385}}
```

The angle of rotation is

```
φ = Arg[evalComplex]
```

```
0.897598
```

Here is the trick: We split the complex eigenvector into its real and imaginary parts :

```
{ξ, η} = { Re[evecComplex], Im[evecComplex]}
```

```
{{0.157243, 0.104828, -0.681385},
 {0.392232, -0.588348, 0}}
```

Remember, we already set the axis of rotation equal to $\underset{\rightharpoondown}{\varsigma}$. We test the three vectors $\underset{\rightharpoondown}{\xi}$, $\underset{\rightharpoondown}{\eta}$, $\underset{\rightharpoondown}{\varsigma}$ for orthogonality :

```
{ξ, η, ς}.Transpose[{ξ, η, ς}] // ChopInteger //
 MatrixForm
```

$$\begin{pmatrix} 0.5 & 0 & 0 \\ 0 & 0.5 & 0 \\ 0 & 0 & 14 \end{pmatrix}$$

They are orthogonal in **x,y,z** space. They are not normalized, but this is not important. The following profound equality shows how rotation can be expressed as a complex eigenequation. It is just the eigenvalue equation with $\mathbf{M_{rot}}$ is the operator, $\underset{\rightharpoondown}{\xi} + i\, \underset{\rightharpoondown}{\eta}$ is a complex eigenvector for it, and the corresponding complex eigenvalue **Exp[ⅈ ϕ]** .

```
M_rot . (ξ + ⅈ η) == Exp[ⅈ ϕ] (ξ + ⅈ η)
```

```
True
```

We look with new eyes, and reinterpret. On the left above, $\underset{\rightharpoondown}{\xi}$ and $\underset{\rightharpoondown}{\eta}$ are two real vectors in **x,y,z** space, perpendicular to each other and to the rotation axis $\underset{\rightharpoondown}{\varsigma}$. Matrix $\mathbf{M_{rot}}$ rotates each one by angle ϕ about $\underset{\rightharpoondown}{\varsigma}$. On the right, $\underset{\rightharpoondown}{\xi} + i\, \underset{\rightharpoondown}{\eta}$ is a triple of complex numbers ($\xi_1 + i\, \eta_1$, etc.) , which may be seen as points in the complex plane. Each of these points rotates in the complex plane by angle ϕ when multiplied by **Exp[ⅈ ϕ]**.

We are dealing with two different spaces here, and therefore two different kinds of "rotation". On the left, it is the rotation in three-dimensional **xyz** space. On the right, it is rotation in the complex plane. Incredibly clever, that **Eigensystem** operator.

13.7 End Notes

13.7.1 Explicit RotoReflection derivation

In a rotoreflection, every point on the rotation axis is reflected through the origin. In symbols,

RRMat.axis == (-1)*axis.

This says that the axis vector is an eigenvector of the rotoreflection matrix, with eigenvalue **-1**. We proceed as before:

```
RRMat =
  AxialRotoReflection[{a, b, c}, α] /. a^2 + b^2 + c^2 → 1;
RRMat // MatrixForm // Size[7]
```

$$\begin{pmatrix} -a^2 + (1 - a^2)\,\text{Cos}[\alpha] & -a\,b\,(1 + \text{Cos}[\alpha]) - c\,\text{Sin}[\alpha] & -a\,c\,(1 + \text{Cos}[\alpha]) + b\,\text{Sin}[\alpha] \\ -a\,b\,(1 + \text{Cos}[\alpha]) + c\,\text{Sin}[\alpha] & -b^2 + (1 - b^2)\,\text{Cos}[\alpha] & -b\,c\,(1 + \text{Cos}[\alpha]) - a\,\text{Sin}[\alpha] \\ -a\,c\,(1 + \text{Cos}[\alpha]) - b\,\text{Sin}[\alpha] & -b\,c\,(1 + \text{Cos}[\alpha]) + a\,\text{Sin}[\alpha] & -c^2 + (1 - c^2)\,\text{Cos}[\alpha] \end{pmatrix}$$

Find the symbolic **Eigensystem**:

```
{time, {evalsRR, evecsRR}} =
  Eigensystem[RRMat] // Timing;
time
```

```
0.857566
```

The vectors are horrifying; don't even look. Look instead at the eigenvalues :

```
(evalsRR /. {-a^2 → -1 + b^2 + c^2} // ExpandAll) /.
  Sqrt[-u_^2] → ⅈ u // TrigToExp
```

$$\{-1,\ e^{-i\alpha},\ e^{i\alpha}\}$$

This time the real eigenvalue is **-1**, and the angle appears as before. Make a numerical rotoreflection matrix with input parameters that are easy to remember.

```
rotMatN =
  AxialRotoReflection[{1., 2., 3.}, -2 π / 5];
rotMatN // MatrixForm
```

$$\begin{pmatrix} 0.215516 & 0.575539 & -0.788865 \\ -0.949544 & -0.0649879 & -0.306827 \\ 0.227857 & -0.815188 & -0.532494 \end{pmatrix}$$

The rest of the derivation is the same as for the reflection matrix.

14. Introduction to the character table

Preliminaries

14.1 Introduction

This chapter is full of unproved results; in fact, it is an advertisement for many theorems about groups that lie just ahead. By describing the character table at this point we give can perhaps rouse a little curiosity about these theorems. We provide forward links to them as we go; if you are reading this chapter as review the links should prove especially useful. We cannot describe the character table without the new term *representation* and an adjective for it, *irreducible*.

A *representation* of an abstract group is a list of concrete mathematical objects that multiply like the defining multiplication table of the abstract group. We will work mainly with matrix representations, but permutation representations also exist. For the formal definition in Chapter 25 (Representations), click here.

Block diagonalization is a way of splitting a large representation into two or more smaller ones. But block diagonalization has a limit; eventually the blocks cannot be reduced any farther. Every group has a small number such representations, called *irreducible representations*. The main purpose of the character table is to give you key information about the irreducible representations.

Irreducible representations are also called symmetry species; or just, *species*.

Even more briefly, we sometimes just say *rep*.

This is a very qualitative explanation of irreducibility, but it should serve to get us started. For the precise definition of *irreducible* in Chapter 28 (Reducible-Reps), click here.

14.2 The standard character table

In this book you can call up character tables in two ways; **(1)** as a typeset box, or **(2)** as a nested **`List`** structure. Here is the typeset version for group $\mathbf{D_{2d}}$:

```
BoxedCharacterTable["D2d"]
```

W.M. McClain, *Symmetry Theory in Molecular Physics with Mathematica*, DOI 10.1007/b13137_14,

D2d	1	2	1	2	2	← Class populations
	E	S_4	C_2	C_2'	σ_d	↓ Basis functions
A_1	1	1	1	1	1	$\{x^2+y^2,\ z^2\}$
A_2	1	1	1	-1	-1	$\{(x^2-y^2)\ z,\ Iz\}$
B_1	1	-1	1	1	-1	$\{x^2-y^2\}$
B_2	1	-1	1	-1	1	$\{z,\ xy\}$
E	2	0	-2	0	0	$\left\{\begin{matrix}x \\ y\end{matrix},\ \begin{matrix}xz \\ yz\end{matrix},\ \begin{matrix}Ix \\ Iy\end{matrix}\right\}$

And here is its homely sister, the nested **List** structure (spruced up a bit by the **Column** wrapper):

```
CharacterTable["D2d"];
Size[8][Column[%]]
```

```
{1, 2, 1, 2, 2}
{E, S4, C2, C'2, σd}
{A1, {1, 1, 1, 1, 1}, {{x^2 + y^2}, {z^2}}}
{A2, {1, 1, 1, -1, -1}, {{(x^2 - y^2) z}, {Iz}}}
{B1, {1, -1, 1, 1, -1}, {{x^2 - y^2}}}
{B2, {1, -1, 1, -1, 1}, {{z}, {x y}}}
{E, {2, 0, -2, 0, 0}, {{x, y}, {x z, y z}, {Ix, Iy}}}
```

It is not so pretty, but it is easier to lift parts from it.

On your own

If you want to see all the character tables in this book, just issue

```
Map[BoxedCharacterTable,GroupCatalog].
```

Perhaps interactive readers should do so at this point. Then you can check the following description against any of the examples you like.

14.3 Some facts about the character table

14.3.1 The essentials

Taking $\mathbf{D_{2\,d}}$ as an example, the most essential parts of the character table are brought forth by

```
SquareEssentials["D2d"]
```

D2d	E	S_4	C_2	C'_2	σ_d
A_1	1	1	1	1	1
A_2	1	1	1	-1	-1
B_1	1	-1	1	1	-1
B_2	1	-1	1	-1	1
E	2	0	-2	0	0

We omit the broad right column, which is discussed below in its own Section. We now discuss several features visible in the **SquareEssentials**.

14.3.2 The columns

As will be proved in Chapter 22 (Classes), all groups may be divided into nonoverlapping *classes* of elements. The class names appear at the head of the columns of the character table. The first class always consists of just one element, the identity element **E** , and it is always called class **E**. Continuing along the top row of this example, the second class is called $\mathbf{S_4}$. The S in its name tells you that it consists only of rotoreflections; the third class is $\mathbf{C_2}$, and the C in its name says that it consists only of rotations. Similarly for class $\mathbf{C'_2}$. The last class is σ_d, and the σ in its name tells you it consists of reflections.

14.3.3 The rows

There is one row for each *irreducible representation* that the group possesses. The names appear down the left column of the table. The first four rows are numerical representations (1×1 matrices), but the **E** row is a 2×2 representation. The full rationale behind the species names will be given in Chapter 28, but you may click here for a peek ahead if you just have to.

14.3.4 The body

The tabulated quantity is the "character" of the representation matrices. The character of a class is the trace, or **Spur**, or diagonal sum of matrix members. *Spur* is a Afrikaans word meaning footprint or scent. It will be proved that in every representation of a group, all group elements in the same class have the same smell, or footprint. This is why the character belongs to a whole class. We demonstrate on rep **T** of group **T**. First, look at its **SquareEssentials** :

```
SquareEssentials["T"]
```

T	E	C3	C32	C2
A1	1	1	1	1
Ea	1	$-\frac{1}{2}-\frac{i\sqrt{3}}{2}$	$-\frac{1}{2}+\frac{i\sqrt{3}}{2}$	1
Eb	1	$-\frac{1}{2}+\frac{i\sqrt{3}}{2}$	$-\frac{1}{2}-\frac{i\sqrt{3}}{2}$	1
T	3	0	0	-1

Construct rep **T** with an operator **MakeRepPoly[polyList,group]** from Chapter 27 (MakeRep). You cannot possibly understand this operator yet, so don't worry about how it does its job. Just take it on faith that it returns a representation of the group that has some relation to the functions **x**, **y**, and **z**.

```
xyzRepT = MakeRepPoly[{x, y, z}, "T"];
% // MatrixList // Size[6]
```

$$\left\{\begin{pmatrix}1&0&0\\0&1&0\\0&0&1\end{pmatrix},\begin{pmatrix}0&1&0\\0&0&1\\1&0&0\end{pmatrix},\begin{pmatrix}0&-1&0\\0&0&1\\-1&0&0\end{pmatrix},\begin{pmatrix}0&-1&0\\0&0&-1\\1&0&0\end{pmatrix},\begin{pmatrix}0&1&0\\0&0&-1\\-1&0&0\end{pmatrix},\begin{pmatrix}0&0&-1\\1&0&0\\0&-1&0\end{pmatrix},\right.$$
$$\left.\begin{pmatrix}0&0&1\\-1&0&0\\0&-1&0\end{pmatrix},\begin{pmatrix}0&0&-1\\-1&0&0\\0&1&0\end{pmatrix},\begin{pmatrix}0&0&1\\1&0&0\\0&1&0\end{pmatrix},\begin{pmatrix}1&0&0\\0&-1&0\\0&0&-1\end{pmatrix},\begin{pmatrix}-1&0&0\\0&1&0\\0&0&-1\end{pmatrix},\begin{pmatrix}-1&0&0\\0&-1&0\\0&0&1\end{pmatrix}\right\}$$

Given a bunch of matrices and a claim that they represent a group, you do know already how to test this claim:

```
MultiplicationTable[xyzRep, Range[12]] ==
 MultiplicationTable[CartRep["T"], Range[12]]

True
```

Since the multiplication table for this bunch of matrices closes, and is identical to the multiplication table of the Cartesian symmetry operators of group "T", the claim is True. Now we invoke the Classify operator from Chapter 22 (Classes) to split the rep into classes. Again take it on faith that this is what it is doing.

```
byClasses = Classify[xyzRepT];
Size[6][Column[MatrixList /@ %]]
```

$$\left\{\begin{pmatrix} 1 & 0 & 0 \\ 0 & 1 & 0 \\ 0 & 0 & 1 \end{pmatrix}\right\}$$

$$\left\{\begin{pmatrix} -1 & 0 & 0 \\ 0 & -1 & 0 \\ 0 & 0 & 1 \end{pmatrix}, \begin{pmatrix} -1 & 0 & 0 \\ 0 & 1 & 0 \\ 0 & 0 & -1 \end{pmatrix}, \begin{pmatrix} 1 & 0 & 0 \\ 0 & -1 & 0 \\ 0 & 0 & -1 \end{pmatrix}\right\}$$

$$\left\{\begin{pmatrix} 0 & -1 & 0 \\ 0 & 0 & -1 \\ 1 & 0 & 0 \end{pmatrix}, \begin{pmatrix} 0 & -1 & 0 \\ 0 & 0 & 1 \\ -1 & 0 & 0 \end{pmatrix}, \begin{pmatrix} 0 & 1 & 0 \\ 0 & 0 & -1 \\ -1 & 0 & 0 \end{pmatrix}, \begin{pmatrix} 0 & 1 & 0 \\ 0 & 0 & 1 \\ 1 & 0 & 0 \end{pmatrix}\right\}$$

$$\left\{\begin{pmatrix} 0 & 0 & -1 \\ -1 & 0 & 0 \\ 0 & 1 & 0 \end{pmatrix}, \begin{pmatrix} 0 & 0 & -1 \\ 1 & 0 & 0 \\ 0 & -1 & 0 \end{pmatrix}, \begin{pmatrix} 0 & 0 & 1 \\ -1 & 0 & 0 \\ 0 & -1 & 0 \end{pmatrix}, \begin{pmatrix} 0 & 0 & 1 \\ 1 & 0 & 0 \\ 0 & 1 & 0 \end{pmatrix}\right\}$$

There is one class on each row. The first class is class **E**, with trace **3**. Then comes a class of three matrices, all having trace **-1**. (Add them up yourself!) Then comes a class of four, all having trace **0**. The fourth class also has four elements with trace **0**. The machine can do it too:

```
Map[Spur, byClasses, {2}]
```

```
{{3}, {-1, -1, -1}, {0, 0, 0, 0}, {0, 0, 0, 0}}
```

The trace is the same for each matrix in any given class, as advertised.

14.3.5 One-dimensional species

Some species consist of **1x1** "matrices"; that is, just numbers. Their names are always anchored on an **A** or a **B**, and they are particularly easy to interpret. Look, for instance, at species $\mathbf{A_2}$ of group $\mathbf{D_{2\,d}}$. An object of species $\mathbf{A_2}$ changes sign when transformed by any one of the $\mathbf{C_2'}$ rotations, or any of the $\sigma_\mathbf{d}$ reflections, but is unaltered by transforms from classes **E** , $\mathbf{S_4}$, and $\mathbf{C_2}$. This accounts for the entries **{1,1,1,-1,-1}** on row $\mathbf{A_2}$.

```
BoxedCharacterTable["D2d"]
```

D2d	1	2	1	2	2	← Class populations
	E	S_4	C_2	C_2'	σ_d	↓ Basis functions
A_1	1	1	1	1	1	$\{x^2+y^2,\ z^2\}$
A_2	1	1	1	-1	-1	$\{(x^2-y^2)\,z,\ Iz\}$
B_1	1	-1	1	1	-1	$\{x^2-y^2\}$
B_2	1	-1	1	-1	1	$\{z,\ xy\}$
E	2	0	-2	0	0	$\left\{\begin{matrix} x \\ y \end{matrix},\ \begin{matrix} xz \\ yz \end{matrix},\ \begin{matrix} Ix \\ Iy \end{matrix}\right\}$

This table says that the polynomial $(\mathbf{x^2 - y^2})$ **z** transforms like rep $\mathbf{A_2}$. Verify:

```
(x² - y²) z /. SymmetryRules["D2d"]
─────────────────────────────────── // Simplify;
            (x² - y²) z
OnePerClass[%, "D2d"]
{1, 1, 1, -1, -1}
```

Yes, this is the **A_2** row of the table.

14.3.6 The trivial representation

The whole first row is always a one-dimensional species, with name anchor **A**. It is just a row of **1**'s, and it is called the "trivial" representation. Objects that "belong to" the first species are called totally symmetric objects, because all the transforms of the group simply multiply them by **1** (i.e., leave them unchanged.) Ground state wave functions are examples of totally symmetric objects.

14.3.7 Multi-dimensional species

D2d	1	2	1	2	2	← Class populations
	E	S_4	C_2	C_2'	σ_d	↓ Basis functions
A_1	1	1	1	1	1	$\{x^2+y^2,\ z^2\}$
A_2	1	1	1	-1	-1	$\{(x^2-y^2)\,z,\ Iz\}$
B_1	1	-1	1	1	-1	$\{x^2-y^2\}$
B_2	1	-1	1	-1	1	$\{z,\ xy\}$
E	2	0	-2	0	0	$\left\{\begin{matrix}x\\y\end{matrix},\ \begin{matrix}xz\\yz\end{matrix},\ \begin{matrix}Ix\\Iy\end{matrix}\right\}$

Look now at species **E**, the last row. In the standard display above, each pair is presented as a column. So the pairs are {**x** and **y**}, {**xz** and **yz**}, and {**Ix** and **Iy**}. This is weird, I admit, but there is a reason for it. If you want them in a **List**, see the **TabulatedBases** operator below.

The first thing to learn is that species **E** has nothing to do with class **E**. They have the same name due to an unfortunate historical accident, and we must live with it.

The species anchor **E** indicates a two-dimensional representation (or sometimes, two complex conjugate one-dimensional representations). These reps are always based on two functions that transform into a linear combination of themselves, and you need the full rep matrix to see exactly how. Nevertheless, the character is meaningful and useful. Click below for a little more detail.

Other groups have even larger multi-dimensional representations; the anchors of their names are **T** , **F**, **G**, or **H**, with dimensions, respectively, of 3, 3, 4, and 5.

These all occur in the icosahedral group "I" :

```
BoxedCharacterTable["I"]
```

I	1	12	12	20	15	← Class populations
	E	C5	C52	C3	C2	↓ Basis functions
A	1	1	1	1	1	$\{x^2 + y^2 + z^2\}$
T1	3	$\frac{1}{2} + \frac{\sqrt{5}}{2}$	$\frac{1}{2} - \frac{\sqrt{5}}{2}$	0	-1	$\left\{\begin{matrix} x & Ix \\ y, & Iy \\ z & Iz \end{matrix}\right\}$
F2	3	$\frac{1}{2} - \frac{\sqrt{5}}{2}$	$\frac{1}{2} + \frac{\sqrt{5}}{2}$	0	-1	$\left\{\begin{matrix} 3x^2 y - y^3 + 6xyz \\ x^3 - 3xy^2 - 3x^2 z + 3y^2 z \\ 3x^2 z + 3y^2 z - 2z^3 \end{matrix}\right\}$
G	4	-1	-1	1	0	$\left\{\begin{matrix} 2x^3 y + 2xy^3 + 3x^2 yz - y^3 z - 12xyz^2 \\ x^4 - y^4 - x^3 z + 3xy^2 z - 6x^2 z^2 + 6y^2 z^2 \\ x^4 - 6x^2 y^2 + y^4 + 3x^3 z + 3xy^2 z - 4xz^3 \\ 4x^3 y - 4xy^3 - 3x^2 yz - 3y^3 z + 4yz^3 \end{matrix}\right\}$
H	5	0	0	-1	1	$\left\{\begin{matrix} \frac{1}{2\sqrt{3}}\left(-x^2 - y^2 + 2z^2\right) \\ \frac{1}{2}\left(x^2 - y^2\right) \\ xy \\ yz \\ xz \end{matrix}\right\}$

If related functions were presented as rows, this table would be way too long to fit on the page.

14.3.8 The species dimension number

Since class **E** has only one member, the unit matrix, its character is the sum of **n** **1**'s, where **n** is the dimension of the representation. Obviously, the dimension of any species may be read as the character of its class **E**. The rep dimensions in our tabulated groups, as it happens, can be any integer from **1** to **5**.

14.3.9 The table is square

In every group, the number of classes is equal to the number of species, so the essential inner part of every character table is square. This is not a quick or easy theorem. It will come as part of Chapter 33 (TheGreatOrthogonality).

14.4 The rightmost column; bases for reps

The broad rightmost column contains information needed for constructing the irreducible representations. They are easily found by the use of projection operators, but many standard books present character tables that have lots of blanks in this column. Take this as proof that projection operators have a high barrier in work by hand. But because our **MakeRep** operators can use these basis sets explicitly, in this book every species has at least one tabulated basis set.

The number of basis functions is the dimension of the species. For many species, several alternate sets of basis functions are given. If only one is given, it is in double braces to conform to the requirement that it be a **List** of **Lists**.

You can extract all the basis information in usable format with

```
TabulatedBases["D2d"];
Size[7][Column[%]]
```

```
{A1, {{x^2 + y^2}, {z^2}}}
{A2, {{(x^2 - y^2) z}, {Iz}}}
{B1, {{x^2 - y^2}}}
{B2, {{z}, {x y}}}
{E, {{x, y}, {x z, y z}, {Ix, Iy}}}
```

or you can extract (for instance) just the $\mathbf{A_1}$ basis information with

```
TabulatedBases["D2d"][[1, 2]]
```

```
{{x^2 + y^2}, {z^2}}
```

The basis information is always a **List** of **Lists**, though in one-dimensional species the inner list has only one member.

1) For species $\mathbf{A_1}$ we see function $\mathbf{x^2 + y^2}$ given as a basis, because it always transforms into itself under the operations of the group. Similarly, so does the second listed basis function, $\mathbf{z^2}$.

$$\left(\frac{(x^2 + y^2)\ /.\ \text{SymmetryRules["D2d"]}}{(x^2 + y^2)}\right) //\ \text{Simplify}$$

```
{1, 1, 1, 1, 1, 1, 1, 1}
```

Try the other one-dimensional species yourself.

2) For species **E** the basis functions come in pairs, because they transform into linear combinations of themselves. Here is a particularly simple example :

```
{x, y} /. SymmetryRules["D2d"]
```

```
{{x, y}, {y, -x}, {-y, x}, {-x, -y},
 {-x, y}, {x, -y}, {y, x}, {-y, -x}}
```

14.5 Various character table operators

You have already seen **BoxedCharacterTable** and **CharacterTable**. Other character table operators bring down named parts of the table.

```
$DefaultGroup = "C3v";
```

```
ClassPopulations[]
```

```
{1, 2, 3}
```

```
ClassNames[]
```

```
{E, C3, σv}
```

```
ElementNames[]
```

```
{E, C3a, C3b, σva, σvb, σvc}
```

```
Classes[]
```

```
{{E}, {C3a, C3b}, {σva, σvb, σvc}}
```

```
SpeciesNames[]
```

```
{A1, A2, E}
```

```
Column[TabulatedBases[]]
```

$\{A1, \{\{z\}, \{x^2 + y^2\}, \{z^2\}\}\}$
$\{A2, \{\{Iz\}\}\}$
$\{E, \{\{x, y\}, \{\frac{1}{2}(x^2 - y^2), x\,y\}, \{x\,z, y\,z\}, \{Ix, Iy\}\}\}$

> **On your own**
> Change the **$DefaultGroup** at the top, and run them all again.

These are the basic facts about character tables. In the following chapters you will see them proved rigorously.

15. The operator MakeGroup

Preliminaries

15.1 An operator for constructing groups

The group generation procedure suggested previously in Chapter 07 (FundamentalThm) is so simple and logical that we ought to be able to automate it. We describe here the construction of an operator that will carry out the whole group construction procedure with a single ENTER.

We call it **MakeGroup[elList]**, where **elList** will be a short list of group elements (usually, one, two, or three of them). The elements may be either matrices or permutations. This operator returns the entire group implied by the short list. We first give two short demonstrations of the operator :

First we let **MakeGroup** make a group of matrices. Start with just one matrix, **C_z[2 π / 3]**:

```
gp1 = MakeGroup[{Cz[2 π / 3] }];
Map[MatrixForm, gp1]
```

$$\left\{\begin{pmatrix} -\frac{1}{2} & -\frac{\sqrt{3}}{2} & 0 \\ \frac{\sqrt{3}}{2} & -\frac{1}{2} & 0 \\ 0 & 0 & 1 \end{pmatrix}, \begin{pmatrix} -\frac{1}{2} & \frac{\sqrt{3}}{2} & 0 \\ -\frac{\sqrt{3}}{2} & -\frac{1}{2} & 0 \\ 0 & 0 & 1 \end{pmatrix}, \begin{pmatrix} 1 & 0 & 0 \\ 0 & 1 & 0 \\ 0 & 0 & 1 \end{pmatrix}\right\}$$

From that one matrix, **MakeGroup** generated a list of three matrices. Are they a group? After Chapter 13, where we constructed the **RecognizeMatrix** operator, we are all set up to answer this :

```
Map[RecognizeMatrix, gp1] // Column
```

$\left\{\text{rotation, 3-fold, } \left\{\{0, 0, 1\}, \frac{2\pi}{3}\right\}\right\}$
$\left\{\text{rotation, 3-fold, } \left\{\{0, 0, -1\}, \frac{2\pi}{3}\right\}\right\}$
{identity}

Yes, this is group **C_3**, the group of rotations by ±1/3 turn. Now we let **MakeGroup** make a permutation group. Start with one permutation of length 4 :

```
perms4 = MakeGroup[{{2, 4, 1, 3}}]
```

```
{{1, 2, 3, 4}, {2, 4, 1, 3}, {3, 1, 4, 2}, {4, 3, 2, 1}}
```

Is this list of permutations a group? The only axioms in question are closure

DOI 10.1007/b13137_15,

and inverse elements, so the quickest test is just to make a multiplication table :

```
MultiplicationTable[
  Permute, perms4, {"E", b, c, d}];
% // MatrixForm
```

$$\begin{pmatrix} E & b & c & d \\ b & d & E & c \\ c & E & d & b \\ d & c & b & E \end{pmatrix}$$

It closes, and every element has an inverse, so **perms4** is a group. The **MakeGroup** operator also has a verbose mode. Watch it operate on two permutations of length 6 :

```
MakeGroup[{{2, 3, 4, 1, 5, 6}, {2, 3, 4, 1, 6, 5}},
 Report → True]
```

```
{{1, 2, 3, 4, 5, 6}, {1, 2, 3, 4, 6, 5},
 {2, 3, 4, 1, 5, 6}, {2, 3, 4, 1, 6, 5}, {3, 4, 1, 2, 5, 6},
 {3, 4, 1, 2, 6, 5}, {4, 1, 2, 3, 5, 6}, {4, 1, 2, 3, 6, 5}}
```

It cycled until the number of elements stopped increasing.

15.2 MakeGroup, step by step

Now we show how **MakeGroup** was constructed. As always, the first job is to carry out an example by hand. So we start with two matrices :

```
step1 = {Cz[2 π / 4], Cx[2 π / 2] };
% // MatrixList // Size[6]
```

$$\left\{\begin{pmatrix} 0 & -1 & 0 \\ 1 & 0 & 0 \\ 0 & 0 & 1 \end{pmatrix}, \begin{pmatrix} 1 & 0 & 0 \\ 0 & -1 & 0 \\ 0 & 0 & -1 \end{pmatrix}\right\}$$

In Chapter 08 (MultiplicatnTbls) we used **Outer**, operating at level 1 :

```
step2 = Outer[Dot, step1, step1, 1];
% // MatrixForm // Size[6]
```

$$\begin{pmatrix} \begin{pmatrix} -1 & 0 & 0 \\ 0 & -1 & 0 \\ 0 & 0 & 1 \end{pmatrix} & \begin{pmatrix} 0 & 1 & 0 \\ 1 & 0 & 0 \\ 0 & 0 & -1 \end{pmatrix} \\ \begin{pmatrix} 0 & -1 & 0 \\ -1 & 0 & 0 \\ 0 & 0 & -1 \end{pmatrix} & \begin{pmatrix} 1 & 0 & 0 \\ 0 & 1 & 0 \\ 0 & 0 & 1 \end{pmatrix} \end{pmatrix}$$

As you see, **Outer** has organized its output as a matrix of matrices, containing all possible two-matrix products, **A.A**, **A.B**, **B.A**, and **B.B** . But we need them in as a **List** of matrices, so we apply **Flatten** at level **1** :

```
step3 = Flatten[step2, 1];
% // MatrixList // Size[6]
```

$$\left\{\begin{pmatrix} -1 & 0 & 0 \\ 0 & -1 & 0 \\ 0 & 0 & 1 \end{pmatrix}, \begin{pmatrix} 0 & 1 & 0 \\ 1 & 0 & 0 \\ 0 & 0 & -1 \end{pmatrix}, \begin{pmatrix} 0 & -1 & 0 \\ -1 & 0 & 0 \\ 0 & 0 & -1 \end{pmatrix}, \begin{pmatrix} 1 & 0 & 0 \\ 0 & 1 & 0 \\ 0 & 0 & 1 \end{pmatrix}\right\}$$

Neither of the two original matrices is in the product list. We don't want to lose them, so we join them in :

```
step4 = Join[step1, step3 ];
% // MatrixList // Size[6]
```

$$\left\{\begin{pmatrix} 0 & -1 & 0 \\ 1 & 0 & 0 \\ 0 & 0 & 1 \end{pmatrix}, \begin{pmatrix} 1 & 0 & 0 \\ 0 & -1 & 0 \\ 0 & 0 & -1 \end{pmatrix}, \begin{pmatrix} -1 & 0 & 0 \\ 0 & -1 & 0 \\ 0 & 0 & 1 \end{pmatrix}, \begin{pmatrix} 0 & 1 & 0 \\ 1 & 0 & 0 \\ 0 & 0 & -1 \end{pmatrix}, \begin{pmatrix} 0 & -1 & 0 \\ -1 & 0 & 0 \\ 0 & 0 & -1 \end{pmatrix}, \begin{pmatrix} 1 & 0 & 0 \\ 0 & 1 & 0 \\ 0 & 0 & 1 \end{pmatrix}\right\}$$

This is the end of the first cycle. We started with two matrices in **step1** and now in **step4** we have six matrices. If we run the **step4** matrices through the same cycle of operations, we will learn something new. The operations in steps 5, 6 and 7 are the same as in steps 2, 3, and 4 :

```
step5 = Outer[Dot, step4, step4, 1];
step6 = Flatten[step5, 1];
step7 = Join[step6, step4];
% // Dimensions
```

```
{42, 3, 3}
```

Indeed, it made 42 matrices. This seems like too many. There must be repeats, and we need to winnow them out. The operator that does this is **Union**. Here is a simple demo :

```
Union[{c, b, a, b, c}]
```

```
{a, b, c}
```

Applied to a single list with repeats in it, **Union** took out the repeats and put the remaining list in canonical order. (For a fuller picture of **Union**, call up its thumbnail description.) What happens if we use **Union** on our list of 42 matrices ?

```
step8 = Union[step7];
% // MatrixList // Size[6]
```

$$\left\{\begin{pmatrix} -1 & 0 & 0 \\ 0 & -1 & 0 \\ 0 & 0 & 1 \end{pmatrix}, \begin{pmatrix} -1 & 0 & 0 \\ 0 & 1 & 0 \\ 0 & 0 & -1 \end{pmatrix}, \begin{pmatrix} 0 & -1 & 0 \\ -1 & 0 & 0 \\ 0 & 0 & -1 \end{pmatrix},\right.$$
$$\left.\begin{pmatrix} 0 & -1 & 0 \\ 1 & 0 & 0 \\ 0 & 0 & 1 \end{pmatrix}, \begin{pmatrix} 0 & 1 & 0 \\ -1 & 0 & 0 \\ 0 & 0 & 1 \end{pmatrix}, \begin{pmatrix} 0 & 1 & 0 \\ 1 & 0 & 0 \\ 0 & 0 & -1 \end{pmatrix}, \begin{pmatrix} 1 & 0 & 0 \\ 0 & -1 & 0 \\ 0 & 0 & -1 \end{pmatrix}, \begin{pmatrix} 1 & 0 & 0 \\ 0 & 1 & 0 \\ 0 & 0 & 1 \end{pmatrix}\right\}$$

So the list of 42 contained just 8 different matrices. That's more like it. Note, however, that **Union** uses an ordering in which the unit matrix comes last. Look at the **MultiplicationTable** of these 8 matrices:

```
MultiplicationTable[Dot, step8,
  {"a", "b", "c", "d", "e", "f", "g", "E"}];
% // MatrixForm // Size[7]
```

$$\begin{pmatrix} E & g & f & e & d & c & b & a \\ g & E & e & f & c & d & a & b \\ f & d & E & b & g & a & e & c \\ e & c & g & a & E & b & f & d \\ d & f & b & E & a & g & c & e \\ c & e & a & g & b & E & d & f \\ b & a & d & c & f & e & E & g \\ a & b & c & d & e & f & g & E \end{pmatrix}$$

It closes, so these eight matrices form a group. If we cycled them through steps 5, 6, 7, and 8 again, the output would repeat the input. So we now see how to construct the general **MakeGroup** operator: we must nest operations 5 through 8 to make a compound operator called **OneMatCycle**, and then run **OneMatCycle** repeatedly until it produces two successive results that are the same.

```
Clear[OneMatCycle];
OneMatCycle[Lst_] := Reverse[Union[
   Join[Lst, Flatten[Outer[Dot, Lst, Lst, 1], 1]]]]
```

The innermost operation, in red, is **Outer**. Then in alternating red and black colors, **Outer** is wrapped with **Flatten**, **Flatten** is wrapped with **Join**, **Join** is wrapped with **Union**, and finally **Union** is wrapped with **Reverse** (which reverses list order, to make the unit matrix come first). This is the *functional style* of programming, so natural to *Mathematica*.

First we apply **OneMatCycle** to the two matrices we called **step1** :

```
cycle1 = OneMatCycle[step1];
% // MatrixList // Size[6]
```

$$\left\{\begin{pmatrix} 1 & 0 & 0 \\ 0 & 1 & 0 \\ 0 & 0 & 1 \end{pmatrix}, \begin{pmatrix} 1 & 0 & 0 \\ 0 & -1 & 0 \\ 0 & 0 & -1 \end{pmatrix}, \begin{pmatrix} 0 & 1 & 0 \\ 1 & 0 & 0 \\ 0 & 0 & -1 \end{pmatrix}, \begin{pmatrix} 0 & -1 & 0 \\ 1 & 0 & 0 \\ 0 & 0 & 1 \end{pmatrix}, \begin{pmatrix} 0 & -1 & 0 \\ -1 & 0 & 0 \\ 0 & 0 & -1 \end{pmatrix}, \begin{pmatrix} -1 & 0 & 0 \\ 0 & -1 & 0 \\ 0 & 0 & 1 \end{pmatrix}\right\}$$

Two matrices made six, just as before. Cycle it again :

```
cycle2 = OneMatCycle[cycle1];
% // MatrixList // Size[6]
```

$$\left\{\begin{pmatrix} 1 & 0 & 0 \\ 0 & 1 & 0 \\ 0 & 0 & 1 \end{pmatrix}, \begin{pmatrix} 1 & 0 & 0 \\ 0 & -1 & 0 \\ 0 & 0 & -1 \end{pmatrix}, \begin{pmatrix} 0 & 1 & 0 \\ 1 & 0 & 0 \\ 0 & 0 & -1 \end{pmatrix}, \begin{pmatrix} 0 & 1 & 0 \\ -1 & 0 & 0 \\ 0 & 0 & 1 \end{pmatrix},\right.$$
$$\left.\begin{pmatrix} 0 & -1 & 0 \\ 1 & 0 & 0 \\ 0 & 0 & 1 \end{pmatrix}, \begin{pmatrix} 0 & -1 & 0 \\ -1 & 0 & 0 \\ 0 & 0 & -1 \end{pmatrix}, \begin{pmatrix} -1 & 0 & 0 \\ 0 & 1 & 0 \\ 0 & 0 & -1 \end{pmatrix}, \begin{pmatrix} -1 & 0 & 0 \\ 0 & -1 & 0 \\ 0 & 0 & 1 \end{pmatrix}\right\}$$

Six matrices made eight, as before. Once more:

```
cycle3 = OneMatCycle[cycle2];
% // MatrixList // Size[6]
```

$$\left\{\begin{pmatrix}1&0&0\\0&1&0\\0&0&1\end{pmatrix},\begin{pmatrix}1&0&0\\0&-1&0\\0&0&-1\end{pmatrix},\begin{pmatrix}0&1&0\\1&0&0\\0&0&-1\end{pmatrix},\begin{pmatrix}0&1&0\\-1&0&0\\0&0&1\end{pmatrix},\begin{pmatrix}0&-1&0\\1&0&0\\0&0&1\end{pmatrix},\begin{pmatrix}0&-1&0\\-1&0&0\\0&0&-1\end{pmatrix},\begin{pmatrix}-1&0&0\\0&1&0\\0&0&-1\end{pmatrix},\begin{pmatrix}-1&0&0\\0&-1&0\\0&0&1\end{pmatrix}\right\}$$

Eight matrices made the same eight again, and we surely have a group. Now we can carry the automation to its final stage.

We want to call **OneMatCycle** on a short list of group elements, and keep applying it to its own output until it starts repeating itself. If we use the **While** construction, we can do this without knowing the number of times OneMatCycle is to be used. The **While** operator requires a little external setup:

```
Clear[ListA, ListB];
ListA = {};
ListB = step1;
```

Remember, **step1** is our original list of two matrices. We will use **While** :

While[*test*, *body*] evaluates *test*, then *body*, repetitively, until *test* first fails to give True.

Below, the test is **Length[ListA]=!=Length[ListB]** (where **=!=** means **is not equal to**) and the body is all the rest.

```
(*Loop entrance*)
While[Length[ListA] =!= Length[ListB],
 (*old answer becomes new input:*)
   ListA = ListB;
 (*generate next answer:*)
   ListB = OneMatCycle[ListA];
 Print[Length[ListA],
  " mats made ", Length[ListB], " mats"];
 (* go back up and apply the test;
 exit if false*)
 (*endWhile*) ]
```

```
2 mats made 6 mats

6 mats made 8 mats

8 mats made 8 mats
```

```
ListB // MatrixList // Size[6]
```

$$\left\{\begin{pmatrix}1&0&0\\0&1&0\\0&0&1\end{pmatrix},\begin{pmatrix}1&0&0\\0&-1&0\\0&0&-1\end{pmatrix},\begin{pmatrix}0&1&0\\1&0&0\\0&0&-1\end{pmatrix},\begin{pmatrix}0&1&0\\-1&0&0\\0&0&1\end{pmatrix},\begin{pmatrix}0&-1&0\\1&0&0\\0&0&1\end{pmatrix},\begin{pmatrix}0&-1&0\\-1&0&0\\0&0&-1\end{pmatrix},\begin{pmatrix}-1&0&0\\0&1&0\\0&0&-1\end{pmatrix},\begin{pmatrix}-1&0&0\\0&-1&0\\0&0&1\end{pmatrix}\right\}$$

It looks like it worked.

On your own

Copy the operator **OneMatCycle** and turn it into **OnePermCycle**, making a permutation group rather than a matrix group. A single word change accomplishes this. Then use it on a short list of permutations until you have a closed permutation group. A good starting point is

```
PermList1 =
  {{2, 3, 4, 1, 5, 6}, {2, 3, 4, 1, 6, 5}};
```

In three cycles, this makes a group of eight permutations.
If you get tangled up, click to the End Notes.

In the **Symmetry`** package operator **MakeGroup**, we have constructed a test that distinguishes matrix input from permutation input, and uses either **Dot** or **Permute**, as appropriate. So it can make either matrix groups of permutation groups. Open up the package to read the details. It also politely declines invalid input:

```
MakeGroup[{{1, 2, 3}, {4, 5}}]
```

```
Input is neither matrix list nor permutation list
```

The rest of this chapter is devoted to some tricky programming issues, only tangentially relevant to Symmetry Theory. Some people may want to stop here and go on to Chapter 16.

15.3 MakeGroup with roundoff

15.3.1 The problem

The operator above worked beautifully with infinite precision input. However, often we will have matrices of **Real** numbers, known only to machine accuracy, and an ugly little issue raises its head. When you compute the same **Real** matrix by two different routes, you may get two answers that differ insignificantly in the last one or two digits of some of the elements (on a Macintosh, in digits 15 and 16). If you use the **Equal** operator (**==**) to ask whether the two matrices are equal, *Mathematica* will normally say **True**, because it reports equality if the two sides are equal to within machine roundoff error.

However, we must apply **Union** to eliminate duplicates. Unfortunately, **Union** has a little quirk that prevents it from working quite the way we want. First, we do an example using exact elements, just to show the answer expected. Applying **OneMatCycle** twice,

```
Map[MatrixForm,
  OneMatCycle[OneMatCycle[{Cz[2 π / 3]}]]] // Size[7]
```

$$\left\{\begin{pmatrix} 1 & 0 & 0 \\ 0 & 1 & 0 \\ 0 & 0 & 1 \end{pmatrix}, \begin{pmatrix} -\frac{1}{2} & \frac{\sqrt{3}}{2} & 0 \\ -\frac{\sqrt{3}}{2} & -\frac{1}{2} & 0 \\ 0 & 0 & 1 \end{pmatrix}, \begin{pmatrix} -\frac{1}{2} & -\frac{\sqrt{3}}{2} & 0 \\ \frac{\sqrt{3}}{2} & -\frac{1}{2} & 0 \\ 0 & 0 & 1 \end{pmatrix}\right\}$$

It worked; that is the three-member cyclic group **C3**. Now we demonstrate the undesirable quirk using the approximate rotation angle **2π/3.** (with a decimal point behind the 3) instead of the exact angle **2π/3** (with no decimal point). We get a different answer :

```
Map[MatrixForm,
   OneMatCycle[OneMatCycle[{Cz[2 π / 3.]}]]] //
  ChopInteger // Size[7]
```

$$\left\{\begin{pmatrix} 1 & 0 & 0 \\ 0 & 1 & 0 \\ 0 & 0 & 1 \end{pmatrix}, \begin{pmatrix} -0.5 & -0.866025 & 0 \\ 0.866025 & -0.5 & 0 \\ 0 & 0 & 1 \end{pmatrix}, \begin{pmatrix} -0.5 & -0.866025 & 0 \\ 0.866025 & -0.5 & 0 \\ 0 & 0 & 1 \end{pmatrix}, \begin{pmatrix} -0.5 & 0.866025 & 0 \\ -0.866025 & -0.5 & 0 \\ 0 & 0 & 1 \end{pmatrix}\right\}$$

We get four matrices, not three. The last three seem to be identical. Clearly, **Union** has failed to do the desired job.

15.3.2 The fix, part 1: A homemade Union operator

How does **Union** really work? Here we make up a homemade **Union** operator that takes a list of symbols as input, returning a sorted list, with repeats removed, as its output. Let the example input be the list **{a,b,b,c,b,d}**.

```
sorted = Sort[{a, b, b, c, b, d}]
```

```
{a, b, b, b, c, d}
```

Now we partition the sorted list into pairs, with an offset of 1. This changes a list into a list of pairs, each pair consisting of an element and its successor.

```
pairedWithSuccessors = Partition[sorted, 2, 1]
```

```
{{a, b}, {b, b}, {b, b}, {b, c}, {c, d}}
```

Now we **Select** from this list of pairs only pairs that are not self-pairs. This eliminates the repeats. The comparison function for **Select** is

```
neq[{a_, b_}] :=
 Union[Flatten[{Chop[a - b]}]] =!= {0}
```

which, as we will see, is so clever that it works for both symbols and matrices of any size. The **Select** operation itself is then

```
differentPairs = Select[pairedWithSuccessors, neq]
```

```
{{a, b}, {b, c}, {c, d}}
```

Now **Transpose** produces two lists, each one being *almost* what we want. However, the last element is missing from the first list, and the first element is missing from the second list.

```
twoLists = Transpose[differentPairs]
```

```
{{a, b, c}, {b, c, d}}
```

We patch this up by prepending the first element of the first list onto the second list.

```
Prepend[twoLists[[2]], twoLists[[1, 1]]]
```

```
{a, b, c, d}
```

This is the desired output, constructed by a perfectly general method. We have cleverly written the comparison function **neq** so that it works on either symbols or matrices, even on matrices of arbitrary size. We test it on two matrices that are close, but not precisely equal :

```
ϵ = 1. * 10^-16;
```

```
twoCloseMats = {(1-ϵ  3), (1 3)};
                ( 2   4)  (2 4)
Map[MatrixForm, %]
```

$\left\{\begin{pmatrix}1. & 3\\2 & 4\end{pmatrix}, \begin{pmatrix}1 & 3\\2 & 4\end{pmatrix}\right\}$

```
neq[twoCloseMats]
```

```
False
```

So it is **False** that they are **NotEqual**; or in other words, it is **True** that their difference is zero, within the criterion used by the **Chop** operator.

15.3.3 RoundSort, an improved Sort for matrices

The homemade **Union** operator above uses the built-in **Sort** operator, and this leads to a second independent difficulty. When sorting matrices or other compound objects, **Sort** begins sorting on the first element of the first row, element **[[1,1]]**. If there is a tie in this element, it breaks the tie using the second element of the first row, element **[[1,2]]**, etc. Thus we see

```
Sort[{(1 3), (1 2)}] // Map[MatrixForm, #] &
      (2 4)  (3 4)
```

$\left\{\begin{pmatrix}1 & 2\\3 & 4\end{pmatrix}, \begin{pmatrix}1 & 3\\2 & 4\end{pmatrix}\right\}$

But suppose the list includes a matrix similar to $\begin{pmatrix}1 & 3\\2 & 4\end{pmatrix}$ but with its **[[1,1]]** element insignificantly smaller than **1**. Then there is no tie on element **[[1,1]]**, and we get a different, and possibly undesired, sorted order :

```
ϵ = 1.*10^-16;
Sort[{(1-ϵ  3), (1 3), (1 2)}] //
      ( 2   4)  (2 4)  (3 4)
 Map[MatrixForm, #] &
```

$\left\{\begin{pmatrix}1. & 3\\2 & 4\end{pmatrix}, \begin{pmatrix}1 & 2\\3 & 4\end{pmatrix}, \begin{pmatrix}1 & 3\\2 & 4\end{pmatrix}\right\}$

The first and third matrices ought to have been neighbors in sorted order because they differ only insignificantly. Instead, **Sort** seized on the tiny defect in **1-ϵ** to list this matrix as the first matrix. Clearly, we need a different kind of **Sort** operator that does a little rounding before it sorts, so that insignificant numerical differences in corresponding elements are ignored. We have written such an operator and stored it in the **Symmetry`** package. It is called **Round‐Sort** :

```
RoundSort[list,(sigfigs)] rounds the list to sigfigs
  significant figures, then sorts the rounded list and
  returns the original unrounded list in sorted order.
  The default for sigfigs is 12. Use it for sorting
  matrices when you want to ignore numerical dither.
```

RoundSort$\left[\left\{\begin{pmatrix}1-\epsilon & 3\\ 2 & 4\end{pmatrix}, \begin{pmatrix}1 & 3\\ 2 & 4\end{pmatrix}, \begin{pmatrix}1 & 2\\ 3 & 4\end{pmatrix}\right\}\right]$ **//**
Map[MatrixForm, #] &

$$\left\{\begin{pmatrix}1 & 2\\ 3 & 4\end{pmatrix}, \begin{pmatrix}1. & 3\\ 2 & 4\end{pmatrix}, \begin{pmatrix}1 & 3\\ 2 & 4\end{pmatrix}\right\}$$

With **RoundSort** we can be sure that matrices that differ only insignificantly will always appear as neighbors in the sorted order. This is a prerequisite for our homemade **RoundUnion** operator to work correctly.

```
RoundUnion[list,(sigfigs)] does the same thing
  as Union[list], except that the elements of
  the list are rounded to the specified number
  of sigfigs (default 12) before the Union
  operation is performed.  It is used mainly to
  ignore numerical dither in lists of matrices.
```

RoundUnion$\left[\left\{\begin{pmatrix}1-\epsilon & 3\\ 2 & 4\end{pmatrix}, \begin{pmatrix}1 & 3\\ 2 & 4\end{pmatrix}, \begin{pmatrix}1 & 2\\ 3 & 4\end{pmatrix}\right\}\right]$ **//**
Map[MatrixForm, #] &

$$\left\{\begin{pmatrix}1 & 2\\ 3 & 4\end{pmatrix}, \begin{pmatrix}1. & 3\\ 2 & 4\end{pmatrix}\right\}$$

15.3.4 The total fix: procedure MakeGroup

Now we show how to construct group $\mathbf{C_3}$ using **Real** matrices. We begin by applying **OneMatCycle** twice to the matrix **C**$_z$**[2 π / 3.]** :

```
approxMats =
  OneMatCycle[OneMatCycle[{Cz[2π/3.]}]];
approxMats // Size[7]
```

```
{{{1., 6.10623×10^-16, 0.}, {-6.10623×10^-16, 1., 0.}, {0., 0., 1.}},
 {{-0.5, -0.866025, 0.}, {0.866025, -0.5, 0.}, {0., 0., 1.}},
 {{-0.5, -0.866025, 0}, {0.866025, -0.5, 0}, {0, 0, 1}},
 {{-0.5, 0.866025, 0.}, {-0.866025, -0.5, 0.}, {0., 0., 1.}}}
```

That is group $\mathbf{C_3}$, but without removal of near-equal matrices. In **MakeGroup**, there is a limit of **200** matrices to prevent runaway operation when the group does not close. The next step is to **RoundSort** to ensure that similar matrices appear as neighbors : (To see the actual outputs below, read this chapter in interactive mode on a computer)

```
listRS = RoundSort[approxMats];
```

Each matrix is paired with its successor :

```
dbls = Partition[listRS, 2, 1];
```

and **Union** is implemented by retaining only pairs that are not equal :

```
difs = Select[dbls, neq];
```

Finally, the successor-pairs are unpaired, and one complete sequence is constructed :

```
Prepend[Transpose[difs][[2]], difs[[1, 1]]] //
   ChopInteger // Map[MatrixForm, #] & // Size[7]
```

$$\left\{\begin{pmatrix} -0.5 & -0.866025 & 0 \\ 0.866025 & -0.5 & 0 \\ 0 & 0 & 1 \end{pmatrix}, \begin{pmatrix} -0.5 & 0.866025 & 0 \\ -0.866025 & -0.5 & 0 \\ 0 & 0 & 1 \end{pmatrix}, \begin{pmatrix} 1 & 0 & 0 \\ 0 & 1 & 0 \\ 0 & 0 & 1 \end{pmatrix}\right\}$$

There is a group of **Real** matrices, with insignificant roundoff errors handled appropriately. This procedure is kept in the **Symmetry`** package as part of the operator **MakeGroup**. See the text of the **Round** operators in the package.

15.4 End notes

15.4.1 `OnePermCycle` for permutations

We want to make a permutation group, starting with the short list

```
PermList1 = {{2, 3, 4, 1, 5, 6}, {2, 3, 4, 1, 6, 5}} ;
```

We will multiply everything by everything, adding all new results to the growing list of elements, until we reach closure. The operator is

```
Clear[OnePermCycle];
OnePermCycle[Lst_] := Reverse[Union[Join[Lst,
    Flatten[Outer[Permute, Lst, Lst, 1], 1] ]]]
```

where all we did was change **Dot** to that blue **Permute**. Let it do its job :

```
PermList2 = OnePermCycle[PermList1]
```

```
{{3, 4, 1, 2, 6, 5}, {3, 4, 1, 2, 5, 6},
 {2, 3, 4, 1, 6, 5}, {2, 3, 4, 1, 5, 6}}
```

```
PermList3 = OnePermCycle[PermList2]
```

```
{{4, 1, 2, 3, 6, 5}, {4, 1, 2, 3, 5, 6},
 {3, 4, 1, 2, 6, 5}, {3, 4, 1, 2, 5, 6}, {2, 3, 4, 1, 6, 5},
 {2, 3, 4, 1, 5, 6}, {1, 2, 3, 4, 6, 5}, {1, 2, 3, 4, 5, 6}}
```

```
PermList4 = OnePermCycle[PermList3]
```

```
{{4, 1, 2, 3, 6, 5}, {4, 1, 2, 3, 5, 6},
 {3, 4, 1, 2, 6, 5}, {3, 4, 1, 2, 5, 6}, {2, 3, 4, 1, 6, 5},
 {2, 3, 4, 1, 5, 6}, {1, 2, 3, 4, 6, 5}, {1, 2, 3, 4, 5, 6}}
```

```
PermList3 == PermList4
```

```
True
```

It closed, so the group is complete.

16. Product groups

Preliminaries

16.1 An introductory example (group C_{3h})

Multiplication to closure is not the only way to construct groups. In this chapter we show how one can sometimes combine two groups already known to make a new, larger group. We begin with an example. Look at the tabulated Cartesian matrices that represent groups C_3 and C_h :

```
CartRep["Ch"] // MatrixList // Size[7]
```

$$\left\{\begin{pmatrix}1&0&0\\0&1&0\\0&0&1\end{pmatrix}, \begin{pmatrix}1&0&0\\0&1&0\\0&0&-1\end{pmatrix}\right\}$$

```
CartRep["C3"] // MatrixList // Size[7]
```

$$\left\{\begin{pmatrix}1&0&0\\0&1&0\\0&0&1\end{pmatrix}, \begin{pmatrix}-\frac{1}{2}&-\frac{\sqrt{3}}{2}&0\\\frac{\sqrt{3}}{2}&-\frac{1}{2}&0\\0&0&1\end{pmatrix}, \begin{pmatrix}-\frac{1}{2}&\frac{\sqrt{3}}{2}&0\\-\frac{\sqrt{3}}{2}&-\frac{1}{2}&0\\0&0&1\end{pmatrix}\right\}$$

Take the **Outer** product of C_h with C_3 and **Flatten** appropriately. So far, the operator ⊗ has been a merely symbolic multiplication of individual group elements, but now we extend its meaning to products of whole groups. In the preliminaries we define **MatListProduct[𝒜, ℬ]** to be a **Flatten**ed **Outer** product using **Dot**, with ⊗ as its "infix alias". The **Outer** product was used in Chapter 8 (MultiplicationTbls). This means you may call **MatList⸜Product[𝒜, ℬ]** as 𝒜⊗ ℬ.

```
ChxC3 = CartRep["Ch"] ⊗ CartRep["C3"];
% // MatrixList // Size[5]
```

$$\left\{\begin{pmatrix}1&0&0\\0&1&0\\0&0&1\end{pmatrix}, \begin{pmatrix}-\frac{1}{2}&-\frac{\sqrt{3}}{2}&0\\\frac{\sqrt{3}}{2}&-\frac{1}{2}&0\\0&0&1\end{pmatrix}, \begin{pmatrix}-\frac{1}{2}&\frac{\sqrt{3}}{2}&0\\-\frac{\sqrt{3}}{2}&-\frac{1}{2}&0\\0&0&1\end{pmatrix}, \begin{pmatrix}1&0&0\\0&1&0\\0&0&-1\end{pmatrix}, \begin{pmatrix}-\frac{1}{2}&-\frac{\sqrt{3}}{2}&0\\\frac{\sqrt{3}}{2}&-\frac{1}{2}&0\\0&0&-1\end{pmatrix}, \begin{pmatrix}-\frac{1}{2}&\frac{\sqrt{3}}{2}&0\\-\frac{\sqrt{3}}{2}&-\frac{1}{2}&0\\0&0&-1\end{pmatrix}\right\}$$

The result is a list of **6** matrices. The first three are just C_3 again; the second three are similar, but with a **-1** in the lower right. Is this group C_{3h}?

```
ChxC3 == CartRep["C3h"]
```

```
True
```


DOI 10.1007/b13137_16,

Yes. So group C_{3h} is the *direct product* of C_h and C_3 . The order of the product group is the product of the orders of the two parent groups. The structure of the **Outer** operator ensures that this is generally true. (Here 6 = 2 × 3.)

16.2 The product group theorem

Research mathematicians do not state theorems and then prove them. Instead, they start with a curious observation based on a concrete calculation. They do more calculations to uncover the origin of the curious behavior, and slowly a proof begins to emerge. When they finally have a general proof, then, as a last step, they state the theorem.

When they write the textbook, they leave out the curious observation and the hard work of exploring blind alleys. They just state the theorem, and then give the proof, leaving the impression that God sent it to them in a dream.

Eschewing this convention, we present the Product Group Theorem in a little more realistic way. We begin with the curious observation that the outer product of C_h and C_3 is group C_{3h}. Can every pair of groups make a product group? More precisely, given a group

$$\mathcal{A} = \{A_1, A_2, \ldots, A_m\}$$

and another group

$$\mathcal{B} = \{B_1, B_2, \ldots, B_n\},$$

is it always true that the set of *all ordered products* $A_m \otimes B_n$ is also a group?

We begin by examining the four axioms. Take the easy ones first :

Axiom 2: (Unit element)
Both groups contain E, so $E \otimes E = E$ will be present.

Axiom 4: (Associativity)
$\mathcal{A}$ and $\mathcal{B}$ are groups under ⊗, so ⊗ is associative.

Axiom 1: (Closure)
If the group consists only of products like $A_m \otimes B_n$, then the product $(A_m \otimes B_n) \otimes (A_r \otimes B_s)$ is by closure necessarily equal to some $A_t \otimes B_u$. But by associativity the product may be rewritten without parentheses as $A_m \otimes B_n \otimes A_r \otimes B_s$. Then **if** we can switch B_n and A_r we can rewrite the product as $(A_m \otimes A_r) \otimes (B_n \otimes B_s)$. Now $A_m \otimes A_r$ will be some A_t, and $B_n \otimes B_s$ will be some B_u, and the product would be of the required form, $A_t \otimes B_u$. So the closure axiom would be satisfied.

But the switch of $\mathbf{B_n}$ and $\mathbf{A_r}$ is not generally valid, and the answer to the question, as originally stated, must be *no*. But we can turn it into a *yes* by imposing the requirement that all the elements of $\mathcal{A}$ and $\mathcal{B}$ must commute. This must be stated prominently in the theorem, when we get to it. But there is one more axiom to check first.

Axiom 3: (Inverse)

Group $\mathcal{A}$ contains $\mathbf{A_i}$ and $\mathbf{A_i^{-1}}$, and group $\mathcal{B}$ contains $\mathbf{B_k}$ and $\mathbf{B_k^{-1}}$, so for every $\mathbf{A_i} \otimes \mathbf{B_k}$ in the product, there is also a product $\mathbf{A_i^{-1}} \otimes \mathbf{B_k^{-1}}$. If they are mutually inverse, they will yield the unit element when multiplied in either order, and the inverse axiom will be satisfied :

$$(\mathbf{A_i} \otimes \mathbf{B_k}) \otimes \left(\mathbf{A_i^{-1}} \otimes \mathbf{B_k^{-1}}\right) \text{ and } \left(\mathbf{A_i^{-1}} \otimes \mathbf{B_k^{-1}}\right) \otimes (\mathbf{A_i} \otimes \mathbf{B_k}) \text{ will be } \mathbf{E}.$$

There is no reason to think this is true as it stands. But we have already found that the two groups must commute, so we can reverse the second product in each expression;

$$(\mathbf{A_i} \otimes \mathbf{B_k}) \otimes \left(\mathbf{B_k^{-1}} \otimes \mathbf{A_i^{-1}}\right) \quad \text{or} \quad \left(\mathbf{A_i^{-1}} \otimes \mathbf{B_k^{-1}}\right) \otimes (\mathbf{B_k} \otimes \mathbf{A_i})$$

and then

$$\mathbf{A_i} \otimes \left(\mathbf{B_k} \otimes \mathbf{B_k^{-1}}\right) \otimes \mathbf{A_i^{-1}} \quad \text{or} \quad \mathbf{A_i^{-1}} \otimes \left(\mathbf{B_k^{-1}} \otimes \mathbf{B_k}\right) \otimes \mathbf{A_i}$$

Taking innermost first, both expressions multiply to an $\mathbf{E}$, so all four axioms are satisfied. Now, finally we know what the theorem should be :

If	$\mathcal{A}$ is a group and $\mathcal{B}$ is a group ,
and if	every element of $\mathcal{A}$ commutes with every element of $\mathcal{B}$
then	$\mathcal{A} \otimes \mathcal{B}$ is a group.

16.3 How to recognize product groups

The character tables of groups are arranged to help you spot product groups. For instance, look at

```
BoxedCharacterTable["C3h"]
```

C3h	1	1	1	1	1	1	← Class populations Note : ϵ = Exp[- 2 π i / 3]
	E	C3	C32	σh	S3	S35	↓ Basis functions
A′	1	1	1	1	1	1	{ $x^2 + y^2$, z^2 , Iz }
E′a	1	ϵ	ϵ *	1	ϵ	ϵ *	{ x + i y }
E′b	1	ϵ *	ϵ	1	ϵ *	ϵ	{ x - i y }
A″	1	1	1	-1	-1	-1	{ z }
E″a	1	ϵ	ϵ *	-1	-ϵ	-ϵ *	{ (x + i y) z }
E″b	1	ϵ *	ϵ	-1	-ϵ *	-ϵ	{ (x - i y) z }

There are an even number of classes, and the upper left quadrant of the table looks like the character table of group **C3**. The first class in the right half is **σh**, suggesting that the group is **Ch ⊗ C3**. Also, the class names in the right half are plausibly **σh** times the class names in the left half. But the ultimate test is

```
CartRep["Ch"] ⊗ CartRep["C3"] == CartRep["C3h"]
```

```
True
```

That clinches it. We did not even have to **Sort** to get a match. You can apply these ideas to any character table to spot product groups.

16.4 Products with Ch vs. products with Ci

The **Cnh** and **Dnh** groups contains a small surprise when **n** is even. We demonstrate using a **Dnh** group:

```
BoxedCharacterTable["D4h"]
```

D4h	1	2	1	2	2	1	2	1	2	2	← Class populations
	E	C4	C_2	C_2'	C_2''	inv	S4	σh	σv	σd	↓ Basis functions
A1g	1	1	1	1	1	1	1	1	1	1	{ $x^2 + y^2$, z^2 }
A2g	1	1	1	-1	-1	1	1	1	-1	-1	{ $x y (x^2 - y^2)$, Iz }
B1g	1	-1	1	1	-1	1	-1	1	1	-1	{ $x^2 - y^2$ }
B2g	1	-1	1	-1	1	1	-1	1	-1	1	{ x y }
Eg	2	0	-2	0	0	2	0	-2	0	0	{ {x z, y z} , {Ix, Iy} }
A1u	1	1	1	1	1	-1	-1	-1	-1	-1	{ $x y (x^2 - y^2) z$ }
A2u	1	1	1	-1	-1	-1	-1	-1	1	1	{ z }
B1u	1	-1	1	1	-1	-1	1	-1	-1	1	{ x y z }
B2u	1	-1	1	-1	1	-1	1	-1	1	-1	{ $(x^2 - y^2) z$ }
Eu	2	0	-2	0	0	-2	0	2	0	0	{ {x, y} }

By the ideas in the last section, it looks like $\mathbf{D_{4\,h}}$ is $\mathbf{C_i} \otimes \mathbf{D_4}$. Test it:

```
CartRep["Ci"] ⊗ CartRep["D4"] == CartRep["D4h"]
```

```
True
```

This same group can also be constructed as $\mathbf{C_h} \otimes \mathbf{D_4}$, but you have to **Sort** to match them up :

```
Sort[CartRep["Ch"] ⊗ CartRep["D4"]] ==
 Sort[CartRep["D4h"]]
```

```
True
```

> **On your own**
> Repeat all the above using a $\mathbf{C_4}$ instead of $\mathbf{D_4}$. Actually, it is true for any group $\mathbf{C_n}$, with **n** even. Try it with $\mathbf{C_6}$.

Also in the Platonic groups, both $\mathbf{C_i} \otimes \mathbf{T}$ and $\mathbf{C_h} \otimes \mathbf{T}$ give $\mathbf{T_h}$,
both $\mathbf{C_i} \otimes \mathbf{O}$ and $\mathbf{C_h} \otimes \mathbf{O}$ give $\mathbf{O_h}$,
and both $\mathbf{C_i} \otimes \mathbf{I}$ and $\mathbf{C_h} \otimes \mathbf{I}$ give $\mathbf{I_h}$.

> **On your own**
> The first two, for groups **T** and **O** are quite easy to prove, with the aid of our tables and operators. Do it.
>
> The third one, for group **I**, is not so easy. The statement is true, but only if the icosahedron is in the right orientation. To see the details, click here, or look in the End Notes of this chapter.

My guess is that Schoenflies himself constructed all these group as $\mathbf{C_h} \otimes \mathbf{C_n}$ or as $\mathbf{C_h} \otimes \mathbf{D_n}$ because he named them $\mathbf{C_{nh}}$ or $\mathbf{D_{nh}}$. But all modern tables are remarkably consistent in ordering the even-**n** groups as $\mathbf{C_i} \otimes \mathbf{C_n}$ or $\mathbf{C_i} \otimes \mathbf{D_n}$. Historians of mathematics, why is this? Do all modern tables trace back to a single post-Schoenflies author who for some reason preferred $\mathbf{C_i}$ over $\mathbf{C_h}$?

16.5 Some careful checking, and a scandal

16.5.1 Product groups with C_i and C_h

If the product group theorem is the governing authority for making product groups, we should be able to show that **Imat** from $\mathbf{C_i}$ and σ_h from $\mathbf{C_h}$ commute with every matrix.

$$\left\{\begin{pmatrix} a & b & c \\ d & e & f \\ g & h & i \end{pmatrix}.\text{Imat}\,,\ \text{Imat}.\begin{pmatrix} a & b & c \\ d & e & f \\ g & h & i \end{pmatrix}\right\} \text{ // MatrixList // Size[7]}$$

$$\left\{\begin{pmatrix} -a & -b & -c \\ -d & -e & -f \\ -g & -h & -i \end{pmatrix}, \begin{pmatrix} -a & -b & -c \\ -d & -e & -f \\ -g & -h & -i \end{pmatrix}\right\}$$

So **Imat** is a universal commuter, and group $\mathbf{C_i}$ can make a product group with any other group.

Surely, you may think, the same thing will be true for σ_h. Try it:

$$\left\{\begin{pmatrix} a & b & c \\ d & e & f \\ g & h & i \end{pmatrix}.\sigma_h\,,\ \sigma_h.\begin{pmatrix} a & b & c \\ d & e & f \\ g & h & i \end{pmatrix}\right\} \text{ // MatrixList // Size[7]}$$

$$\left\{\begin{pmatrix} a & b & -c \\ d & e & -f \\ g & h & -i \end{pmatrix}, \begin{pmatrix} a & b & c \\ d & e & f \\ -g & -h & -i \end{pmatrix}\right\}$$

This is a shock. Matrix σ_h is not a universal commuter! How can we get away with using it to make product groups, in violation of the theorem?

It just happens to work in the cases where we use it, because it is used only with groups $\mathbf{C_n}$ and $\mathbf{D_n}$, and they have a special matrix pattern that does commute with σ_h. It is

$$\left\{\begin{pmatrix} a & b & 0 \\ d & e & 0 \\ 0 & 0 & i \end{pmatrix}.\sigma_h\,,\ \sigma_h.\begin{pmatrix} a & b & 0 \\ d & e & 0 \\ 0 & 0 & i \end{pmatrix}\right\} \text{ // MatrixList // Size[7]}$$

$$\left\{\begin{pmatrix} a & b & 0 \\ d & e & 0 \\ 0 & 0 & -i \end{pmatrix}, \begin{pmatrix} a & b & 0 \\ d & e & 0 \\ 0 & 0 & -i \end{pmatrix}\right\}$$

For the truly shocking scandal, read on.

16.5.2 Product groups with C_v

The $\mathbf{C_{nv}}$ groups with odd **n** are not product groups, but those with even **n** are indeed product groups and are tabulated as $\mathbf{C_v \otimes C_n}$. Group $\mathbf{C_v}$ is $\{\mathbf{Emat}, \sigma_y\}$ so you might assume that matrix σ_y commutes with all rotations. But in group theory, never assume. Look at this:

$$\left\{\begin{pmatrix} a & b & 0 \\ d & e & 0 \\ 0 & 0 & i \end{pmatrix}.\sigma_y,\ \sigma_y.\begin{pmatrix} a & b & 0 \\ d & e & 0 \\ 0 & 0 & i \end{pmatrix}\right\} \text{ // MatrixList // Size[7]}$$

$$\left\{\begin{pmatrix} a & -b & 0 \\ d & -e & 0 \\ 0 & 0 & i \end{pmatrix},\ \begin{pmatrix} a & b & 0 \\ -d & -e & 0 \\ 0 & 0 & i \end{pmatrix}\right\}$$

This is *not* commutation, and there is no commonly seen matrix pattern that makes it OK anyway. So why can we construct product groups with it?

It is because of a variant of the product group theorem, which uses *pseudo-commuters*. Look at the way σ_y commutes with rotation matrices:

$$\begin{pmatrix} \text{Cos}[\alpha] & -\text{Sin}[\alpha] & 0 \\ \text{Sin}[\alpha] & \text{Cos}[\alpha] & 0 \\ 0 & 0 & 1 \end{pmatrix}.\sigma_y === \sigma_y.\begin{pmatrix} \text{Cos}[\alpha] & -\text{Sin}[\alpha] & 0 \\ \text{Sin}[\alpha] & \text{Cos}[\alpha] & 0 \\ 0 & 0 & 1 \end{pmatrix}$$

```
False
```

But we can turn this `False` into a `True` with a little trick; just negate the trigonometric angle when you swap the multiplication order :

$$\begin{pmatrix} \text{Cos}[\alpha] & -\text{Sin}[\alpha] & 0 \\ \text{Sin}[\alpha] & \text{Cos}[\alpha] & 0 \\ 0 & 0 & 1 \end{pmatrix}.\sigma_y === \sigma_y.\begin{pmatrix} \text{Cos}[-\alpha] & -\text{Sin}[-\alpha] & 0 \\ \text{Sin}[-\alpha] & \text{Cos}[-\alpha] & 0 \\ 0 & 0 & 1 \end{pmatrix}$$

```
True
```

This is what we mean by *pseudo-commutation*. It is not true commutation, but it is close enough that we do not abandon the word altogether. The product group proof for pseudocommuters is a great deal like that for commuters; so similar, in fact, that we relegate it to the End Notes, where it may be ignored by the many. But if you are one of the few, click here. The theorem it leads to is this:

If	$\mathcal{A}$ is a group and $\mathcal{B}$ is a group ,
and if	$\mathcal{A}$ is such that **A[-n α]** is the inverse of **A[+n α]**,
and if	$\mathbf{A[n\,\alpha] \otimes B_u == B_u \otimes A[-n\,\alpha]}$
then	the outer product $\mathcal{A} \otimes \mathcal{B}$ is a group.

Actually, $\mathbf{C_v} \otimes \mathbf{D_n}$ is an alternate construction for $\mathbf{D_{nh}}$.

```
Sort[CartRep["Cv"] ⊗ CartRep["D6"]] ==
 Sort[CartRep["D6h"]]
```

```
True
```

But don't try to call this group $\mathbf{D_{6\,v}}$. Our Great Father Schönflies did not use $\mathbf{D_{n,v}}$ at all, and no one will know what you are talking about.

16.6 End notes

16.6.1 Construction of Schönemann's I and $\mathrm{I_h}$ (numerical)

Group **I** may be generated by two rotations, one about the **z** axis and one in the **zx** plane at angle α to the **z** axis, where the **Sin** and **Cos** of α are

$$\{\cos\alpha,\ \sin\alpha\} = \left\{\sqrt{\frac{2}{5+\sqrt{5}}},\ \sqrt{\frac{1}{10}\left(5+\sqrt{5}\right)}\right\};$$

One axis must be a twofold and the other a fivefold. The standard convention says to make the **z** axis fivefold. However, the matrices generated this way do not make a product group with group $\mathbf{C_h}$. To see the product group you must make the **z** axis twofold. This is similar to the irregular choice for group **T**, in which **z** is a also twofold axis, ignoring the threefold axis on the diagonal of the enclosing cube.

Group **I** is easily generated from two rotation matrices, in either the nonstandard Schönemann orientation (**z** is a twofold axis), or in standard orientation (**z** is a fivefold axis). Below we do it in Schönemann orientation. For standard orientation, just interchange the axes. The big difference is that σ_h is a symmetry operation for the icosahedron in Schönemann's orientation, but not in standard orientation.

```
matA = AxialRotation[{0, 0, 1}, π] // N;
MatrixForm[%]
```

$$\begin{pmatrix} -1. & 0. & 0. \\ 0. & -1. & 0. \\ 0. & 0. & 1. \end{pmatrix}$$

```
matB = AxialRotation[{-sinα, 0, cosα}, 2 π / 5] // N;
MatrixForm[%]
```

$$\begin{pmatrix} 0.809017 & -0.5 & -0.309017 \\ 0.5 & 0.309017 & 0.809017 \\ -0.309017 & -0.809017 & 0.5 \end{pmatrix}$$

Then group **I** in Schönemann's orientation is given by

```
groupI = MakeGroup[{matA, matB}, Report → True];
Dimensions[groupI]
```

```
{60, 3, 3}
```

The next step is even quicker:

```
groupIh = (CartRep["Ch"] ⊗ groupI) // RoundSort ;
Dimensions[groupIh]
```

```
{120, 3, 3}
```

```
groupIi = (CartRep["Ci"] ⊗ groupI) // RoundSort ;
Dimensions[groupIi]
```

```
{120, 3, 3}
```

Is **groupIh** the same as **groupIi** ?

```
(groupIh - groupIi) // Chop // Union
```

```
{{{0, 0, 0}, {0, 0, 0}, {0, 0, 0}}}
```

Is it a group? We need to check only for closure. This takes a few seconds.

```
mtIh =
  MultiplicationTable[Dot, groupIh, Range[120]];
% // Dimensions
```

```
{120, 120}
```

The table is the right size, but were there any unrecognized products?

```
FreeQ[mtIh, "?"]
```

```
True
```

So using Schönemann's orientation for **I** the multiplication table is free of question marks. If **I** is generated in standard position, the table is mostly question marks.

16.6.2 Pseudocommutation theorem for $C_v \otimes C_n$

The product group theorem can also be proved under the assumption that group $\mathcal{A}$ is the rotation group $\mathbf{C_n}$ and group $\mathcal{B}$ is the vertical reflection group $\mathbf{C_v}$ given by **{Emat, σ_x}**. The elements of group $\mathcal{A}$ (namely, $\mathbf{C_n}$) are constructed by the formula

```
A[m_ , n_] := RotationMatrix3Dz[2 π m / n]
```

as **m** runs from **0** to **n-1**.

Lemma

Prove: Matrix **A[-m,n]** is in group $\mathbf{C_n}$, even though it is not used explicitly in the group construction.

The proof is simple. If **A[m,n]** is present by construction in group $\mathbf{C_n}$, it has a unique inverse that is also present somewhere in the group. But **A[m,n]** has only one inverse, and it is returned by **A[-m,n]**:

```
A[m, n].A[-m, n] // TrigExpand
```

```
{{1, 0, 0}, {0, 1, 0}, {0, 0, 1}}
```

so **A[-m,n]** is present in the group.

Main theorem

Now prove that the matrices generated by $\mathbf{C_n} \otimes \mathbf{C_v}$ form a group. We do only the hard parts:

Axiom 1 (Closure) for pseudocommutation:
If all products of the form **A[m, n]** $\otimes$ $\mathbf{B_q}$ form a group, then the product $\left(\mathbf{A[r, n]} \otimes \mathbf{B_q}\right) \otimes (\mathbf{A[s, n]} \otimes \mathbf{B_t})$ is by closure necessarily equal to some **A[u, n]** $\otimes$ $\mathbf{B_v}$. But by associativity the product may be rewritten without parentheses as **A[r, n]** $\otimes$ $\mathbf{B_q}$ $\otimes$ **A[s, n]** $\otimes$ $\mathbf{B_t}$. Then **if** $\mathbf{B_q}$ $\otimes$ **A[s, n]** is identical to **A[-s,n]** $\otimes$ $\mathbf{B_q}$ we can rewrite the product as

$$\mathbf{A[r, n]} \otimes \mathbf{A[-s, n]} \otimes \mathbf{B_q} \otimes \mathbf{B_t}.$$

Now (**A[r, n]** $\otimes$ **A[-s, n]**) will be some **A[u, n]**, and $\mathbf{B_q} \otimes \mathbf{B_s}$ will be some $\mathbf{B_v}$, so the product will be of the required form **A[u, n]** $\otimes$ $\mathbf{B_v}$.

Axiom 3 (Inverse axiom) for pseudocommutation:
Group $\mathcal{A}$ contains **A[m, n]** and its inverse **A[-m, n]**, and group $\mathcal{B}$ contains $\mathbf{B_j}$ and $\mathbf{B_j^{-1}}$, so for every **A[m, n]** $\otimes$ $\mathbf{B_j}$ in the product group, there is also a

product $\mathbf{A[m,\ n]} \otimes \mathbf{B}_j^{-1}$. The product of these two products may be taken two ways:

$$\left(\mathbf{A[m,\ n]} \otimes \mathbf{B}_j\right) \otimes \left(\mathbf{A[m,\ n]} \otimes \mathbf{B}_j^{-1}\right)$$

or

$$\left(\mathbf{A[m,\ n]} \otimes \mathbf{B}_j^{-1}\right) \otimes \left(\mathbf{A[m,\ n]} \otimes \mathbf{B}_j\right).$$

But we have already found the two groups must pseudo-commute, so we reverse and negate the angle of the second factor in each expression;

$$\left(\mathbf{A[m,\ n]} \otimes \mathbf{B}_j\right) \otimes \left(\mathbf{B}_j^{-1} \otimes \mathbf{A[-m,\ n]}\right) == \mathbf{A[m,\ n]} \otimes \mathbf{A[-m,\ n]} == \mathbf{E}$$

or

$$\left(\mathbf{A[m,\ n]} \otimes \mathbf{B}_j^{-1}\right) \otimes \left(\mathbf{B}_j \otimes \mathbf{A[-m,\ n]}\right) == \mathbf{A[m,\ n]} \otimes \mathbf{A[-m,\ n]} == \mathbf{E}$$

Therefore, every element in the product group possesses an inverse, and the Inverse Axiom is satisfied. Now all four axioms are satisfied. As was to be proved, the matrices generated by $\mathbf{C_n} \otimes \mathbf{C_v}$ form a group. Click back to the main text to reread the statement.

17. Naming the point groups

Preliminaries

17.1 Molecules and point groups

In Chapter 11 (EulerRotations) and Chapter 12 (LieRotations) we studied the construction of individual rotation and rotoreflection matrices that are the elements of all point groups, and in Chapter 13 (RecognizeMatrix) we made operators that recognize the matrix types automatically, and in Chapter 14 (MakeMatrixGroup) we used them to generate whole groups. Now we are ready to understand the rationale behind the naming of the Point Groups.

The point groups are all you need to analyze the symmetry of individual finite molecules, because molecules, like the point groups, also have a unique point, their center of mass. If a molecule has any symmetry, its center of mass will be the unique origin point where all its symmetry elements intersect. Therefore, all geometric symmetries of molecules are described by point groups.

There are infinitely many point groups; indeed, there are infinitely many axial rotation groups alone. They may be sorted out and given systematic names in several different ways, each way based on the minimal lists of elements needed to generate the groups.

This does not lead to a single unique best way to classify them and name them, because often a point group can be generated in several different ways. In the sections below, we study two different systems in common use, the Schönflies system (favored by chemists and spectroscopists) and the International system (favored by crystallographers and solid state physicists).

17.2 Schönflies names

The first satisfactory naming system for point groups was proposed by the German mathematician Schönflies. He provided a name for each possible point group, proving exhaustively that he had left no point group unnamed. We omit his massive proof (see Bibliography).

The names that Schönflies made up have taken root in chemistry, where the symmetries of molecules are invariably described by his names. Also, in molecular spectroscopy excited states are always named using the Schönflies

W.M. McClain, *Symmetry Theory in Molecular Physics with Mathematica*,
DOI 10.1007/b13137_17,

names. His system is not unique or compellingly perfect, but it is not going to go away. It can be explained quite concisely.

In every naming system for symmetry groups, one must first name the symmetry elements (the individual transform matrices). Here is the way Schönflies named them :

Schönflies Name	Symmetry Element
E (our **Emat**)	identity
i (our **Imat**)	inversion
C_n	n – fold rotation axis*
C'_2	2 – fold rotation axis, perpendicular to a higher C_n axis
σ_v	vertical mirror plane, containing the z and x axes
σ_h	horizontal mirror plane, containing the x and y axes
σ_d	dihedral** mirror plane
S_n (n even)	n – fold rotoreflection axis* (this name for even n)
C_{nh} (n odd)	n – fold rotoreflection axis* (this name for odd n)
i. C_n	n – fold rotoinversion axis (not used in the Schönflies system)

* The C_n or S_n axis of highest order, called the "main" axis, defines the "vertical".

** Dihedral means vertical, bisecting the angles between σ_v planes or C'_2 axes.

Once the symmetry elements are named, one may specify groups by giving short list of elements that generate the entire group. One then invents a name for the group by combining (in some attractive way) the names on the list. But there is no unique minimal list, so quite different names for the same group may arise in different naming systems.

Schönflies followed a systematic plan for all groups based on a single n-fold axis, plus mirrors and perpendicular twofold axes. But when he got to groups related to the Platonic solids (tetrahedron, cube, octahedron, dodecahedron, and icosahedron), he made up special shorter names.

Here is the way Schönflies picked the minimal list of generating elements, and the way that he collapsed the list of element names into a group name :

Generating Elements	Schönflies Group Name
i, the inversion	C_i
σ_v or σ_h, the reflection	C_v or C_h
S_n axis, n even*	$S_{n\,(n\,even)}$
C_n axis	C_n
$S_{n\,(n\,odd)}$ or (C_n and σ_h)	C_{nh}
C_n (n odd) and a σ_v plane	C_{nv}
C_n (n even), a σ_v plane, and a σ_d plane	C_{nv}
C_n axis and a $\perp$ C_2' axis	D_n
D_n elements, and a σ_h plane	D_{nh}
(1) D_n elements and a σ_d plane **or** **(2)** an S_{2n} axis and a σ_d plane	D_{nd}
tetrahedron, rotations only	T
cube or octohedron, rotations only	O
dodecahedon or icosahedron, rotations only	I or Y
rotations of a tetrahedron, plus inversion**	T_h
all the symmetries of a tetrahedron	T_d
all the symmetries of a cube or octohedron	O_h
all symmetries of a dodecahedron or icosahedron	I_h or Y_h

- * Except that S_2 is inversion, usually named as C_i

  ** The tetrahedron itself does not have this symmetry

In Chapter 18 (CartesianRepresentations) we will verify these names by using them to construct the groups.

17.3 International names

17.3.1 Rotoinversion

There is a very fundamental difference between the Schönflies and International naming systems. Where the Schönflies system makes exclusive use of the rotoREFLECTION axis, the International system makes exclusive use of the roto-INVERSION axis. This is the root source of some confusion that we will try to clear up.

We have shown that products of rotations and rotoreflections can never produce anything *fundamentally* new, but that word *fundamentally* betrays a small difficulty. Let's bring it into the open. One can make a symmetry element called the *rotoinversion* by combining a rotation with an inversion. The rotoinversion about the **z** axis is given by

```
Clear[iCz];
iCz[ϕ_] := Imat.Cz[ϕ];
```

Look at **z**-axis rotoinversion (**iCz**) and rotoreflection ($\mathbf{S_z}$) together :

```
{ iCz[ϕ], Sz[ϕ]} // MatrixList // Size[7]
```

$$\left\{\begin{pmatrix} -\text{Cos}[\phi] & \text{Sin}[\phi] & 0 \\ -\text{Sin}[\phi] & -\text{Cos}[\phi] & 0 \\ 0 & 0 & -1 \end{pmatrix}, \begin{pmatrix} \text{Cos}[\phi] & -\text{Sin}[\phi] & 0 \\ \text{Sin}[\phi] & \text{Cos}[\phi] & 0 \\ 0 & 0 & -1 \end{pmatrix}\right\}$$

Clearly they are different, but is the rotoinversion a "fundamentally" new kind of element? Not really. With a little thought, it is easy to see what it is :

```
{iCz[ϕ], Sz[ϕ + π]} // MatrixList // Size[7]
```

$$\left\{\begin{pmatrix} -\text{Cos}[\phi] & \text{Sin}[\phi] & 0 \\ -\text{Sin}[\phi] & -\text{Cos}[\phi] & 0 \\ 0 & 0 & -1 \end{pmatrix}, \begin{pmatrix} -\text{Cos}[\phi] & \text{Sin}[\phi] & 0 \\ -\text{Sin}[\phi] & -\text{Cos}[\phi] & 0 \\ 0 & 0 & -1 \end{pmatrix}\right\}$$

So **rotoinversion** about **z** by ϕ is the same **rotoreflection** about **z** by $\phi + \pi$. The isotropy of space assures us that this will be the same in any axial direction. But this kind of reasoning is a little too grand for some people, and we check it out directly for an arbitrary axis, **{a,b,c}**. Let **iC** be rotoinversion

```
iC[ϕ_] := AxialRotation[{a, b, c}, ϕ].Imat
```

and let **σC** be rotoreflection :

```
σC[ϕ_] := AxialRotation[{a, b, c}, ϕ].
  AxialReflection[{a, b, c}]
```

If the theorem is true for arbitrary axis, we will see zeroes below :

```
(iC[ϕ] - σC[ϕ + π]) // Simplify
{{0, 0, 0}, {0, 0, 0}, {0, 0, 0}}
```

As Mussolini used to say, "I think I may assert, without fear of contradiction, that ..."

> Rotoinversion by ϕ about any axis $\underset{\rightharpoondown}{\mathbf{a}}$ is identical to rotoreflection about $\underset{\rightharpoondown}{\mathbf{a}}$ by $\phi + \pi$.

The table below summarizes this discussion. We just add another entry to a table you have seen before :

Special Case	General Case
unit	rotation[0]
reflection	rotoreflection[0]
inversion	rotoreflection[π]
rotoinversion[ϕ]	rotoreflection[$\phi + \pi$]

17.3.2 International naming system

The International system (or Hermann-Mauguin system) is very frugal in naming symmetry elements. There is no indication within the element name itself as to whether it is vertical, horizontal, or canted at some angle. An **n**-fold rotation axis is indicated by the integer **n**. With a bar over it, it indicates an **n**-fold rotoinversion axis. A rotation with a perpendicular mirror is indicated as **n/m** or $\frac{\mathbf{n}}{\mathbf{m}}$.

International Name	Symmetry Element
n	n – fold rotation axis
$\bar{\text{n}}$	n – fold rotoinversion axis*
m	mirror plane
$\frac{\mathbf{n}}{\mathbf{m}}$ or n / m	n –fold rotation and $\perp$ mirror

*The inversion bar, if it appears, always appears over the first rotation axis in the group name, making it into a rotoinversion axis. No group needs more than one rotoinversion axis, just as Schönflies never used more than one rotoreflection axis (S) in a group name.

The group names in the International system are very subtle. The first axis named may be thought of as vertical. In the table below the "m" is a literal m standing for "mirror", but "*n*" is an integer.

Generating Elements	International Name
i, the inversion	$\bar{1}$
n – fold rotation axis	n
n – fold rotoinversion axis	$\bar{n}$
n – fold vertical axis and a horizontal mirror*	$\frac{n}{m}$ or n/m
n – fold vertical rotation and a vertical mirror plan	$n\,m$
n – fold vertical rotation and two vertical mirrors	nmm
n – fold rotation and a $\perp$ 2 – fold rotation	$n2$ (with $n > 2$)
vertical 2 – fold and canted 3 – fold	23

* two elements, NOT combined to make a single rotoreflection

17.4 Schönflies-International correspondences

Usually these correspondences are important only when dealing with crystals. It may be proved that a 3D crystal cannot form if its unit cell has rotations or rotoreflections of order other than 1, 2, 3, 4, or 6. This excludes many groups; indeed, it excludes all but 32 of them, which are known as the "crystal point groups" Here they are, with Schönflies names on the red stripes and the Hermann-Mauguin (or International) names directly below on the blue stripes.

```
CrystalPointGroupsTable
```

triclinic	C_1	C_i					
	1	$\bar{1}$					
monoclinic	C_h	C_2	C_{2h}				
	m	2	2/m				
rhombic	C_{2v}	D_2	D_{2h}				
	mm	222	mmm				
trigonal	C_3	S_6	C_{3v}	D_3	D_{3h}		
	3	$\bar{3}$	3m	32	$\bar{6}$m2		
tetragonal	C_4	S_4	C_{4h}	C_{4v}	D_{2d}	D_4	D_{4h}
	4	$\bar{4}$	4/m	4mm	$\bar{4}$2m	422	4/mmm
hexagonal	D_{3d}	C_{3h}	C_6	C_{6h}	C_{6v}	D_6	D_{6h}
	$\bar{3}$m	$\bar{6}$	6	6/m	6mm	622	6/mmm
cubic	T	T_h	T_d	O	O_h		
	23	m3	$\bar{4}$3m	432	m3m		

Fig. 17.1 Proposed flag for the International Crystallographic Union. Schönflies names are on the red background; the Hermann-Maugin names are on blue.

It may seem odd to you that $D_{4\,d}$ and $D_{6\,d}$ are missing. But take a look at the third symmetry elements from each of these groups:

```
RecognizeMatrix[CartRep["D4d"][[3]]]
```

$\left\{\text{rotoreflection, 8-fold, }\left\{\{0,\ 0,\ -1\},\ \frac{\pi}{4}\right\}\right\}$

```
RecognizeMatrix[CartRep["D6d"][[3]]]
```

$\left\{\text{rotoreflection, 12-fold, }\left\{\{0,\ 0,\ -1\},\ \frac{\pi}{6}\right\}\right\}$

Because 8-fold and 12-fold rotoreflection axes cannot fit into a crystal lattice, these groups are excluded, in spite of their tempting Schönflies names. The International names begin with the axis of highest symmetry, so $D_{4\,d}$ would start with $\bar{8}$ and $D_{6\,d}$ would start with $\overline{12}$, making it obvious they could not be crystal groups. Maybe this is one reason that crystallographers like that system.

17.5 Some interesting name correspondences

17.5.1 Group names C2v and mm

The **CrystalPointGroupsTable** says that **C2v** is the same as **mm**. This sounds a little odd, because the Schönflies name says it comes from a rotation and a reflection, whereas the International name says it comes from two perpendicular reflections. Let's make sure there is no contradiction.

```
groupC2v = MakeGroup[{Cz[2 π / 2], σx}];
groupmm = MakeGroup[{σx, σy}];
groupC2v == groupmm
```

True

All is well.

17.5.2 Group names C6v and 6mm

The Schönflies name **C6v** implies a vertical **C6** axis and vertical mirror plane. Our orientation convention requires that it be σ_x.

```
groupC6v = MakeGroup[{Cz[2 π / 6], σx}];
Map[RecognizeMatrix, groupC6v] // Sort // Column
```

{identity}
{reflection, {0, 1, 0}}
$\{\text{reflection}, \{\frac{1}{2}, -\frac{\sqrt{3}}{2}, 0\}\}$
$\{\text{reflection}, \{\frac{1}{2}, \frac{\sqrt{3}}{2}, 0\}\}$
{reflection, {1, 0, 0}}
$\{\text{reflection}, \{\frac{\sqrt{3}}{2}, -\frac{1}{2}, 0\}\}$
$\{\text{reflection}, \{\frac{\sqrt{3}}{2}, \frac{1}{2}, 0\}\}$
{rotation, 2-fold, {{0, 0, 1}, π}}
$\{\text{rotation, 3-fold}, \{\{0, 0, -1\}, \frac{2\pi}{3}\}\}$
$\{\text{rotation, 3-fold}, \{\{0, 0, 1\}, \frac{2\pi}{3}\}\}$
$\{\text{rotation, 6-fold}, \{\{0, 0, -1\}, \frac{\pi}{3}\}\}$
$\{\text{rotation, 6-fold}, \{\{0, 0, 1\}, \frac{\pi}{3}\}\}$

Now we can check that this is the same as **6mm**; but wait- a vertical C_6 axis and vertical mirror plane is **6m** , which we just did above. The second **m** has to contain the **6**-fold axis and be perpendicular to the first **m**, which means that it has to be σ_y. Indeed, σ_y (i.e., **{reflection,{0,1,0}}**) came out automatically as one of the group members, so we would get the same group if we included it in the original short list. But why do so?

This is a mystery known only to the old International Commission, long since gone to its eternal reward. Obviously the name **6mm** generates group **C6v**, so it is formally correct; but it could be shortened to **6m**. Still, **6mm** is official, and we stick with it.

17.5.3 Group names D2h and mmm

The Schönflies name says this group is generated by a vertical **C2**, a perpendicular **C2**, and a σ_h. We generate it :

```
Cz[π]
```

```
{{-1, 0, 0}, {0, -1, 0}, {0, 0, 1}}
```

```
D2hCalc =
  MakeGroup[{Cz[π], Cy[π], σz}, Report → True];
MatrixList[%] // Size[7]
```

$$\left\{\begin{pmatrix}-1&0&0\\0&-1&0\\0&0&-1\end{pmatrix},\begin{pmatrix}-1&0&0\\0&-1&0\\0&0&1\end{pmatrix},\begin{pmatrix}-1&0&0\\0&1&0\\0&0&-1\end{pmatrix},\begin{pmatrix}-1&0&0\\0&1&0\\0&0&1\end{pmatrix},\begin{pmatrix}1&0&0\\0&-1&0\\0&0&-1\end{pmatrix},\begin{pmatrix}1&0&0\\0&-1&0\\0&0&1\end{pmatrix},\begin{pmatrix}1&0&0\\0&1&0\\0&0&-1\end{pmatrix},\begin{pmatrix}1&0&0\\0&1&0\\0&0&1\end{pmatrix}\right\}$$

The International name **mmm** says it is generated by three perpendicular mirrors, which are certainly there. We just check that **mmm** actually does the whole job:

```
mmmGroup = MakeGroup[{σx, σy, σz}, Report → True];
MatrixList[%] // Size[7]
```

$$\left\{\begin{pmatrix}-1&0&0\\0&-1&0\\0&0&-1\end{pmatrix},\begin{pmatrix}-1&0&0\\0&-1&0\\0&0&1\end{pmatrix},\begin{pmatrix}-1&0&0\\0&1&0\\0&0&-1\end{pmatrix},\begin{pmatrix}-1&0&0\\0&1&0\\0&0&1\end{pmatrix},\begin{pmatrix}1&0&0\\0&-1&0\\0&0&-1\end{pmatrix},\begin{pmatrix}1&0&0\\0&-1&0\\0&0&1\end{pmatrix},\begin{pmatrix}1&0&0\\0&1&0\\0&0&-1\end{pmatrix},\begin{pmatrix}1&0&0\\0&1&0\\0&0&1\end{pmatrix}\right\}$$

```
D2hCalc == mmmGroup
```

```
True
```

All is well. The name **mmm** cannot be shortened, because **mm** is **C2v**, as we saw above.

18. Tabulated representations of groups

Preliminaries

18.1 Cartesian representation tables

18.1.1 Introduction

In this chapter we discuss the package operator **CartesianRepresentationTable**, or **CartRep** for short, first defined back in Chapter 8 (MultiplicationTables). All the group names returned by **GroupCatalog** have a **CartRep** kept in the **Symmetry`** package:

```
GroupCatalog
```

```
{C1, C2, C2h, C2v, C3, C3h, C3v, C4, C4h, C4v, C5, C6, C6h,
 C6v, Ch, Ci, D2, D2d, D2h, D3, D3d, D3h, D4, D4d, D4h, D5,
 D5d, D5h, D6, D6d, D6h, I, O, Oh, S4, S6, S8, T, Td, Th}
```

The official group names are written with subscripts raised to the main line, and must be enclosed in quotes when fed to an operator. These 41 point groups include all 32 groups that describe the unit cells of crystals, plus other symmetries possessed by molecules, but which cannot serve as a unit cell symmetries.

In this chapter we present operators that construct from scratch the groups $\mathbf{C_n}$, $\mathbf{D_n}$, $\mathbf{D_{nd}}$, $\mathbf{S_{n\,(n\,even)}}$, **T**, **O**, and **I**. The other groups may be constructed as product groups; indeed, their names are based on this fact. In detail,

$$\mathbf{C_{nh}} = \mathbf{C_n} \otimes \mathbf{C_h} \text{ or } \mathbf{C_n} \otimes \mathbf{C_i}$$

$$\mathbf{C_{nv}} = \mathbf{C_n} \otimes \mathbf{C_2'}$$

$$\mathbf{D_{nh}} = \mathbf{D_n} \otimes \mathbf{C_h} \text{ or } \mathbf{D_n} \otimes \mathbf{C_i}$$

$$\mathbf{I_h} = \mathbf{I} \otimes \mathbf{C_h} \text{ or } \mathbf{I} \otimes \mathbf{C_i}$$

$$\mathbf{O_h} = \mathbf{O} \otimes \mathbf{C_h} \text{ or } \mathbf{O} \otimes \mathbf{C_i}$$

$$\mathbf{T_h} = \mathbf{T} \otimes \mathbf{C_h}$$

$$\mathbf{T_d} = \mathbf{T} \otimes \mathbf{C_i}$$

Our group-making operators return their results in infinite precision for all the tabulated groups. But usually they will not work for **n** higher than **6**, because (with some exceptions) *Mathematica* does not evaluate expressions like $\mathbf{C_z[2\,\pi\,/\,n]}$ in infinite precision for **n** higher than **6**. But if you just include a decimal after the fold-number **n** (like $\mathbf{C_z[2\,\pi\,/\,7.]}$) all operators return a correct answer in machine precision.

W.M. McClain, *Symmetry Theory in Molecular Physics with Mathematica*,
DOI 10.1007/b13137_18, © Springer Science+Business Media, LLC 2009

Most of the tabulated representations are generated automatically in infinite precision from a few basic matrices every time you load the **Symmetry`** package. (Infinite precision groups with five-fold axes are an exception, for trivial reasons, discussed in the End Notes.) The formulas are based on " Cayley diagrams" for each group, constructed by a method explained in Duffey. They do not contain errors: any typo in the basic matrices or in the Cayley formulas would produce massive errors in the **MultiplicationTable** of the group, and these are not seen.

The Cayley method is fast but not very informative. In this chapter, we follow the Schönflies classification scheme of Chapter 17 (Naming Groups), as made more specific by each group's **Presentation** information, brought up by commands like

```
Presentation["O"]
```

```
a.a.a.a == b.b.b == a.b.a.b.a.b.a.b == e
```

This example says that group **O** is the closure of any two matrices, **a** and **b** that obey the relations shown. The first, **a.a.a.a==e**, says that **a** is a four-fold rotation, while **b.b.b==e** says that **b** is a three-fold rotation. The angle between the **a** and **b** axes is implicit in the relation **(a.b).(a.b).(a.b).(a.b)==e**. This is a beautiful orientation-free way of specifying the group **O** uniquely.

Our concrete presentation matrices will agree with the orientation rules (next Section). We then feed them to **MakeGroup**, developed in Chapter 15. Then we check that the generated group is identical to the tabulated group. This is our basic check on the correctness of the tables.

After you understand the construction scheme, you may wonder why we go to the trouble of tabulating all the representations. Why not just give people the tools, and let them do it on their own? It is because there is no unique or systematic way to order the elements of the groups, and the ordering is important to many uses of the group representations. The ordering can be established only by standard tables.

You may not want to read this whole chapter at one sitting because it is long and systematic, intended as a reference on the construction of any point group you may become interested in, tabulated or not. We cannot go above six-fold axes in infinite precision, and there are times when you may want higher groups. This chapter is a manual on how to construct them in machine precision.

Read until you get the idea, then have a look at the Platonic groups (which are a little different), then skip on to the next chapter. You can come back to any particular group whenever you need to.

18.1.2 Orientation conventions

Cartesian transform matrices must refer to absolute orientations in space, and the orientations must be chosen by convention. Once and for all, here they are :

1. The horizontal plane is the **x,y** plane. Planar molecules always lie in this plane.

2. If there is a unique axis of highest symmetry, it is the **z** axis.

3. When there are rotation axes in addition to **z**, one must lie in the **z,x** plane.

4. When there are vertical reflection planes (i.e., planes containing the **z** axis), one must be the **z,x** plane. But if this conflicts with rule 3 (as in the D_{nd} groups), then rule **3** takes precedence.

5. In the Platonic groups, there is no unique highest axis.
(a) In the **T** groups, **z** is two-fold and **{1,1,1}** is three-fold.
(b) In the **O** groups, **z** and **x** are fourfold axes.
(c) The **I** groups take **z** as two-fold, plus a five-fold in the **z,x** plane, at angle α to the **z** axis such that

$$\text{Sin}[\alpha] = \sqrt{\frac{2}{5+\sqrt{5}}} \quad \text{and} \quad \text{Cos}[\alpha] = \sqrt{\frac{1}{2} + \frac{1}{2\sqrt{5}}}\,.$$

Nearly everyone agrees on the first two conventions. Agreement on the others is not so universal, but we will try to stick to them in this book.

Molecule definitions also have to follow these rules if you want to use the tools of this book on them. For instance, never define a planar molecule to lie in the **z,x** plane or the **y,z** plane. Planar molecules must obey the group (or subgroup) C_h, which uses the **x,y** plane. This is why we do not include any group C_v among our tabulated groups; by these conventions no molecule can belong to C_v.

18.1.3 An example of a Cartesian Representation

The command **CartesianRepresentation["gpNm"]**, where **"gpNm"** is any group name from **GroupCatalog**, brings up the entire matrix group. For instance,

```
CartesianRepresentation["C3"] // GridList
```

$$\left\{\begin{pmatrix}1&0&0\\0&1&0\\0&0&1\end{pmatrix},\ \begin{pmatrix}-\frac{1}{2}&-\frac{\sqrt{3}}{2}&0\\\frac{\sqrt{3}}{2}&-\frac{1}{2}&0\\0&0&1\end{pmatrix},\ \begin{pmatrix}-\frac{1}{2}&\frac{\sqrt{3}}{2}&0\\-\frac{\sqrt{3}}{2}&-\frac{1}{2}&0\\0&0&1\end{pmatrix}\right\}$$

The abbreviation **CartRep**, which we usually use, does exactly the same job.

These are 3×3 Cartesian transform matrices that "represent" the group (obey its multiplication table). Any one of them can be dotted into a Cartesian point **{x,y,z}** on its right, producing the new position of the point after the symmetry operation. Therefore, as will be fully explained in Chapter 25 (Representations), the **CartReps** are "**{x,y,z}** representations".

18.1.4 Auxiliary operators

There are several auxiliary operators that help you use the **CartReps**. First, every matrix of every group has a specific name:

```
ElementNames["C3"] // FullForm
```

```
List["E", "C3", "C32"]
```

The names are strings, and you must not omit the quotes when you feed them to an operator. The names are in the same order as the matrices returned by **CartRep**.

It is often easier to specify operations using matrix names rather than the matrices themselves. Here is pseudo-symbolic multiplication of matrices **C3** and **C32** from group C_3:

```
"C3"."C32" /. NamesToRepMats["C3"] /.
 RepMatsToNames["C3"]
```

```
E
```

The thumbnail sketch of **NamesToRepMats** is

- **NamesToRepMats["g"]** returns rules that replace the name of each element of group **"g"** by its Cartesian representation matrix. The inverse rules are given by **RepMatsToNames["g"]**.

18.2 Very simple groups

18.2.1 One element

For completeness, we begin with group $\mathbf{C_1}$, which consists only of the identity matrix **E**. Even the most unsymmetric objects remain unchanged (are symmetric) under **E**. Issue **CartRep["C1"]** to see the actual matrix.

18.2.2 Two-element groups

Next in simplicity come the two-element groups $\mathbf{C_i}$, $\mathbf{C_2}$, $\mathbf{C_h}$, and $\mathbf{C_v}$. They are all isomorphic to each other, so to an abstract group theorist they are all the same. But geometrically, they are very different.

A two-element group can have only elements **E** and **X**. In its multiplication table the entries $\begin{pmatrix} \mathbf{E} & \mathbf{X} \\ \mathbf{X} & \square \end{pmatrix}$ are fixed by the format of the table. The fourth entry must be either **E** or **X**. If it were **X** the table would be $\begin{pmatrix} \mathbf{E} & \mathbf{X} \\ \mathbf{X} & \mathbf{X} \end{pmatrix}$, which is not a Latin square. So it must be $\begin{pmatrix} \mathbf{E} & \mathbf{X} \\ \mathbf{X} & \mathbf{E} \end{pmatrix}$.

1. Group $\mathbf{C_i}$ contains only the unit **Emat** and the inversion matrix **Imat**. Issue **CartRep["Ci"]** to see the matrices. The inversion is not used in the Schönflies system of naming, but is used extensively in International (every axis with a bar over it is a rotation times in inversion.) There are simple molecules that belong to this group and to no higher group; for example, here is **C2H2Br2Cl2staggered**.

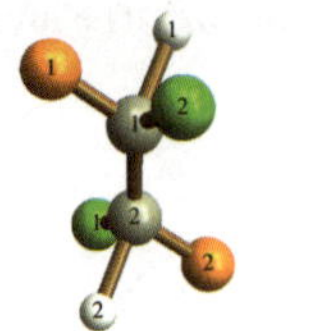

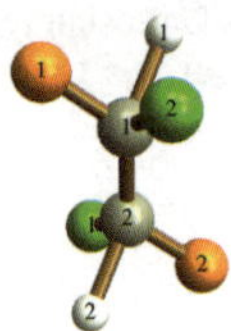

Fig. 18.1 Stereo pair for a molecule that possesses only inversion symmetry

2. Group $\mathbf{C_2}$ contains only **E** and a $\mathbf{C_2}$ axis, which by convention must be the **z** axis. Issue **CartRep["C2"]** to see the matrices. Many protein and nucleic acid dimers have only a $\mathbf{C_2}$ axis. The smallest molecule with *only* a $\mathbf{C_2}$ axis is a deuterated borane **transHB3H2D2**:

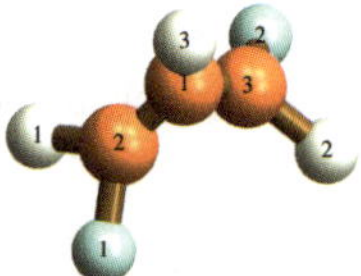

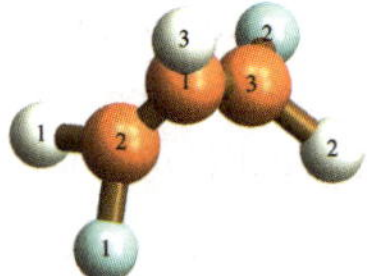

Fig. 18.2 Stereo pair for a molecule that possesses only a twofold rotation

4. Group $\mathbf{C_h}$ has a horizontal reflecting plane. Issue **CartRep["Ch"]** to see the matrices. All triatomic molecules **ABC** with three different atoms belong to $\mathbf{C_h}$, and to no higher group. The orientation rules say all three atoms should be in the horizontal plane.

5. Group $\mathbf{C_v}$ has only a vertical reflection plane. This group is not in the catalog because the vertical reflection plane can be at any angle θ to the **z,x** plane, and this causes trouble with some of our auxiliary operators. If a reflection plane contains the point **{Cos[θ],Sin[θ],0}** then its reflection matrix is

Cos[2 θ]	Sin[2 θ]	0
Sin[2 θ]	-Cos[2 θ]	0
0	0	1

On your own
Prove this using the **ReflectionMatrix** operator.

18.3 Groups with a principal axis

18.3.1 Introduction

In the next three sections we enumerate all the point group types that have a unique axis of highest symmetry. This axis may be either a rotation or rotoreflection. Only a limited number of other elements may be added to the principal axis to make a generator list for a finite point group. All possible cases are listed systematically below.

18.3.2 Groups C_n, C_{nh}, and C_{nv}

Groups C_n

As always, look first at the presentation:

```
Presentation["C3"]
```

```
a.a.a == e
```

This says that C_3 is generated by a list containing only one matrix, a three-fold rotation. Rule 2 says it has to be the **z** axis. Generalizing, C_n needs an **n**-fold rotation about **z** :

```
MakeCn[int_] :=
 Module[{op},
  If[int == 5, op = PentaDot, op = Dot];
  MakeGroup[op, {Cz[2 π / int]}]]
```

The five-fold **z** axis presents a difficulty in infinite precision calculations, so case **5** alone uses a special operator called **PentaDot**, defined in the **Symmetry`** package. It is the usual **Dot** operator with some special simplifications for five-fold **z** axes. Click to the End Notes for a full discussion. The three-fold rotation group is the briefest:

```
groupC3 = MakeCn[3];
% // MatrixList // Size[6]
```

$$\left\{\begin{pmatrix} -\frac{1}{2} & -\frac{\sqrt{3}}{2} & 0 \\ \frac{\sqrt{3}}{2} & -\frac{1}{2} & 0 \\ 0 & 0 & 1 \end{pmatrix}, \begin{pmatrix} -\frac{1}{2} & \frac{\sqrt{3}}{2} & 0 \\ -\frac{\sqrt{3}}{2} & -\frac{1}{2} & 0 \\ 0 & 0 & 1 \end{pmatrix}, \begin{pmatrix} 1 & 0 & 0 \\ 0 & 1 & 0 \\ 0 & 0 & 1 \end{pmatrix}\right\}$$

Is this the same as the tabulated group? The ordering of the matrices is unlikely to be the same, so always **Sort** before asking the equality question.

```
Sort[groupC3] == Sort[CartRep["C3"]]
```

```
True
```

Make sure that the special case **5** works smoothly. Look at just the last two matrices:

```
groupC5 = MakeCn[5];
Take[% // Sort, -2] // GridList // Size[5]
```

$$\left\{\begin{pmatrix} \frac{1}{4}\left(-1+\sqrt{5}\right) & -\frac{1}{2}\sqrt{\frac{1}{2}\left(5+\sqrt{5}\right)} & 0 \\ \frac{1}{2}\sqrt{\frac{1}{2}\left(5+\sqrt{5}\right)} & \frac{1}{4}\left(-1+\sqrt{5}\right) & 0 \\ 0 & 0 & 1 \end{pmatrix}, \begin{pmatrix} \frac{1}{4}\left(-1+\sqrt{5}\right) & \frac{1}{2}\sqrt{\frac{1}{2}\left(5+\sqrt{5}\right)} & 0 \\ -\frac{1}{2}\sqrt{\frac{1}{2}\left(5+\sqrt{5}\right)} & \frac{1}{4}\left(-1+\sqrt{5}\right) & 0 \\ 0 & 0 & 1 \end{pmatrix}\right\}$$

Those radicals under radicals are the source of the difficulty. But the generated group is correct:

```
Sort[groupC5] == Sort[CartRep["C5"]]
```

```
True
```

Groups C_{nh}

The $\mathbf{C_{nh}}$ groups are product groups $\mathbf{C_n} \otimes \mathbf{C_h}$, so they are just rotations with a horizontal mirror. The presentation is

```
Presentation["C3h"]
```

```
a.a.a == b.b == a.b.Inverse[a].b == e
```

By Rule 1, matrix b must be the horizontal mirror σ_{xy}. Check the presentation :

```
aMat = RotationMatrix3Dz[2 π / 3];
bMat = σxy;
aInv = Inverse[aMat];
aMat.bMat.aInv.bMat == Emat
```

```
True
```

It works. So the general operator for these groups is

```
MakeCnh[int_] :=
```

```
  Module[{op},
   If[int == 5, op = PentaDot, op = Dot];
   MakeGroup[op, {Cz[2 π / int], σxy}]]
```

Check it :

```
groupC3h = MakeCnh[3];
Sort[groupC3h] == Sort[CartRep["C3h"]]
```

```
True
```

Groups C_{nv}

The Schönflies name C_{nv} implies a product group $C_n \otimes C_v$, as detailed previously in Chapter 16, ProductGroups. The presentation confirms this :

```
Presentation["C3v"]
```

```
a.a.a == b.b == a.b.a.b == e
```

By Rules 2 and 4, a is a **z**-rotation and b is σ_{zx}.

```
MakeCnv[int_] :=
  Module[{op},
   If[int == 5, op = PentaDot, op = Dot];
   MakeGroup[op, {Cz[2 π / int], σzx}]]
```

Check it :

```
groupC3v = MakeCnv[3];
Sort[groupC3v] == Sort[CartRep["C3v"]]
```

```
True
```

18.3.3 The S_n groups (n even)

The axial rotoreflection groups S_n are based on powers of a roto-reflection (NOT a roto-inversion) about the **z** axis. The only subtlety is that the name S_n is used only for even **n**. You will see why.

```
Presentation["S4"]
```

```
a.a.a.a == e
```

The presentation agrees.

```
Clear[MakeSn];
MakeSn[int_] :=
```

```
MakeGroup[Dot,
   {AxialRotoReflection[{0, 0, 1}, 2 π / int]}] /;
  EvenQ[ChopInteger[int]]
```

That trailing fragment **/;EvenQ[ChopInteger[int]]** is a **Condition** prevents evaluation of the definition unless it is `True`. Try the operator:

```
Sort[MakeSn[4]] == Sort[ CartRep["S4"]]
```

```
True
```

To understand the restriction to even integers, look at the group generated by

```
groupS3 = MakeGroup[Dot,
    {AxialRotoReflection[{0, 0, 1}, 2 π / 3]}];
% // GridList // Size[6]
```

$$\left\{\begin{pmatrix} -\frac{1}{2} & -\frac{\sqrt{3}}{2} & 0 \\ \frac{\sqrt{3}}{2} & -\frac{1}{2} & 0 \\ 0 & 0 & -1 \end{pmatrix}, \begin{pmatrix} -\frac{1}{2} & -\frac{\sqrt{3}}{2} & 0 \\ \frac{\sqrt{3}}{2} & -\frac{1}{2} & 0 \\ 0 & 0 & 1 \end{pmatrix}, \begin{pmatrix} -\frac{1}{2} & \frac{\sqrt{3}}{2} & 0 \\ -\frac{\sqrt{3}}{2} & -\frac{1}{2} & 0 \\ 0 & 0 & -1 \end{pmatrix}, \begin{pmatrix} -\frac{1}{2} & \frac{\sqrt{3}}{2} & 0 \\ -\frac{\sqrt{3}}{2} & -\frac{1}{2} & 0 \\ 0 & 0 & 1 \end{pmatrix}, \begin{pmatrix} 1 & 0 & 0 \\ 0 & 1 & 0 \\ 0 & 0 & -1 \end{pmatrix}, \begin{pmatrix} 1 & 0 & 0 \\ 0 & 1 & 0 \\ 0 & 0 & 1 \end{pmatrix}\right\}$$

It is based on the rotoreflection by a third of a turn, but the cycle has six members! Furthermore, half the matrices have **-1** in position **3,3**. That makes it look like a product with group $\mathbf{C_h}$. Is it?

```
Sort[groupS3] == Sort[CartRep["C3"] ⊗ CartRep["Ch"]]
```

```
True
```

Indeed it is. Schönflies used product group names as much as possible, (product group calculations are so much quicker if you work by hand) so instead of calling it $\mathbf{S_3}$, he called it $\mathbf{C_{3\,h}}$.

On your own

This operator happens to work in infinite precision for **8** and **12**. Try it.

Make a big numerical S_n group and **Map** the operator **RecognizeMatrix** on it. Note that rotation matrices alternate with rotoreflection matrices.

Then try an odd integer, as in **MakeS$_n$[3]**. You will see it refuse to evaluate because the **Condition** is not **True**.

Then try a higher even integer, as in **MakeS$_n$[10.]**. The decimal tells it to work numerically, so you will see a numeric version of group S_{10}.

18.3.4 Groups D_n, D_{nh}, and D_{nd}

Groups D_n

The D_n groups are always pure rotation groups, with an **n**-fold rotation around the principal axis and another **n** two-fold rotations about horizontal axes.

```
Presentation["D3"]
```

```
a.a.a == b.b == a.b.a.b == e
```

By Rule 2, a is a three-fold rotation about **z**. By Rule 3, the b axis must lie in the **z,x** plane, but where? You could work it our from a.b.a.b == e, but you know from the **D** in the name that it must be the **x** axis.

```
Clear[MakeDn];
MakeDn[int_] :=
 Module[{op},
  If[int == 5, op = PentaDot, op = Dot];
  MakeGroup[op, {Cz[2 π / int], Cx[2 π / 2]}]]
```

Click back to the C_n groups if you need to review **PentaDot**. Try it out:

```
Sort[MakeDn[3]] == Sort[CartRep["D3"]]
```

```
True
```

On your own
The $\mathbf{D_{n,h}}$ groups are products of group $\mathbf{C_h}$ with the groups $\mathbf{D_n}$. Show this symbolically, following the models above.

Groups $\mathbf{D_{nh}}$

The name implies a product group $\mathbf{D_n} \otimes \mathbf{C_h}$. This is true, and the minimal presentation agrees:

```
Presentation["D3h"]
```

```
a.a.a == b.b == c.c == a.b.a.b == b.c.b.c == a.c.Inverse[a].c == e
```

By Rule 2, a is a three-fold rotation about **z**. By Rule 3, b is a two-fold about **x**, and by Rule 1, c is the horizontal reflection plane σ_{xy}.

```
Clear[MakeDnh];
MakeDnh[int_] :=
 Module[{op},
  If[int == 5, op = PentaDot, op = Dot];
  MakeGroup[op, {Cz[2 π / int], Cx[2 π / 2], σxy}]]

Sort[MakeDnh[3]] == Sort[CartRep["D3h"]]
```

```
True
```

Groups $\mathbf{D_{nd}}$

The name implies a product group $\mathbf{D_n} \otimes \mathbf{C_d}$, which is the way Schönflies looked at it. But this was the tragic error which led crystallographers to the reject Schönflies notation. Here is a quick demonstration. Ask if the $\mathbf{D_{nd}}$ groups contain the inversion matrix **Imat**, or not :

```
Map[FreeQ[CartRep[#], Imat] &,
  {"D2d", "D3d", "D4d", "D5d", "D6d"}]
```

```
{True, False, True, False, True}
```

All the odd-**n** groups contain inversion center, while the even-**n** groups do not! The International system splits the $\mathbf{D_{nd}}$ groups into two series with quite different names. There is the odd series $\{\mathbf{D_{3d}}, \mathbf{D_{5d}}, ..\} = \{\bar{3}\,\mathbf{m}, \bar{5}\,\mathbf{m}, ..\}$ and the even series $\{\mathbf{D_{2d}}, \mathbf{D_{4d}}, ..\} = \{\bar{4}\,2\,\mathbf{m}, \bar{8}\,2\,\mathbf{m}, ..\}$. Remember that a bar over an integer means roto-*inversion* axis, a concept that Schönflies did not use. For more on this, click to the End Notes.

Where does the inversion center come from? It is easy to see from Fig. 18.3 (below) that all the odd-**n** groups have a reflection in the **y, z** plane, while the even-**n** groups do not. It is the product of this reflection times the 2-fold rotation about **x** that produces the inversion center:

```
Imat == AxialReflection[{1, 0, 0}].Cx[π]
```

```
True
```

Here are the pictures :

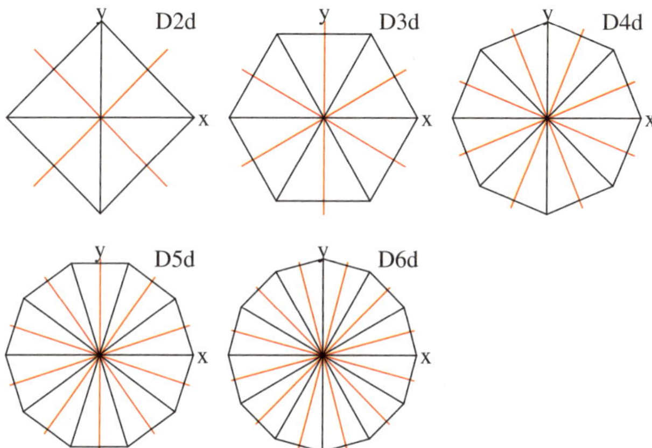

Fig. 18.3 A top view of the elements of the tabulated $D_{n,d}$ groups. The horizontal C_2 axes are black, and the edges of the σ_v planes are red.

The black **C2** axes are connected by a polygonal rim for visual clarity. By convention, one of them is the **x**-axis. The edges of the dihedral reflecting planes are in red. These figures have **2n**-fold symmetry, not **n**-fold. This means that the angle between any two adjacent C_2 axes is $2\pi/(2n)$, and therefore that the angle between a C_2 axis and the adjacent dihedral plane is $2\pi/(4n)$.

Generate group D_{nd}

For a discussion of the **Presentation** for D_{nd}, click to the End Notes.

We follow Schönflies in constructing these groups as $D_n \otimes C_d$. His way has at least the advantage of doing them all in a unified way, as an **n**-fold rotation about **z**, a two-fold rotation about **x**, and a dihedral reflection matrix at $\pi/(2n)$. The dihedral mirror is in a different position in each group, so it is constructed by the operator **DihedralReflMat[n]**. Flip to (or click to) the End Notes to see it developed. Here we just use it in the group generator :

```
Clear[MakeDnd];
MakeDnd[int_] := Module[{op},
  If[int == 5 || int == 10, op = PentaDot, op = Dot];
  MakeGroup[op,
   {Cz[2 π / int], Cx[π], DihedralReflMat[int]}]  ]
```

Click back to the axial rotation groups if you need to review **PentaDot**. Watch it work on the simplest case :

```
groupD2d = MakeDnd[2];
% // NeatList // Size[6]
```

$$\left\{\begin{pmatrix}-1&0&0\\0&-1&0\\0&0&1\end{pmatrix},\begin{pmatrix}-1&0&0\\0&1&0\\0&0&-1\end{pmatrix},\begin{pmatrix}0&-1&0\\-1&0&0\\0&0&1\end{pmatrix},\begin{pmatrix}0&-1&0\\1&0&0\\0&0&-1\end{pmatrix},\begin{pmatrix}0&1&0\\-1&0&0\\0&0&-1\end{pmatrix},\begin{pmatrix}0&1&0\\1&0&0\\0&0&1\end{pmatrix},\begin{pmatrix}1&0&0\\0&-1&0\\0&0&-1\end{pmatrix},\begin{pmatrix}1&0&0\\0&1&0\\0&0&1\end{pmatrix}\right\}$$

Make sure it agrees with the tabulated group :

```
Sort[groupD2d] == Sort[CartRep["D2d"]]

True
```

Run **MakeDnd** for **2** through **6** and compare to the tables:

```
TestDnd[int_] :=
 Module[{gpName, byTable, byConstruction},
  gpName = "D" <> ToString[int] <> "d";
  byTable = CartRep[gpName] // Sort;
  byConstruction = MakeDnd[int] // Sort;
 byTable == byConstruction ]

Map[TestDnd, {2, 3, 4, 5, 6}]

{True, True, True, True, True}
```

We said in the introduction that our group-making operators work with decimal numbers beyond 6, returning numeric results. Try it, looking at only two mats:

```
MakeDnd[9.];
%[[{3, 4}]] // GridList // Size[7]
```

{

-0.939693	-0.34202	0
-0.34202	0.939693	0
0	0	-1.

,

-0.939693	-0.34202	0
0.34202	-0.939693	0
0	0	1.

}

Numeric, as advertised. The crown ethers have high $\mathbf{D_{nd}}$ symmetries, so this is not completely academic.

On your own
It happens that `MakeD`$_{nd}$ works in infinite precision for the non-catalog group $D_{10\,d}$. You can generate it just like $D_{5\,d}$, above.

18.4 Platonic groups

There are infinitely many groups in all the uniaxial types enumerated above, because the axis order can go to infinity. After these, there are exactly seven more point groups. They are all based on the Platonic solids.

The Platonic solids are bounded by identical equilateral polygons. There are exactly five of them: the tetrahedron (4 triangular faces), the cube (6 square faces), the octahedron (8 triangular faces), the dodecahedron (12 pentagonal faces), and the icosahedron (20 triangular faces).

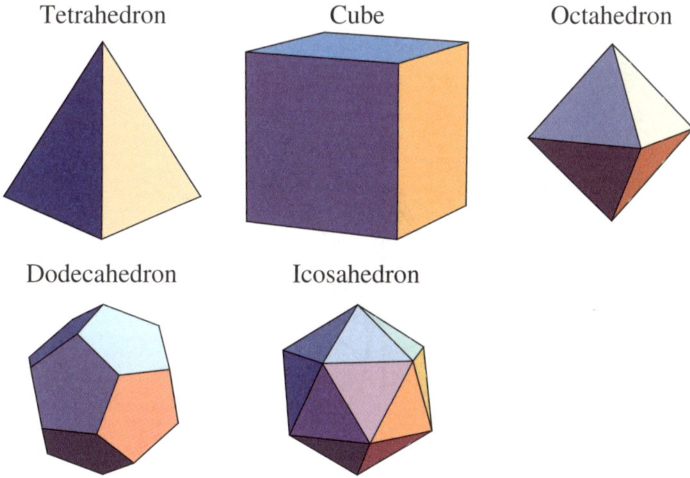

Fig. 18.4 The five Platonic solids. Three-D space does not support any others.

They are called "Platonic" because Plato based a prescientific theory of atoms on them. However, it may be that they were first discovered in Scotland.

Like the Platonic solids, the Platonic groups contain several axes of the same highest order. Before calculating **`CartRep`** of a Platonic group, its solid must be given a standard orientation according to Rules 5a, b, and c at the top of this chapter.

18.4.1 Group T (tetrahedral rotations)

Schönflies defined group **T** as all the rotations of the tetrahedron, with no attempt to base the name on a minimal list. The **Presentation** says that the minimal list consists of only two rotations, a two-fold and a three-fold:

```
Presentation["T"]
```

```
a.a.a == b.b == a.b.a.b.a.b == e
```

This agrees with the International name **23** But it is up to us to derive from classical geometry the angle between these axes.

Following Rule 5a, look at a tetrahedron inscribed in a unit cube, with one vertex at **{1,1,1}** and the others at alternating corners of a cube of edge length **2**, centered squarely on the coordinate axes. You will see that all the coordinate axes are two-fold symmetry axes, and that a slanted three-fold rotation axis runs through each of the four apices and out though the centroid of the opposite face. The red axes below are the two we use as to generate the group.

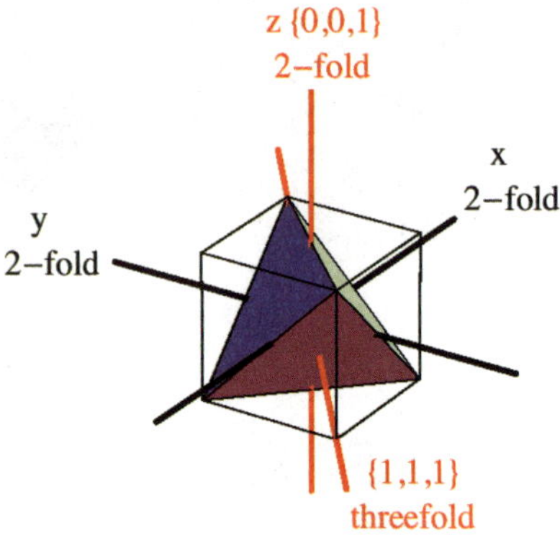

Fig. 18.5 The two generating axes of group T, in red.

So the minimal generating list is

```
aMatT = AxialRotation[{1, 1, 1}, 2 π / 3];
bMatT = AxialRotation[{0, 0, 1}, π];
```

Feed them to **MakeGroup**. Watch it close in two rounds of calculation:

```
groupT =
  MakeGroup[Dot, {aMatT, bMatT}, Report → True] //
    Sort // Reverse;
```

```
2 elements made 6 elements

6 elements made 12 elements

12 elements made 12 elements
```

It is a very simple and beautiful group :

```
groupT // NeatList // Size[6]
```

$$\left\{\begin{pmatrix}1&0&0\\0&1&0\\0&0&1\end{pmatrix},\begin{pmatrix}1&0&0\\0&-1&0\\0&0&-1\end{pmatrix},\begin{pmatrix}0&1&0\\0&0&1\\1&0&0\end{pmatrix},\begin{pmatrix}0&1&0\\0&0&-1\\-1&0&0\end{pmatrix},\right.$$
$$\begin{pmatrix}0&0&1\\1&0&0\\0&1&0\end{pmatrix},\begin{pmatrix}0&0&1\\-1&0&0\\0&-1&0\end{pmatrix},\begin{pmatrix}0&0&-1\\1&0&0\\0&-1&0\end{pmatrix},\begin{pmatrix}0&0&-1\\-1&0&0\\0&1&0\end{pmatrix},$$
$$\left.\begin{pmatrix}0&-1&0\\0&0&1\\-1&0&0\end{pmatrix},\begin{pmatrix}0&-1&0\\0&0&-1\\1&0&0\end{pmatrix},\begin{pmatrix}-1&0&0\\0&1&0\\0&0&-1\end{pmatrix},\begin{pmatrix}-1&0&0\\0&-1&0\\0&0&1\end{pmatrix}\right\}$$

On your own
`Map` the operator **`RecognizeMatrix`** onto these matrices You will see that it is a pure rotation group.

There are three two-fold axes and four three-fold axes. Why does each three-fold axis yield two groups elements, while the two-fold axes yield only one?

18.4.2 Group T_h

From its name, it seems that Schönflies constructed it as a product group of **`T`** times **C_h**. But the minimal list is a three-fold rotation about **`{1,1,1}`** plus a horizontal reflection:

```
groupTh = MakeGroup[Dot,
   {AxialRotation[{1, 1, 1}, 2 π / 3], σz},
   Report → True];
```

```
2 elements made 6 elements

6 elements made 15 elements

15 elements made 24 elements

24 elements made 24 elements
```

Is it the same as the tabulated group?

```
Sort[groupTh] == Sort[CartRep["Th"]]
```

```
True
```

Its **24** matrices all have six **0**'s and three **±1**'s, like those of **T**. Call them up with

```
groupTh // NeatList // Size[6]
```

$$\left\{\begin{pmatrix}-1&0&0\\0&-1&0\\0&0&-1\end{pmatrix},\begin{pmatrix}-1&0&0\\0&-1&0\\0&0&1\end{pmatrix},\begin{pmatrix}-1&0&0\\0&1&0\\0&0&-1\end{pmatrix},\begin{pmatrix}-1&0&0\\0&1&0\\0&0&1\end{pmatrix},\right.$$
$$\begin{pmatrix}0&-1&0\\0&0&-1\\-1&0&0\end{pmatrix},\begin{pmatrix}0&-1&0\\0&0&-1\\1&0&0\end{pmatrix},\begin{pmatrix}0&-1&0\\0&0&1\\-1&0&0\end{pmatrix},\begin{pmatrix}0&-1&0\\0&0&1\\1&0&0\end{pmatrix},\begin{pmatrix}0&0&-1\\-1&0&0\\0&-1&0\end{pmatrix},$$
$$\begin{pmatrix}0&0&-1\\-1&0&0\\0&1&0\end{pmatrix},\begin{pmatrix}0&0&-1\\1&0&0\\0&-1&0\end{pmatrix},\begin{pmatrix}0&0&-1\\1&0&0\\0&1&0\end{pmatrix},\begin{pmatrix}0&0&1\\-1&0&0\\0&-1&0\end{pmatrix},\begin{pmatrix}0&0&1\\-1&0&0\\0&1&0\end{pmatrix},$$
$$\begin{pmatrix}0&0&1\\1&0&0\\0&-1&0\end{pmatrix},\begin{pmatrix}0&0&1\\1&0&0\\0&1&0\end{pmatrix},\begin{pmatrix}0&1&0\\0&0&-1\\-1&0&0\end{pmatrix},\begin{pmatrix}0&1&0\\0&0&-1\\1&0&0\end{pmatrix},\begin{pmatrix}0&1&0\\0&0&1\\-1&0&0\end{pmatrix},$$
$$\left.\begin{pmatrix}0&1&0\\0&0&1\\1&0&0\end{pmatrix},\begin{pmatrix}1&0&0\\0&-1&0\\0&0&-1\end{pmatrix},\begin{pmatrix}1&0&0\\0&-1&0\\0&0&1\end{pmatrix},\begin{pmatrix}1&0&0\\0&1&0\\0&0&-1\end{pmatrix},\begin{pmatrix}1&0&0\\0&1&0\\0&0&1\end{pmatrix}\right\}$$

18.4.3 Group T_d

Group $\mathbf{T_d}$ comprises all the symmetries of the tetrahedron, rotations and reflections. It may be $\mathbf{T \otimes C_d}$, but the **Presentation** says the minimal list is a fourfold and a threefold:

```
Presentation["Td"]
```

```
a.a.a.a == b.b.b == a.b.a.b.a.b.a.b == e
```

The fourfold is a rotoreflection, not a rotation. The threefold is the usual cube diagonal. It closes in three rounds of calculation :

```
groupTd = MakeGroup[Dot, {
    AxialRotoReflection[{0, 0, 1}, π / 2],
    AxialRotation[{1, 1, 1}, 2 π / 3]}, Report → True];
```

```
2 elements made 6 elements

6 elements made 22 elements

22 elements made 24 elements

24 elements made 24 elements
```

```
Sort[groupTd] == Sort[CartRep["Td"]]
```

```
True
```

If you run the command

```
Map[RecognizeMatrix, groupTd] // Sort
```

you will see that T_d has six reflecting planes, each bisecting the tetrahedron and containing one of its six edges. These are the dihedral planes that give the name its subscript **d**. If you run the next command, you will see 24 beautifully simple matrices:

```
groupTd // NeatList // Size[6]
```

18.4.4 Group O (octahedral rotations)

Group **O** is the group of all the rotations of the octahedron; or equivalently, the group of all the rotations of the cube. Here is an odd fact :

```
Presentation["O"] == Presentation["Td"]
```

```
True
```

Explicitly,

```
Presentation["O"]
```

```
a.a.a.a == b.b.b == a.b.a.b.a.b.a.b == e
```

It is a fourfold and a threefold. But orientation rule **5b**, above, says it must include four-fold rotations about the **z** and **x** axes. So is it also possible to construct it from two perpendicular fourfold rotations?

```
groupO = MakeGroup[Dot,
    {Cx[2 π / 4], Cz[2 π / 4]} // N, Report → True];
```

```
2 elements made 6 elements

6 elements made 21 elements

21 elements made 24 elements

24 elements made 24 elements
```

Does this agree with the tabulated group ?

```
Sort[groupO] == Sort[CartRep["O"]]
```

```
True
```

18.4.5 Group O_h

Group $\mathbf{O_h}$ consists of all the symmetries of the octahedron; or equivalently, all the symmetries of the cube. It is the product of groups **O** and $\mathbf{C_h}$.

```
Sort[CartRep["O"] ⊗ CartRep["Ch"]] == Sort[CartRep["Oh"]]
```

```
True
```

But its short list has only two elements in it :

```
Presentation["Oh"]
```

```
a.a.a.a == b.b == a.b.a.b.a.b.a.b.a.b.a.b == e
```

This must mean a vertical four-fold rotation and a reflecting plane (because two rotations cannot generate a reflection, and the cube has many reflections). But which plane? Looking at the octahedron, you can see that one of them is perpendicular to **{1,0,1}**, so we use that:

```
aMat = RotationMatrix3Dz[2 π / 4];
bMat = ReflectionMatrix[{1, 0, 1}];
MatrixPower[aMat, 4] == MatrixPower[bMat, 2] ==
 MatrixPower[aMat.bMat, 6] == Emat
```

```
True
```

That works. So now make the group and test it:

```
groupOh = MakeGroup[Dot, {aMat, bMat}];
Sort[groupOh] == Sort[CartRep["Oh"]]
```

```
True
```

Again, it is beautifully simple. All elements are similar to the first 6:

```
Take[groupOh, 6] // Sort // Reverse // GridList //
 Size[6]
```

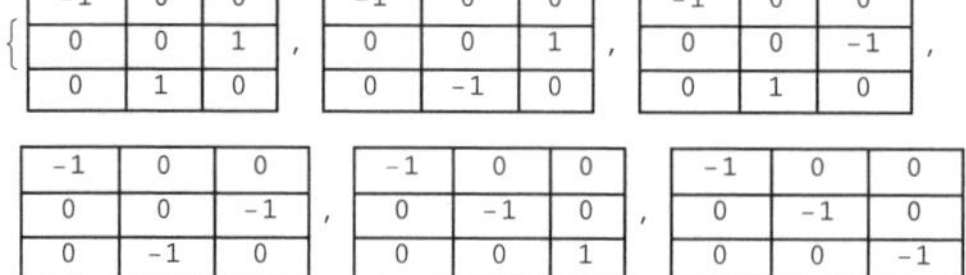

$$\left\{\begin{pmatrix}-1&0&0\\0&0&1\\0&1&0\end{pmatrix},\begin{pmatrix}-1&0&0\\0&0&1\\0&-1&0\end{pmatrix},\begin{pmatrix}-1&0&0\\0&0&-1\\0&1&0\end{pmatrix},\begin{pmatrix}-1&0&0\\0&0&-1\\0&-1&0\end{pmatrix},\begin{pmatrix}-1&0&0\\0&-1&0\\0&0&1\end{pmatrix},\begin{pmatrix}-1&0&0\\0&-1&0\\0&0&-1\end{pmatrix}\right\}$$

18.4.6 Group I (icosahedral rotations)

Group **I** is the group of all the rotations of the icosahedron; or equivalently, of the dodecahedron. The International name of this group is **52**, implying that it is generated by a fivefold axis and a twofold axis.

The icosahedron is the least familiar of the Platonic solids. Strangely, the Greeks never understood the icosahedron completely. The climax of Euclid's books on geometry is the construction of the icosahedron (Book XIII, proposition 18) but he failed to see the one fact that makes the icosahedron easily understandable. Schönemann finally discovered it in 1873, some 2200 years after Euclid. The icosahedron is built around a skeleton of three interlocking Golden Rectangles, and using this he was the first to list the Cartesian coordinates of their vertices. An excellent modern exposition has been given by Coxeter. We give a quick sketch of it just below.

Modern construction of the icosahedron

The sides **a** and **b** of the Golden rectangles obey the classic Greek "golden" proportion $\frac{a+b}{a} == \frac{a}{b}$, where **a** is the long side and **b** is the short side. The golden ratio γ is **a/b**. Setting $a = \gamma$ and $b = 1$,

```
Clear[γ];
γSolns = Solve[(γ + 1)/γ == γ/1, γ]
```

$$\left\{\left\{\gamma \to \frac{1}{2}\left(1-\sqrt{5}\right)\right\}, \left\{\gamma \to \frac{1}{2}\left(1+\sqrt{5}\right)\right\}\right\}$$

We want the positive solution :

```
γ = 1/2 (1 + √5);
```

We draw three perpendicular golden rectangles, intersecting at the origin. One is in the **xy** plane (long in the **x**-direction), one is in the **yz** plane (long in the **y** direction), and one is in the **zx** plane (long in the **z** direction). Each rectangle has length **2**γ and width **2**.

```
threeGoldenRectangles = {
{{γ, 1, 0}, {γ, -1, 0}, {-γ, -1, 0}, {-γ, 1, 0}},
{{0, γ, 1}, {0, γ, -1}, {0, -γ, -1}, {0, -γ, 1}},
{{1, 0, γ}, {1, 0, -γ}, {-1, 0, -γ}, {-1, 0, γ}}};
```

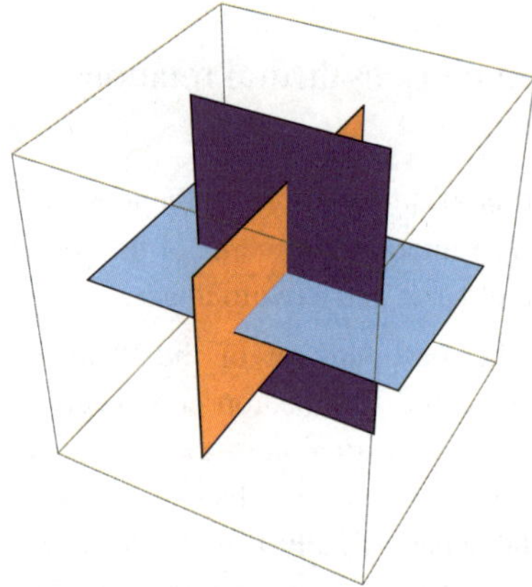

Fig. 18.6 Three perpendicular golden rectangles; the heart of the icosahedron

It is easy to calculate (with *Mathematica*) that each vertex is equidistant from its five closest neighbors.

```
vertices = Flatten[threeGoldenRectangles // N, 1];
DMofIcos = DistanceMatrix[vertices];
MatrixForm[%] // PaddedForm[#, {4, 2}] & //
 Size[5.5]
```

$$\begin{pmatrix}
0.00 & 2.00 & 3.80 & 3.24 & 2.00 & 2.00 & 3.24 & 3.24 & 2.00 & 2.00 & 3.24 & 3.24 \\
2.00 & 0.00 & 3.24 & 3.80 & 3.24 & 3.24 & 2.00 & 2.00 & 2.00 & 2.00 & 3.24 & 3.24 \\
3.80 & 3.24 & 0.00 & 2.00 & 3.24 & 3.24 & 2.00 & 2.00 & 3.24 & 3.24 & 2.00 & 2.00 \\
3.24 & 3.80 & 2.00 & 0.00 & 2.00 & 2.00 & 3.24 & 3.24 & 3.24 & 3.24 & 2.00 & 2.00 \\
2.00 & 3.24 & 3.24 & 2.00 & 0.00 & 2.00 & 3.80 & 3.24 & 2.00 & 3.24 & 3.24 & 2.00 \\
2.00 & 3.24 & 3.24 & 2.00 & 2.00 & 0.00 & 3.24 & 3.80 & 3.24 & 2.00 & 2.00 & 3.24 \\
3.24 & 2.00 & 2.00 & 3.24 & 3.80 & 3.24 & 0.00 & 2.00 & 3.24 & 2.00 & 2.00 & 3.24 \\
3.24 & 2.00 & 2.00 & 3.24 & 3.24 & 3.80 & 2.00 & 0.00 & 2.00 & 3.24 & 3.24 & 2.00 \\
2.00 & 2.00 & 3.24 & 3.24 & 2.00 & 3.24 & 3.24 & 2.00 & 0.00 & 3.24 & 3.80 & 2.00 \\
2.00 & 2.00 & 3.24 & 3.24 & 3.24 & 2.00 & 2.00 & 3.24 & 3.24 & 0.00 & 2.00 & 3.80 \\
3.24 & 3.24 & 2.00 & 2.00 & 3.24 & 2.00 & 2.00 & 3.24 & 3.80 & 2.00 & 0.00 & 3.24 \\
3.24 & 3.24 & 2.00 & 2.00 & 2.00 & 3.24 & 3.24 & 2.00 & 2.00 & 3.80 & 3.24 & 0.00
\end{pmatrix}$$

The minimum vertex-vertex distance is 2.00, and it occurs five times in each row. That is, each vertex is surrounded by five equidistant nearest neighbor vertices. This is the definition of the icosahedron.

Below, we draw the shortest vertex-vertex lines to show the icosahedron more clearly.

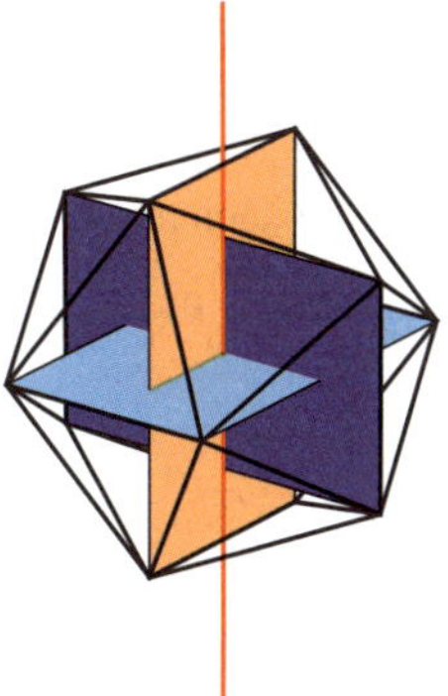

Fig. 18.7 Three interlocked golden rectangles with nearest vertices joined. It's the icosahedron!

This icosahedron is not in standard orientation because its **z** axis is a twofold axis. The fivefold axes all lie on the diagonals of the golden rectangles. To put the figure in standard orientation, pick a five-fold axis in the **zx** plane and rotate about **y** by an angle α to bring it to the vertical.

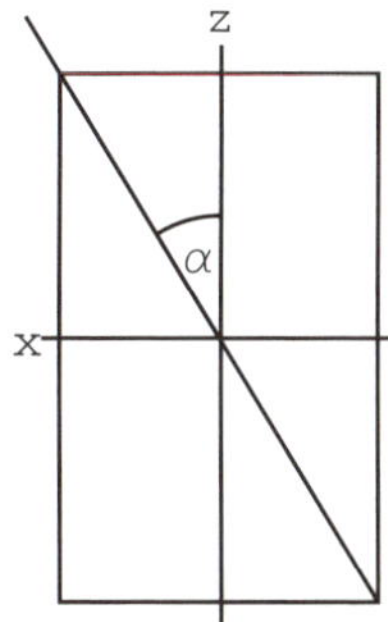

Fig. 18.8 Golden rectangle, with angle α defined. Width is 2; height is 2γ.

We can write exact expressions for the **Sin** and **Cos** of α :

$$\{\sin\alpha,\ \cos\alpha\} = \left\{\frac{1}{\sqrt{1+\gamma^2}},\ \frac{\gamma}{\sqrt{1+\gamma^2}}\right\} \ //\ \text{FullSimplify}$$

$$\left\{\sqrt{\frac{2}{5+\sqrt{5}}},\ \sqrt{\frac{1}{10}\left(5+\sqrt{5}\right)}\right\}$$

Icosahedron with vertical five-fold

Two rotation matrices (about the **z** axis and the diagonal axis in the figure above) are all we need to generate group **I**. If you want the icosahedron to have a vertical five-fold (as it is usually depicted) use

```
matA = AxialRotation[{0, 0, 1}, 2 π / 5] // Simplify;
MatrixForm[%]
```

$$\begin{pmatrix} \frac{1}{4}\left(-1+\sqrt{5}\right) & -\frac{1}{2}\sqrt{\frac{1}{2}\left(5+\sqrt{5}\right)} & 0 \\ \sqrt{\frac{5}{8}+\frac{\sqrt{5}}{8}} & \frac{1}{4}\left(-1+\sqrt{5}\right) & 0 \\ 0 & 0 & 1 \end{pmatrix}$$

plus a twofold about the diagonal axis.

```
matB = AxialRotation[{-sinα, 0, cosα}, π] //
   FullSimplify;
MatrixForm[
 %]
```

$$\begin{pmatrix} -\frac{1}{\sqrt{5}} & 0 & -\frac{2}{\sqrt{5}} \\ 0 & -1 & 0 \\ -\frac{2}{\sqrt{5}} & 0 & \frac{1}{\sqrt{5}} \end{pmatrix}$$

Icosahedron with vertical two-fold

But to follow Rule 5c, we must take the two-fold as the **z** axis and the five-fold as the diagonal axis of the figure. Also, to calculate group $\mathbf{I_h}$ as a product of **I** and $\mathbf{C_h}$, you must have a vertical two-fold. So we need

```
aMat = AxialRotation[{Sqrt[2/(5 + Sqrt[5])], 0, Sqrt[1/2 + 1/(2 Sqrt[5])]},
    2 π / 5] // FullSimplify;
% // GridForm
```

$$\text{aMat} = \text{AxialRotation}\left[\left\{\sqrt{\frac{2}{5+\sqrt{5}}},\ 0,\ \sqrt{\frac{1}{2}+\frac{1}{2\sqrt{5}}}\right\},\ 2\pi/5\right]$$

$\frac{1}{2}$	$\frac{1}{4}\left(-1-\sqrt{5}\right)$	$\frac{1}{4}\left(-1+\sqrt{5}\right)$
$\frac{1}{4}\left(1+\sqrt{5}\right)$	$\frac{1}{4}\left(-1+\sqrt{5}\right)$	$-\frac{1}{2}$
$\frac{1}{4}\left(-1+\sqrt{5}\right)$	$\frac{1}{2}$	$\frac{1}{4}\left(1+\sqrt{5}\right)$

```
bMat = Cz[π];
% // GridForm
```

-1	0	0
0	-1	0
0	0	1

These matrices are the basis on which **CartRep["I"]** is calculated. The presentation of group **I**, the largest group we tabulate (60 elements) is no more difficult than any other point group:

```
Presentation["I"]
```

```
a.a.a.a.a == b.b == a.b.a.b.a.b == e
```

So **a** is the fivefold and **b** is the twofold, and the third matrix power of **a.b** must be unity. This is the acid test of whether we have the angle between the two rotation axes correct. Here it is:

```
MatrixPower[aMat.bMat, 3] // FullSimplify
```

```
{{1, 0, 0}, {0, 1, 0}, {0, 0, 1}}
```

Construction of group I

As in many infinite precision calculations that involve $\sqrt{5}$, we must make sure that all expressions equal in value are simplified to the same form. We insure this with a special **Dot** operator (like **PentaDot**) which we call **IcosaDot**, defined in the **Symmetry`** package. Read about in the End Notes.

Get ready to be amazed.

```
{timeI1, groupI} = MakeGroup[IcosaDot,
    {aMat, bMat}, Report → True] // Timing;
timeI1 Second
```

```
2 elements made 6 elements

6 elements made 19 elements

19 elements made 52 elements

52 elements made 60 elements

60 elements made 60 elements
```

```
8.78538 Second
```

```
Sort[groupI] == Sort[CartRep["I"]]
```

```
True
```

Even the multiplication table works if you give it **IcosaDot**, and it takes only 15 seconds or so (MacBook Pro). We suppress the display:

```
{timeI2, mtI} = MultiplicationTable[
    IcosaDot, groupI, Range[60]] // Timing;
timeI2 Second
```

```
13.4219 Second
```

All this is infinite precision. Did it recognize every infinite precision product?

```
FreeQ[mtI, "?"]
```

```
True
```

> **On your own**
> Try this one:
> **Map[RecognizeMatrix,N[groupI]]//Sort//Chop//Column**
> You will find only rotation matrices.

> **On your own**
> Repeat this construction using the simple **Dot** operator and a machine precision minimal list, and see how much faster it is.

18.4.7 Group I_h

We do not tabulate this 120 element group, because it slows down the loading of the **Symmetry`** package. If you ever need it, construct it and send it to disk using **groupIh >> fileIh**. Here is the construction; it is almost instantaneous :

```
groupIh = CartRep["I"] ⊗ CartRep["Ch"];
Dimensions[groupIh]
```

```
{120, 3, 3}
```

Interactive readers can see the whole thing, arranged by classes, by issuing

```
Map[RecognizeMatrix, N[groupIh]] // Chop // Sort //
  Column // Size[7]
```

Checking for closure takes about a minute :

```
{timeIh, mtIh} = MultiplicationTable[
    IcosaDot, groupIh, Range[120]] // Timing;
timeIh Second
```

```
53.5 Second
```

Did it close?

```
FreeQ[groupIh, "?"]
```

```
True
```

Coda

If you ever find a point group that cannot be generated by the methods of this chapter, be sure to notify the American Mathematical Society. They will know what to do with you.

18.5 End Notes

18.5.1 Dihedral reflection planes for groups D_{nd}

Look back to Fig. 18.3 (click here) to see where the dihedral mirror planes lie in the D_{nd} groups. Just below that figure we argued that the dihedral plane adjacent to the **x**-axis lies at angle **2π/(4n)**. Here we make an operator that returns the matrix for this mirror.

If we insist that all our tabulated Cartesian representations be in infinite precision (and we do) then the plane for $D_{4\,d}$ presents a special problem.

The dihedral plane at angle θ contains the point **{Cos[θ],Sin[θ],0}**. So a vector perpendicular to this plane is given by **{Sin[θ],-Cos[θ],0}**. The first dihedral reflection plane (the one needed for D_{nd} generation lies) at **π/(2n)**. This means that **n=4** needs **{Sin[π/8],Cos[π/8]}**, and these are not evaluated in infinite precision. However, the trig functions for **π/4** are evaluated in infinite precision, so we can get exact expressions for **π/8** using the half-angle formulas

$$\text{SinHalf[x\_] := FullSimplify}\left[(-1)^{\text{Floor}\left[\frac{x}{2\pi}\right]}\sqrt{\frac{1-\text{Cos[x]}}{2}}\right];$$

$$\text{CosHalf[x\_] := FullSimplify}\left[(-1)^{\text{Floor}\left[\frac{x+\pi}{2\pi}\right]}\sqrt{\frac{1+\text{Cos[x]}}{2}}\right];$$

Here are the functions of **π/8** in infinite precision:

```
{SinHalf[π / 4], CosHalf[π / 4]}
```

$$\left\{\frac{\sqrt{2-\sqrt{2}}}{2}, \frac{\sqrt{2+\sqrt{2}}}{2}\right\}$$

Do a numerical check :

```
{N[SinHalf[π / 4]] == N[Sin[π / 8]],
 N[CosHalf[π / 4]] == N[Cos[π / 8]]}
```

```
{True, True}
```

The axis perpendicular to the first dihedral plane is therefore

```
DihedralReflAxis[int_] :=
 {SinHalf[π/int], -CosHalf[π/int], 0}
```

Put this into the definition of the first **n**-fold dihedral reflection matrix

```
Clear[DihedralReflMat];
DihedralReflMat[int_] :=
 Map[ToRadicals, Map[FullSimplify,
    AxialReflection[DihedralReflAxis[int]]] /.
   noEighth]
```

(The replacement rule **noEighth** is discussed in another End Note.) For example, the plane for the **2**-fold case lies at **π/4** (see Fig. 18.3), and the reflec tion matrix is

```
DihedralReflMat[2]
```

```
{{0, 1, 0}, {1, 0, 0}, {0, 0, 1}}
```

This exchanges **x** and **y** coordinates, as it should. Try other examples for yourself, checking with **RecognizeMatix**.

18.5.2 Schönflies vs. International for the D_{nd} groups

This name implies a minimal list of three elements (two matrices for D_n plus the matrix σ_d), but the presentation says there is a simpler way that uses only two elements:

```
Presentation["D2d"]
```

```
a.a.a.a == b.b == a.b.a.b == e
```

```
aMat = AxialRotoReflection[{0, 0, 1}, 2 π / 4];
```

```
bMat = DihedralReflMat[2];
MatrixPower[aMat, 4] ==
 MatrixPower[bMat, 2] == aMat.bMat.aMat.bMat == Emat

True

groupD2d =
  MakeGroup[Dot, {aMat, bMat}, Report → True];

2 elements made 6 elements

6 elements made 8 elements

8 elements made 8 elements

Dimensions[CartRep["D2d"]]

{8, 3, 3}

Sort[groupD2d] == Sort[CartRep["D2d"]]

True

Presentation["D3d"]

a.a.a.a.a.a == b.b == a.b.a.b == e

aMat = Cz[2 π / 6].σh;
bMat = DihedralReflMat[3];
MatrixPower[aMat, 6] ==
 MatrixPower[bMat, 2] == aMat.bMat.aMat.bMat == Emat

True

groupD3d =
  MakeGroup[Dot, {aMat, bMat}, Report → True];

2 elements made 6 elements

6 elements made 12 elements

12 elements made 12 elements

Dimensions[CartRep["D3d"]]

{12, 3, 3}

Sort[groupD3d] == Sort[CartRep["D3d"]]

True
```

On your own
Redo the whole calculation above, using **aMat = C$_z$[2 π / 6].Imat** instead of **aMat = C$_z$[2 π / 6].σ$_h$**. It does make a **D$_{3d}$** group, but it is not the tabulated group.

18.5.3 Five-fold z-axis: PentaDot[mt1,mt2]

Vexing problems can arise when you try to calculate radical expressions with infinite precision. Take the multiplication table **C$_5$** for example:

```
MultiplicationTable[Dot, CartRep["C5"], Range[5]] //
 MatrixForm
```

$$\begin{pmatrix} 1 & 2 & 3 & 4 & 5 \\ 2 & ? & ? & ? & 1 \\ 3 & ? & ? & 1 & ? \\ 4 & ? & 1 & ? & ? \\ 5 & 1 & ? & ? & ? \end{pmatrix}$$

There is nothing wrong with the group; if you ask it to use a machine precision representation **N[CartRep["C5"]]**, it works well :

```
MultiplicationTable[Dot,
  N[CartRep["C5"]], Range[5]] // MatrixForm
```

$$\begin{pmatrix} 1 & 2 & 3 & 4 & 5 \\ 2 & 3 & 4 & 5 & 1 \\ 3 & 4 & 5 & 1 & 2 \\ 4 & 5 & 1 & 2 & 3 \\ 5 & 1 & 2 & 3 & 4 \end{pmatrix}$$

The problem is in the way **MultiplicationTable** recognizes identical products. Some of the products are generated as different forms that are numerically equal, and it correctly assigns a **?** to the unfamiliar products. Here are two of the troublemakers:

$$\textbf{formA} = \frac{1}{2}\sqrt{\frac{1}{2}\left(5-\sqrt{5}\right)}\,;\quad \textbf{formB} = \sqrt{\frac{5}{8}-\frac{\sqrt{5}}{8}}\,;$$

```
N[formA] == N[formB]
```

```
True
```

Here is the most powerful simplification in *Mathematica*'s core:

```
Map[ToRadicals, Map[FullSimplify, {formA, formB}]]
```

$$\left\{\frac{1}{2}\sqrt{\frac{1}{2}\left(5-\sqrt{5}\right)},\ \sqrt{\frac{5}{8}-\frac{\sqrt{5}}{8}}\right\}$$

> **On your own**
> Try **Simplify** alone, and then **FullSimplify** alone. The latter is a bit surprising the first time you see it. This explains why **ToRadicals** has to be applied last.

To put them in the same form, *Mathematica* would have to make an arbitrary choice, which by design it almost never does. So we make the arbitrary choice for it, as a little **Rule** called **noEighth**:

noEighth =

$$\left\{\sqrt{\frac{5}{8}+\frac{\sqrt{5}}{8}} \to \frac{1}{2}\sqrt{\frac{1}{2}\left(5+\sqrt{5}\right)},\ \sqrt{\frac{5}{8}-\frac{\sqrt{5}}{8}} \to \frac{1}{2}\sqrt{\frac{1}{2}\left(5-\sqrt{5}\right)}\right\};$$

Fortunately, these two ambiguities are the only ones that arise when working with axial five-fold groups. So in the **Symmetry`** package we define a special version of the **Dot** operator, called **PentaDot**, that contains this choice of form :

```
(*NOT EVALUATABLE*)
PentaDot[mat1_, mat2_] := ToRadicals[
    FullSimplify[Dot[mat1, mat2]]] /. noEighth;
```

The cell above is not evaluatable because it repeats a definition already made in the **Symmetry`** package. With this fancy operator, the table closes in infinite precision :

```
MultiplicationTable[PentaDot, CartRep["C5"]] //
 MatrixForm
```

$$\begin{pmatrix} 1 & 2 & 3 & 4 & 5 \\ 2 & 3 & 4 & 5 & 1 \\ 3 & 4 & 5 & 1 & 2 \\ 4 & 5 & 1 & 2 & 3 \\ 5 & 1 & 2 & 3 & 4 \end{pmatrix}$$

It is also important to use **PentaDot** in the **MakeGroup** operator when generating five-fold axial groups.

On your own
Make the multiplication tables for groups $\mathbf{D_5}$, $\mathbf{D_{5\,h}}$, and $\mathbf{D_{5\,d}}$ using both **Dot** and **PentaDot**.

Use **MakeGroup[PentaDot,{... }]** to generate some of these groups

18.5.4 Five-fold slanted axis: IcosaDot[mt1,mt2]

Another simplification problem, similar to that in the five-fold axial groups, occurs in the icosahedral groups **I** and $\mathbf{I_h}$. Again, the root of the problem is five-fold axes. The two expressions below are ambiguous in these groups :

$$\text{formC} = \left(1-\sqrt{5}\right)\sqrt{\frac{1}{10}\left(5+\sqrt{5}\right)};$$

$$\text{formD} = \frac{-1}{5}\left(5-\sqrt{5}\right)\sqrt{\frac{1}{2}\left(5+\sqrt{5}\right)};$$

```
N[formC] == N[formD]
```

```
True
```

Part of the cure is the rule list

$$\text{noTenth} = \left\{\left(1-\sqrt{5}\right)\sqrt{\frac{1}{10}\left(5+\sqrt{5}\right)} \to \frac{1}{5}\left(-5+\sqrt{5}\right)\sqrt{\frac{1}{2}\left(5+\sqrt{5}\right)},\right.$$
$$\left(-1+\sqrt{5}\right)\sqrt{\frac{1}{10}\left(5+\sqrt{5}\right)} \to \frac{1}{5}\left(5-\sqrt{5}\right)\sqrt{\frac{1}{2}\left(5+\sqrt{5}\right)},$$
$$\left(-5+\sqrt{5}\right)\sqrt{\frac{1}{10}\left(5+\sqrt{5}\right)} \to \left(1-\sqrt{5}\right)\sqrt{\frac{1}{2}\left(5+\sqrt{5}\right)},$$
$$\left.\left(5-\sqrt{5}\right)\sqrt{\frac{1}{10}\left(5+\sqrt{5}\right)} \to \left(-1+\sqrt{5}\right)\sqrt{\frac{1}{2}\left(5+\sqrt{5}\right)}\right\};$$

Try it out:

```
{formC, formD} /. noTenth
```

$$\left\{\frac{1}{5}\left(-5+\sqrt{5}\right)\sqrt{\frac{1}{2}\left(5+\sqrt{5}\right)},\ -\frac{1}{5}\left(5-\sqrt{5}\right)\sqrt{\frac{1}{2}\left(5+\sqrt{5}\right)}\right\}$$

Now the radicals are the same, but there is a problem with the external sign. So we write a function **noLeadingMinus** that moves any leading minus sign inside the first factor of the expression :

```
noLeadingMinus[f_] := Module[{g},
  g = Simplify[f] // Expand // Factor;
  If[Dimensions[g] == {3} && g[[1]] < 0,
   Apply[Times, {-g[[1]], -g[[2]], g[[3]]}],
   g]]
```

Try it :

$$\text{noLeadingMinus}\left[-\left(1-\sqrt{5}\right)\sqrt{\frac{1}{2}\left(5+\sqrt{5}\right)}\,\right]$$

$$\left(-1+\sqrt{5}\right)\sqrt{\frac{1}{2}\left(5+\sqrt{5}\right)}$$

It works. These two corrections are certainly not elegant or general, but they are all that is needed for the icosahedral groups. We put them together :

```
(*NOT EVALUATABLE*)
IcosaDot[a_, b_] :=
 noLeadingMinus[Dot[a, b] /. noTenth]
```

IcosaDot can be used just like **PentaDot**, with either **MultiplicationTable** or **MakeGroup**.

18.5.5 Scotland, Scotland, über alles

From the Wikipedia's article on Platonic solids: "There is evidence that these figures were known long before the time of the Greeks. The neolithic people of Scotland constructed stone models of all five solids at least 1000 years before Plato (Atiyah and Sutcliffe 2003). These models are kept at the Ashmolean Museum in Oxford."

Check this out at http://en.wikipedia.org/wiki/Platonic_solid.

19. Visualizing groups

Preliminaries

19.1 Constellations vs. stereograms

19.1.1 Introduction; the constellation operators

In Chapter 15 (MakeGroup) we developed tools for generating every possible point group, and in Chapter 18 (CartesianRepresentations) we actually did it (well, not all of them).

In books with paper pages one often sees a conventional diagram called a stereogram, a kind of polar projection of the symmetry elements of a group. Stereograms are a systematic way of presenting symmetries semi-visually, but they are not 3D-pictorial and one suspects that few people actually use them. In this electronic book, we think we have a better way. We will visualize the point groups by making geometric objects that have all the symmetries of any given point group (and no extra symmetry from higher groups).

In mathematical group theory, the set of points generated by applying all the transforms of a group to some one point is called the *orbit* of the point, under the given group. In cyclic groups the orbit points really are placed like points on an orbit around the principal axis. But for other kinds of groups they are scattered, like a symmetric set of stars in space. With apologies to the mathematical community, we will call such a set a *constellation* instead of an *orbit* .

In the **Preliminaries** of this chapter we define two special graphics operators; together, they can make figures showing the symmetries of the finite point groups. The names of the special operators are **Constellate** and **MakeLines**.

W.M. McClain, *Symmetry Theory in Molecular Physics with Mathematica*,
DOI 10.1007/b13137_19, © Springer Science+Business Media, LLC 2009

19.1.2 Constellate

In the operator **Constellate[pt,matList]**, parameter **pt** is a three dimensional point, and **matList** is a list of 3-by-3 transform matrices, usually a group of such matrices. A typical command would be

```
{constellation,distanceMatrix}=
                        Constellate[pt, matList]
```

where **constellation** is the list of points created as the constellation of point **pt** by the transform matrices of **matList**, and where **distanceMatrix** is a matrix of all the distances between the points of the constellation. Both **constellation** and **distanceMatrix** are used by **MakeLines** to create the final figure.

19.1.3 MakeLines

A typical command would be

```
blueObjects = MakeLines[constellation,
                  distanceMatrix,{1,3},"Blue"];
```

It uses the outputs of **Constellate**, plus a list of integers and a color. In this example the list **{1,3}** tells it to draw two kinds of lines: first, the **1** says to draw in all the shortest lines in the distance matrix; then the **3** says to draw in all lines of the third longest length. Similarly, **{2}** would tell it to draw in lines of only the second length, etc.

The output of **MakeLines** is a **Graphics3D** object that includes all the constellation points and all the specified lines, all in the specified color. When we want lines of different lengths to be in different colors, we use **MakeLines** several times, combining all its graphic object outputs in one final list, which is fed to a rendering operator.

When all desired graphics objects are collected together in a list, they are fed to StereoView.

19.2 Constellations with a principal axis

19.2.1 Groups C_n and S_n

We show one construction in the open; the rest will be closed. We use group `C6`. The one point **`{1,0,0.15}`** constellates into six blue points and the shortest distance within the constellation is rendered as a blue line (six times). The point **`{0,0,1}`** constellates only onto itself, and is shown in red. Axes **`x`**, **`y`**, and **`z`** are shown in black.

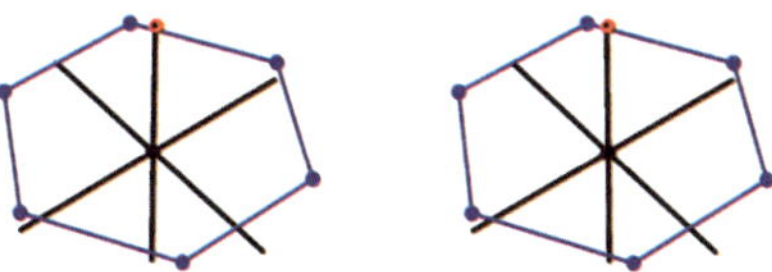

Fig. 19.1 Stereogram for group `C6`

Let them fuse into a 3D image. The red dot lies on the only symmetry element that this group has (the z axis) so it makes a very defective constellation (namely, just itself). The most general constellation, a hexagon, is made by any point not on the **`z`** axis,

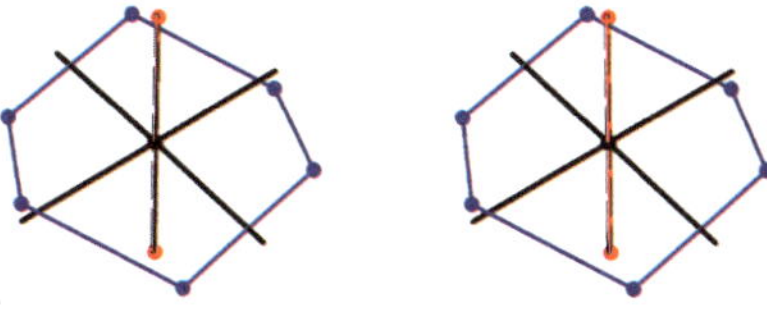

Fig. 19.2 Stereogram for group `S6`

The general constellation is a buckled hexagon, its points alternating equally above and below the **x,y** plane. The red dot now makes a special two-point constellation. This is roto-reflection.

19.2.2 Groups C_{nh} and C_{nv}

Group $\mathbf{C_{nh}}$ adds a horizontal reflection plane to $\mathbf{C_n}$, so the constellation now has two layers.

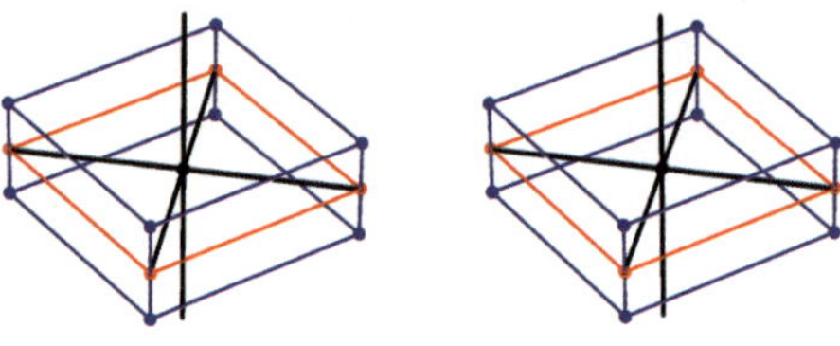

Fig. 19.3 Stereogram for group $C_{4\,h}$

Group $\mathbf{C_{nv}}$ adds vertical dihedral planes to $\mathbf{C_n}$, starting with one in the z,x plane. This doubles the points horizontally.

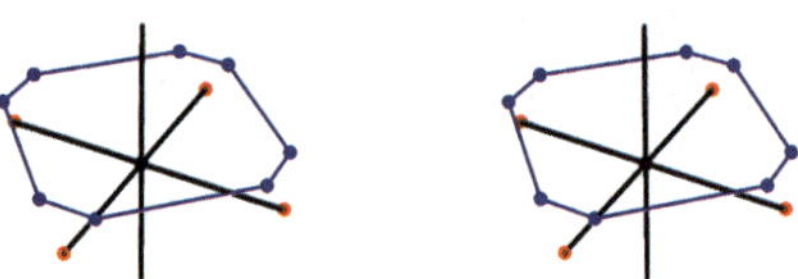

Fig. 19.4 Stereogram for group $C_{4\,v}$

19.2.3 Groups D_n

These groups have an **n**-fold central axis and one or more horizontal **2**-fold rotation axes. By our convention, one of the axes must be the **x** axis.

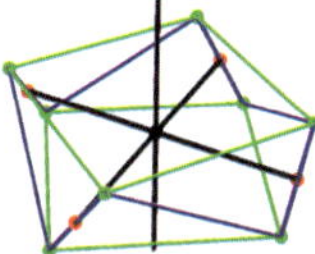
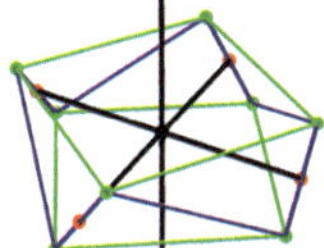

Fig. 19.5 Stereogram for group D_4

There is a simplified red constellation and a general blue constellation. The individual eye-views are hard to interpret, but in stereo the picture is quite clear. The red dots lie directly on the horizontal C_2 axes, so they have no rotation partners with respect to these axes. However, the pairs of blue dots linked by blue lines are C_2 rotation partners. The green lines outline the top and bottom faces of a pseudoprism. A pseudoprism is a figure which is neither prism (eclipsed top and bottom faces) nor antiprism (staggered top and bottom faces), but somewhere between, with top and bottom faces twisted a little, but not to a symmetric position. The top face may be rotated (with respect to the bottom face) a little in the right-handed sense, or a little in the left-handed sense.

The lateral faces are not plane parallelograms, but rather parallelograms broken into similar triangles, with a dihedral angle between them. Clearly this can be done in either of two ways (depending on which diagonal you pick as the dihedral intersection). Above we chose the way that makes the solid a convex solid.

19.2.4 Groups D_{nh}

If you add a horizontal reflection plane to group D_2, you have group D_{2h}.

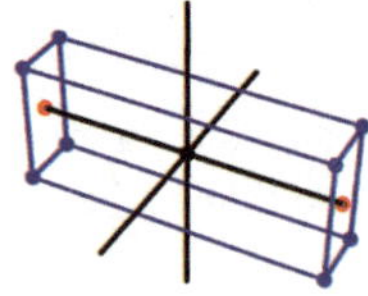
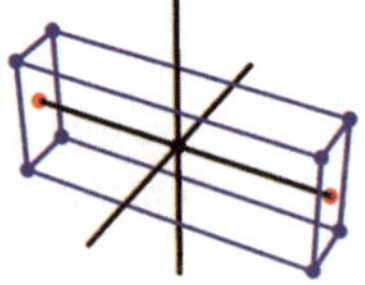

Fig. 19.6 Stereogram for group D_{2h}

Group D_{2h} is the symmetry group of a brick. In the general case, group D_{nh} has an n-fold rotational symmetry about the central vertical axis, a horizontal reflection plane, and a set of horizontal C_2 axes. It is quite instructive to look at this with a larger integer. Making the appropriate trivial change, the figure transforms itself to

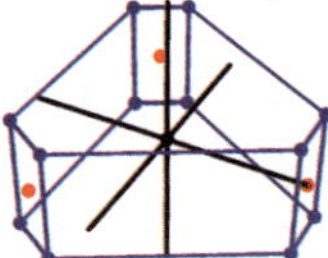
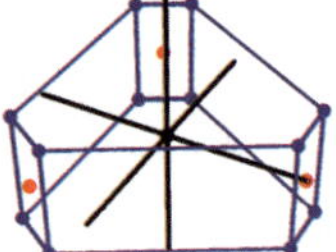

Fig. 19.7 Stereogram for group D_{3h}

The group order of D_{3h} is 12, and the 12 blue dots show the most general constellation of this group. The three red dots lie on the three horizontal two-fold axes, forming a less general constellation of their own.

19.2.5 Groups D_{nd}

The generating elements of D_{nd} are an n-fold rotation about z, a 2-fold rotation about x, and a vertical reflecting plane at angle $2\pi/(4n)$. The vertical plane bisects the angle between two adjacent C_2 axes. The picture for D_{4d} is

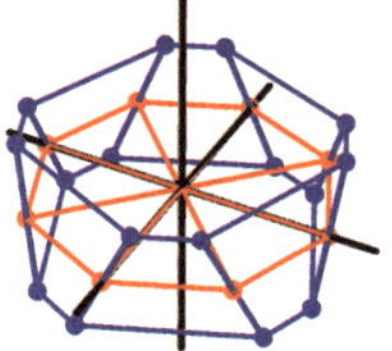
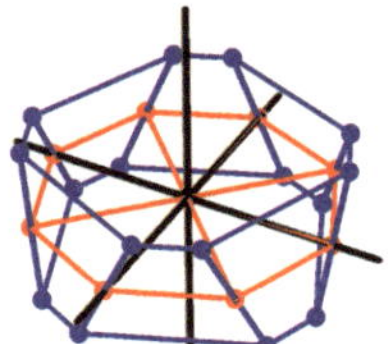

Fig. 19.8 Stereogram for group D_{4d}

The D_{nd} groups are quite different for odd and even **n**. You can see easily that there is no inversion center above, but there is one below :

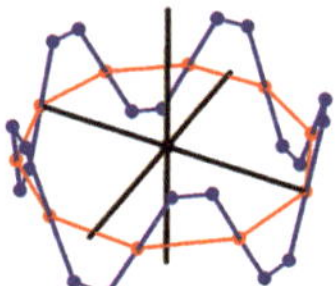
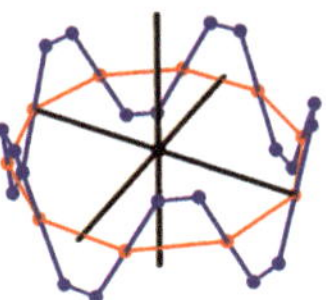

Fig. 19.9 Stereogram for group D_{5d}. Note the inversion center.

Now maybe you see why crown ethers belong to this group.

19.3 Constellations based on Platonic solids

19.3.1 Group T (rotations of the tetrahedron)

Group **T** consists of all the rotations of the tetrahedron.

The tetrahedron and the octahedron both belong to group **T**. But of course they are really of higher symmetry, possessing additional reflections not required in group **T**. But it is easy to generate the tetrahedral and octahedral vertices as constellations of group **T**. In the next cell we make a two complementary tetrahedra enclosing an octahedron :

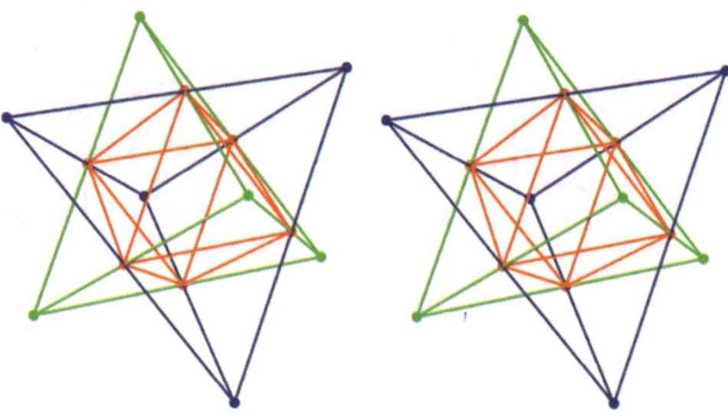

Fig. 19.10 Three special constellations of group T

The blue is the constellation of **{1,1,1}**; the green, of **{1,1,-1}**; the red, of **{1,0,0}**. None is general, having have fewer than 12 points. Below we construct a general constellation that belongs to group **T**, and to no higher group. Every coordinate axis is a twofold axis, and every cube diagonal is a threefold axis. But it has no reflection planes. First, we take a stereo view from an asymmetric perspective. The individual eye-views are difficult to interpret, but in stereo they are quite easy.

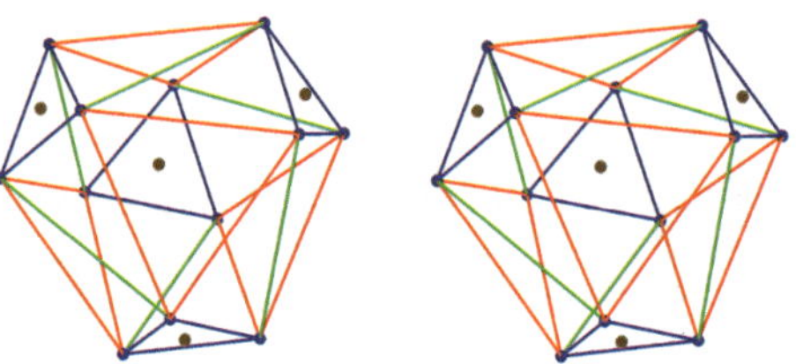

Fig. 19.11 General stereogram for group T.

There are 12 blue points, so this is the most general constellation of group **T**. It has 12 triangular faces, so it has the same connectivity as a dodecahedron. The brown dots are at tetrahedral vertices, just to help the eye organize this picture. Each blue vertex is five-coordinate, but the edges are of three different lengths, whereas in a true dodecahedron all edges are of equal length.

If the blue lines shrink to zero, the figure becomes a tetrahedron. If the red lines shrink to zero, it becomes another tetrahedron. If the green lines shrink to zero, it becomes an octahedron.

Below is a view down the x-axis (left) and a view down the cube diagonal (right). Note the two-fold and three-fold symmetries of the 2D figures.

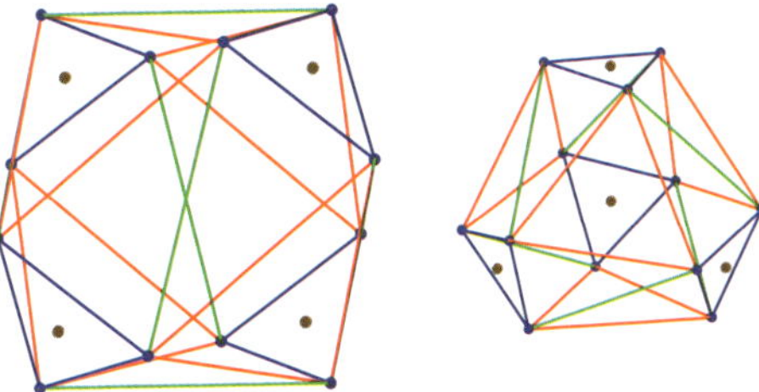

Fig. 19.12 Two different views of the general `T` constellation.

19.3.2 Group T_h A non-tetrahedral group

This group has a **`T`**-name because it is closely related to the tetrahedron, but the tetrahedron itself does not have this symmetry. However, both the cube and the octahedron may be constructed as constellations of group $\mathbf{T_h}$. We draw them together in conjugate position :

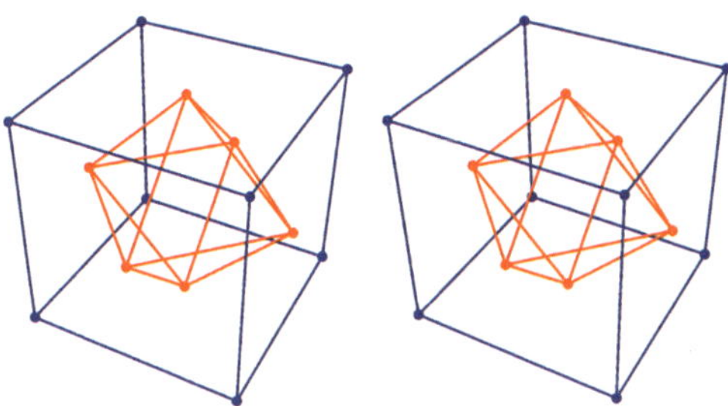

Fig. 19.13 Conjugacy of the cube and octahedron, using group $\mathbf{T_h}$.

The $\mathbf{C_3}$ axes of group **`T`** (the diagonals of the bounding box) become $\mathbf{S_6}$ axes in group $\mathbf{T_h}$. Here is a view directly down one of the $\mathbf{S_6}$ axes. Both the cube and the octahedron do possess $\mathbf{S_6}$ symmetry (around a body diagonal).

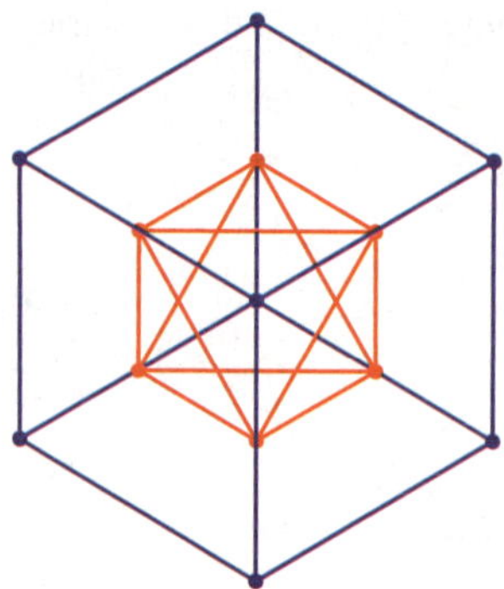

Fig. 19.14 Three-fold rotational symmetry of the cube and octahedron.

Group $\mathbf{T_h}$ has 24 elements, and things are starting to get complicated. Everything we have seen so far are special symmetric constellations; now we construct the most symmetric 24-point constellation of group $\mathbf{T_h}$. It is made of 18 squares and 8 triangles, all equilateral with each other. Each vertex is 4-coordinate. The individual eye-views below are too busy to interpret easily, but in stereo it looks quite simple.

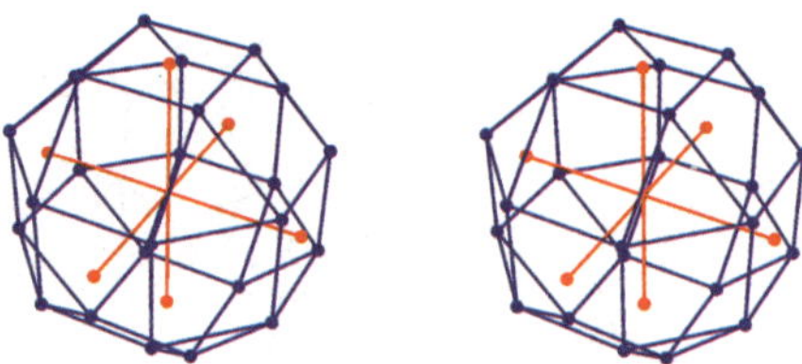

Fig. 19.15 General constellation of group $\mathbf{T_h}$, all edges of equal length.

If the classical geometers gave this solid a name, we have not been able to find it. (It is not the cuboctahedron.) It's a nice one though. But molecules that belong to $\mathbf{T_h}$ do not need to look as simple as this constellation. Below, we draw a 24-point constellation with three different edge lengths that belongs to $\mathbf{T_h}$ and to no higher group.

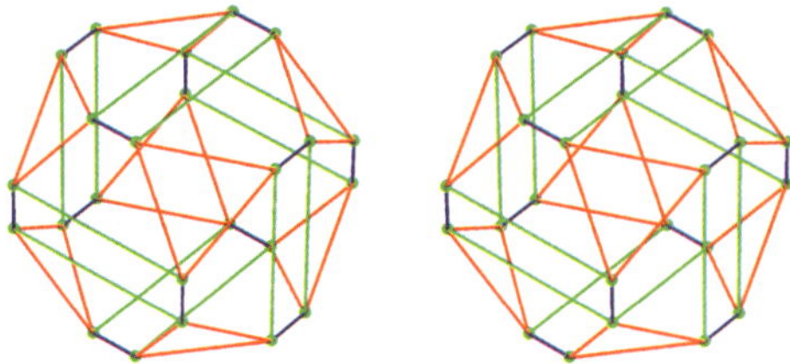

Fig. 19.16 General constellation of group T_h, with three different edge lengths.

This object has the same connectivity as the anonymous equilateral T_h polytope constructed above, but it has three different edge lengths.

19.3.3 Group T_d All symmetries of the tetrahedron

Schönfliess defined group T_d as all the symmetries of the tetrahedron.

We construct the most general constellation of group T_d, with 24 vertices and three different edge lengths.

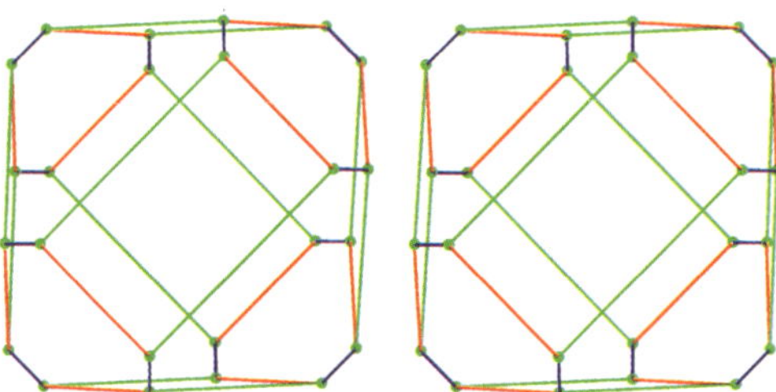

Fig. 19.17 General constellation of group T_d, with three different edge lengths.

The symmetry of group T_d is so high that the general constellation is quite symmetric. It is a cube with alternate vertices frustrated in two different ways, deeply enough that the original cube edges are completely consumed. The six rectangular faces are all that is left of the original cube faces. The remaining eight six-sided faces divide into two types. Four of them relate like the faces of a tetrahedron, while the other four relate like the faces of the conjugate tetrahedron.

19.3.4 Group O (rotations of the cube and octahedron)

Group O is defined by Schönfliess as all the rotations of the octahedron. We draw the general 24-point constellation with three different edge lengths :

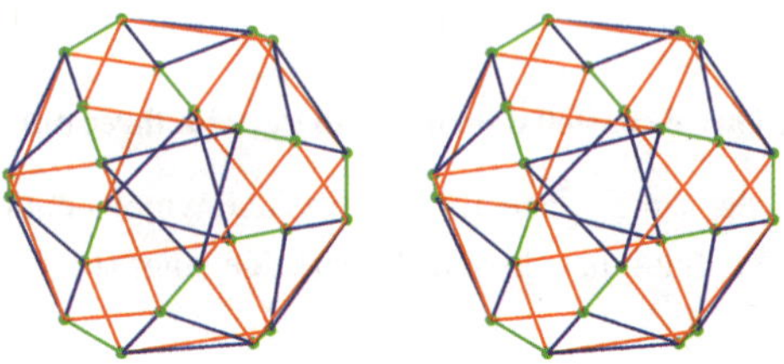

Fig. 19.18 General constellation of group O, with three different edge lengths.

A special symmetric constellation of this group was named by the classical geometers as the cuboctahedron. All its edges are the same length. It has 12 vertices and 12 faces; six of its faces are squares parallel to the faces of a cube; the other eight faces are triangles parallel to the faces of the conjugate octahedron.

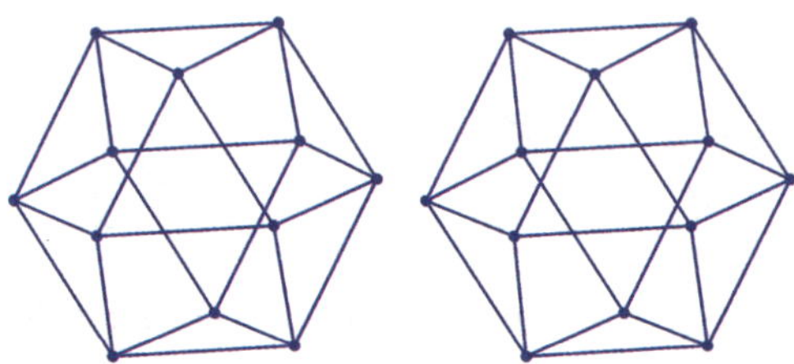

Fig. 19.19 The cuboctahedon, as a constellation of point {1,1,0} in group O.

19.3.5 Group O_h (all symmetries, cube and octahedron)

First, we display the cube and an inscribed octahedron, just to prove that they are both constellations of O_h.

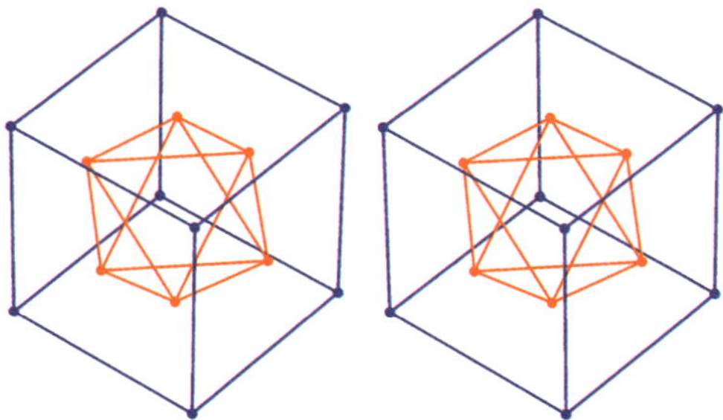

Fig. 19.20 The cube and the octahedron also belong to group `Oh`.

These simple figures are far short of the maximum number of points. To see the full complexity of group `Oh`, we must generate a constellation with 48 points.

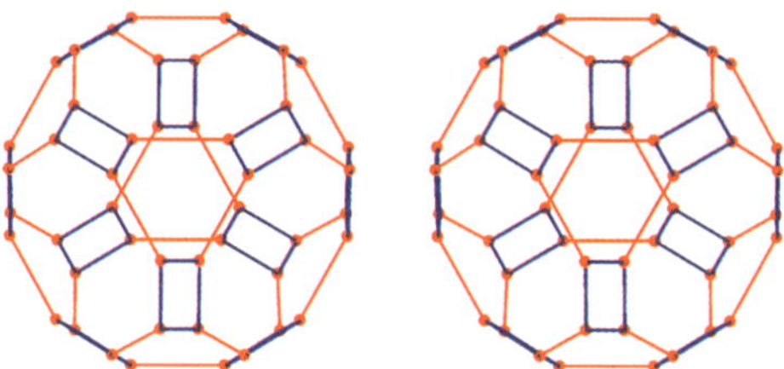

Fig. 19.21 The most general constellation of group `Oh`. All 48 points were generated from a single point.

It is perhaps notable that `Oh` possesses an inversion, though it starts from a four-fold rotation and a single reflection.

19.3.6 Group I (rotations of dodecahedron and icosahedron)

Construction of a five-fold upright group I

The International name of this group could be either **25** or **35**, indicating that it is generated by a fivefold axis and a two-fold or a three-fold axis. Taking the five-fold axis as the **z**-axis, there is a threefold axis in the **zx** plane given by

```
threefoldAxis = {-2 (3 + √5), 0, 7 + 3 √5};
```

Therefore, a minimal generator list for this group consists of the matrices

```
C5z = Cz[2 π / 5.];
C3q = AxialRotation[threefoldAxis, 2 π / 3.];
```

The decimals above make the whole thing numerical and very fast.

```
groupI = MakeGroup[Dot, {C5z, C3q}];
```

This group is not the same as our tabulated group `I`, but it works just as well.

> **On your own**
> It would be interesting to alter the graphics commands to work with the new two-fold upright version of group `I` in our tables. I would do it, but this book is taking too long already.

The icosahedron and dodecahedron

Both the icosahedron and the dodecahedron can be drawn as constellations of this group. The icosahedron is the constellation of any point on a fivefold axis :

Fig. 19.22 The icosahedron as a constellation of any five-fold axis point in group I.

We know that one vertex of the dodecahedron lies on the threefold axis that was used to construct group `I`, so the dodecahedron is the constellation of any point on one of the threefold axes of the group :

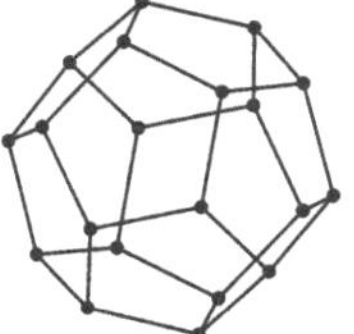

Fig. 19.23 Dodecahedron as a constellation of any three-fold axis point in group I.

The geodesic dome; the C_{60} molecule fullerene

The icosahedron has only **12** vertices and the dodecahedron has only **20**, so these figures are far from the general **60**-vertex constellation of this group. We now present the most symmetric general constellation of group **I**, which is none other than Buckminster Fuller's fabulous geodesic dome. Its **60** vertices are all connected to three other vertices, and all its edges are the same length. Its faces are either pentagons or hexagons. Each pentagonal face is surrounded by five hexagons; each hexagonal face is surrounded by three hexagons alternating with three pentagons. It is easily generated as a constellation of group **I**, of a single carefully chosen point. (We use $\left\{4,\ 0,\ \sqrt{5}+2\right\}$.)

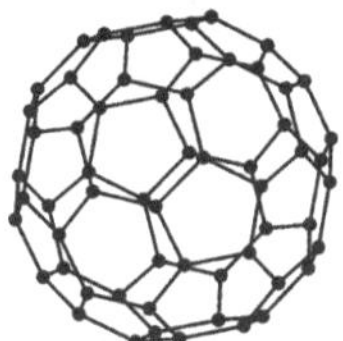

Fig. 19.24 The molecule buckminsterfullerene, C_{60}, as a constellation of group I.

There it is; the first fundamentally new architectural framework found since the middle ages; superbly light and strong. Buckminster Fuller used half of this structure to enclose sports arenas and other huge spaces. The molecule C_{60} forms in reasonable yield from carbon vapor under the right conditions.

By changing the constellation point to anything not on a symmetry element we get the most general constellation for group **I**, with two different edge lengths. Below, the constellation point is **{5,1,4}**.

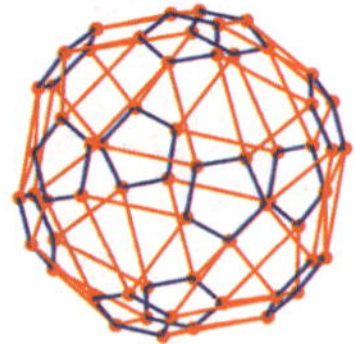 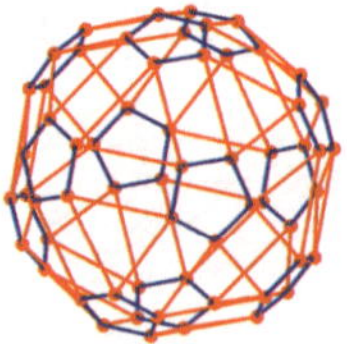

Fig. 19.25 The most general constellation of group I.

It still has all **60** vertices; the fivefold axes are still in the center of pentagons, and the threefold axes are still in the center of triangles, but new rectangles have appeared to fill the gaps.

19.3.7 Group I_h (all symmetries, dodeca- and icosa- hedron)

Here, at the very end of our catalog of point group constellations, we come to group $\mathbf{I_h}$ with 120 elements. The constellation is becoming too crowded to see much, even in stereo, so we will show only the top hemisphere of the constellations we create.

But first, we must create the group. We construct it as the product of the group **I** we used above, with an inversion center. Being numerical, it's instantaneous:

```
groupInv = {Emat, Imat};
groupIh = groupI ⊗ groupInv;
Dimensions[groupIh]
```

```
{120, 3, 3}
```

It has an inversion center and many reflecting planes, which **I** did not have.

```
Map[RecognizeMatrix, groupIh] // Chop // Sort;
Take[%, 3]
```

```
{{identity}, {inversion}, {reflection, {0, 1., 0}}}
```

Mathematicians know of finite groups with unbelievably large numbers of elements. One, called "the monster", has a precisely known number of elements said to be greater than the number of elementary particles in the universe. So don't get overwhelmed by $\mathbf{I_h}$.

Make the constellation for $\mathbf{I_h}$ of an arbitrarily chosen point, **{1,0.3,1.2}**, but show only the top half for visual clarity;

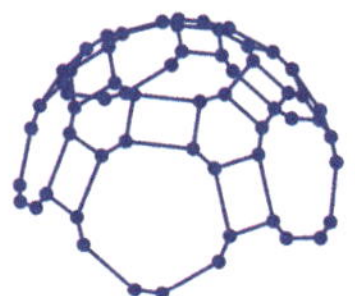
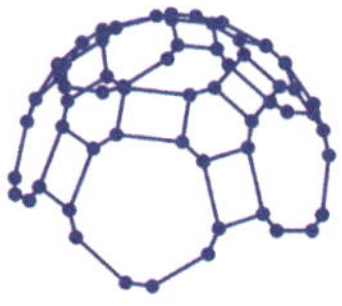

Fig. 19.26 Top half of the most general constellation of group $\mathtt{I_h}$.

The reflection planes of group $\mathtt{I_h}$ have done their job. The pentagons of the asymmetric geodesic dome of group $\mathtt{I}$ have expanded to become alternating decagons, and the triangles have expanded to become alternating hexagons.
No known molecule has a symmetry this complex. Academic chemists may take this as a challenge.

19.4 That's it, folks. What Next?

This completes the catalog of all possible kinds of point groups, and therefore the catalog of all symmetries that molecules can possess. We know how to generate the Cartesian representation of any point group, and how to use the representations to visualize pseudo-molecules that possess all the group symmetries. So this is the end of Part I of this book.

But we still know nothing about the concepts of *subgroup*, *class*, *species*, or *character* which make groups so useful in molecular physics. We begin to develop those concepts in the next chapter.

20. Subgroups

Preliminaries

20.1 Definition of subgroup

We begin with the fundamental definition of *subgroup*, and with two immediate modifications of that concept.

> **Subgroup**
> If $\mathcal{G}$ is a group, and if $\mathcal{H}$ is a subset of $\mathcal{G}$ that obeys all the group axioms, then $\mathcal{H}$ is a subgroup of $\mathcal{G}$.

The next definition is little more than a linguistic annoyance, a result of the way mathematicians define subsets. (Every **set** is deemed to be a **subset** of itself.)

> **Trivial subgroups**
> Every group has two trivial subgroups :
> **(1)** The unit element by itself, and
> **(2)** The entire group.

With the trivia out of the way, we can say what we nearly always mean when we speak of subgroups :

> **Proper subgroup**
> A subgroup that is not a trivial subgroup is a proper subgroup.

In other words, proper subgroups are smaller than the full group, but larger than the unit element alone.

All this is very nice, but why would a chemist or a molecular physicist bother with it? It is because of the following very practical fact: When a molecule is perturbed, the group generally changes. But there are constraints on how the group can change; the old and new groups are always related by the group-subgroup relation.

If you start with a highly symmetric molecule and exchange one of its symmetric atoms for an atom of another kind (with similar bonding) it is said that the symmetry is "broken" or "lowered" to a subgroup of the original group. Often the lowered symmetry will have new fundamental properties, such as a dipole moment or a transition moment lacking in the original. Thus the symmetry

W.M. McClain, *Symmetry Theory in Molecular Physics with Mathematica*,
DOI 10.1007/b13137_20, © Springer Science+Business Media, LLC 2009

analysis of molecular properties becomes a powerful tool for structural chemistry.

On the other hand, if you start with a molecule that has only one atom that prevents its belonging to some group, and if you change that atom appropriately, the symmetry will rise to a higher group that has the lower group as a subgroup.

The rest of this chapter is devoted to detailed examples of such simple substitutions, and their consequences in terms of group-subgroup relations between the old and new symmetries.

20.2 A simple but nontrivial example (group C_{3h})

20.2.1 A three-element subgroup of C_{3h}

We look immediately at an example of a proper subgroup. As the main group $\mathcal{G}$ we pick group $\mathbf{C_{3h}}$. Remembering that it is a product group $\mathbf{C_3} \otimes \mathbf{C_h}$ we know right away that both $\mathbf{C_3}$ and $\mathbf{C_h}$ will be subgroups. To verify, partition its table into 3×3 subtables using a special display operator, **NiceTable**, defined in the preliminaries :

```
NiceTable[MultiplicationTable["C3h"], 3, 3]
```

E	C3	C32	σh	S3	S35
C3	C32	E	S3	S35	σh
C32	E	C3	S35	σh	S3
σh	S3	S35	E	C3	C32
S3	S35	σh	C3	C32	E
S35	σh	S3	C32	E	C3

Group $\mathbf{C_3}$ is in the upper left. We check this :

```
NiceTable[MultiplicationTable["C3"], 3, 3]
```

E	C3	C32
C3	C32	E
C32	E	C3

Whenever a group is ordered so that one of its subgroups occurs together at the head of the list, it is very easy to spot that subgroup. The subgroup table is just an upper left section of the whole group table.

As a counter-example of a subset that is not subgroup, look at the upper left 4-by-4 sub-table of the $\mathbf{C_{3h}}$ table :

```
NiceTable[ MultiplicationTable["C3h"], 4, 4]
```

E	C3	C32	σh	S3	S35
C3	C32	E	S3	S35	σh
C32	E	C3	S35	σh	S3
σh	S3	S35	E	C3	C32
S3	S35	σh	C3	C32	E
S35	σh	S3	C32	E	C3

The upper left is not a subgroup because it is not a Latin Square, and therefore not a group.

20.2.2 A two-element subgroup of C_{3h}

We know that $\mathbf{C_h}$ is a subgroup, but even if we did not, it is easy to spot any subgroups of order 2 : just look for **E**'s on the diagonal. This is the sign of a self-inverse element, and therefore the sign of a 2-element subgroup. In the $\mathbf{C_{3h}}$ case, the reflection **σh** is such an element. To make this crystal clear, we permute the elements of $\mathbf{C_{3h}}$ so that elements **E** and **σh** are the first two elements. If you need to refresh yourself on permutations, click here. Note that we permute the names as well as the elements, so that the each name stays with its element.

```
perm = {1, 4, 2, 3, 5, 6};
names = ElementNames["C3h"][[perm]];
mats = CartRep["C3h"][[perm]];
NiceTable[
 MultiplicationTable[Dot, mats, names], 2, 2]
```

E	σh	C3	C32	S3	S35
σh	E	S3	S35	C3	C32
C3	S3	C32	E	S35	σh
C32	S35	E	C3	σh	S3
S3	C3	S35	σh	C32	E
S35	C32	σh	S3	E	C3

The top left is the table for group $\mathbf{C_h}$.

```
NiceTable[ MultiplicationTable["Ch"], 2, 2]
```

E	σh
σh	E

How many other subgroups are there in $\mathbf{C_{3h}}$?

On your own
Guessing that **{E,C3,C32,S3,S35}** might be a subgroup, follow the model above, permuting these five elements and their names to the front of the group, and make a multiplication table. Partition it using **Nice\
Table[…,5,5]** .

If this does not give a subgroup, try to find others.

Now you have tried very hard to find other subgroups of $\mathbf{C_{3\,h}}$, and you have not found any. So now you are ready to conclude that $\mathbf{C_3}$ and $\mathbf{C_h}$ are the only subgroups of $\mathbf{C_{3\,h}}$.*

Preview of coming attractions
A person who knows Lagrange's Theorem (coming in the next chapter) would say instantly that it is impossible for $\mathbf{C_{3\,h}}$ (six elements) to have a five-element subgroup. How could he possibly know this? You will see.

- * If you are deeply offended by this, read on.

20.3 Finding all the subgroups of a group

Think for a moment about the assertion at the end of the last section, and what level of credence you should give it. Searching for something and failing to find it is not a proof that it is not there, unless you prove formally that you have exhaustively examined all the possibilities. So the claim that we have found all the subgroups of $\mathbf{C_{3\,h}}$, as it stands, is mere persiflage.

Making certain that you have exhausted all possible candidates for subgroup is tedious, even with the help of Lagrange's Theorem. No one has ever found any algorithm for subgroup extraction other than exhaustive examination, though many have tried. Therefore, we need a tabulation of subgroups, and this is provided in one of the earliest chemistry books to use group theory, Wilson, Decius, and Cross *Molecular Vibrations*. WD&C tabulate many group-subgroup relations in their Appendix X-8, (Table X-14). To tell the frozen truth, our **Subgroups** operator relies upon the near-divine authority of WD&C, who in turn rely on many years of hand calculation by 19th century mathematicians. Just a couple of examples :

```
Subgroups["C2v"]
```

```
{C2, 2 Cv}
```

Why does it say **2 Cv** ? Look at the **ElementNames** of this group:

```
ElementNames["C2v"]
```

```
{E, C2, σvzx, σvyz}
```

There are two ways to choose a vertical reflection plane to make a subgroup: **{E,σvzx}** and **{E,σvyz}**. Sometimes the repeats are surprisingly numerous.

```
Subgroups["D6h"]
```

```
{D6, 2 D3h, C6v, C6h, 2 D3d, D2h, C6, C3h, 2 D3,
 2 C3v, S6, D2, 2 C2v, 3 C2h, C3, 3 C2, Cv, Ch, Cd, Ci}
```

Interactive readers can see the subgroups of all our tabulated point groups by processing the following cell :

```
Transpose[
  {GroupCatalog, Map[Subgroups, GroupCatalog]}];
% // Column
```

20.4 An orientation issue

There is a tricky issue that you must be aware of in dealing with subgroups. Taking $\mathbf{D_{3\,h}}$ as the main group, there are a lot of subgroups:

```
Subgroups["D3h"]
```

```
{C3h, D3, C3v, C2v, C3, C2, 2 Ch}
```

The tricky issue will appear if we look at subgroup $\mathbf{C_{2\,v}}$. It has four elements, all of which ought to be in $\mathbf{D_{3\,h}}$. But when we ask to see the common elements

```
Intersection[CartRep["D3h"], CartRep["C2v"]]
```

```
{{{1, 0, 0}, {0, -1, 0}, {0, 0, 1}},
 {{1, 0, 0}, {0, 1, 0}, {0, 0, 1}}}
```

there are only two. Do Wilson, Decius, and Cross have a mistake in their subgroup tables? Don't bet on it. Group $\mathbf{C_{2\,v}}$ is pretty simple:

```
Column[RecognizeMatrix /@ CartRep["C2v"]]
```

```
{identity}
{rotation, 2-fold, {{0, 0, 1}, π}}
{reflection, {0, 1, 0}}
{reflection, {1, 0, 0}}
```

It consists of a twofold rotation and two reflections perpendicular to the rotation axis, and to each other. Now look at D_{3h} :

```
Column[RecognizeMatrix /@ CartRep["D3h"]]
```

{identity}

$\{\text{rotation, 3-fold, } \{\{0, 0, 1\}, \frac{2\pi}{3}\}\}$

$\{\text{rotation, 3-fold, } \{\{0, 0, -1\}, \frac{2\pi}{3}\}\}$

$\{\text{rotation, 2-fold, } \{\{1, 0, 0\}, \pi\}\}$

$\{\text{rotation, 2-fold, } \{\{\frac{1}{\sqrt{3}}, 1, 0\}, \pi\}\}$

$\{\text{rotation, 2-fold, } \{\{-\frac{1}{\sqrt{3}}, 1, 0\}, \pi\}\}$

$\{\text{reflection, } \{0, 0, 1\}\}$

$\{\text{rotoreflection, 3-fold, } \{\{0, 0, 1\}, \frac{2\pi}{3}\}\}$

$\{\text{rotoreflection, 3-fold, } \{\{0, 0, -1\}, \frac{2\pi}{3}\}\}$

$\{\text{reflection, } \{0, 1, 0\}\}$

$\{\text{reflection, } \{\frac{\sqrt{3}}{2}, -\frac{1}{2}, 0\}\}$

$\{\text{reflection, } \{\frac{\sqrt{3}}{2}, \frac{1}{2}, 0\}\}$

There are three twofold rotation axes, but none is the **z** axis. One twofold axis is the **x**-axis. To build a C_{2v} subgroup on this, we need two reflections, one along **z** and the other along **y**. And indeed they are there: `{reflection,{0,0,1}}`and `{reflection,{0,1,0}}`. So C_{2v} really is present inside D_{3h}, but not in standard orientation.

20.5 Symmetry breaking in PCl_5 (group D_{3h})

20.5.1 Phosphorous pentachloride

Consider the molecule phosphorous pentachloride, PCl_5. Below, you can see that it is a triangular bi-pyramid. The two axial bonds are a little longer than the three equatorial bonds.

Isotopically pure

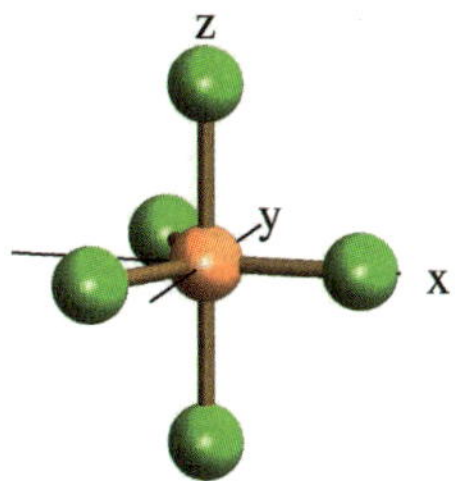

Fig. 20.1 The structure of $P(^{35}Cl)_5$. The axial bonds are longer than the equatorial bonds.

What are the symmetry elements of the isotopically pure **PCl$_5$**?

(1) The vertical axis **Cl - P - Cl** is a threefold rotation axis.
(2) Each of the three **P - Cl** bonds in the **xy** plane is a twofold axis.
(3) There is a horizontal reflection plane.
(4) The vertical axis is also a three-fold rotoreflection axis.
(5) There are three vertical reflection planes.

Compare these to the elements of group **D_{3h}** :

```
ElementNames["D3h"]
```

```
{E, C3a, C3b, C2a, C2b, C2c, σh, S3a, S3b, σva, σvb, σvc}
```

There is an exact match, so isotopically pure **PCl$_5$** belongs to group **D_{3h}**. But if we break its symmetry by replacing a **^{35}Cl** atom by a **^{37}Cl** atom, the group must become one of the following :

```
Subgroups["D3h"]
```

```
{C3h, D3, C3v, C2v, C3, C2, 2 Ch}
```

There are two possibilities : **(1)** Axial substitution, or **(2)** Equatorial substitution. We discuss these now in turn.

20.5.2 Axial isotope substitution

Axial substitution gives the molecule seen below. The bond distances and bond angles have not changed.

Axial isotope

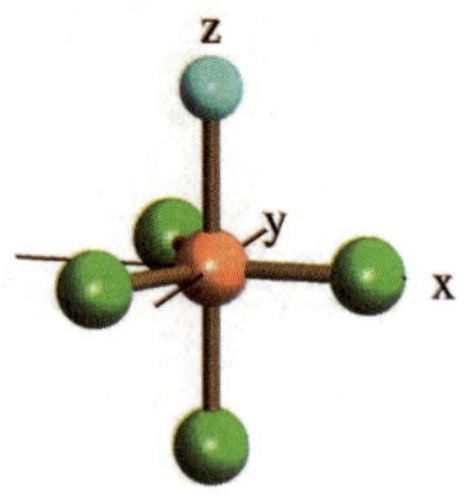

Fig. 20.2 **Axial** $P(^{37}Cl)(^{35}Cl)_4$

What are the symmetry elements?

The principal threefold rotation axis remains;

The three vertical reflection planes remain.

But the following elements are destroyed:

The three twofold horizontal axes;

The horizontal reflection plane;

The threefold rotoreflection axis.

With only a threefold rotation axis and three equivalent vertical reflection planes remaining, this is group $\mathbf{C_{3v}}$:

```
ElementNames["C3v"]
```

```
{E, C3a, C3b, σva, σvb, σvc}
```

Good; $\mathbf{C_{3v}}$ was one of the subgroups of $\mathbf{D_{3h}}$.

20.5.3 Equatorial substitution

What happens if the changed atom is equatorial?

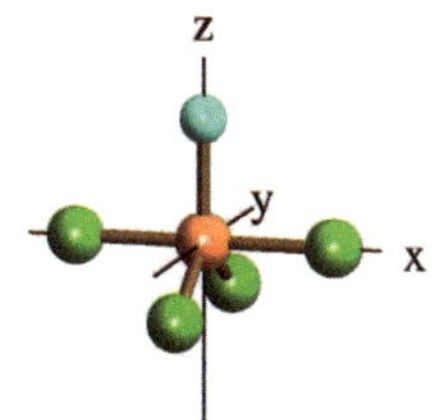

Fig. 20.3 Equatorial $P(^{37}Cl)(^{35}Cl)_4$. Left, naïve orientation; Right, standard.

On the left in Fig.20.3 you see PCl_5, changed to include one heavy equatorial atom. But the threefold vertical axis has been destroyed and cannot serve legally as the **z** axis (Rule 2). The rotation axis of highest order is the **x** axis in this picture, so we must turn it vertical, as you see on the right. There is no other rotation axis, so Rule 3 does not apply. Rule 4 says that one reflection plane must be the **z,x** plane, and it is. There are two vertical reflection planes, and we could legally have used the other just as well.

(1) The new **z** axis is a twofold rotation axis. There is no other rotation.

(2) There are two reflecting planes (**z,x** and **z,y**) intersecting along **z**.

Compare this to

```
ElementNames["C2v"]
```

```
{E, C2, σvzx, σvyz}
```

There is an exact match. Good; C_{2v} was also was one of the subgroups of D_{3h}.

20.5.4 Two-atom substitutions $P(^{37}Cl)_2(^{35}Cl)_3$

Here are three easy problems. Click to the answers only to check yourself.

Problem 1 Two heavy axial atoms

What is the group of PCl_5 with two heavy axial isotopes?

Two axial isotopes

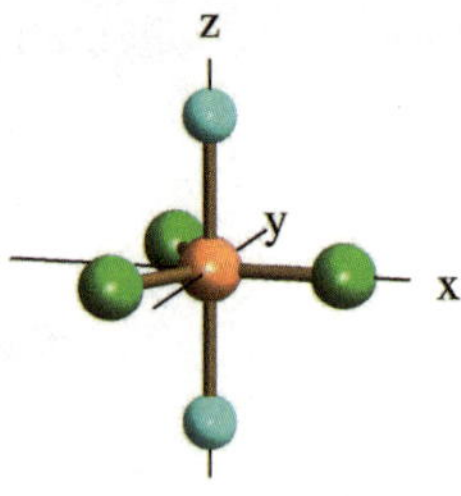

Fig. 20.4 Diaxial $P(^{37}Cl)_2(^{35}Cl)_3$

Click here to check your answer.

Problem 2 : Two heavy equatorial atoms

What is the group of PCl_5 with two heavy equatorial isotopes?

```
Show[npcl2]
```

Two equatorial
isotopes

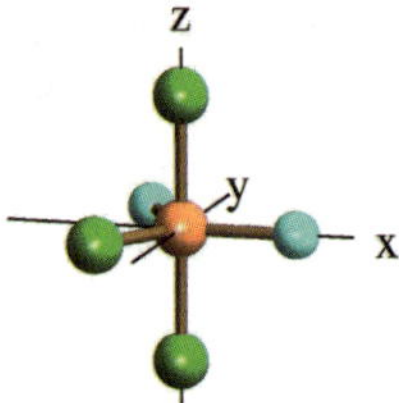

Click here to check your answer.

Problem 3 : One axial and one equatorial

Axial–equatorial
standing up

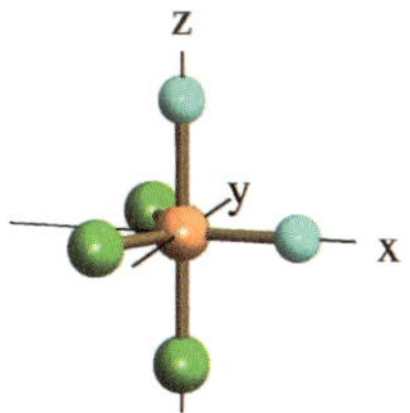

With one heavy axial atom it became C_{3v}; with one heavy equatorial atom it became C_{2v}. But what is it with both?

Click here to check your answer.

20.6 Definition of "symmetry breaking"

We have now seen examples where isotopic substitution has **(1)** lowered the symmetry, **(2)** left the symmetry unchanged, and **(3)** raised the symmetry. So now we can make a precise statement of what "symmetry breaking" really means.

> **Symmetry breaking**
> When we say that a high symmetry is "broken", we usually mean that a perturbation has destroyed a symmetry element. Therefore, when the remaining symmetry elements are used to generate a group, the new group can only be a subgroup of the old group. It cannot switch to a new unrelated symmetry.
>
> In cases where substitution raises the symmetry (introduces new symmetry elements) the old group is a subgroup of the new group.

20.7 End Notes

Answer 1 : two heavy isotopes, diaxial

Two heavy axial isotopes restore the full symmetry of isotopically pure molecule, $\mathbf{D_{3\,h}}$ Click back to the top for a blow-by-blow analysis.

Answer 2 : two heavy isotopes, diequatorial

It helps a little to rotate by 1/3 turn about **z**, because that makes the **x**-axis into a twofold rotation axis (left figure). But the vertical axis has lost its rotation symmetry. Turning the **x** axis vertical, it is like Fig.20.XXX, except that the al atoms are below the **xy** plane instead of above. So the group is again $\mathbf{C_{2\,v}}$.

Two equatorial isotopes
zx reflection

Two equatorial
standard orientation

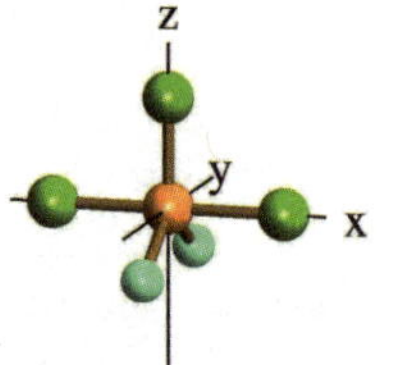

Answer 3 : one axial, one equatorial

This molecule has one reflection plane and no rotation axis. If you orient it so its plane is **zx**, it belongs to group $\mathbf{C_v}$; but Rule 1 says to put the reflection plane horizontal when it is the only symmetry, so it is group $\mathbf{C_h}$.

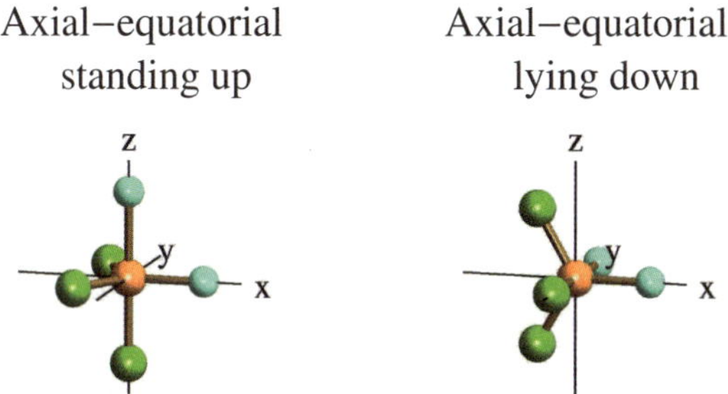

21. Lagrange's Theorem

Preliminaries

21.1 Purpose

Our goal in this chapter is Lagrange's theorem, a surprising and powerfully useful result about the order of subgroups. It is the earliest result in abstract group theory (ca. 1770), and its beauty and simplicity motivated a great deal of work in this field, before it even was a field. We prove it in 21.5.0, below. But this theorem depends on the concept of **cosets**, and on a **coset lemma**. Therefore we must do these first.

21.2 Definition of coset

First, a little extension of the meaning of ⊗ :

> **An element times a group, a ⊗ H**
> If **H** is a group of elements $\{h_1, h_2, \ldots\}$ and **a** is an element that multiplies with them, then **a ⊗ H** means $\{a \otimes h_1, a \otimes h_2, \ldots\}$.
>
> In *Mathematica*-talk, this means that the operation ⊗ is `Listable`.

Now we can define the *coset*. Actually, there are two kinds, the *left coset* and the *right coset* :

> **Cosets, left and right**
> The set of elements in **a ⊗ H** is the **left coset** of **H** by **a**;
> similarly, **H ⊗ a** is the **right coset** of **H** by **a**.

We will work below only with *left* cosets, but everything is similar if you switch to *right* cosets.

W.M. McClain, *Symmetry Theory in Molecular Physics with Mathematica*,
DOI 10.1007/b13137_21,

21.3 An operator for cosets

We do not know for sure how Lagrange discovered his great coset theorem, but it is likely that once he had the idea of cosets, he did examples on a number of groups that he knew about. We can do such experiments very easily and quickly using *Mathematica.*

We define an operator that make it easy to construct cosets of a subgroup within a group. It is called

```
MakeCoset[groupName,subgroupName,elementName]
```

It converts names to matrices, constructs the coset using the matrices, and then converts the coset matrices back to names :

```
Clear[MakeCoset];
MakeCoset[gpNm_, sbgpNm_, elNm_] := Module[
    (*gpN=groupName,
    sbgpN=subgroupName,elN=elementName*)
    {elMat, sMats, leftCosetMats, cosetNames},
    (*Turn the element name into a matrix : *)
    elMat = elNm /. NamesToRepMats[gpNm];
    (*Pull out the subgroup matrices: *)
    sMats = CartRep[sbgpNm];
    (*Make the left coset matrices: *)
    leftCosetMats = Map[elMat.# &, sMats, 1];
    (*Turn the coset matrices into names: *)
    cosetNames =
     leftCosetMats /. RepMatsToNames[gpNm] ];
```

Note the trailing **1** in the **Map** operation. This makes it **Map** at level **1**, so that **Dot** will act on matrices, not matrix elements.

Technical note:

In the **Symmetry`** tables, we sometimes give the same matrix a different name in different groups, and that could mess up the experiment. So when we convert matrices to names (or vice-versa) we use the names from the *group*, not the *subgroup*.

21.3.1 Make all the cosets of a subgroup

We test this operator immediately. We have been using group $\mathbf{D_{3\,h}}$ as an example, and we continue here. First look at the names of its elements:

```
ElementNames["D3h"]
```

```
{E, C3a, C3b, C2a, C2b, C2c, σh, S3a, S3b, σva, σvb, σvc}
```

There are a lot of subgroups to choose from:

```
Subgroups["D3h"]
```

```
{C3h, D3, C3v, C2v, C3, C2, 2 Ch}
```

We pick $\mathbf{C_3}$ because its principal axis, as it sits inside $\mathbf{D_{3\,h}}$, is the standard **z** axis, (If we picked $\mathbf{C_{2\,v}}$ we would run into trouble because its principal axis, as it sits within $\mathbf{D_{3\,h}}$, is the **x**-axis. Click to review.)

```
ElementNames["D3h"]
```

```
{E, C3a, C3b, C2a, C2b, C2c, σh, S3a, S3b, σva, σvb, σvc}
```

First, make a coset based on an arbitrary element of $\mathbf{D_{3\,h}}$ (namely, **"C2c"**):

```
MakeCoset["D3h", "C3", "C2c"]
```

```
{C2c, C2b, C2a}
```

So **MakeCoset** seems to work. Since subgroup $\mathbf{C_3}$ has three elements, there will always be three elements in any coset of $\mathbf{C_3}$.

The experiment is to make as many cosets as possible, reporting the list of cosets made and the list of elements left over. We base it on a **While** operator, which cycles *while* some criterion is met, and stops when it is not met.

Look at the setup below. The empty **cosets** list will gradually fill, and the **unused** list will gradually empty as the operations cycle under **While**. Every cycle takes the first element from the **unused** list and makes its coset. That coset is appended to the **cosets** list and its members are deleted from the **unused** list. The cycle stops when there are not enough **unused** elements to make one more coset.

```
(*Setup*)
cosets = {};
unused = ElementNames["D3h"];

(*Start the cycle*)
While[Length[unused] > 2,
```

```
newCoset = MakeCoset["D3h", "C3", unused[[1]]];
cosets = Append[cosets, newCoset];
unused = Complement[unused, Flatten[cosets]] ];
{cosets, unused}
```

```
{{{E, C3a, C3b}, {C2a, C2c, C2b},
  {S3a, S3b, σh}, {σva, σvc, σvb}}, {}}
```

On this trial, the 12 elements of $\mathbf{D_{3h}}$ were neatly partitioned into mutually exclusive cosets, with nothing left over. And strangely, when Lagrange tried this on all the groups he knew of, the same thing happened.

(1) No element ever appeared in two cosets, and

(2) Nothing was ever left over.

The first observation became the Coset Lemma, and the second became Lagrange's Theorem. The proofs follow.

On your own

Try some other groups and subsets.

But be aware of the orientation issue. To work with every subgroup, **MakeCoset** would have to lift its subgroup elements out of the tabulated main group, rather than the tabulated subgroup. By suitable use of **RecognizeMatrix** you might be able to make this happen. If you enjoy creative programming, try it.

21.4 The coset lemma

21.4.1 Statement and proof

Coset Lemma
If two cosets of a proper subgroup have one element in common, then they have all their elements in common.

For proof, go through the experiment using general symbols instead of concrete examples, as we did in the experiments.

Let the group be **{E,A,B,C,... ,R,S,T,... ,Z}** and let the subgroup b **{E,R,S,T}**. First, construct the coset of **E** and set it aside (it will be the subgroup itself.) Then construct cosets with two elements, **A** and **B**, that are in the main group but outside the subgroup. That is, construct

A⊗{E,R,S,T} and **B⊗{E,R,S,T}**.

Suppose these two cosets are different, except for one element that they have in common; i.e., suppose for some subgroup element **R**, it is true that **A⊗R = B⊗S**. All group elements have an inverse, so we can right multiply by $\mathbf{R}^{-1}$:

$$\mathbf{A} = \mathbf{B} \otimes \mathbf{S} \otimes \mathbf{R}^{-1}$$

But **S** and **R** and $\mathbf{R}^{-1}$ are all in the subgroup. So also $\mathbf{S} \otimes \mathbf{R}^{-1}$ must be in the subgroup; call it element **T**. Then

$$\mathbf{A} = \mathbf{B} \otimes \mathbf{T}$$

Using this form of **A** in the calculation of the **A** coset , it becomes

(B ⊗ T) ⊗{E,R,S,T}

Now the power of using general symbols kicks in. We regroup as

B⊗ (T⊗{E,R,S,T}))

By the rearrangement theorem, **T⊗{E,R,S,T}**contains the same elements as **{E,R,S,T}**, but in a different order. So the **A** and **B** coset lists are the same except for order. But order does not matter in a set, so the two co*sets* are identical.

Therefore, if two cosets have one element in common, they have all their elements in common. Which was to be shown.

21.4.2 A diagram for the lemma

Below is a diagram of the situation described the Coset Lemma, with big group $\mathcal{G}$ and subgroup $\mathcal{S}$, and cosets **A⊗**$\mathcal{S}$ and **B⊗**$\mathcal{S}$. If there are other cosets, consider them to be shown similarly. The main point is, by the Coset Lemma, the cosets have no points in common.

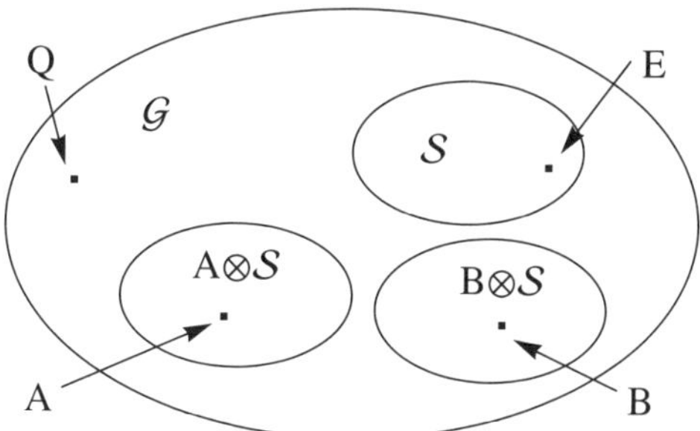

But this diagram also raises a question. If subgroup $\mathcal{S}$ contains **m** elements, then all its cosets also contain **m** elements, and if it is possible to form **n** cosets, the

total number of elements in cosets is **m*n**. But suppose that the number of elements in big group $\mathcal{G}$ is **m*n+1**, with orphan element **Q** left out of all cosets.

Lagrange's experiments never found such a **Q**, but that does not prove that somewhere out there is a peculiar group that does possess an orphan **Q**.

- Linguistic note: Long ago, the word "factored" meant "separated into parts". Today it means almost exclusively "separated into parts that multiply together to make the whole". But in Lagrange's time it had not yet drifted toward this meaning. So when Lagrange divided a group into cosets, he said he had "factored" it. This usage persists to this day in many statements of LaGrange's Theorem. Don't let it confuse you.

21.5 Lagrange's theorem proved

Lagrange's theorem
If **G** is a group and **S** is a subgroup of **G**, the group order of **S** is an exact divisor of the group order of **S**.

The formal proof is pretty easy in the light of the Coset Lemma. Let $\mathcal{G}$ be a group of order **g**, and let $\mathcal{S}$ be a subgroup of $\mathcal{G}$, of order **s**.

Suppose that when $\mathcal{G}$ has been partitioned into as many cosets as possible, there is one element **Q** left over. But nothing prevents us from making the coset of **Q**, and when we do it will contain **s** elements, just like any other coset. This means it must contain **Q** plus other elements already in one of the cosets. But the coset lemma says this is impossible, so the existence of leftover element **Q** is a contradiction. There are never any leftover elements.

Let **c** be the number of cosets. Since every coset contains **s** elements, it must be that **c*s = g**, the total number of elements. Thus integer **c** is **g/s**. In other words, the number of subgroups **s** is an exact divisor of the order of the main group **g**, which was to be shown.

21.5.1 Check the Symmetry` package tabulations

If we divide the order of the main group by a list of the orders of all its sub-groups, we should get a list of integers. Try it :

```
GroupOrder["D6d"]
Map[GroupOrder, Subgroups["D6d"]]
```

$$\left\{2,\ 2,\ 4,\ 3,\ 4,\ 4,\ 6,\ 6,\ 6,\ 8,\ \frac{24}{2 + C2},\ 12\right\}$$

On Your Own
It almost works. Can you fix it? It may need a **Module**.

21.5.2 Groups of prime order: a corollary

A corollary of Lagrange's theorem
Groups of prime order have no proper subgroups.

Thus it is useless to look for the proper subgroups of group C_5, for instance, because it has five elements, and **5** is a prime number. Prime numbers, by definition, have no divisors other than themselves and **1**.

21.5.3 A caution

Suppose you find a group with **28** elements. In examining it for subgroups, what orders might be expected?

```
Divisors[28]
```

```
{1, 2, 4, 7, 14, 28}
```

Any of these divisors *might* be the order of a subgroup, but it does not *have* to be. There is no requirement that each divisor has to correspond to a subgroup, except for the two trivial subgroups with orders **28** and **1**.

22. Classes

Preliminaries

22.1 Qualitative meaning of classes

In this chapter we will construct operator named **MakeClass** that will separate the elements of any group of elements into "classes" of *similar* elements. We demonstrate it immediately :

```
Classify[CartRep["C3v"]] /. RepMatsToNames["C3v"]
```

```
{{E}, {σvc, σvb, σva}, {C3a, C3b}}
```

The identity **E**, the three **σv** reflections, and the two **C3** rotations constitute the three *classes* of group **C3v**. And indeed,the elements in each class seem intuitively "similar" to each other.

How did it perform this trick? Be assured **Classify** did not use the name similarities to classify them. It would have separated rotations from reflections no matter what the names were. Also, it did not use any specific geometric properties to classify them. All groups have classes, even those with no geometrical meaning.

22.2 Graphics of two "similar" elements

In Chapter 4 we showed a before- and after- picture of the ammonia molecule in its standard position, undergoing a σ_v reflection (in the **zx** plane). Also we showed the potential due to the three H atoms of ammonia, a function with C_{3v} symmetry. Now we show a sequence of operations that performs a reflection in a plane that is not used directly :

W.M. McClain, *Symmetry Theory in Molecular Physics with Mathematica*,
DOI 10.1007/b13137_22, © Springer Science+Business Media, LLC 2009

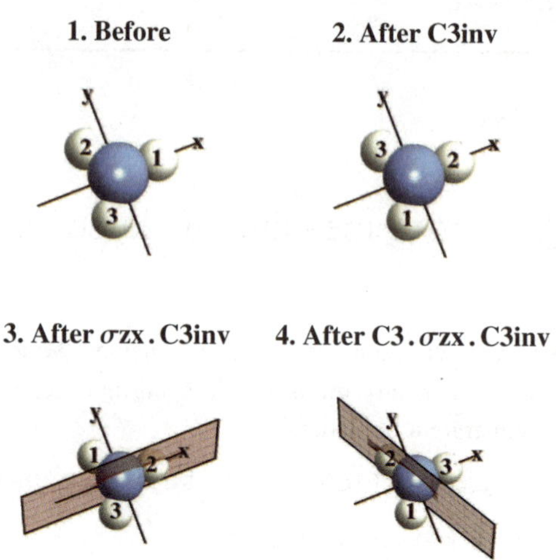

Fig. 22.1 Two similar reflections in group C_{3v}

Starting from original position in the upper left, we first rotate clockwise by **`1/3`** turn (**After C3inv**), then we reflect in the colored **`zx`** plane (**After** σ**zx.-C3inv**), then we back-rotate both the molecule and the colored plane (**After C3.**σ**zx.C3inv**). Comparing H atom numbers in the first and last pictures, the overall operation is simply reflection in the rotated plane.

This is an illustration of a very general principle. In any compound operation of the type $\mathbf{U.V.U^{-1}}$, operation $\mathbf{U^{-1}}$ first transforms the molecule, then **V** performs some other transform, and then **U** undoes the original transform both on the molecule and on the symmetry element used by **V** (above, the reflection plane shown). The final result is the transform of the molecule by a symmetry element *similar* to **V**, but in a different orientation.

> $\mathbf{U.V.U^{-1}}$ is called a "similarity transform" of **V** by **U**

It seems natural to say that **V** and $\mathbf{U.V.U^{-1}}$ are elements in the same "class".

22.3 Finding all the similar elements

22.3.1 All the reflections of group $C_{3\,v}$

Let's repeat this "experiment" keeping the central σ_{va} operator, but using another element in the place of the $C_{3\,a}$ rotation. In fact, it is easy to do it for all the elements in the group at once. We just **Map** the triple product **mat.**$C_{3\,a}$**.** **mat**$^{-1}$ onto every **mat** in the group.

```
mats6 =
 Map[(#.σzx.Inverse[#]) &, CartRep["C3v"]] /.
  RepMatsToNames["C3v"]
```

```
{σva, σvc, σvb, σva, σvc, σvb}
```

They are all reflections, named σ-something. So the operation

$$\mathbf{mat.reflection.mat}^{-1}$$

produces only reflections as **mat** runs over the whole group! Six reflections are generated as **mat** runs over the six elements of the group. But the group contains only three reflections, so there must be repeats. We cut them away with **Union** :

```
Union[mats6]
```

```
{σva, σvb, σvc}
```

So we have created an operator which, when applied to a whole group, selects out just the vertical reflections of the group. Can we do the same thing for rotations?

22.3.2 All the rotations of group $C_{3\,v}$

We do the same thing again, but this time using one of the rotations as the central element of the operator sandwich. This time, we put **Union** right into the operator to eliminate repeats automatically :

```
rotMat = "C3a" /. NamesToRepMats["C3v"];
Map[(#.rotMat.Inverse[#]) &, CartRep["C3v"]] /.
  RepMatsToNames["C3v"] // Union
```

```
{C3a, C3b}
```

It did it again! As **mat** runs over the whole group, the operation

$$\mathbf{mat.rotMat.mat^{-1}}$$

produces all the group elements that are rotations by multiples of **1/3** turn about the **z** axis.

22.4 Definition of class

This is enough to motivate a formal definition :

> **Definition of class**
> The class of element **x** in group **G** is the set of elements generated by
>
> $$\mathbf{m.x.m^{-1}}$$
>
> as **m** runs over all the elements of group **G**. Alternative statement:
>
> $$\mathcal{C}[\mathbf{x}] = \langle \mathbf{m.x.m^{-1}} \rangle \;\forall\; \mathbf{m} \text{ in } \mathbf{G}$$

The script capital $\mathcal{C}$ means “class”. The ∀ symbol is read by mathematicians as “for all”, and in this book the pointed brackets 〈... 〉 encloses a “set”. A set i like a **List**, except that no ordering is implied and no repeats are allowed. The **Symmetry`** package has an operator that creates classes by an algorithm based directly on this definition. We discuss it next.

22.5 A very pregnant little discussion

Now that you know what a similarity transform is, and what classes are, you should click into the MatrixReview and read an amazing little discussion there. It shows that entire theories can be similarity transformed, sometimes with an important simplification. It is summarized with a metaphor:

"... similarity transforms change the costumes of vectors and matrices, but underneath the costumes they remain the same troupe of clowns, and they perform the same show."

If you can't recognize a clown by his costume, how do you recognize him? You have to do it by his *character*, the kind of role he plays in the show. The *character table* divides the clowns into classes on the basis of their characters, and tells how many belong to each class. Every troupe has a unique character table. So if you have a set of *character tables* for different troupes, you can always tell which you are looking at, whatever their costumes happen to be.

22.6 Operator `MakeClass`

The definition of class has been embodied as a **Symmetry`** package operator called **MakeClass**. The thumbnail sketch is

```
MakeClass[el (, grp)] returns the class of element
el within group grp. Element el may be a matrix or
a matrix name, and grp may be a group of matrices
or a group name. The default for grp is
$DefaultGroup.
```

We use this operator to quickly review the results we got above :

```
MakeClass["C3a", "C3v"] /. RepMatsToNames["C3v"]
```

```
{C3a, C3b}
```

```
MakeClass["σva", "C3v"] /. RepMatsToNames["C3v"]
```

```
{σvc, σvb, σva}
```

Fine, it does what it should. Try it now on a bigger group with a lot more possibilities, the octahedral group **O** . We will see a little surprise at the end.

```
ElementNames["O"]
```

```
{E, C3mmm, C3mmp, C3mpm, C3mpp, C3pmm, C3pmp, C3ppm,
 C3ppp, C2x, C2y, C2z, C4xm, C4xp, C4ym, C4yp, C4zm,
 C4zp, C2mxy, C2myz, C2mzx, C2pxy, C2pyz, C2pzx}
```

```
MakeClass["E", "O"] // Map[RecognizeMatrix, #] &
```

```
{{identity}}
```

The class of the identity is just the identity itself. Now make the class of a threefold rotation, and use **RecognizeMatrix** to see what kinds of transforms are in the class :

```
MakeClass["C3mmm", "O"] // Map[RecognizeMatrix, #] &
```

$$\left\{\left\{\text{rotation, 3-fold, }\left\{\{1,-1,1\},\frac{2\pi}{3}\right\}\right\},\left\{\text{rotation, 3-fold, }\left\{\{-1,1,1\},\frac{2\pi}{3}\right\}\right\},\right.$$
$$\left\{\text{rotation, 3-fold, }\left\{\{1,-1,-1\},\frac{2\pi}{3}\right\}\right\},\left\{\text{rotation, 3-fold, }\left\{\{-1,-1,1\},\frac{2\pi}{3}\right\}\right\},$$
$$\left\{\text{rotation, 3-fold, }\left\{\{-1,1,-1\},\frac{2\pi}{3}\right\}\right\},\left\{\text{rotation, 3-fold, }\left\{\{1,1,1\},\frac{2\pi}{3}\right\}\right\},$$
$$\left.\left\{\text{rotation, 3-fold, }\left\{\{1,1,-1\},\frac{2\pi}{3}\right\}\right\},\left\{\text{rotation, 3-fold, }\left\{\{-1,-1,-1\},\frac{2\pi}{3}\right\}\right\}\right\}$$

They are all threefold rotations. Try the class generated by a fourfold rotation :

```
MakeClass["C4xm", "O"] // Map[RecognizeMatrix, #] &
```

$\{\{\text{rotation, 4-fold, }\{\{0,0,1\},\frac{\pi}{2}\}\},\{\text{rotation, 4-fold, }\{\{0,-1,0\},\frac{\pi}{2}\}\},$
$\{\text{rotation, 4-fold, }\{\{0,1,0\},\frac{\pi}{2}\}\},\{\text{rotation, 4-fold, }\{\{0,0,-1\},\frac{\pi}{2}\}\},$
$\{\text{rotation, 4-fold, }\{\{1,0,0\},\frac{\pi}{2}\}\},\{\text{rotation, 4-fold, }\{\{-1,0,0\},\frac{\pi}{2}\}\}\}$

They are all four-folds, as you might expect. Now generate the class of the two-fold rotation named **C2x** :

```
MakeClass["C2x", "O"] // Map[RecognizeMatrix, #] &
```

```
{{rotation, 2-fold, {{0, 0, 1}, π}},
 {rotation, 2-fold, {{0, 1, 0}, π}}, {rotation, 2-fold, {{1, 0, 0}, π}}}
```

There are only three members; that is a bit of a surprise. The two-fold named **C2mxy** , among others, is missing above. What is its class?

```
MakeClass["C2mxy", "O"] // Map[RecognizeMatrix, #] &
```

```
{{rotation, 2-fold, {{0, -1, 1}, π}}, {rotation, 2-fold, {{0, 1, 1}, π}},
 {rotation, 2-fold, {{-1, 1, 0}, π}}, {rotation, 2-fold, {{-1, 0, 1}, π}},
 {rotation, 2-fold, {{1, 0, 1}, π}}, {rotation, 2-fold, {{1, 1, 0}, π}}}
```

The first class of 2-folds rotates around the coordinate axes; the second, around dihedral axes that lie between two coordinate axes. The two classes are "pretty similar", but not similar enough. To put them in the same class, the group would have to contain rotations around the coordinate axes by **π/4**.

In group theory, never trust simplistic statements. It just takes too many words to cover all possibilities, so some things people say informally (like qualitative explanations of "similar") are mostly true, but not precisely true.

22.7 Table lookup operators for classes

The group tables in the **Symmetry`** package know all about classes. If you know the name of a group, and if it is tabulated, there is a shortcut around **MakeClass** The tetrahedral group **Td** is a good illustration :

```
Classes["Td"]
```

```
{{E}, {C3a, C3b, C3c, C3d, C3e, C3f, C3g, C3h},
 {C2a, C2b, C2c}, {S4a, S4b, S4c, S4d, S4e, S4f},
 {σda, σdb, σdc, σdd, σde, σdf}}
```

The 24 elements of the group form five classes. They need names, which have to be tabulated, like all conventions :

```
ClassNames["Td"]
```

```
{E, C3, C2, S4, σd}
```

How many elements are there in each class?

```
ClassPopulations["Td"]
```

```
{1, 8, 3, 6, 6}
```

This kind of data is available for every tabulated group. Change the group name and go through all the operations again. You will see.

22.8 Theorems about classes

The little calculation-experiments above illustrate several points that are true for the classes of every group. These points are stated as the five theorems below. The proofs are easy; you start them all by writing down the definition of class and staring at it, while you think about how to use the special condition mentioned in the theorem. Then start to calculate.

Proving these theorems for yourself will help to fix the definition of class in your mind, which you MUST do. Class is a central concept in the physical applications of group theory, and will reappear many times throughout the rest of this book.

> **1. Every element belongs to the class that it generates.**
> Write the definition down and think through the details of what happens when the definition is set equal to the middle element in the triple product. The proof is a one-liner.

> **2. In every group, the unit element E forms a class by itself.**
> Again, the proof is a one-liner, based on a special thing that happens when the triple product is evaluated with an **E** in the center.

> **3. No element belongs to two classes.**
> This one is just a little harder. Write two equations expressing the idea that element **Q** is a member of two different classes. Then eliminate Q and work until you find a contradiction. If it takes more than a few minutes, click here. But remember: No pain, no gain.

4. Classes are invariant under similarity transforms.
Another statement, using more words:
Let $\mathbf{a_i}$ be any member of class $\mathbf{A}$. Then subject $\mathbf{a_i}$ to a similarity transform by $\mathbf{g}$, any member of the group. The result will be in class $\mathbf{A}$.

Put the definition of class together with Theorem 1, above. But to see it argued in detail, click here.

5. Every class of a product group is the outer product of two classes of the generating groups.

Click back to the definition of the product group. The proof is a straightforward manipulation of symbols, but if you gotta have it, click here.

22.9 Class character, and other class properties

In Appendix A3, the MatrixReview, it is shown that if two matrices $\mathbf{M}$ and $\mathbf{N}$ are in the same class, then $\mathbf{M}$ and $\mathbf{N}$ have several properties in common. So all the matrices in the class have these properties in common.

The proofs of these common properties are not difficult, but we prefer to keep them all together in the MatrixReview. We quote the most important one :

6. Every matrix in a class has the same trace.

The trace of a transform matrix is called its **character**. It is very likely you can prove this for yourself. Write the reputed property

$$\mathbf{Trace[M]=Trace}\left[\mathbf{Q.M.Q^{-1}}\right]$$

in index notation, and rearrange the order of summation. Use the fact that $\mathbf{Q}$ is unitary. It really is not that hard. But if you just can't wait to see the proof, click here.

It is hard to overestimate the importance of this theorem. Applied group theory revolves around the "character table" for the group, which gives the values of the class characters *of all the inequivalent irreducible representations of the group* for all classes of the group. The italicized words have not yet been explained in this book, so this statement may not be crystal clear to you. But the full explanation lies in the chapters just ahead. Take this as motivation for learning what *inequivalent irreducible representations* are.

7. Every matrix in a class has the same eigenvalues.

We have already used the eigenvalues of transform matrices as a key ingredient in the operator **RecognizeMatrix**. The proof that eigenvalues are a common property of the class is not quite as elementary as usual. That is why it lives in the MatrixReview as one of a chain of theorems about class invariant properties.

We explored the uses of the eigenvalues very thoroughly in Chapter 13, and used them in the operator **RecognizeMatrix.** So this is also a theorem of great importance.

But note that Theorems 6 and 7 do not say that different classes must have different traces or different eigenvalues. It is easy to prove this by example. Group **O** contains elements **C2x** and **C2mxy** , twofold rotations that belong to different classes.

Element **C2x** belongs to class **C2**

```
C2Mat = CartRep["C2x", "O"]
```

```
{{1, 0, 0}, {0, -1, 0}, {0, 0, -1}}
```

but element **C2mxy** belongs to class **C2d** :

```
C2dMat = CartRep["C2mxy", "O"]
```

```
{{0, -1, 0}, {-1, 0, 0}, {0, 0, -1}}
```

Look at their eigenvalues:

```
Map[Eigenvalues, {C2Mat, C2dMat}]
```

```
{{-1, -1, 1}, {-1, -1, 1}}
```

Look at their traces:

```
Map[Spur, {C2Mat, C2dMat}]
```

```
{-1, -1}
```

So sometimes different classes have the same eigenvalues and the same trace.

22.10 The Classify operator

In your work with molecules you may well need to work with a group for which no character table exists (very likely, a permutation group), so you will need to divide the group into classes on your own. For this purpose, it will be very useful to have an operator that "classifies" groups automatically.

Classify is just an extension of the **MakeClass** operator. We make a **classList** which is initially empty, and a **freeList** which initially has the whole group in it. We take the first element from **freeList** and construct its class. All the elements in the class are then removed from **freeList** and (grouped as a sublist) put into the **classList**. This procedure recycles until **freeList** is empty. To see the text of it, click here and you will go right into the **Symmetry`** package to have a look.

We try it on our familiar example, group $\mathbf{C_{3\,v}}$, using the tabulated **xyz** representation of the group :

```
classesC3v =
 Classify[CartRep["C3v"]] /. RepMatsToNames["C3v"]
```

```
{{E}, {σvc, σvb, σva}, {C3a, C3b}}
```

Lovely; it works. We give it one more test. Group $\mathbf{T_d}$ has lots of classes:

```
ClassPopulations["Td"] // Sort
```

```
{1, 3, 6, 6, 8}
```

Now let's see if **Classify** can produce the same answer by calculation, working on the standard tabulated matrices for group $\mathbf{T_d}$. This table is kept under the name CartRep**["Td"]** :

```
TdClassified =
 Classify[CartRep["Td"]] /. RepMatsToNames["Td"]
```

```
{{E}, {C2a, C2c, C2b}, {S4d, S4c, S4b, S4f, S4e, S4a},
 {σde, σdd, σda, σdb, σdf, σdc},
 {C3c, C3b, C3a, C3e, C3d, C3h, C3f, C3g}}
```

```
Map[Length, TdClassified] // Sort
```

```
{1, 3, 6, 6, 8}
```

It agrees.

22.11 Isomorphisms among order 6 point groups

Several point groups in the standard tables are isomorphic to each other. We would like to find them by examining their class structure. We begin by looking at the names of all the tabulated point groups :

```
groupNames = GroupCatalog
```

```
{C1, C2, C2h, C2v, C3, C3h, C3v, C4, C4h, C4v, C5, C6, C6h,
 C6v, Ch, Ci, D2, D2d, D2h, D3, D3d, D3h, D4, D4d, D4h, D5,
 D5d, D5h, D6, D6d, D6h, I, O, Oh, S4, S6, S8, T, Td, Th}
```

What is the order of each of these groups? That is, how many elements does each of these groups possess? To make the result more readable we group names and orders together :

```
allpairs = Transpose[{orders, groupNames}] // Sort
```

```
{{1, C1}, {2, C2}, {2, Ch}, {2, Ci}, {3, C3}, {4, C2h},
 {4, C2v}, {4, C4}, {4, D2}, {4, S4}, {5, C5}, {6, C3h},
 {6, C3v}, {6, C6}, {6, D3}, {6, S6}, {8, C4h}, {8, C4v},
 {8, D2d}, {8, D2h}, {8, D4}, {8, S8}, {10, D5}, {12, C6h},
 {12, C6v}, {12, D3d}, {12, D3h}, {12, D6}, {12, T},
 {16, D4d}, {16, D4h}, {20, D5d}, {20, D5h}, {24, D6d},
 {24, D6h}, {24, O}, {24, Td}, {24, Th}, {48, Oh}, {60, I}}
```

The groups that have order 6 are :

```
order6pairs = Select[allpairs, MatchQ[#, {6, _}] &]
```

```
{{6, C3h}, {6, C3v}, {6, C6}, {6, D3}, {6, S6}}
```

```
order6names = Transpose[order6pairs]〚2〛
```

```
{C3h, C3v, C6, D3, S6}
```

The populations of the classes of these groups are

```
Column[Transpose[
  {order6names, ClassPopulations /@ order6names}]]
```

```
{C3h, {1, 1, 1, 1, 1, 1}}
{C3v, {1, 2, 3}}
{C6, {1, 1, 1, 1, 1, 1}}
{D3, {1, 2, 3}}
{S6, {1, 1, 1, 1, 1, 1}}
```

The five groups of order 6 appear to be of two different types: One type has six classes (each class containing a single element, clearly); the other type has three classes. We look more closely at the two groups that have three classes; namely, $\mathbf{C_{3\,v}}$ and $\mathbf{D_3}$:

```
MultiplicationTable["C3v"] // MatrixForm
```

```
E    C3a  C3b  σva  σvb  σvc
C3a  C3b  E    σvb  σvc  σva
C3b  E    C3a  σvc  σva  σvb
σva  σvc  σvb  E    C3b  C3a
σvb  σva  σvc  C3a  E    C3b
σvc  σvb  σva  C3b  C3a  E
```

```
MultiplicationTable["D3"] // MatrixForm
```

```
E    C3a  C3b  C2a  C2b  C2c
C3a  C3b  E    C2b  C2c  C2a
C3b  E    C3a  C2c  C2a  C2b
C2a  C2c  C2b  E    C3b  C3a
C2b  C2a  C2c  C3a  E    C3b
C2c  C2b  C2a  C3b  C3a  E
```

These tables have the same form. To make this perfectly explicit, we recalculate the two tables using exactly the same list of element names, and ask if the two tables are equal :

```
sixNames = {"E", "A", "B", "C", "D", "F"};
mtC3v = MultiplicationTable[
   Dot, CartRep["C3v"], sixNames];
mtD3 = MultiplicationTable[Dot,
   CartRep["D3"], sixNames];
mtC3v == mtD3
```

```
True
```

To an abstract group theorist, groups **C3v** and **D3** are "the same", because their multiplication tables are the same. But to a symmetrician, they are quite different. In group $\mathbf{C_{3\,v}}$, the classes are $\mathbf{E}$, $\mathbf{C_3}$, and $\boldsymbol{\sigma_v}$, while in group $\mathbf{D_3}$ the classes are $\mathbf{E}$, $\mathbf{C_3}$, and $\mathbf{C_2}$. Classes $\mathbf{E}$ and $\mathbf{C_3}$ are the same in both groups, but $\boldsymbol{\sigma_v}$ is a class of vertical reflections, while class $\mathbf{C_2}$ is a class of horizontal twofold rotations. These are two very different symmetry groups which happen to be isomorphic.

Now we turn to the other three groups of order **6**, each of which has **6** classes (each class containing a single element). They were **C6**, **S6**, and **C3h**. We now show that these are all cyclic groups of order 6 :

```
MultiplicationTable[Dot, CartRep["C6"], sixNames] //
  MatrixForm
```

$$\begin{pmatrix} E & A & B & C & D & F \\ A & B & C & D & F & E \\ B & C & D & F & E & A \\ C & D & F & E & A & B \\ D & F & E & A & B & C \\ F & E & A & B & C & D \end{pmatrix}$$

The cyclic nature of the group is apparent in the pure counter-diagonals of the multiplication table. (They mean that each row is a one-place rotation of the one above it.) If you do the same thing with either group **S6** or **C3h**, you will not see the pure counter-diagonals. However, this is only because the conventional ordering of these groups does not follow cyclic order. If we reorder them appropriately (starting always with the unit element, and going cyclically from there) and then compute their tables, the pure counter-diagonals appear :

```
S6Reordered = CartRep["S6"][[{1, 6, 2, 4, 3, 5}]];
```

```
Column[RecognizeMatrix /@ S6Reordered]
```

{identity}
$\left\{\text{rotoreflection, 6-fold, }\left\{\{0, 0, 1\}, \frac{\pi}{3}\right\}\right\}$
$\left\{\text{rotation, 3-fold, }\left\{\{0, 0, 1\}, \frac{2\pi}{3}\right\}\right\}$
{inversion}
$\left\{\text{rotation, 3-fold, }\left\{\{0, 0, -1\}, \frac{2\pi}{3}\right\}\right\}$
$\left\{\text{rotoreflection, 6-fold, }\left\{\{0, 0, -1\}, \frac{\pi}{3}\right\}\right\}$

```
mtS6 =
  MultiplicationTable[Dot, S6Reordered, sixNames];
        mtS6 // MatrixForm
```

$$\begin{pmatrix} E & A & B & C & D & F \\ A & B & C & D & F & E \\ B & C & D & F & E & A \\ C & D & F & E & A & B \\ D & F & E & A & B & C \\ F & E & A & B & C & D \end{pmatrix}$$

So group **S6** is a cyclic group, based on a sixfold rotoreflection.. Now do **C3h**.

```
C3hReordered = CartRep["C3h"][[{1, 5, 3, 4, 2, 6}]];
```

```
Column[RecognizeMatrix /@ C3hReordered]
```

{identity}
$\left\{\text{rotoreflection, 3-fold, }\left\{\{0, 0, 1\}, \frac{2\pi}{3}\right\}\right\}$
$\left\{\text{rotation, 3-fold, }\left\{\{0, 0, -1\}, \frac{2\pi}{3}\right\}\right\}$

```
{reflection, {0, 0, 1}}
{rotation, 3-fold, {{0, 0, 1}, 2π/3}}
{rotoreflection, 3-fold, {{0, 0, -1}, 2π/3}}
```

```
MultiplicationTable[Dot, C3hReordered, sixNames] //
 MatrixForm
```

$$\begin{pmatrix} E & A & B & C & D & F \\ A & B & C & D & F & E \\ B & C & D & F & E & A \\ C & D & F & E & A & B \\ D & F & E & A & B & C \\ F & E & A & B & C & D \end{pmatrix}$$

So they are both sixfold cyclic groups, but they differ in the nature of what is cycling. In group $\mathbf{C_6}$, it is a sixfold rotation. In group $\mathbf{S_6}$ it is a sixfold roto-reflection. In group $\mathbf{C_{3h}}$ it is a threefold roto-reflection. The latter group is NOT named $\mathbf{S_3}$ because Schönfliess declared odd indices in $\mathbf{S_n}$ groups to be *verboten.*

On your own
This kind of analysis can be applied to groups of any order. Do another order . Pick an easy one; not too large.

A theorem of your very own
Put the concept of Abelian groups together with the concept of class, and prove something about the classes of Abelian groups. If you succeed, you may name this theorem after yourself.

For a calculational clue, try **MakeClass** on an Abelian group.

Probably you will see it; but still, just in case ...

22.12 End notes

22.12.1 Proof of Theorem 1

Every element belongs to the class that it generates : When **mat** runs over the whole group, **mat** will eventually be the unit matrix **E**, and then $\mathbf{mat.X.mat}^{-1}$ will give **X**.

22.12.2 Proof of Theorem 2

When **X** is **E**, you can write $\mathbf{mat.E.mat}^{-1} = \mathbf{mat.mat}^{-1} = \mathbf{E}$, regardless of what **mat** is. So as **mat** runs over the whole group, **E** is the only element generated.

22.12.3 Proof of theorem 3

We assume that there exists a questionable element **Q** that belongs to the class of **X** and to the class of **Y**, where **X** and **Y** are in different classes. If this is true, there exist elements **M** and **N** such that $\mathbf{Q} = \mathbf{M.X.M}^{-1}$and $\mathbf{Q} = \mathbf{N.Y.N}^{-1}$.

But if so, we can eliminate **Q** and rearrange to yield
$\mathbf{X} = \mathbf{M}^{-1}\mathbf{.N.Y.N}^{-1}\mathbf{.M}$

However, $(\mathbf{M}^{-1}\mathbf{.N}) \mathbf{.} (\mathbf{N}^{-1}\mathbf{.M}) = \mathbf{E}$, so $(\mathbf{M}^{-1}\mathbf{.N})$ and $(\mathbf{N}^{-1}\mathbf{.M})$ are mutually inverse, and may be symbolized by **U** and $\mathbf{U}^{-1}$. Therefore $\mathbf{X} = \mathbf{U.Y.U}^{-1}$ and **X** and **Y** are therefore in the same class, contrary to assumption. Therefore, there exists no element **Q** belonging to two different classes.

22.12.4 Proof of Theorem 4

Let $\mathbf{a_m}$ and $\mathbf{a_n}$ be two different elements of class $\mathbf{A}$. Let $\mathbf{g}$ be any element of the group, and consider $\mathbf{g\,a_m\,g^{-1}}$ and $\mathbf{g\,a_n\,g^{-1}}$. By the definition of class they are still in class $\mathbf{A}$. But is it possible that they are the same? No, because, if $\mathbf{g\,a_m\,g^{-1}} = \mathbf{g\,a_n\,g^{-1}}$ we could left multiply each side by $\mathbf{g^{-1}}$and right multiply each side by $\mathbf{g}$ and we would have $\mathbf{a_m} = \mathbf{a_n}$, which contradicts the assumption that $\mathbf{a_m}$ and $\mathbf{a_n}$ are different. So the similarity transform of any class produces only elements in the class, with no repeats. (The element order will generally change, but that makes no difference in a *set*.) So classes are invariant under similarity transforms, which was to be proved.

22.12.5 Proof of Theorem 5

Prove: If $\mathcal{AB}$ is a product group, the class of element $\mathbf{AB}$ is the class of element $\mathbf{A}$ times the class of element $\mathbf{B}$.

The class of element $\mathbf{AB_i}$ is the set $\left\langle \mathbf{AB_m} \otimes \mathbf{AB_i} \otimes \mathbf{AB_m^{-1}} \text{ for all } \mathbf{m} \right\rangle$. But $\mathbf{AB_m}$ was some product $\mathbf{A}_\alpha \otimes \mathbf{B}_\Gamma$, and $\mathbf{AB_i}$ was some product $\mathbf{A}_\beta \otimes \mathbf{B}_\Delta$. Therefore the class of $\mathbf{AB_i}$ is

$$\left\langle (\mathbf{A}_\alpha \otimes \mathbf{B}_\Gamma) \otimes (\mathbf{A}_\beta \otimes \mathbf{B}_\Delta) \otimes (\mathbf{A}_\alpha \otimes \mathbf{B}_\Gamma)^{-1} \text{ for all } \alpha \text{ and } \Gamma \right\rangle$$

or expressing the inverse in more detail,

$$\left\langle (\mathbf{A}_\alpha \otimes \mathbf{B}_\Gamma) \otimes (\mathbf{A}_\beta \otimes \mathbf{B}_\Delta) \otimes \left(\mathbf{B}_\Gamma^{-1} \otimes \mathbf{A}_\alpha^{-1}\right) \text{ for all } \alpha \text{ and } \Gamma \right\rangle$$

Since $\otimes$ is associative, the parentheses may be removed, leaving

$$\left\langle \mathbf{A}_\alpha \otimes \mathbf{B}_\Gamma \otimes \mathbf{A}_\beta \otimes \mathbf{B}_\Delta \otimes \mathbf{B}_\Gamma^{-1} \otimes \mathbf{A}_\alpha^{-1} \text{ for all } \alpha \text{ and } \Gamma \right\rangle$$

But all $\mathbf{A}$ and $\mathbf{B}$ matrices commute, so commuting the $\mathbf{A}$'s leftward and the $\mathbf{B}$'s rightward, (but never passing an $\mathbf{A}$ through an $\mathbf{A}$ or a $\mathbf{B}$ through a $\mathbf{B}$) we arrive at

$$\left\langle \mathbf{A}_\alpha \otimes \mathbf{A}_\beta \otimes \mathbf{A}_\alpha^{-1} \text{ for all } \alpha \right\rangle \otimes \left\langle \mathbf{B}_\Gamma \otimes \mathbf{B}_\Delta \otimes \mathbf{B}_\Gamma^{-1} \text{ for all } \Gamma \right\rangle$$

But this is just the direct product of the class of $\mathbf{A}_\beta$ and the class of $\mathbf{B}_\Delta$, the factors of element $\mathbf{AB_i}$, which was to be proved.

22.12.6 Proof of Theorem 6

Prove: The trace of the product is the product of the traces.

We want an expression for the trace of $\mathbf{AB_m}$. But $\mathbf{AB_m}$ was some product $\mathbf{A}_\alpha \otimes \mathbf{B}_\Gamma$, and $\mathbf{AB_n}$ was some product $\mathbf{A}_\beta \otimes \mathbf{B}_\Delta$. So we are looking for the trace of

$$\mathbf{A}_\alpha \otimes \mathbf{B}_\Gamma \otimes \mathbf{A}_\beta \otimes \mathbf{B}_\Delta \,;$$

or equivalently, by permitted commutation, the trace of

$$\mathbf{A}_\alpha \otimes \mathbf{A}_\beta \otimes \mathbf{B}_\Gamma \otimes \mathbf{B}_\Delta$$

But suppose element $\mathbf{A}_\alpha$ is represented by matrix $\mathbb{A}^{(\alpha)}_{\_,\_}$; element $\mathbf{B}_\Gamma$ by matrix $\mathbb{B}^{(\Gamma)}_{\_,\_}$; element $\mathbf{A}_\beta$ by matrix $\mathbb{A}^{(\alpha)}_{\_,\_}$; and element $\mathbf{B}_\Delta$ by matrix $\mathbb{B}^{(\Delta)}_{\_,\_}$. Then the desired trace, in index notation, is

$$\mathbb{A}^{(\alpha)}_{p,q}\,\mathbb{A}^{(\beta)}_{q,r}\,\mathbb{B}^{(\Gamma)}_{r,s}\,\mathbb{B}^{(\Delta)}_{s,p}$$

We want an expression for the trace of $\mathbf{AB_m}$. But $\mathbf{AB_m}$ was some product $\mathbf{A}_\alpha \otimes \mathbf{B}_\Gamma$, so in double index notation it is $\mathbf{AB}_{\alpha,\Gamma}$.

Suppose element $\mathbf{A}_\alpha$ is represented by matrix $\mathbb{A}^{(\alpha)}_{\_,\_}$; element $\mathbf{B}_\Gamma$ by matrix $\mathbb{B}^{(\Gamma)}_{\_,\_}$; element $\mathbf{A}_\beta$ by matrix $\mathbb{A}^{(\alpha)}_{\_,\_}$; and element $\mathbf{B}_\Delta$ by matrix $\mathbb{B}^{(\Delta)}_{\_,\_}$. Then the desired trace, in index notation, is

$$\mathbb{A}^{(\alpha)}_{p,q}\quad\mathbb{B}^{(\Gamma)}_{q,p}$$

22.12.7 Classes of Abelian groups

In Abelian groups you are allowed to rearrange multiplications, so

$$\mathbf{Y.X.Y^{-1} = X.\left(Y.Y^{-1}\right) = X.E = X}$$

As $\mathbf{Y}$ runs over the whole group it gives nothing but $\mathbf{X}$, so in an Abelian group every element is a class by itself. This is the theorem.

If you had to read this, it is not your theorem.

23. Symmetry and quantum mechanics

Preliminaries

23.1 Transformation of eigenfunctions

23.1.1 The Hamiltonian symmetry theorem

We have defined geometric and algebraic symmetry, we have shown that the transforms of symmetric objects always come in groups, and we are now ready to think about how these concepts apply to Schrödinger's eigenequation, the basis of quantum chemistry and solid state physics. This equation is

$$\mathcal{H}[\psi_i] = E_i\,\psi_i \tag{23.1}$$

where $\mathcal{H}$ is a Hamiltonian operator for the molecule, ψ_i is its wave function for eigenstate i, and E_i is the energy of eigenstate i. Such solutions are called *stationary* states because *all* their physical properties (not just the energy) are constant in time. Stable molecules in a bottle have time-independent properties because they are all in or very close to an eigenstate; namely, the one that has the lowest energy, the ground state.

We do not yet specify exactly what $\mathcal{H}$ is, because it has numerous levels of approximation, and we want to discuss a feature that applies to all of them. $\mathcal{H}$ and ψ_i are defined in a coordinate space that gives the positions of all the particles that make up the system. Let $\mathcal{T}$ be a coordinate transform defined in this space, and apply it to both sides of Eq.23.1 :

$$\mathcal{T}[\mathcal{H}[\psi_i]] = \mathcal{T}[E_i\,\psi_i] \tag{23.2}$$

The $\mathcal{T}$ transforms everything within its scope of operation, so this is the same as

$$\mathcal{T}[\mathcal{H}][\mathcal{T}[\psi_i]] = \mathcal{T}[E_i]\,\mathcal{T}[\psi_i] \tag{23.3}$$

The term $\mathcal{T}[E_i]$ is nothing but E_i, because E_i has no particle coordinates in it, and cannot be transformed by $\mathcal{T}$ .So we have

$$\mathcal{T}[\mathcal{H}][\mathcal{T}[\psi_i]] = E_i\;\mathcal{T}[\psi_i] \tag{23.4}$$

We are interested in symmetry transforms of the Hamiltonian. By the basic definition of symmetry, these are transforms that leave the Hamiltonian unchanged :

$$\mathcal{T}[\mathcal{H}] = \mathcal{H} \tag{23.5}$$

W.M. McClain, *Symmetry Theory in Molecular Physics with Mathematica*,
DOI 10.1007/b13137_23,

so that Eq.23.4 becomes

$$\mathcal{H}\,[\mathcal{T}\,[\psi_i]\,] = E_i\,\mathcal{T}\,[\psi_i] \tag{23.6}$$

Comparing this to the original Schrödinger equation $\mathcal{H}\,[\psi_i] = E_i\,\psi_i$, we notice that $\mathcal{T}\,[\psi_i]$ is playing the role of ψ_i in the original equation, which means that $\mathcal{T}[\psi_i]$ is an eigenfunction of $\mathcal{H}$ with eigenvalue E_i. This is profound. We encapsulate it as the Hamiltonian symmetry theorem :

> **If**
> $\mathcal{H}[\psi_i] = E_i\,\psi_i$, and if $\mathcal{T}[\mathcal{H}] = \mathcal{H}$
> **then**
> $\mathcal{T}[\psi_i]$ is an eigenfunction of $\mathcal{H}$ with eigenvalue E_i.

Immediate comment:
Yes, **an** eigenfunction; but sometimes, **another linearly independent** eigenfunction.

This theorem has very different consequences depending on whether only one linearly independent eigenfunction exists at the given energy, or whether there are several different such eigenfunctions. There is a word that everybody uses to describe this :

> If an energy level possesses more than one linearly independent eigenfunction, it is said to be *degenerate*. Otherwise, *nondegenerate*.

In the nondegenerate case, transform $\mathcal{T}$ just gives back the same eigenfunction it operated on, possibly with a different phase factor $e^{i\phi}$. But if the level is degenerate, it may return a completely different (a linearly independent) eigenfunction with the same eigenenergy. Thus, in degenerate levels, the symmetry transforms $\mathcal{T}$ of the Hamiltonian can make new eigenfunctions out of old ones. It's like magic.

23.1.2 A figure for the theorem

Benzene is a hexagonal ring molecule $C_6 H_6$ (group $\mathbf{D_{6\,h}}$) with a delocalized electron density function that extends around the ring of C atoms. We can approach it through an exactly soluble idealized model: Let six positively charged centers be arranged hexagonally, and consider the stationary wave functions that a single electron might have in the vicinity of the six attractors. (Such one-electron wave functions are called "orbitals".) Orbitals are the solutions of Schrödinger's eigenequation, and for one particle they can be found as accurately as desired using known numerical techniques for solving differential equations.

The function $\psi_{\mathbf{E1g\,a}}$**[x,y,z]**, defined in the preliminaries, resembles such an eigenfunction in all its essentials. If you want to see the formula, issue **ψE1ga**. It looks complicated to a human, but not to *Mathematica.*

The left side of the next figure is an **x,y** contour plot of $\psi_{\mathbf{E1g\,a}}$. It belongs to a degenerate level, so the theorem says we might generate a linearly independent partner by simply letting a $\mathbf{D_{6\,h}}$ transform work on it.

The right side of the figure is such a transform, to be picked by you. Group $\mathbf{D_{6\,h}}$ has **24** elements, and below you can set the transform number **Tnbr** to be anything from **1** to **24**. The appropriate transform rule will be looked up and applied to $\psi_{\mathbf{E1ga}}$, and the contour plot will be shown on the right. Do the next cell before you start making figures. It will help you to choose and understand the results.

```
Transpose[{Range[24], ElementNames["D6h"]}]
```

```
{{1, E}, {2, C6a}, {3, C6b}, {4, C3a}, {5, C3b},
 {6, C2}, {7, C2pa}, {8, C2pb}, {9, C2pc}, {10, C2ppa},
 {11, C2ppb}, {12, C2ppc}, {13, inv}, {14, S3a},
 {15, S3b}, {16, S6a}, {17, S6b}, {18, σh}, {19, σda},
 {20, σdb}, {21, σdc}, {22, σva}, {23, σvb}, {24, σvc}}
```

Put in values for **Tnbr** and **Enter** until you see what is going on.

```
Tnbr = 3;
left = Show[cpBlue[ψE1ga /. z → 0.1`],
  PlotLabel → "ψE1ga", BaseStyle → {"TR", 9}];
rule = SymmetryRules["D6h"][[Tnbr]];
rcg = RecognizeMatrix[CartRep["D6h"][[Tnbr]]];
rght = Show[cpBlue[ψE1ga /. rule /. z → 0.1`],
  PlotLabel → rcg, BaseStyle → {"TR", 6}];
Show[GraphicsRow[{left, rght}]]
```

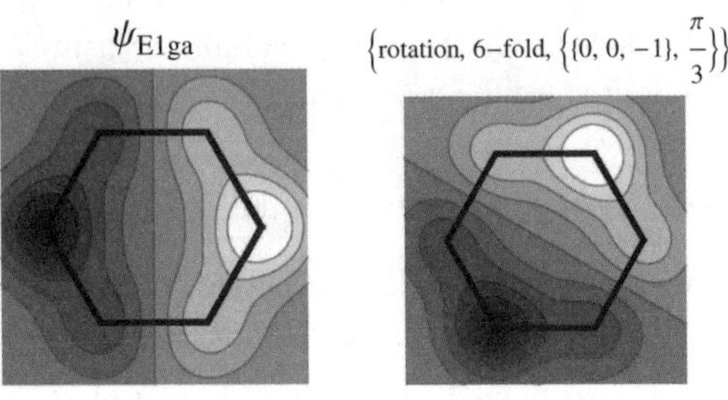

Fig. 23.1 The Hamiltonian symmetry theorem using `Tnbr = 3`, a rotation by $2\pi/6$, showing linear independence.

Some transforms will be linearly *dependent*, and some not. If normalization is maintained, linear dependence on a single real function means multiplication by **±1**, so the whitest atom must stay where it is, or move diametrically across the ring. If some other atom becomes the whitest atom, the two orbitals are linearly *independent*.

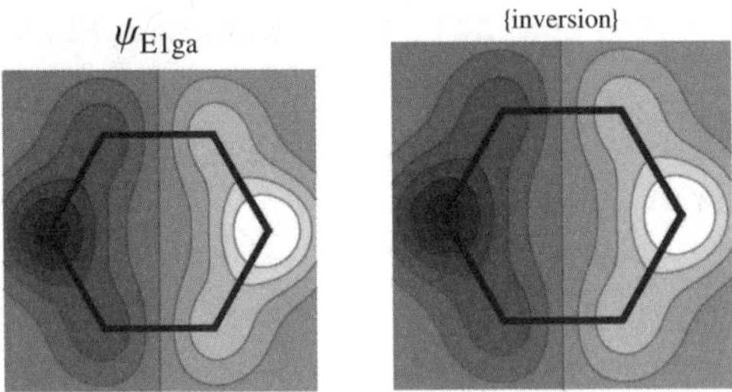

Fig. 23.2 Using `Tnbr = 13` (the inversion) we see linear dependence.

All these orbitals are antisymmetric for reflection in the plane of the molecule (`Tnbr = 18`). This means that the orbitals are exactly zero in the plane, so the diagrams are drawn for a cross section slightly above the plane, at `z = 0.1` . If you think the white spot should have moved across the ring above, you are thinking of 2D inversion. No, it's 3D inversion; the white spot comes across the ring and up from below.

23.2 Delocalized orbitals of benzene

Here is a somewhat untraditional, but very instructive diagram for the π orbitals of benzene. Energy runs vertical on this diagram, so orbitals on the same level all have the same energy. There are an infinity of orbitals on each degenerate level, but we show just three on each level, constructed by rotational transforms, as in the last section.

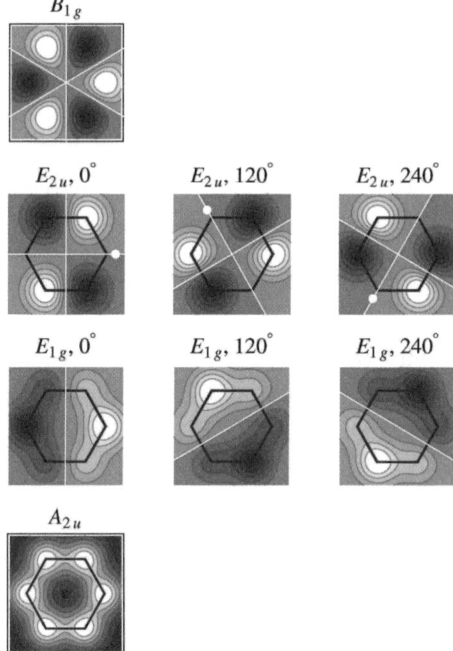

Fig. 23.3 A few benzene orbitals. Top and bottom rows are nondegenerate. The middle two rows are degenerate, but the orbitals shown are not orthogonal. They are related by rotations through 1/3 turn. An extra white dot resolves orientational ambiguity for species E_{2u}.

The orbitals resemble standing waves in a hexagonal pond of water excited by disturbances of the right symmetry.

This does not exhaust the games you can play with degenerate orbitals. Read on.

23.3 Linear combinations of degenerate orbitals

23.3.1 A quick theorem on degenerate orbitals

There are infinitely many wave functions in each degenerate level, because linear combinations of degenerate wave functions always produce other wave functions of the same energy. The proof is quick and easy. If ψ_1 and ψ_2 both have energy **E**, then because $\mathcal{H}$ is a linear operator

$$\mathcal{H}[\mathbf{a}\,\psi_1 + \mathbf{b}\,\psi_2] = \mathbf{a}\,\mathcal{H}[\psi_1] + \mathbf{b}\,\mathcal{H}[\psi_2] = \mathbf{a}\,\mathbf{E}\,\psi_1 + \mathbf{b}\,\mathbf{E}\,\psi_2 = \mathbf{E}\,(\mathbf{a}\,\psi_1 + \mathbf{b}\,\psi_2)$$

Since $\mathbf{a}\,\psi_1 + \mathbf{b}\,\psi_2$ plays the role of ψ in $\mathcal{H}[\psi] = \mathbf{E}\,\psi$, it must be another eigenfunction with the same energy **E**. So the three functions shown in each degenerate level of Figure 23.3 are just the winners of a beauty contest; other linear combinations not shown are just as valid, but look sort of lumpy and asymmetric. Their magnificent continuous flow from one to another comes out only when you see them in a movie, as you will below in Subsection 23.3.3.

23.3.2 Orthogonal basis functions

In our discussion of the character table, we said that degenerate levels named **E** are two dimensional; that is, they need two basis functions to span the function space. An orthogonal pair of basis functions can be extracted from any two linearly independent functions. As an illustration, we take $\psi_{\mathbf{E1g}}[0\,°]$ and $\psi_{\mathbf{E1g}}[120\,°]$ from Fig.23.3. These functions are a bit unsightly, so their full algebraic definition is closeted in the preliminaries as **ψ0N** and **ψ120N**. You can call them forth for inspection if you like. Below, we just plot them to verify that they are what we say :

$\psi E^{0°}_{1g}$

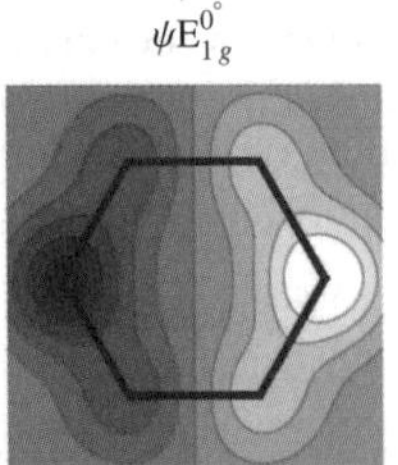

$\psi E^{120°}_{1g}$

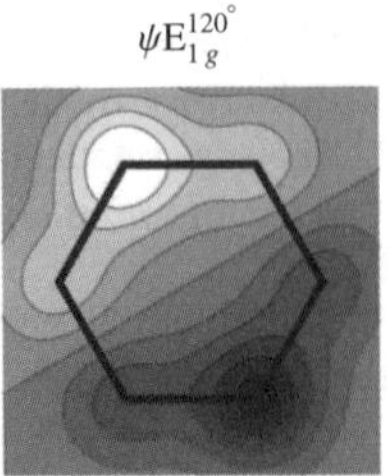

Fig. 23.4 Two linearly independent benzene orbitals from the E_{1g} level. These are the input to `Orthogonalize`, below.

These functions will be fed to the **`Orthogonalize`** operator, which by default implements the Gram-Schmidt orthogonalization procedure. But when orthogonalizing functions, you must tell it what to use for the scalar product of two functions. There are a number of things that are used in different kinds of problems, but in quantum mechanics the scalar product of functions **`f`** and **`g`** is always $\iiint$ **`f[x, y, z]* g[x, y, z] dx dy dz`**, where the integrals run over all space. When both functions are real, this comes down to

```
NumInt[f1_, f2_] := NIntegrate[(f1) (f2),
   {z, -∞, ∞}, {x, -∞, ∞}, {y, -∞, ∞}]

Off[NIntegrate::"slwcon", NIntegrate::"eincr"];
{timeOrth, {ψE1g0, ψE1g90}} =
   Orthogonalize[{ψ0N, ψ120N}, NumInt] //
     Chop[#, 10^-8] & // Timing;
timeOrth Second
{ψE1g0, ψE1g90}
```

```
42.5846 Second
```

$$\{0.917217\, e^{-3.90747-5.59104\,x-2\,x^2-4.84198\,y-2\,y^2-2\,z^2} \left(-1+e^{5.59104\,x}\right)\left(e^{2.79552\,x}+2\,e^{4.84198\,y}+2\,e^{2.79552\,(2\,x+1.73205\,y)}+e^{2.79552\,(x+3.4641\,y)}\right) z,$$
$$1.58867\, e^{-3.90747-2.79552\,x-2\,x^2-4.84198\,y-2\,y^2-2\,z^2}\left(1+e^{5.59104\,x}\right)\left(-1+e^{9.68397\,y}\right) z\}$$

It is more meaningful to look at the contour plots:

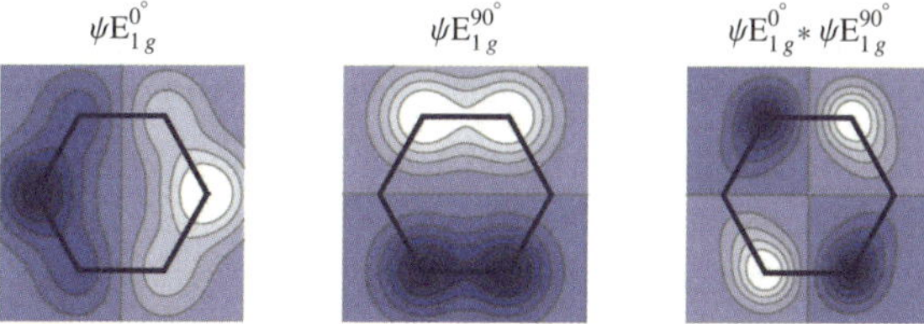

Fig. 23.5 Two orthogonal basis functions for the E_{1g} level, and their product.

It is clear that the first two are orthogonal because their product (third panel) is made of equal but opposite lobes, and so must integrate to **`0`**.

If each degenerate space is expressible as linear combinations of a small number "basis functions" then the most economical display of the space is just the basis functions. The Gram-Schmidt orthogonalization procedure (or the numerical Householder algorithm) provides a quick, practical way of doing this. Here is the result for the benzene orbitals:

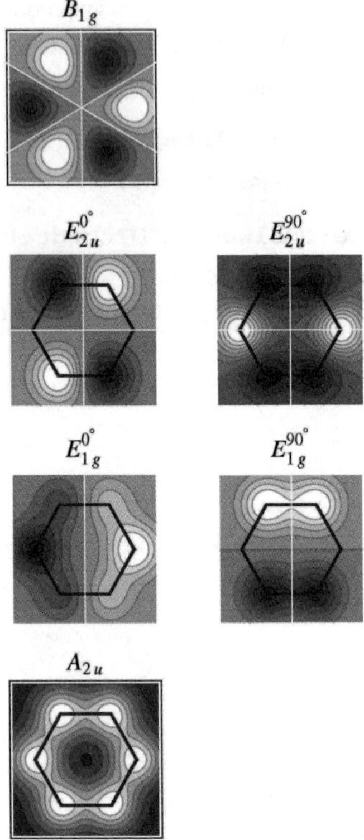

Fig. 23.6 Orthonormal benzene orbitals. The degenerate rows are only twofold degenerate, so only two basis functions are needed. The nodal planes of one partner are reflection planes of the other, ensuring orthogonality.

The two orthogonal shapes in each degenerate level differ by more than a simple reorientation. Nevertheless, the "new" shape on the right is just a linear combination of reoriented shapes.

23.3.3 Movies of the whole degenerate space

In a twofold degenerate level, we can add any two basis functions with coefficients **Cos[α]** and **Sin[α]** and then let **α** run around the circle. This will give us a quick tour of the whole space. With *Mathematica* in front of us, this is not an impractical idea. A convenient function named **E1gPicture[α]** is defined in the preliminaries. Take a look if you are interested.

> **On your own**
> **E1gPicture[0]** shows the basis state with vertical node;
> **E1gPicture[90]** shows the basis state with horizontal node.
> Do these, and try other mixing angles.

Using the function **E1gPicture**, it is easy now to tour the whole space in a movie. Interactive readers can activate the next cell and see the whole thing :

```
tblE1g = Table[E1gPicture[α], {α, 0, 350, 10}];
ListAnimate[tblE1g, 5]
```

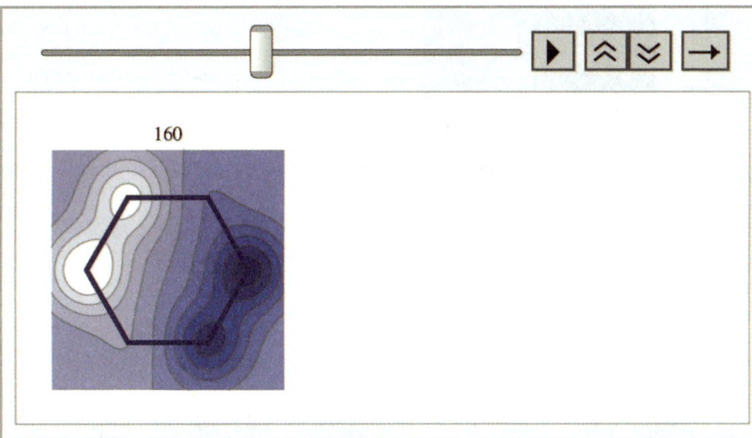

For the book, here is every fifth cell of the movie:

```
Table[E1gPicture[α, 50], {α, 0, 360, 40}]
```

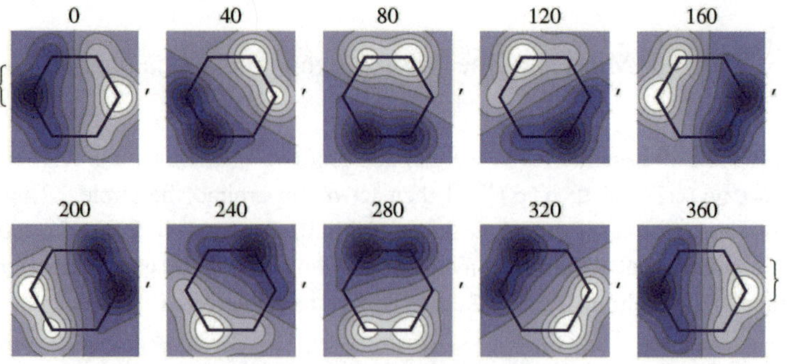

The two-node movie of the $\mathbf{E_{2\,u}}$ function space (below) is a little tricky to grasp intuitively. You may have to watch it for a while to get it.

```
tblE2u = Table[E2uPicture[α], {α, 0, 350, 10}];
ListAnimate[tblE2u, 5]
```

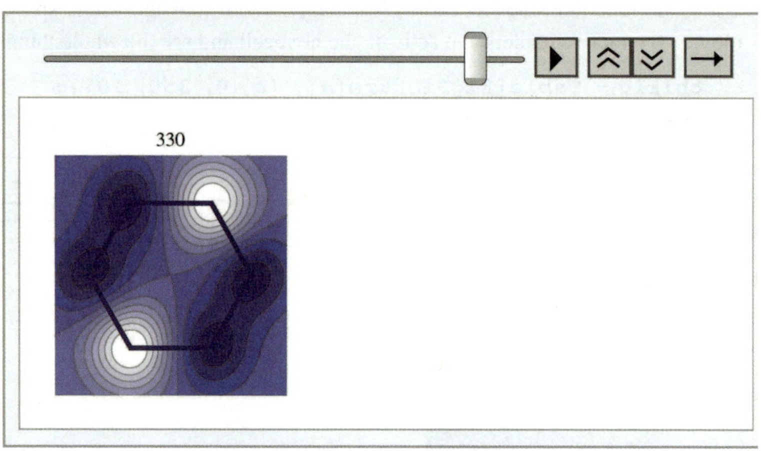


```
Table[E2uPicture[α, 50], {α, 0, 360, 40}]
```

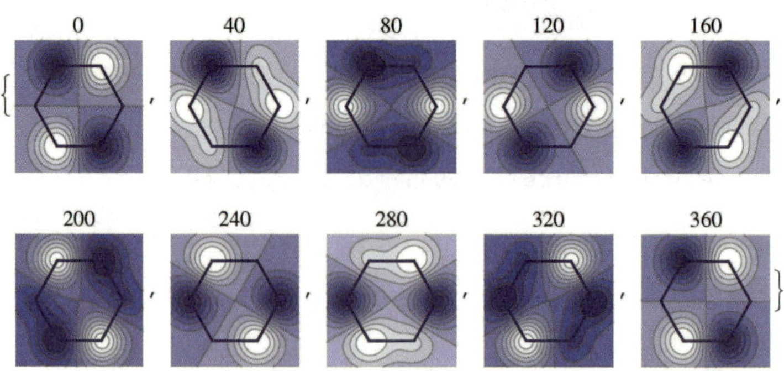

23.3.4 Invariant function spaces in general

The movies you have just seen are examples of a space that is *invariant* to the transforms of group **D6 h**. Technically, this means that if you take any inhabitant of the space (any frame of the movie) and subject to any **D6 h** transform, the result is just another frame of the movie (or perhaps, one that lies between two frames).

Now consider this: Given an arbitrary function, does it belong to an invariant space? The answer is always *yes*, as you will see by the following argument:

You can make an invariant function space for any function by simply applying all the symmetry rules of some group to it. If there are **h** elements in the group, and if **rule1** is the identity,

```
ψ/.{rule1,rule2,... ,ruleh}={ψ,ψ2,... ,ψh}
```

Now think about applying any one of these symmetry rules again to any one of the generated functions; for instance,

```
ψ2/.rule3=(ψ/.rule2)/.rule3 =ψ/.(rule2/.rule3)
```

But **rule2/.rule3** is just **rule2⊗3**, and **2⊗3** is in the group. So this rule has already been applied to **ψ**, and its result is already in the list of transformed functions. So this list spans an invariant space that contains the original arbitrary **ψ2**. There, that was easy. Click to an End note to see this actually carried out. There are some interesting issues that arise, and we show how to deal with them.

23.4 Qualitative meaning of degeneracy

There is qualitative difference between the physics of the single state of a nondegenerate level, and the infinitely many states of a degenerate level. The electron density cloud in a nondegenerate state has a definite unique natural shape. The shape can be perturbed, but it takes energy to do so, and the cloud returns to its natural shape when the perturbation is removed.

A good nondegenerate example would be a ground state He atom bouncing from a solid collision object. Before the collision it is absolutely spherical. We consider only a gentle collision with energy less than the lowest electronic excitation. As the atom comes to rest against the collision object, its former kinetic energy is converted to potential energy stored in a distortion of its spherical shape. But this is not a stationary state, and it immediately begins to recover. To do this, it has to spring away from the collision object. When the electron cloud has recovered its natural shape, the collision energy is fully reconverted to kinetic energy and an "elastic" collision has occurred.

By contrast, the electron density cloud in a degenerate level has no single definite shape. It can take on a continuous range of shapes without any expense of energy, and therefore without any restoring force whatsoever. Sometimes the range of shapes looks like a reorientation, but in other cases the range includes energy-neutral plastic deformations. (Remember the s-p "hybridization" pictures in your freshman chemistry textbook? This is exactly what they were talking about. In the H atom, the s and p orbitals are exactly degenerate and can mix without energy cost, but with a great change in the density cloud).

Because of this, degeneracies allow zero-cost changes in bonding at the saddle point of reactive collision complexes Without such degeneracy, reactive collisions would have to be so energetic that bonds would be fully broken before reforming as new species, and this would rule out many chemical reactions at room temperature. Life would be impossible without degeneracy.

23.5 Accidental degeneracy

We make a very bold and simple statement :

Symmetry is the cause of degeneracy in quantum mechanics.

Throughout this chapter we have seen that geometrical symmetry in the molecule causes algebraic symmetry in the Hamiltonian, and algebraic symmetry in the Hamiltonian causes degeneracy in some of its eigenstates.

There are exceptions to this rule, but they all involve non-quantum parameters (like electric or magnetic field, or bond length) that cause two levels to cross at a particular value of the parameters. Exactly at the crossing point, there is an "accidental" degeneracy that seems very surprising in the symmetry you are using to analyze the problem. But you may rest assured that this degeneracy is the consequence of some higher symmetry that you may not yet have recognized.

A very simple example of this is offered by the orbitals of cyclopropenylCation, $\mathbf{C_3H_3^+}$, discussed Chapter 40 (AtomicOrbitalSALCs). If the three **C** atoms are arrayed as an irregular triangle in the **x,y** plane (group $\mathbf{C_h}$), the three π orbitals of $\mathbf{C_3H_3}$ all have different energy. But the bond lengths are non-quantum parameters in this calculation, and if they are all set equal, two of the orbitals develop an " accidental" degeneracy. Of course, the equilateral triangle belongs to a higher symmetry (**D3 h**), and the degeneracy of the two orbitals is expected and required in **D3 h**. This has spectroscopic consequences, called the Jahn-Teller effect.

But the most famous example of pseudo-accidental degeneracy is the "shell" degeneracy of one-electron atoms, which is elementary and very sophisticated at the same time. In shell **2**, the **2s** orbital is degenerate with any function spanned by **{2px, 2py, 2pz}**. In shell **3**, the **3s** orbital is degenerate with the three **3p**'s and with the five **3d**'s; and so on. Rotational geometry explains why, in any shell, the three **p**'s have the same energy, or why the five **d**'s have the same energy, but it cannot explain why, in shell **n**, the energies of {{one **ns**}, {three **np**'s}, {five **nd**'s}**,…** } are all *exactly* equal. In older quantun books this intra-shell degeneracy was the prime example (indeed, the only example) of non-perturbative "accidental" degeneracy. It is now recognized as a dynamic symmetry, arising from a most unexpected quarter.

23.6 Dynamic symmetry

A dynamic symmetry is a a quantity with a formula that depends on constantly changing quantities, but which itself is constant in time. The only familiar example is the angular momentum, given by $\underline{L} = \underline{r}[t] \times \underline{p}[t]$.

However, with the help of Lie groups and their connection with commuting quantum operators, it was finally recognized that the one-electron atom, by virtue of its pure **1/r** potential, possesses an extra dynamic symmetry that accounts for the shell degeneracies. (With two or more electrons, electron-electron repulsion ruins the pure **1/r** potential and causes shell degeneracy to break.)

This dynamic symmetry occurs also in the classical mechanics of planetary systems, which also obey a **1/r** potential. Isolated mechanical systems obey conservation of energy, linear momentum, and angular momentum. But in planetary systems there is another one. It is easy to describe and understand; it was a fact known to Newton, though he did not know its formula or think of it as a constant of the motion. It is called $\underline{RL}$, the Runge-Lenz vector. It runs through the center of mass and through the perihelion of each planet's orbit. It is a dynamic symmetry because it is computed by the formula

$$\underline{RL}[t\_] = \underline{p}[t] \times \underline{L} - m\,k \left(\frac{\underline{r}[t]}{\left| \underline{r}[t] \right|} \right) \qquad (23.7)$$

where $\underline{\vec{p}}[t]$ is linear momentum of the planet, $\underline{\vec{L}}$ is its angular momentum, and $\underline{\vec{r}}[t]$ is the radial position of the planet from the center of mass, **m** is the reduced mass, and **k** is from the potential **-k/r**. Despite the fact that $\underline{\vec{p}}$ and $\underline{\vec{r}}$ are constantly changing, $\underline{\vec{RL}}$ is rock-solid. In this it is like the angular momentum.

This fact makes the position of the perihelion direction a constant in the frame of the fixed stars. This is why most planets run over the same orbit, year after year.

Well, almost the same orbit. In multi-planet systems every planet feels all the others, so the total potential is not exactly **1/r** , and perihelions do wander a bit. But for all our planets, save one, the wandering is mainly accounted for by Jupiter's attraction. The one exception is Mercury. Mercury shows an extra steady precession of its perihelion that was a mystery right down into the 20th century, when it became a great early triumph of special relativity. Mercury moves faster than any other planet as it rounds its aperihelion (nearest distance to the sun). There it is so fast that it picks up a detectable amount of relativistic mass. Alternatively, you could say that it feels a detectable relativistic perturbation of the sun's **1/r** potential. Either way, it loses the absolute constancy of its Runge-Lenz vector direction, and its orbit undergoes a constant, observable precession in the frame of the fixed stars.

Other dynamic symmetries arise from other potentials that appear on the ultimate microscopic scale of the structure of matter. To mention a few, they are called charm and color, for instance. This dealt with by Lie groups. But this book must stick with simple geometric symmetry.

23.7 Motion reversal symmetry

The character tables of several groups have a feature that is due a physical effect that lies outside the mathematics of group theory, but which has affected the way the Mulliken names of symmetry species.

$$\mathcal{H}\,[\psi[\xi,\eta,\zeta]]\ \mathrm{Exp}[+i\,\omega\,t] = (-i/\hbar)\,\partial_t\ \psi[x,y,z,+t] \qquad (23.8)$$

$$\mathcal{H}\,[\psi^*[x,y,z]]\ \mathrm{Exp}[-i\,\omega\,t] = (+i/\hbar)\,\partial_t\ \psi^*[x,y,z,-t] \qquad (23.9)$$

The space-dependent part of the quantum state (as shown in Figs. 23.3 and 23.6) is not complete. The complete eigenfunction also has an exponential time dependence factor, so that $\Psi[x,y,z,t] = \psi[x,y,z]\ \mathrm{Exp}[i\ \omega\ t]$, where the angular frequency ω is **energy / $\hbar$**. In group theory this shows up whenever a group permits 2-D rotation in either clockwise or counterclockwise sense. The mark of such groups is complex-valued characters.

```
imCh = Select[GroupCatalog,
  Not[FreeQ[CharacterTable[#], Complex]] &]

{C3, C3h, C4, C4h, C5, C6, C6h, S4, S6, S8, T, Th}
```

Here is one of the character tables:

```
BoxedCharacterTable["C6"]
```

C6	1	1	1	1	1	1	← Class populations Note : ϵ = Exp[-2πi/3]
	E	C6	C62	C63	C64	C65	↓ Basis functions
A	1	1	1	1	1	1	$\{(z), (x^2+y^2), (z^2), (Iz)\}$
B	1	-1	1	-1	1	-1	$\{((x \pm i\, y)^3)\}$
E1a	1	$-\epsilon$	ϵ^*	-1	ϵ	$-\epsilon^*$	$\{(x+-i\,y)\}$
E1b	1	$-\epsilon^*$	ϵ	-1	ϵ^*	$-\epsilon$	$\{(x+\ i\,y)\}$
E2a	1	ϵ^*	ϵ	1	ϵ^*	ϵ	$\{(x^2-2\,i\,x\,y-y^2)\}$
E2b	1	ϵ	ϵ^*	1	ϵ	ϵ^*	$\{(x^2+2\,i\,x\,y-y^2)\}$

The **E** species in these groups are not truly two-dimensional; each has a one-D rep supported by a power of **x ± i y**. One relates to clockwise rotation; the other, counterclockwise. But such conjugate one-D species were made into honorary two-D species by Prof. Mulliken, who created the naming scheme. We return a bit toward a strict naming system by appending an **a** or **b** to the name. This helps when counting species. A true 2-D species **E** counts as only one, but **Ea** and **Eb** should count as two. This keeps the character table square, for instance.

23.8 End Notes

23.8.1 Orbitals and water waves

You can see this in a fountain with six small hexagonally arrayed jets of water, alternately covered and uncovered by waves in the pond. When a jet is covered, its water adds to the local wave amplitude, maintaining the wave. When uncovered, it sprays out over a wide area, doing nothing for the wave shape. Waves of any of the various orbital shapes can develop and be maintained, until the wind provokes a change from one mode to another. A steady wind favors the waves with a single nodal line perpendicular to the wind direction.

There is (or used to be) a fountain like this near the old train station in St. Louis.

23.8.2 Orthonormalization

With energies and boundary conditions set appropriately, any one of the panels of Fig. 23.3 can be generated by a numerical solver. However, the solver cannot tell you about the degeneracy of the whole set of solutions. For this, you need three things: one good numerical solution, the Hamiltonian symmetry transforms, and a Gram-Schmidt orthogonalization. Here is the procedure :

First, get your numerical integrator to make a function with a simple nodal pattern, such as the left-most function of each degenerate row of Fig. 23.3. Then transform this function by all the Hamiltonian symmetries, pruning away any duplicates. Finally, feed all the survivors to a Gram-Schmidt orthogonalization procedure, which will return a minimum set of orthonormalized basis functions for the given level.

GramSchmidt will use any inner product (or **S**calar **P**roduct) formula you give it in the option **InnerProduct→fn** . Mathematicians know of many kinds of scalar products, but the only one used in quantum mechanics is

$$\mathbf{SP}_{1,2} == \int_{\mathrm{all\ space}} \left(\psi_1[\underline{\vec{r}}]\right)^* \psi_2[\underline{\vec{r}}]\, d\underline{\vec{r}}$$

Orthogonalize uses Gram-Schmidt by default. It is completely automated, but here is a description of how it works. It accepts a list of functions $\{\psi_1, \ldots, \psi_N\}$, and without further ado, it normalizes ψ_1 and accepts it as the first basis function $\psi_{1\,B}$. Then it calculates $\psi_{2\,B} = \mathbf{Norm}[\psi_2 - \psi_1\, \mathbf{SP}_{1,2}]$ as a possible second basis function. If nonzero, $\psi_{2\,B}$ is accepted and added to the basis list. Functions $\psi_{1\,B}$ and $\psi_{2\,B}$ are necessarily orthogonal because

$$\mathbf{SP}_{1\,B,2\,B} == \int_{\mathrm{all\ space}} \left(\psi_1[\underline{\vec{r}}]\right)^* \left(\psi_2 - \psi_1\, \mathbf{SP}_{1,2}\right) d\underline{\vec{r}} ==$$

$$\int_{\mathrm{all\ space}} \left(\left(\psi_1[\underline{\vec{r}}]\right)^* \psi_2 - \left(\psi_1[\underline{\vec{r}}]\right)^* \psi_1\, \mathbf{SP}_{1,2}\right) d\underline{\vec{r}} == \mathbf{SP}_{1,2} - \mathbf{SP}_{1,2} = 0$$

Then it continues on to ψ_3, attempting to extract from it another nonzero function $\psi_{3\,B}$ orthogonal to both $\psi_{1\,B}$ and $\psi_{2\,B}$. If it succeeds, it adds $\psi_{3\,B}$ to the basis list; if not, it passes on to ψ_4, etc., always hopeful that it may find a new orthogonal function. When it finishes the last function ψ_N it returns a **List** of all the orthonormal basis functions found. Thus may a complete set of numerically exact basis functions be generated, starting from any one numerically exact orbital of the level.

GramSchmidt always returns a list of orthonormal functions, but note that the list returned depends on the order of the functions in the list submitted. This is just reality; there are infinitely many ways to specify the basis functions. To a human, some lists look simpler than others, and we tend to focus on the nice-looking ones. But to a computer, all basis lists look the same aesthetically, and there is no way to choose a "best" one.

23.8.3 The simplest example: parity

Let $\mathcal{P}$ be an operator that changes the sign of the **x** coordinate. Many Hamiltonians are invariant under this operator. The kinetic energy is always invariant to it, and the potential energy of anything with a σ_x reflection plane) is also. For the eigenfunctions ψ of all such Hamiltonians we can write

$$\mathcal{P}[\psi[\mathbf{x}]] = \psi[-\mathbf{x}] = \mathbf{p}\,\psi[\mathbf{x}]$$

where **p** is the eigenvalue of $\mathcal{P}$. Note that $\mathcal{P}^2$ is the identity operator, because two changes of sign on **x** get you back to the original.

$$\mathcal{P}^2[\psi] = \mathcal{P}[\mathcal{P}[\psi]] = \mathcal{P}[\mathbf{p}\,\psi] = \mathbf{p}\,\mathcal{P}[\psi] = \mathbf{p}^2\,\psi = \psi$$

So $\mathbf{p}^2 = 1$, and $\mathbf{p} = \pm 1$. So the most that $\mathcal{P}$ can do is to change the sign of ψ. If $\mathcal{P}[\psi] = \psi$ we say that ψ has *even* parity; if $\mathcal{P}[\psi] = -\psi$, *odd* parity.

Functions of even and odd parity are easy to spot graphically :

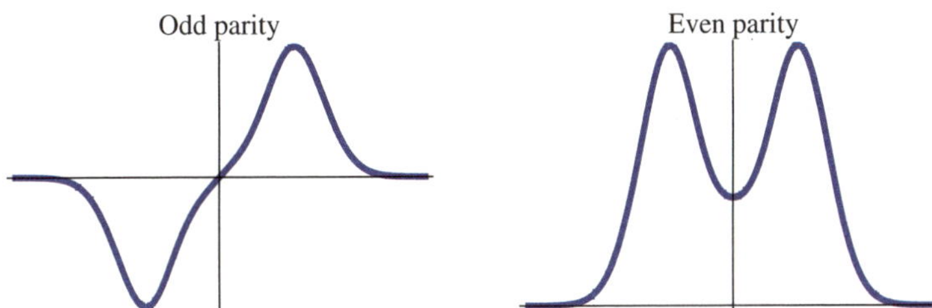

Note that the odd parity function will integrate to zero, because whatever is on the right will cancel with an equal but opposite contribution from the left. If you think you need a formal proof of this, read on.

23.8.4 Construct the invariant space of an arbitrary function

Click back above for the introductory discussion. Now we do a concrete example. Generating the space is easy, but the cleanup requires a little ingenuity. We take as our example function

```
Φ = Sqrt[3] (x^5 y + 2 x^3 y^3 + x y^5) / 2;
gp = "D3";
```

$$\Phi = \frac{\sqrt{3}\,\left(x^5 y + 2 x^3 y^3 + x y^5\right)}{2};$$

First, rescale the function so its first coefficient is **1**. (The function **scale** is defined in the preliminaries.)

```
fn = scale[Φ] // Expand
```

$$\frac{x^5 y}{4} + \frac{x^3 y^3}{2} + \frac{x y^5}{4}$$

Make the raw invariant space:

```
basis1 = fn /. SymmetryRules[gp] // Expand;
% // Size[6]
```

$$\left\{\frac{x^5 y}{4} + \frac{x^3 y^3}{2} + \frac{x y^5}{4},\ -\frac{1}{16}\sqrt{3}\,x^6 - \frac{x^5 y}{8} - \frac{1}{16}\sqrt{3}\,x^4 y^2 - \frac{x^3 y^3}{4} + \frac{1}{16}\sqrt{3}\,x^2 y^4 - \frac{x y^5}{8} + \frac{\sqrt{3}\,y^6}{16},\right.$$
$$\frac{\sqrt{3}\,x^6}{16} - \frac{x^5 y}{8} + \frac{1}{16}\sqrt{3}\,x^4 y^2 - \frac{x^3 y^3}{4} - \frac{1}{16}\sqrt{3}\,x^2 y^4 - \frac{x y^5}{8} - \frac{\sqrt{3}\,y^6}{16},\ -\frac{x^5 y}{4} - \frac{x^3 y^3}{2} - \frac{x y^5}{4},$$
$$-\frac{1}{16}\sqrt{3}\,x^6 + \frac{x^5 y}{8} - \frac{1}{16}\sqrt{3}\,x^4 y^2 + \frac{x^3 y^3}{4} + \frac{1}{16}\sqrt{3}\,x^2 y^4 + \frac{x y^5}{8} + \frac{\sqrt{3}\,y^6}{16},$$
$$\left.\frac{\sqrt{3}\,x^6}{16} + \frac{x^5 y}{8} + \frac{1}{16}\sqrt{3}\,x^4 y^2 + \frac{x^3 y^3}{4} - \frac{1}{16}\sqrt{3}\,x^2 y^4 + \frac{x y^5}{8} - \frac{\sqrt{3}\,y^6}{16}\right\}$$

Apply **scale**, and then hit it with **Union** to get rid of redundancies :

```
basis2 = Map[scale, basis1] // Union
```

$$\left\{x^5 y + 2 x^3 y^3 + x y^5,\right.$$
$$x^6 - \frac{2 x^5 y}{\sqrt{3}} + x^4 y^2 - \frac{4 x^3 y^3}{\sqrt{3}} - x^2 y^4 - \frac{2 x y^5}{\sqrt{3}} - y^6,$$
$$\left.x^6 + \frac{2 x^5 y}{\sqrt{3}} + x^4 y^2 + \frac{4 x^3 y^3}{\sqrt{3}} - x^2 y^4 + \frac{2 x y^5}{\sqrt{3}} - y^6\right\}$$

That's getting better. If you stare at it for a little while, you will see that the two generated functions are just the original function, plus other stuff. Sometimes this happens; sometimes not. But when it does, use the **remove** operator, below. It finds the coefficient that removes the first term of **f1** from **f2**, and then removes the whole thing. The **If** makes sure that **f1** is not removed from **f1** itself:

```
Clear[remove];
remove[f2_, f1_] := Module[{lead, cf},
If[f1 === f2, Return[f1]];
```

```
lead = f1[[1]] ;
cf = Coefficient[f2, lead];
(f2 - cf * f1) // Expand ];
```

Map **remove** onto **basis2**, hit it with a **Union**, and you have it:

```
basis3 = Map[remove[#, fn] &, basis2] // Union
```

$$\left\{x^5 y + 2 x^3 y^3 + x y^5,\ x^6 - \frac{2 x^5 y}{\sqrt{3}} + x^4 y^2 - \frac{4 x^3 y^3}{\sqrt{3}} - x^2 y^4 - \frac{2 x y^5}{\sqrt{3}} - y^6,\ x^6 + \frac{2 x^5 y}{\sqrt{3}} + x^4 y^2 + \frac{4 x^3 y^3}{\sqrt{3}} - x^2 y^4 + \frac{2 x y^5}{\sqrt{3}} - y^6\right\}$$

Pull it all together in one function :

```
Clear[makeInvariantSpace];
makeInvariantSpace[Φ_, gp_] :=
 Module[{fn, basis1, basis2, basis3},
fn = scale[Expand[Φ]];
basis1 = fn /. SymmetryRules[gp] // Expand;
basis2 = Map[scale, basis1] // Union;
basis3 = Map[remove[#, fn] &, basis2] // Union;
Map[scale, basis3] // Union  ]
```

Try it out:

```
invSpace = makeInvariantSpace[Φ, gp]
```

$$\{x^5 y + 2 x^3 y^3 + x y^5,\ x^6 + x^4 y^2 - x^2 y^4 - y^6\}$$

We must test the reputed invariance from first principles, and see if it really is. Since the functions are polynomials, all we have to do is assure ourselves that the group operations create no new polynomial terms. Here is an operator that evaporates the coefficients from any polynomial, returning just a list of its bare terms :

```
bareTerms[poly_] := Module[{termLst, coeffs},
  termLst = Expand[poly] /. Plus → List;
coeffs = termLst /. {x → 1, y → 1, z → 1};
termLst / coeffs]
```

Use it to make a list of terms in both functions before the big group transform :

```
before = Map[bareTerms, invSpace] // Flatten // Union
```

$$\{x^6,\ x^5 y,\ x^4 y^2,\ x^3 y^3,\ x^2 y^4,\ x y^5,\ y^6\}$$

Apply all the symmetry rules of the group, and see if anything new appears:

```
after =
 Map[bareTerms, invSpace /. SymmetryRules[gp] //
     Expand] // Flatten // Union
```

$\{x^6, x^5 y, x^4 y^2, x^3 y^3, x^2 y^4, x y^5, y^6\}$

```
before == after
```

```
True
```

Indeed, that little two-polynomial space was invariant under $\mathbf{D_3}$.

24. Transformation of functions

Preliminaries

24.1 What is a function transform ?

Consider a functional form **f** that depends on the list **{x,y}**, and think of its contour plot over the **x,y** plane. If you want a concrete example, think of

```
f[{x_, y_}] := x^2 y;
```

We make its contour plot with **BlueContourPlot**, defined in the preliminaries. It is just the standard core operator **ContourPlot**, with a few options to make it look prettier.

```
bcp1 =
 BlueContourPlot[f[{x, y}], {x, -2, 2}, {y, -2, 2}]
```

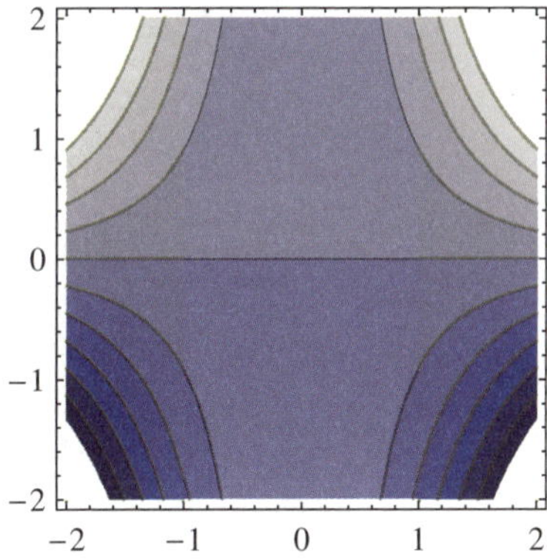

Fig. 24.1 Contour plot of the function f[{x,y}] = x^2 y

Here is our basic question, in concrete form: What function has a contour plot that looks just like this, but rotated **20** degrees counterclockwise? To begin, let's think about how to rotate a single point **{x,y}** by δ degrees counterclockwise. The basic formula is quite familiar. If we let

$$\text{Rmat}[\delta\_] := \begin{pmatrix} \text{Cos}[\delta] & -\text{Sin}[\delta] \\ \text{Sin}[\delta] & \text{Cos}[\delta] \end{pmatrix}$$

we can define a $\mathcal{T}$ that will rotate any **{x,y}** point by **+20** degrees:

```
𝒯[{x_, y_}] := Rmat[20. Degree].{x, y}
```

Try it on a point that lies on the positive **x**-axis:

DOI 10.1007/b13137_24,

```
𝒯[{1, 0}]
```

```
{0.939693, 0.34202}
```

It moved left and up. Point your right thumb at your nose to verify that this is right handed (or counter-clockwise) rotation, as seen from a point on the positive side of the **z** axis.

24.2 An intuitive way (but it's wrong)

Now we do something that usually seems intuitively correct to innocent people. If we want a contour plot to rotate by +20 degrees, we will just rotate all the points by +20 degrees before we compute the function. We apply $\mathcal{T}$ to the general **{x,y}** point before we apply **f** to it :

```
𝒯[bcp] = BlueContourPlot[
  f[𝒯[{x, y}]], {x, -2, 2}, {y, -2, 2}]
```

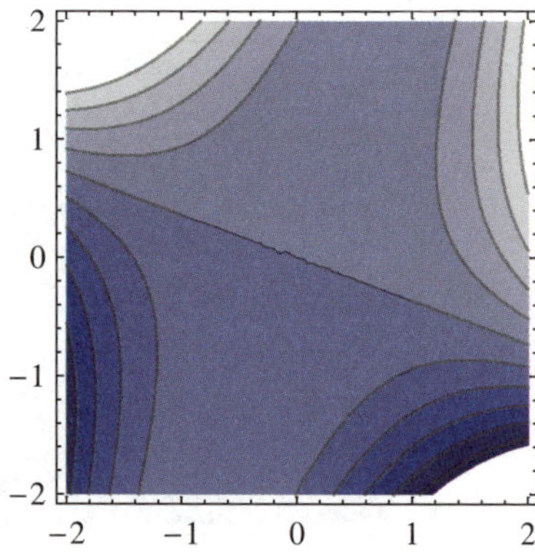

Fig. 24.2 Contours of f[𝒯{x, y}] ≐ (0.93x−0.34y)^2 (0.34x + 0.94y), with • a rotation by +20 °.

It seems to work pretty well. But wait... Do you notice something strange? Positive rotations are *counterclockwise*, but this plot has rotated *clockwise*. What causes this? Maybe we need to think a little more carefully.

24.3 The right way

We begin with a careful statement of what we want :

We have a transform $\mathcal{T}$ that transforms the independent variables **x,y** according to $\mathcal{T}$`[{`x_{old}`, `y_{old}`}] == {`x_{new}`, `y_{new}`}`.

We define the functional form f_{new} to be such that

$$f_{new}[x_{new}, y_{new}] == f_{old}[x_{old}, y_{old}]$$

We already know the meanings of f_{old}, x_{old}, y_{old}, x_{new}, and y_{new}. Therefore, this is a definition of f_{new}.

In words: The height of the contour of f_{new} at the point $\{x_{new}, y_{new}\}$ must be the same as that of f_{old} at the point $\{x_{old}, y_{old}\}$. Thus when the new point rotates counterclockwise, the whole contour will rotate counterclockwise. If we can get rid of x_{old} and y_{old}, rewriting them in terms of x_{new} and y_{new}, then our definition of f_{new} will be in perfectly standard form, with a function of independent variables on the left, and its meaning on the right. So pretending we know the functions $x_{old}[\{x_{new}, y_{new}\}]$ and $y_{old}[\{x_{new}, y_{new}\}]$ we write

$$f_{new}[\{x_{new}, y_{new}\}] == f_{old}[\{x_{old}[\{x_{new}, y_{new}\}], y_{old}[\{x_{new}, y_{new}\}]\}]$$

How do we construct $\{x_{old}, y_{old}\}$ out of $\{x_{new}, y_{new}\}$? We certainly do not use $\mathcal{T}$ because we have already said

$$\{x_{new}, y_{new}\} == \mathcal{T}[\{x_{old}, y_{old}\}]$$

We want the opposite of this. Apply the inverse transform $\mathcal{T}^{-1}$ to both sides, and you have it:

$$\mathcal{T}^{-1}[\{x_{new}, y_{new}\}] == \mathcal{T}^{-1}[\mathcal{T}[\{x_{old}, y_{old}\}]]$$

Whatever $\mathcal{T}$ does, $\mathcal{T}^{-1}$ undoes, so this becomes

$$\mathcal{T}^{-1}[\{x_{new}, y_{new}\}] == \{x_{old}, y_{old}\}$$

or for linear transforms based on matrix $\mathbb{T}$, just use the matrix inverse of $\mathbb{T}$:

$$\{x_{old}, y_{old}\} = \mathbb{T}^{-1}.\{x,y\}$$

Now we use this to eliminate $\{x_{old}, y_{old}\}$ from the equation for f_{new}:

$$f_{new}[\{x, y\}] == f_{old}\left[\mathbb{T}^{-1}.\{x, y\}\right]$$

In our intuitive calculation we used $\mathbb{T}$ instead of $\mathbb{T}^{-1}$, and got it wrong. Now it is easy to fix it. Define the inverse operator just like the direct operator, but with negative angle. In the preliminaries we made $\mathcal{T}^{-1}$ into a symbol. So

```
𝒯⁻¹[{x_, y_}] := Rmat[-20. Degree].{x, y};
```

Just rerun the same **ContourPlot**, but with inverse operator :

```
BlueContourPlot[
  f[𝒯⁻¹[{x, y}]], {x, -2, 2}, {y, -2, 2}]
```

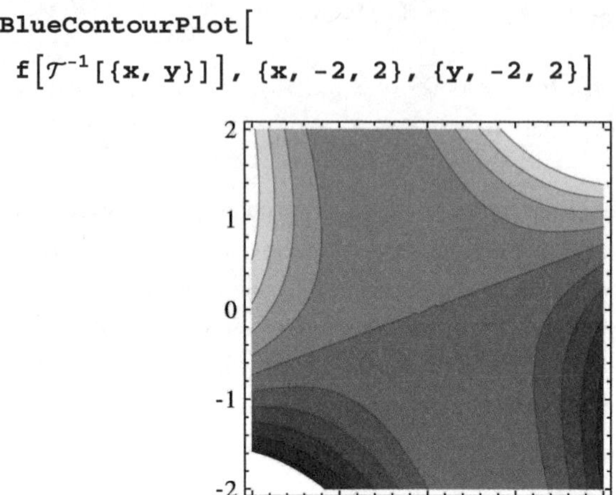

Fig. 24.3 Contours of f[𝒯 {x, y}] ≐ (0.93 x + 0.34 y)² (0.34 x - 0.94 y), with • a rotation by -20 °. Now the plot itself rotates by +20 °.

Cynics will say that the change in the sign of the angle is just a kludge to make it turn the other way. Yes, it does that, but it is also profoundly correct. The inverse formula works even for nonlinear transforms.

Really, we have given a pretty good proof of the following theorem :

> If you want the contour plot of **f[{x,y,...}]** to be transformed by transform $\mathcal{T}$, make a plot of **f[𝒯⁻¹[{x, y, ...}]]**.

The function **f[𝒯⁻¹[{x, y, ...}]]** is just another function of **{x,y,...}**, and it needs a function name of its own. We will call it **𝒯[f]**, and we will write the function and its variables as **𝒯[f][{x,y,...}]**. *Mathematica* supports this kind of function name. It is traditionally written simply as **𝒯f[{x,y,...}]**. But this notation should be abandoned because it does not indicate the scope of the $\mathcal{T}$. It might mean that $\mathcal{T}$ should apply to **f** and **x** and **y** as well. No, we mean it to apply only to **f**.

Rewriting in a form we will pick up and use in the next chapter

```
𝒯[f][{x, y, z}] == f[𝒯⁻¹[{x, y, z}]]
```

The left side is what we want to know, and the right side shows how to construct it.

24.4 End Notes

24.4.1 Kludge

This is an official English word. The Oxford English Dictionary says:
Kludge (kloooj) : An ill-assorted collection of poorly-matching parts, forming a distressing whole; esp.in Computing, a machine, system, or program that has been improvised or "bodged" together; a hastily improvised and poorly thought-out solution to a fault or "bug".

25. Matrix representations of groups

Preliminaries

25.1 Representation theory

We have worked extensively with groups of Cartesian transform matrices that multiply exactly like a corresponding abstract group, and you may have the impression that these are the only matrices that can represent the group. This is far from the truth. The Cartesian representation (technically, "the representation based on `{x, y, z}`") is only one among an infinite possible number of representations. In this chapter we will develop the concept of matrix representation in its full generality.

Click back to the basic definition of symmetry (Chapter 4) and read it again. There are many kinds of symmetric objects in chemistry and physics, each with its own kind of transformation rule. In the Representation Theorem below, we use such general language that we manage to deal with all of them at once. This is wonderful, but unless you already know some examples, it leaves you with an empty feeling. So we follow the general theorem with three kinds of examples, which should restore you.

25.2 Definition of representation

Definition of *representation*
Let the matrices **{𝔸, 𝔹, ... , ℤ}** correspond to the elements **{A, B, ... , Z}** of an abstract group. The matrices are a *representation* of the group if for every relation **A⊗B = C** in the abstract group it is true that 𝔸 **.** 𝔹 = ℂ in the corresponding matrices.

In words, the matrices **{𝔸, 𝔹, ... , ℤ}** *represent* the group in the sense tha every relation in the group multiplication table is also true for **Dot** products of the matrices.

W.M. McClain, *Symmetry Theory in Molecular Physics with Mathematica*,
DOI 10.1007/b13137_25, © Springer Science+Business Media, LLC 2009

25.3 Faithful and unfaithful reps

Note that the implication in the definition runs only one way. *If* a product relation holds in the abstract group, *then* it must also hold in the representation. Sometimes the converse of this is true and sometimes not. Explicitly, the converse is : If a product relation is true among the matrices, then it is also true among the group elements. When the converse holds, the representation is called a *faithful* representation. When not, *unfaithful*.

You have already seen many examples of faithful representations. (All the matrix groups tabulated under the names **CartRep["group"]** are faithful.) So you need to see immediately an example of an unfaithful representation, which is a new idea. We define two small groups :

```
grpA = CartRep["C4"];
```

$$\texttt{grpB} = \left\{\begin{pmatrix} 1 & 0 \\ 0 & 1 \end{pmatrix}, \begin{pmatrix} -1 & 0 \\ 0 & -1 \end{pmatrix}, \begin{pmatrix} 1 & 0 \\ 0 & 1 \end{pmatrix}, \begin{pmatrix} -1 & 0 \\ 0 & -1 \end{pmatrix}\right\};$$

First, we check for closure:

```
MultiplicationTable[Dot, grpA]
```

```
{{1, 2, 3, 4}, {2, 3, 4, 1}, {3, 4, 1, 2}, {4, 1, 2, 3}}
```

```
MultiplicationTable[Dot, grpB]
```

```
{{1, 2, 1, 2}, {2, 1, 2, 1}, {1, 2, 1, 2}, {2, 1, 2, 1}}
```

Neither table contains any question marks, so they both close. We accept this as evidence that they are both groups.

So **grpA** is the point group $\mathbf{C_4}$, the rotations by multiples of 90°, with elements $\{\mathbf{E}, \mathbf{C_4}, \mathbf{C_4^2}, \mathbf{C_4^3}\}$. In **grpB** you will note immediately that this matrix list includes repeats. However you will recall that this is not illegal. We make a truth table for multiplication, taking all possible cases of indices **{i,j,k}** in both groups,, and bracketing together the results for **grpA** and **grpB**.

```
TFtable = Table[{
      grpA⟦i⟧.grpA⟦j⟧ == grpA⟦k⟧,
      grpB⟦i⟧.grpB⟦j⟧ == grpB⟦k⟧},
     {i, 4}, {j, 4}, {k, 4}] // Flatten[#, 2] &;
Dimensions[TFtable]
```

```
{64, 2}
```

Yes, both groups have **4** members, and $\mathbf{4}^3$ is **64**. So we have exhaustive examination. We **Select** all the cases where **grpA** yields a **True**:

```
Select[TFtable, (#[[1]] === True) &] // Union
```

```
{{True, True}}
```

In every case where **grpA** yields a **True**, then **grpB** does also. This means that **grpB** represents **grpA**. But is it a faithful or an unfaithful representation ? If it is faithful, then a **False** in **grpA** will imply a **False** in **grpB**, and there will be no pairs **{False,True}**. Look for them :

```
Select[TFtable, (# === {False, True}) &] // Union
```

```
{{False, True}}
```

No! There are plenty of them, so **grpB** is an *unfaithful* representation of **grpA**.

We mentioned this before in Chapter 14 (CharacterTableIntro). All groups have a representation that is the ultimate in unfaithfulness : Let each group element be represented by a simple **1**. The test is: "If **A⊗B == C**, then it must be true that **1×1 = 1**". Indeed, this is a true syllogism (however scanty the content). So a list **1**'s is indeed a representation, called the *trivial* representation, or the *totally symmetric* representation (because **1** is unchanged (or symmetric) under every transform of the group).

Why do mathematicians make such a seemingly silly definition? It's because we will be interested in counting the different irreducible representations of groups, and the theorems will be prettier if we count the trivial representation as a true representation.

25.4 MorphTest

In order to test the constructions that we make below, we have written an operator called **MorphTest** that tests any list of matrices to see if it is a representation of a given group; and if so, whether it is faithful or unfaithful. It is a straightforward generalization of the test we made above. The thumbnail sketch is

- **MorphTest[slot1,slot2,opts]** returns three possible statements:
  1) **slot1** and **slot2** are isomorphic (faithful to each other),
  2) **slot1** is an unfaithful representation of **slot2**
  3) **slot2** is an unfaithful representations of **slot1**
  slot1 must be a list of group elements
  slot2 is either another list of elements or a group name.
  The opts are **Numeric→False//True** and **TypeSize→0//nbr**.

The defaults are **False** and **0**.
TypeSize→6 (for example) returns both multiplication tables for homomorphic groups in type size **6**; the default **TypeSize → 0** suppresses this.
Numeric → True is used when symbolic identification fails.

Try it out on the unfaithful representation of group C_4 given above :

```
MorphTest[grpB, "C4"]
```

```
slot#1 represents slot#2 unfaithfully (homomorphism)
```

For further detail, read the text of the operator in the **Symmetry`** package.

25.5 Why representations are important

The idea of representation begins to assume its true importance when you learn that any collection of symmetric objects can be used to construct a representation. In fact, many physical applications of group theory follow a pattern that goes like this:

(1) Identify a set of objects that transform linearly among themselves under the action of a group $\mathcal{G}$ of transforms (i.e., identify some objects that span an invariant space). These could be polynomials, infinitesimal rotation matrices, atom orbitals in a symmetric molecule, symmetry adapted combinations of atomic orbitals, molecular orbitals, quantum formulas for spectroscopic operators, or many other things.

(2) Use these objects to make a representation of group $\mathcal{G}$.

(3) Analyze the representation in a way that will be described. This analysis tells you everything that can be said about the symmetry of the objects.

Step 1 is the hard one. Human intelligence is required to identify a suitable group of objects. But once this is done, Step 2 is just calculational work, and can often be automated, as we show in Chapter 27.

Step 3, the analysis of the representation, will be described in Chapter 28 (Reducible Reps), then carried out by an amazing method that involves an arbitrary matrix in Chapter 30 (Schur's Reduction), and finally made fast and automatic in Chapter 35 (RepAnalysis).

25.6 The rep construction recipe

25.6.1 Rep construction in general

We can describe the construction process without specifying what the objects are, or what the transforms are. This is important, because it is the same for all embodiments. Here is the notation we will use:

> **Notation**
> $\{\mathcal{A}, \mathcal{B}, \ldots, \mathcal{Z}\}$ are linear transform operators that constitute group $\mathcal{G}$
> They "know" how to work on objects $\{f_1, .., f_N\}$.

The operators "belong" to a group $\mathcal{G}$ if composition of the operators obeys the multiplication table for $\mathcal{G}$. In detail, if $\mathbf{A} \otimes \mathbf{B} = \mathbf{C}$, then $\mathcal{A}[\mathcal{B}[f_i]] = \mathcal{C}[f_i]$.

> **The invariance requirement**
> The space spanned by $\{f_1, .., f_N\}$ must be *invariant* under each of the transforms $\{\mathcal{A}, \mathcal{B}, \ldots, \mathcal{Z}\}$.
>
> A space is *invariant* under a transform $\mathcal{T}$ if the transformed bases, $\{\mathcal{T}[f_1], .., \mathcal{T}[f_N]\}$, are just linear combinations the original bases $\{f_1, .., f_N\}$, and vice-versa.

Remember, symmetric means that something is unchanged under a transform. Here, the individual f objects change, but the space they span does not. So an *invariant* space is a *symmetric* space (with respect to a transform $\mathcal{T}$). In plain English, operator $\mathcal{T}$ moves points around in the space, but it brings no new points in, and casts no old ones out.

Sometimes $\{\mathcal{T}[f_1], .., \mathcal{T}[f_N]\}$ is just a permutation of $\{f_1, .., f_N\}$. This is invariance, since permutations are just a special kind of linear combination (having only 0 and 1 as coefficients).

With these understandings we can state the general representation recipe. It is really just a statement of the invariance requirement, viewed from a new perspective :

For every operator $\mathcal{T}_\Lambda$, find its representation matrix $\mathbb{T}^{(\Lambda)}$ by solving

$$\mathcal{T}_\Lambda[\underline{\mathbf{f}}] = \underline{\mathbf{f}}.\mathbb{T}^{(\Lambda)}$$

or in index notation

$$\mathcal{T}_\Lambda[\mathbf{f}_k] = \sum_{h=1}^{nf} \mathbf{f}_h \, \mathbb{T}^{(\Lambda)}_{h,k} \qquad (25.1)$$

where the **f** objects lie in a space invariant under all the transforms $\mathcal{T}_\Lambda$, and **nf** is the number of **f** objects (i.e., the dimension of vector space $\underline{\mathbf{f}}$).

Note that matrix $\mathbb{T}$ will be of size **nf** × **nf**.

It may seem odd to you that $\mathbb{T}$ multiplies leftward into the vector of **f** objects. Usually our matrices multiply rightward into vectors. But this is not a mistake; we really do mean leftward multiplication. You will see exactly why when you follow a detailed construction, below.

25.6.2 Check that the recipe works

We do a *Mathematica*-assisted constructive proof for the case of 2 completely general basis objects. That is, we will do the constructions required by Eq. 25.1 and test whether it produces a representation, or not. Then we will argue that the case of **N** basis objects is essentially the same.

Take the invariant linear transforms to be

```
Arule = {f1 → a f1 + b f2, f2 → c f1 + d f2};
Brule = {f1 → r f1 + s f2, f2 → t f1 + u f2};
```

No 2-D transform can be more general than this. We need *Mathematica* operators that impose these rules :

```
Clear[𝒜, ℬ];
𝒜[u_] := u /. Arule;
ℬ[u_] := u /. Brule;
```

For example:

```
𝒜[2 f1 + 3 f2] // Collect[#, {f1, f2}] &
```

```
(2 a + 3 c) f1 + (2 b + 3 d) f2
```

Operator $\mathcal{A}$ has moved the point, but it is still in the $\{\mathbf{f}_1, \mathbf{f}_2\}$ space. To get the matrices, Eq. 25.1 says that we must solve for the matrix objects in

```
𝒜[{f1, f2}] = {f1, f2}.𝔸
```
, etc.

In the MatrixReview we show that this is accomplished by

```
𝔸 = MatrixOfCoefficients[
    𝒜[{f1, f2}], {f1, f2}] // Transpose;
𝔹 = MatrixOfCoefficients[
    ℬ[{f1, f2}], {f1, f2}] // Transpose;
{𝔸, 𝔹} // MatrixList
```

$$\left\{\begin{pmatrix} a & c \\ b & d \end{pmatrix}, \begin{pmatrix} r & t \\ s & u \end{pmatrix}\right\}$$

And so forth, with all the other elements of the group. Remember, the key property that we want is *composition*. In other words, we want the compound transform 𝒜 ℬ (first ℬ , then 𝒜) to be represented by the matrix product **𝔸.𝔹**. To make sure that this is true, apply the substitution operators, working rightward :

```
form1 = 𝒜[ℬ[{f1, f2}]] // Expand
```

$\{a\, f_1\, r + b\, f_2\, r + c\, f_1\, s + d\, f_2\, s,\ a\, f_1\, t + b\, f_2\, t + c\, f_1\, u + d\, f_2\, u\}$

Then apply the compounded matrix **𝔸.𝔹**, working leftward :

```
form2 = {f1, f2}.(𝔸.𝔹) // Expand
```

$\{a\, f_1\, r + b\, f_2\, r + c\, f_1\, s + d\, f_2\, s,\ a\, f_1\, t + b\, f_2\, t + c\, f_1\, u + d\, f_2\, u\}$

Are they the same? Don't trust your eyes. You are too excited; you have too much invested in the outcome, and you might make a mistake. Let the machine do it:

```
form1 == form2
```

True

Success ! Yes, **𝒜 [ℬ[{f1, f2, ..}]]** (operating rightward) has exactly the same effect as **{f1, f2, ..}.𝔸.𝔹** (operating leftward). So the whole table of all compounded transforms in the group can be written as a table of matrix products, taken in the same order.

On your own

1. Try the last test above with matrices **𝔸** and **𝔹** reversed.

2. Try it with a variant of Eq. 25.1 that uses rightward matrices. (Just click back and remove the **Transpose** from the equations for **𝔸** and **𝔹**, evaluating on down to the result. Then does **𝔸.𝔹** or **𝔹.𝔸** give a **True** ?)

The construction has been illustrated, but the formal argument continues: The operators belong to group $\mathcal{G}$, so if the abstract group elements obey **A⊗B = C** then the operators obey $\mathcal{A}[\mathcal{B}[\square]] = \mathcal{C}[\square]$. Therefore

$$\mathcal{C}[\{f_1, f_2\}] == \{f_1, f_2\}.\mathbb{A}.\mathbb{B}, \tag{25.2}$$

But Eq. 25.1 says that a matrix $\mathbb{C}$ must stand in the position occupied by $\mathbb{A}.\mathbb{B}$ so it must be that

$$\text{If } \mathbf{A}\otimes\mathbf{B}=\mathbf{C}, \quad \text{then} \quad \mathbb{A}.\mathbb{B}=\mathbb{C} \tag{25.3}$$

which was to be proved.

Looking back, we used nothing that was unique to the case of **2** basis objects, and it seems clear that the same construction will work for the case of **N** objects, with any **N** you like.

> **On your own**
> Using **MultiplicationTable["group"]**, find a non-Abelian table and select a pair of elements where multiplication order makes a difference. Then use **CartRep["element","group"]** to pull out the representation matrices, and verify that if the table says **A⊗B=C**, then $\mathbb{A}.\mathbb{B}=\mathbb{C}$.

In Chapter 27 (MakeReps), we will carry out the rep construction for several kinds of basis objects, and automate the constructions.

26. Similar representations

Preliminaries

26.1 Given two reps, are they similar?

A new understanding of **congruent** will help us below:

We will call two groups of matrices **congruent** if they have the same group order and the same matrix size.

In the MatrixReview it is shown that if a group $\mathcal{A}$ is similarity transformed into a congruent group $\mathcal{B}$ then they both obey the same multiplication table (click to review the theorem). Now think about the converse of this: *If two congruent groups of matrices, $\mathcal{A}$ and $\mathcal{B}$ obey the same table, then they are similarity transforms of each other.* As always, converses are not automatically true. In fact, this one is quite false. The proof is simple: If $\mathcal{A}$ is faithful and $\mathcal{B}$ is unfaithful, they both obey the abstract group multiplication table. But $\mathcal{B}$ must contain an illicit relationship (the essence of unfaithfulness) that is untrue in $\mathcal{A}$ and that relation is maintained in any similarity transform of $\mathcal{B}$ Since $\mathcal{A}$ does not have it, it cannot be a similarity transform of $\mathcal{B}$

But there is a little stronger question that remains unanswered. If $\mathcal{A}$ and $\mathcal{B}$ are faithful representations of some group, is there always a similarity transform that will change one into the other? The formal statement is

Let $\mathcal{A}$ and $\mathcal{B}$ be different faithful representations of some abstract group. This implies that their elements are in one-to one correspondence $\mathbb{A}_i$ to $\mathbb{B}_i$, for all i.

Then does there exist a matrix $\mathbb{P}$ such that for every member $\mathbb{A}_i$ of $\mathcal{A}$ and its corresponding member $\mathbb{B}_i$ of $\mathcal{B}$ it is true that

$$\mathbb{A}_i = \mathbb{P} . \mathbb{B}_i . \mathbb{P}^{-1}$$

(26.1)

W.M. McClain, *Symmetry Theory in Molecular Physics with Mathematica*,
DOI 10.1007/b13137_26, © Springer Science+Business Media, LLC 2009

This question will launch us into a surprisingly rich development that will occupy the next several chapters, and give us the crowning achievement of finite group theory, the Great Orthogonality.

Many applied group theory books skip over the development of the Great Orthogonality and present it as *fait accompli.* They do this because one cannot conduct numerical experiments in a paper paged book, so they cannot easily show how the ideas developed and led logically to the theorem. But in doing so they skip a development that is one of the most beautiful and surprising in all of mathematics. Our guided tour in the coming chapters will avoid many bypaths and pitfalls, and will include several firm pushes in the right direction. We hope to show that the Great Orthogonality was the result of human curiosity and logic, piqued by some surprising mathematical experiments.

26.2 Universal pseudocommuter matrices

26.2.1 Strengthen the problem

We begin our work with a slightly altered form of Eq. 26.1 . Multiplying from the right by matrix $\mathbb{P}$, it becomes

$$\mathbb{A}_i . \mathbb{P} = \mathbb{P} . \mathbb{B}_i \tag{26.2}$$

Matrix $\mathbb{P}$ needs a name. We will call it a *universal pseudocommuter* matrix for $\mathcal{A}$ and $\mathcal{B}$ The "universal" part just means that it is a pseudocommuter for all the elements of the group. The "pseudo" part means that its defining property is similar to commutation, but just a little different. If $\mathbb{B}_i$ and $\mathbb{A}_i$ in Eq. 26.2 were the same, then $\mathbb{P}$ would be a true commuter. We will be very interested in the true commuter equation, but that comes later.

Eq. 26.2 is a little stronger than 26.1 because there is no presupposition that $\mathbb{P}$ has an inverse, and we can work with the trivial solution $\mathbb{P} = \mathbf{0}$, which plays an important role. Therefore our basic question becomes: When is there a nontrivial solution for $\mathbb{P}$, and can we compute it from the given $\mathcal{A}$ and $\mathcal{B}$?

In Eq. 26.2 both sides are linear in $\mathbb{P}$, so any multiple of a solution $\mathbb{P}$ is also a solution. Thus $\mathbb{P}$ is somewhat like an eigenvector (Any multiple of an eigenvector is still an eigenvector. In fact, if we required $\mathbb{B}_i$ to be diagonal, Eq. 26.2 would be the eigensystem equation... but we don't.)

Because the group axioms insure that $\mathbb{A}_i^{-1}$ always exists, we lose no generality in rewriting Eq. 26.2 as

$$\mathbb{P} = \mathbb{A}_i^{-1}.\mathbb{P} \,.\, \mathbb{B}_i \tag{26.3}$$

If some matrix $\mathbb{P}$ solves this equation for every element $\mathbf{i}$ of the two reps, then we still have a valid equation if we sum over the whole group. If the group contains $\mathbf{h}$ elements, this gives

$$\mathbf{h}\,\mathbb{P} = \sum_{i=1}^{h} \mathbb{A}_i^{-1}.\mathbb{P} \,.\, \mathbb{B}_i \tag{26.4}$$

This is a very promising equation because it contains a lot of information from both groups. So all we need is to solve for $\mathbb{P}$. But how?

26.2.2 A bright idea: successive approximations

Eq. 26.4, if divided on both sides by $\mathbf{h}$, is of the form $\mathbb{P} = \mathbf{f}[\mathbb{P}]$, and with this form you can sometimes get a solution for the unknown $\mathbb{P}$ by the method of successive approximations. Let $\mathbb{P}_1$ be a first guess for the target matrix $\mathbb{P}$. It need not be very good. Then the first improvement is

$$\mathbb{P}_2 = \left(\frac{1}{h}\right) \sum_{i=1}^{h} \mathbb{A}_i^{-1}.\mathbb{P}_1.\, \mathbb{B}_i \tag{26.5}$$

With $\mathbb{P}_2$ in hand we continue to iterate, so that after $\mathbf{n}$ rounds we have

$$\mathbb{P}_{n+1} = \left(\frac{1}{h}\right) \sum_{i=1}^{h} \mathbb{A}_i^{-1}.\mathbb{P}_n.\, \mathbb{B}_i \tag{26.6}$$

Sometimes the successive approximations $\mathbb{P}_1$, $\mathbb{P}_2$, $\mathbb{P}_3$, ..., $\mathbb{P}_n$, $\mathbb{P}_{n+1}$ oscillate wildly, giving nothing useful; but other times they settle in toward a target matrix $\mathbb{P}$, and when they do you have found a solution to the problem.

26.2.3 A much brighter idea

However, before implementing this rather wildcat scheme, we scout around a little. How can we use the fact that our matrices form a group? We remember that any sum over the whole group lets one use the rearrangement theorem. Choosing corresponding elements **k** from both groups, this theorem lets us rewrite Eq. 26.5 as

$$\mathbb{P}_2 = \left(\frac{1}{h}\right) \sum_{i=1}^{h} \left(\mathbb{A}_i . \mathbb{A}_k^{-1}\right)^{-1} . \mathbb{P}_1 . \left(\mathbb{B}_i . \mathbb{B}_k^{-1}\right) \tag{26.7}$$

This multiplication reorders both groups in the same way, and since we sum over all the elements, the sum is the same. The first factor in the sum is

$$\left(\mathbb{A}_i . \mathbb{A}_k^{-1}\right)^{-1} = \left(\mathbb{A}_k^{-1}\right)^{-1} . \mathbb{A}_i^{-1} = \mathbb{A}_k . \mathbb{A}_i^{-1} \tag{26.8}$$

so making this substitution, and moving the sum as far inward as possible,

$$\mathbb{P}_2 = \mathbb{A}_k . \left(\frac{1}{h} \sum_i \mathbb{A}_i^{-1} . \mathbb{P}_1 . \mathbb{B}_i\right) . \mathbb{B}_k^{-1} \tag{26.9}$$

But the parenthesis is just the right side of the original Eq. 26.5 , so

$$\mathbb{P}_2 = \mathbb{A}_k . \mathbb{P}_2 . \mathbb{B}_k^{-1} \tag{26.10}$$

or

$$\mathbb{P}_2 . \mathbb{B}_k = \mathbb{A}_k . \mathbb{P}_2 \tag{26.11}$$

where **k** can be chosen as any index in the group.

This is an astonishing result. It is exactly the pseudo commuter definition, Eq. 26.2, but with sides switched. It says there is no need for iteration because $\mathbb{P}_2$, as calculated by Eq. 26.5, is already the universal pseudocommuter! Furthermore, $\mathbb{P}_1$ was completely arbitrary, so it might as well absorb the factor of **1/h** and be called $\mathbb{Q}$.

A universal pseudocommuter $\mathbb{P}$ for groups $\mathcal{A}$ and $\mathcal{B}$ is

$$\mathbb{P} = \sum_{i=1}^{h} \mathbb{A}_i^{-1} . \mathbb{Q} . \mathbb{B}_i \tag{26.12}$$

where $\mathbb{Q}$ is any matrix that makes the formula evaluatable.

When $\mathbb{A}$ and $\mathbb{B}$ are the same size, nothing prevents the simplest choice of $\mathbb{Q}$, a unit matrix. Then the formula becomes

$$\mathbb{P} = \sum_{i=1}^{h} \mathbb{A}_i^{-1} . \mathbb{B}_i \quad \text{(if } \mathbb{A} \text{ and } \mathbb{B} \text{ are the same size)} \tag{26.13}$$

We anticipated that the formula for $\mathbb{P}$ would have an arbitrary multiplicative constant in it. But the general formula contains a whole arbitrary multiplicative *matrix*. Can this possibly be correct? We must try a numerical example immediately.

26.2.4 A nice square example

Here we will make two very different-looking reps of the same size that are actually similarity transforms of each other. Then we use them to construct their universal pseudocommuter, and then carry out the similarity transform from one to another, just to verify it.

Group **T** (all rotations of the tetrahedron) has a very pretty Cartesian representation, so let's use that as our group.

In the list of operations below we call up a group $\mathcal{G}$ and transform $\mathcal{G}$ into $\mathcal{A}$ using an arbitrary tangling matrix $\mathbb{T}_{\mathcal{A}}$:

```
𝒢 = CartRep["T"];
len = Length[𝒢];
𝕋𝒜 = {{6, -10, -5}, {7, -5, -3}, {9, 9, 3}};
𝒜 = Similarity[𝒢, 𝕋𝒜];
MorphTest[𝒜, "T"]
```

```
Faithful, or Isomorphic
```

Do it again, using a different tangling matrix, and calling the result $\mathcal{B}$:

```
𝕋ℬ = {{-9, -2, 10}, {4, -1, 10}, {-2, -3, -10}};
ℬ = Similarity[𝒢, 𝕋ℬ];
MorphTest[ℬ, "T"]
```

```
Faithful, or Isomorphic
```

We will need the inverses of all the $\mathcal{B}$ matrices :

```
ℬ⁻¹ = Map[Inverse, ℬ];
```

Now make the universal pseudocommuter $\mathbb{P}_{\mathcal{A}\mathcal{B}}$ for $\mathcal{A}$ and $\mathcal{B}$, using an arbitrarily chosen matrix $\mathbb{Q}_1$:

```
ℚ1 = {{8, -9, 8}, {1, 1, -2}, {3, 0, -4}};
```

Make the pseudocommuter :

$$\mathbb{P}_{\mathcal{A}\mathbb{Q}\mathcal{B}} = \sum_{i=1}^{len} \text{Table}\left[\mathcal{B}^{-1}[\![i]\!].\mathbb{Q}_1.\mathcal{A}[\![i]\!]\right]$$

$$\left\{\left\{\frac{268513}{45}, -\frac{159901}{20}, \frac{1173613}{540}\right\}, \left\{\frac{295666}{45}, -\frac{1571857}{180}, \frac{1318429}{540}\right\}, \left\{-\frac{48272}{9}, \frac{1288259}{180}, -\frac{1083103}{540}\right\}\right\}$$

This ought to transform $\mathcal{A}$ into $\mathcal{B}$. Will it really do it ?

```
Similarity[𝒜, ℙ𝒜ℚℬ] == ℬ
```

```
True
```

However unlikely this may seem, it works.

> **On your own**
> Go back and try this with different $\mathbb{Q}$ matrices. In particular, try the unit matrix.

26.2.5 A strange rectangular example

By using the word "congruent" we encouraged you to think that reps $\mathcal{A}$ and $\mathcal{B}$ are of the same dimension. But go back and read Section 26.2 again, this time thinking that reps $\mathcal{A}$ and $\mathcal{B}$ are of different dimensions. It still makes sense if the pseudocommuter $\mathbb{P}$ is rectangular, with the height of $\mathcal{A}$ and the width of $\mathcal{B}$:

$$\mathbb{A}_i \cdot \mathbb{P} = \mathbb{P} \cdot \mathbb{B}_i$$

$$\begin{pmatrix} \square & \square & \square \\ \square & \square & \square \\ \square & \square & \square \end{pmatrix} \cdot \begin{pmatrix} \square & \square \\ \square & \square \\ \square & \square \end{pmatrix} = \begin{pmatrix} \square & \square \\ \square & \square \\ \square & \square \end{pmatrix} \cdot \begin{pmatrix} \square & \square \\ \square & \square \end{pmatrix} \tag{26.14}$$

$$\text{or} \quad \begin{pmatrix} \square & \square \\ \square & \square \\ \square & \square \end{pmatrix} = \begin{pmatrix} \square & \square \\ \square & \square \\ \square & \square \end{pmatrix}$$

Nothing in the development prevents the usual conclusion, though now $\mathbb{Q}$ must be rectangular :

$$\mathbb{P} = \sum_i \mathbb{A}_i^{-1} \cdot \mathbb{Q} \cdot \mathbb{B}_i$$

$$\begin{pmatrix} \square & \square \\ \square & \square \\ \square & \square \end{pmatrix} = \sum_i \begin{pmatrix} \square & \square & \square \\ \square & \square & \square \\ \square & \square & \square \end{pmatrix}_i \cdot \begin{pmatrix} \square & \square \\ \square & \square \\ \square & \square \end{pmatrix} \cdot \begin{pmatrix} \square & \square \\ \square & \square \end{pmatrix}_i \tag{26.15}$$

$$\text{or} \quad \begin{pmatrix} \square & \square \\ \square & \square \\ \square & \square \end{pmatrix} = \sum_i \begin{pmatrix} \square & \square \\ \square & \square \\ \square & \square \end{pmatrix}_i$$

This is even harder to believe than the original. To test it, make two reps; one 3×3, the other 2×2 :

```
𝒜33 = CartRep["C3h"];
% // GridList // Size[5]
```

$$\left\{\begin{pmatrix}1&0&0\\0&1&0\\0&0&1\end{pmatrix}, \begin{pmatrix}-\frac{1}{2}&-\frac{\sqrt{3}}{2}&0\\\frac{\sqrt{3}}{2}&-\frac{1}{2}&0\\0&0&1\end{pmatrix}, \begin{pmatrix}-\frac{1}{2}&\frac{\sqrt{3}}{2}&0\\-\frac{\sqrt{3}}{2}&-\frac{1}{2}&0\\0&0&1\end{pmatrix}, \begin{pmatrix}1&0&0\\0&1&0\\0&0&-1\end{pmatrix}, \begin{pmatrix}-\frac{1}{2}&-\frac{\sqrt{3}}{2}&0\\\frac{\sqrt{3}}{2}&-\frac{1}{2}&0\\0&0&-1\end{pmatrix}, \begin{pmatrix}-\frac{1}{2}&\frac{\sqrt{3}}{2}&0\\-\frac{\sqrt{3}}{2}&-\frac{1}{2}&0\\0&0&-1\end{pmatrix}\right\}$$

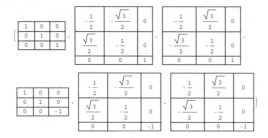

Make a complex 2×2 representation of $\mathbf{C_{3\,h}}$:

```
ℬ22 = MakeRepPoly[{(x + ⅈ y), (x - ⅈ y)}, "C3h"] //
      ComplexExpand;
% // GridList // Size[5]
```

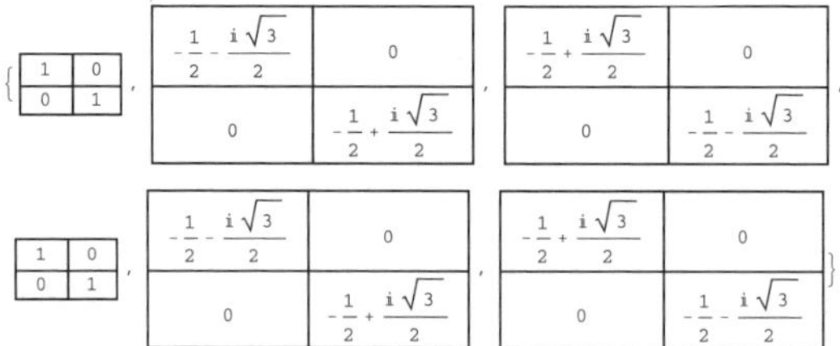

Just check it :

```
MorphTest[ℬ22, "C3h"]
```

```
slot#1 represents slot#2 unfaithfully (homomorphism)
```

Make the rectangular pseudocommuter (Eq. 26.12) using $\mathbb{Q}_{32}$:

```
𝒜33⁻¹ = Map[Inverse, 𝒜33];
ℚ32 = {{10, -8}, {4, 10}, {8, 9}};
ℙ32 = Σ_{i=1}^{6} (𝒜33⁻¹[[i]].ℚ32.ℬ22[[i]]) // ExpandAll;
% // GridForm // Size[8]
```

30 - 12 i	-24 + 30 i
12 + 30 i	30 + 24 i
0	0

Test it on all the elements of the group :

```
Test[i_] := (Expand[𝒜33⟦i⟧.ℙ32] == Expand[ℙ32.ℬ22⟦i⟧]);
Table[Test[i], {i, 6}]
```

```
{True, True, True, True, True, True}
```

Well, perhaps you still don't *have to* believe it (this is only an example), but at least you must take it seriously. That algorithm for the universal pseudocommuter is a very hardy algorithm.

> **On your own**
> Further examples won't prove anything, but by working them out you insure that you really understand the process. So do a few other examples.

26.3 Universal true commuter matrices

Nothing in Section 26.2 (Universal pseudocommuter matrices) excludes the case where reps $\mathcal{A}$ and $\mathcal{B}$ are the same. Go back and read through it again thinking $\mathcal{A}$ and $\mathcal{B}$ are the same. It all still works. Therefore, without any further work, we can say

> A universal commuter $\mathbb{C}$ for rep $\mathcal{A}$ is given by
> $$\mathbb{C} = \sum_{i=1}^{h} \mathbb{A}_i^{-1}.\mathbb{Q}.\mathbb{A}_i \qquad (26.16)$$
> where $\mathbb{Q}$ is any matrix that makes the formula evaluatable. (But if $\mathbb{Q}$ is a unit matrix the result is trivial.)

If $\mathbb{Q}$ is the unit matrix, the summands are all unit matrices, and $\mathbb{C}$ comes out as a constant matrix (**h** times the unit matrix). Constant matrices commute with all matrices, so this is a correct but trivial result, and it does not lead to the place we want to go. This will be discussed further in Chapter 30 (SchursReduction) where we make great use of the universal commuter.

On your own
Make a numerical test of universal commuter construction (Eq. 26.16). Then pick an example group and confirm that the answer really is a universal commuter.

Does Eq. 26.16 work for lists of matrices that are not a group?

Click here for a relevant End note.

26.4 A whiff of Schur's Second Lemma

Let's try to crash the computer by asking it to do something impossible. Let's ask it to find the similarity transform matrix for a pair of reps that are not related by a similarity transform.

No two different irreducible reps of a group can be similarity transformed into each other, because similarity transforms preserve the characters of the rep, and different irreducible reps have different characters.

```
BoxedCharacterTable["D6h", 6]
```

D6h	1	2	2	1	3	3	1	2	2	1	3	3	← Class populations
	E	C6	C3	C_2	C'_2	C''_2	inv	S3	S6	σh	σd	σv	↓ Basis functions
A1g	1	1	1	1	1	1	1	1	1	1	1	1	$\{(x^2+y^2), (z^2)\}$
A2g	1	1	1	1	-1	-1	1	1	1	1	-1	-1	$\{(-3x^5y+10x^3y^3-3xy^5), (Iz)\}$
B1g	1	-1	1	-1	1	-1	1	-1	1	-1	1	-1	$\{(x^3z-3xy^2z)\}$
B2g	1	-1	1	-1	-1	1	1	-1	1	-1	-1	1	$\{(-3x^2yz+y^3z)\}$
E1g	2	1	-1	-2	0	0	2	1	-1	-2	0	0	$\{\begin{pmatrix} xz \\ yz \end{pmatrix}, \begin{pmatrix} Ix \\ Iy \end{pmatrix}\}$
E2g	2	-1	-1	2	0	0	2	-1	-1	2	0	0	$\{\begin{pmatrix} \frac{x^2}{2}-\frac{y^2}{2} \\ xy \end{pmatrix}\}$
A1u	1	1	1	1	1	1	-1	-1	-1	-1	-1	-1	$\{(-3x^5yz+10x^3y^3z-3xy^5z)\}$
A2u	1	1	1	1	-1	-1	-1	-1	-1	-1	1	1	$\{(z)\}$
B1u	1	-1	1	-1	1	-1	-1	1	-1	1	-1	1	$\{(3x^2y-y^3)\}$
B2u	1	-1	1	-1	-1	1	-1	1	-1	1	1	-1	$\{(-x^3+3xy^2)\}$
E1u	2	1	-1	-2	0	0	-2	-1	1	2	0	0	$\{\begin{pmatrix} x \\ y \end{pmatrix}\}$
E2u	2	-1	-1	2	0	0	-2	1	1	-2	0	0	$\{\begin{pmatrix} \frac{x^2z}{2}-\frac{y^2z}{2} \\ xyz \end{pmatrix}\}$

Reps $\mathbf{E_{2g}}$ and $\mathbf{E_{1u}}$ of group $\mathbf{D_{6h}}$ are both 2-dimensional and have simple basis functions that can be read from the character table. So let's construct concrete $\mathbb{A}_{E2g}$ and $\mathbb{A}_{E1u}$ matrices, and see what happens when we ask it to compute the non-existent pseudocommuter.

We construct the universal pseudocommuter all in one cell below, but interactive readers can take it apart and watch it work step by step.

$\mathcal{A}_{E1u}$ = **MakeRepPoly[{x, y}, "D6h"];**

$\mathcal{B}_{E2u}$ = **MakeRepPoly**$\left[\left\{\left(\frac{x^2 - y^2}{2}\right), \; x\, y\right\}, \; \text{"D6h"}\right]$;

$\mathcal{A}^{-1}_{E1u}$ = **Map[Inverse,** $\mathcal{A}_{E1u}$**];**

$\mathbb{Q}_{22}$ = **{{9, -6}, {1, -2}};**

$\mathbb{P}_{E1u,E2u} == \sum_{i=1}^{24} \left(\mathcal{A}^{-1}_{E1u}[\![i]\!] . \mathbb{Q}_{22} . \mathcal{B}_{E2u}[\![i]\!]\right)$ **// Simplify**

$\mathbb{P}_{E1u,E2u}$ == {{0, 0}, {0, 0}}

Well, at least the computer did not crash. And indeed, Eq. 26.2 is obeyed with all-zeroes in $\mathbb{P}_{E1u,E2u}$, though every element produces only **0 == 0**.

As we said before, this is a very hardy algorithm. When asked for the impossible, it produces something that is at least formally correct. This numerical surprise is the clue that leads to Schur's famous Second Lemma.

On your own

There are many congruencies among the **CartReps** of various point groups. Do all groups of matrices that are congruent but non-similar return an all-zero pseudocommuter ?

After your own experiments, try to state what you think Schur's Second Lemma will be. Then click here.

26.5 End notes

26.5.1 Congruent, non-similar reps

Here is a table from which you can pick **CartRep** congruencies i.e. congruencies among the Cartesian representations. All **CartRep** matrices are 3×3, so just pick groups with the same group order :

```
Map[{#, GroupOrder[#]} &, GroupCatalog]
```

```
{{C1, 1}, {C2, 2}, {C2h, 4}, {C2v, 4}, {C3, 3}, {C3h, 6},
 {C3v, 6}, {C4, 4}, {C4h, 8}, {C4v, 8}, {C5, 5}, {C6, 6},
 {C6h, 12}, {C6v, 12}, {Ch, 2}, {Ci, 2}, {D2, 4}, {D2d, 8},
 {D2h, 8}, {D3, 6}, {D3d, 12}, {D3h, 12}, {D4, 8},
 {D4d, 16}, {D4h, 16}, {D5, 10}, {D5d, 20}, {D5h, 20},
 {D6, 12}, {D6d, 24}, {D6h, 24}, {I, 60}, {O, 24}, {Oh, 48},
 {S4, 4}, {S6, 6}, {S8, 8}, {T, 12}, {Td, 24}, {Th, 24}}
```

Groups $\mathbf{C_{2\,h}}$ and $\mathbf{C_{2\,v}}$ are congruent; try them.

```
𝒜 = CartRep["C2h"];
Map[Spur, 𝒜]
```

```
{3, -1, -3, 1}
```

```
ℬ = CartRep["C2v"];
Map[Spur, ℬ]
```

```
{3, -1, 1, 1}
```

Reps $\mathcal{A}$ and $\mathcal{B}$ have different character vectors, so they cannot be similar. Will they give an all-zero pseudocommuter?

```
𝒜^-1 = Map[Inverse, 𝒜] // ChopInteger;
ℚ = RandomMat[Dimensions[𝒜]⟦2⟧];
ℙ = Apply[Plus,
    Table[𝒜^-1⟦i⟧.ℚ.ℬ⟦i⟧, {i, Length[𝒜]}]] // Chop;
% // MatrixForm
```

$$\begin{pmatrix} 0 & -20 & 0 \\ 0 & 12 & 0 \\ 0 & 0 & 0 \end{pmatrix}$$

It is not all-zero, but does it obey $\mathbb{A}_i . \mathbb{P} = \mathbb{P} . \mathbb{B}_i$?

```
Map[#.ℙ &, 𝒜] == Map[ℙ.# &, ℬ]
```

```
True
```

Wow! Originally, we found $\mathbb{P}$ to be a pseudo-commuter when $\mathcal{A}$ and $\mathcal{B}$ were different irreducible representations of the *same* group. But here, the groups are *different*, and $\mathbb{P}$ is not all zeroes, but the pseudocommuter property still holds! But before you get too excited, go back and try $\mathbf{C_{4\,h}}$ and $\mathbf{C_{4\,v}}$.

Perhaps after muddying the waters for you, we should reclarify them: the all-zero $\mathbb{P}$ is guaranteed only if the congruent reps are *different irreducible representations* of the *same* group.

27. The MakeRep operators

Preliminaries

27.1 Four `MakeRep` operators

Chapter 25 (Representations) gave the general formula for constructing representations, and a proof that it works in principle. The actual work of constructing various kinds of representations was left for this chapter.

You will learn in Chapter 37 (ProjectionOperators) that some of the greatest and neatest applications of group theory involve the use of projection operators. But some projections require a concrete representation belonging to the given species. This is a huge barrier to hand calculation, because whole representations are not tabulated anywhere, and no one (except, maybe, great geniuses and idiots *savante*) can construct them quickly and accurately out of their head. But the **MakeRep** operators developed in this chapter deliver them to the common man in a single stroke.

In Chapter 25 we said that representation matrices are found by solving for $\mathbb{T}^{(\mathbf{R})}$ in an equation that expresses the invariance of objects **f** under transforms $\mathcal{T}_{\mathbf{R}}$ of group $\mathcal{G}$:

The master representation equation

$$\mathcal{T}_{\mathbf{R}}[\mathbf{f}_k] = \sum_{h=1}^{N[f]} \mathbf{f}_h \, \mathbf{T}^{(\mathbf{R})}_{h,k} \qquad \text{for } k = \{1, \ldots, N[f]\} \tag{27.1}$$

where the given quantities are

$\mathcal{T}_{\mathbf{R}}$ is the transform for element **R**
$\mathbf{f}_h$ is basis object number **h**
N[f] is the number of basis objects

and the unknown to be found is
$\mathbf{T}^{(\mathbf{R})}_{h,k}$ (or $\mathbb{T}^{(\mathbf{R})}$), the desired rep matrix for element **R**

The solution method depends on the nature of the **f** objects and their transforms. There is no neat little algorithm that takes care of all cases.

In Section 27.2 we present an algorithm **MakeRepPoly** that works when the **f** objects are polynomials in **{x,y,z}**.

DOI 10.1007/b13137_27,

This chapter is too long to print it all. The next three Sections differ from Section 27.2 (**MakeRepPoly**) only in the way the master equation is solved. You can come back to them on CD when you need them.

In Section 27.3 (closed; please read from the CD) we present **MakeRepRot** that works when the **f** objects are infinitesimal rotation matrices **{Ix, Iy, Iz}**.

In Section 27.4 (closed; please read from the CD) we present **MakeRepAtomic**. Chemists often need to make representations on the basis of the way atoms or atomic orbitals are swapped around in a molecule by symmetry transforms, and this operator makes a matrix representation on the basis of such swapping.

In Section 27.5 we present **MakeRepMolec** (closed; please read from the CD). This is used when the basis set consists of molecular orbitals; i.e., linear combinations of atomic orbitals.

27.2 Polynomial basis (polynomials in x, y, z)

27.2.1 The tabulated basis functions

In Chapter 14 (CharacterTableIntro) we discussed the broad right column of the character table in some detail. Now we will start work with the polynomial bases listed there, so click back if you need to review.

27.2.2 A single polynomial (in one step)

When the basis object is a single polynomial **p[x,y,z]**, the "matrix" $\mathbb{T}^{(R)}$ is of size **1×1**; just a number, $\mathbf{T}^{(R)}$. Then the master equation 27.1 becomes

$$\mathcal{T}_R[\mathtt{p[x,\ y,\ z]}] = \mathtt{p[x,\ y,\ z]} * \mathbf{T}^{(R)} \tag{27.2}$$

where the transform $\mathcal{T}_R$ is performed by replacement rules working on **x**, **y**, **z**. Solving is trivial :

$$\mathbf{T}^{(R)} = \frac{\mathtt{p[x,y,z]/.SymmetryRulesInverse[R]}}{\mathtt{p[x,y,z]}}$$

Are you surprised that we use the **SymmetryRulesInverse** rather than **SymmetryRules** ? If so, click back to the definition of functional transform.

```
BoxedCharacterTable["O"]
```

O	1	8	3	6	6	← Class populations
	E	C3	C42	C4	C2	↓ Basis functions
A1	1	1	1	1	1	$\{(x^2+y^2+z^2)\}$
A2	1	1	1	-1	-1	$\{(x\,y\,z)\}$
E	2	-1	2	0	0	$\left\{\begin{pmatrix} -x^2-y^2+2z^2 \\ \sqrt{3}\,x^2-\sqrt{3}\,y^2 \end{pmatrix}\right\}$
T1	3	0	-1	1	-1	$\left\{\begin{pmatrix} x \\ y \\ z \end{pmatrix}, \begin{pmatrix} Ix \\ Iy \\ Iz \end{pmatrix}\right\}$
T2	3	0	-1	-1	1	$\left\{\begin{pmatrix} x\,y \\ y\,z \\ x\,z \end{pmatrix}\right\}$

In the **A2** row of the **O** character table, the only basis object given is the polynomial **x * y * z**. Further, the **Listable** property of the replacement operator lets us do all the rules at once:

```
repA2ofO = (x*y*z /. SymmetryRulesInverse["O"]) / (x*y*z)
```

```
{1, 1, 1, 1, 1, 1, 1, 1, 1, 1, 1, 1, -1,
 -1, -1, -1, -1, -1, -1, -1, -1, -1, -1, -1}
```

As you see, each number $\mathbf{T}^{(R)}$ is just a **1** or a **-1**. Is this a representation of group **O**? Try MorphTest on it:

```
MorphTest[repA2ofO, "O"]
```

```
slot#1 represents slot#2 unfaithfully (homomorphism)
```

Yes, it is a representation, but since it consists of only two different numbers, it has to be unfaithful. Make sure that the package operator **MakeRepPoly** also produces this result :

```
(MakeRepPoly[{x y z}, "O"] // Flatten) == repA2ofO
```

```
True
```

27.2.3 A list of polynomials (step by step)

Raw materials

When the basis is a list of polynomials, the master equation 27.1 becomes

```
{p1, …, pN} /.
SymmetryRulesInverse[R] =
{p1, …, pN}.ᴿG
```
(27.3)

Now $^{R}\mathbb{G}$ is a matrix, and we need an algorithm that solves for it. As usual in algorithm development, we first do an example, step-by-step. We do species $\mathbf{E_5}$ of group $\mathbf{D_{6\,d}}$.

```
TabulatedBases["D6d"][[9]]
```

$$\{E5, \{\{x\,z, y\,z\}, \{x^5 - 10\,x^3\,y^2 + 5\,x\,y^4, 5\,x^4\,y - 10\,x^2\,y^3 + y^5\}, \{Ix, Iy\}\}\}$$

Three basis pairs are tabulated; we take the second :

```
group = "D6d";
basis = {x^5 - 10 x^3 y^2 + 5 x y^4, 5 x^4 y - 10 x^2 y^3 + y^5};
```

Transform the basis functions

The next step is to transform the **basis** by applying to them the inverse symmetry rules for every element of the group. Below, interactive readers can remove the semicolon and see them all:

```
allBasisTfs =
  basis /. SymmetryRulesInverse[group] // ExpandAll;
Dimensions[%]
```

```
{24, 2}
```

The group has 24 members, so we got one transformed basis pair for every element. To keep things simple, we will do the rep matrix for only one group element. Picking element #2, its transformed basis pair, which goes on the left side of the master equation Eq. 26.3, is

```
rule2 = SymmetryRulesInverse[group][[2]];
left = basis /. rule2 // ExpandAll
```

$$\left\{-\frac{1}{2}\sqrt{3}\,x^5 + \frac{5\,x^4\,y}{2} + 5\sqrt{3}\,x^3\,y^2 - 5\,x^2\,y^3 - \frac{5}{2}\sqrt{3}\,x\,y^4 + \frac{y^5}{2},\right.$$
$$\left.-\frac{x^5}{2} - \frac{5}{2}\sqrt{3}\,x^4\,y + 5\,x^3\,y^2 + 5\sqrt{3}\,x^2\,y^3 - \frac{5\,x\,y^4}{2} - \frac{\sqrt{3}\,y^5}{2}\right\}$$

Now make the right side. Use an **Array** of **a[i,j]** as the symbolic matrix to be found. We will need the **a[i,j]** both as a matrix and as a list :

```
dim = Length[basis]
```

```
2
```

```
aMat = Array[a, {dim, dim}];
aList = % // Flatten
```

```
{a[1, 1], a[1, 2], a[2, 1], a[2, 2]}
```

Carefully making **aMat** operate leftward, the right side is

```
right = basis.aMat
```

$$\{(x^5 - 10 x^3 y^2 + 5 x y^4)\, a[1, 1] + (5 x^4 y - 10 x^2 y^3 + y^5)\, a[2, 1],\; (x^5 - 10 x^3 y^2 + 5 x y^4)\, a[1, 2] + (5 x^4 y - 10 x^2 y^3 + y^5)\, a[2, 2]\}$$

We get one representation equation for each polynomial :

```
repEqs = Thread[left == right]
```

$$\left\{-\frac{1}{2}\sqrt{3}\, x^5 + \frac{5 x^4 y}{2} + 5\sqrt{3}\, x^3 y^2 - 5 x^2 y^3 - \frac{5}{2}\sqrt{3}\, x y^4 + \frac{y^5}{2} == (x^5 - 10 x^3 y^2 + 5 x y^4)\, a[1, 1] + (5 x^4 y - 10 x^2 y^3 + y^5)\, a[2, 1],\right.$$
$$\left. -\frac{x^5}{2} - \frac{5}{2}\sqrt{3}\, x^4 y + 5 x^3 y^2 + 5\sqrt{3}\, x^2 y^3 - \frac{5 x y^4}{2} - \frac{\sqrt{3}\, y^5}{2} == (x^5 - 10 x^3 y^2 + 5 x y^4)\, a[1, 2] + (5 x^4 y - 10 x^2 y^3 + y^5)\, a[2, 2]\right\}$$

Use SolveAlways to make the representation matrix

Can we solve these equations? First we try our old friend, the **Solve** operator:

```
vars = Variables[basis]
```

```
{x, y}
```

```
Solve[repEqs, aList, vars]
```

```
{{}}
```

The empty bracket means that we did not pose a proper problem for **Solve**. We asked it to solve two equations for four variables (**aList**), with the elimination of three variables (**{x,y,z}**). When you put it this way, you can see why **Solve** considered the problem ill-posed. But a related operator that goes by the name of **SolveAlways[eq,vars]** is exactly set up for this situation, The **Always** in its name means "for all values of the **vars**". This operator considers the given equation to be an identity, true for all values of **x,y,z** , and tries to find values of the parameters in **aList** that make it so. This is exactly what we want:

```
aRules = SolveAlways[repEqs, vars] // Flatten
```

$$\left\{a[1,1]\to-\frac{\sqrt{3}}{2},\ a[1,2]\to-\frac{1}{2},\ a[2,1]\to\frac{1}{2},\ a[2,2]\to-\frac{\sqrt{3}}{2}\right\}$$

Those are the right answers, in a simple **List**. As a matrix they are

```
mat = aMat /. aRules;
% // GridForm
```

$-\frac{\sqrt{3}}{2}$	$-\frac{1}{2}$
$\frac{1}{2}$	$-\frac{\sqrt{3}}{2}$

Check the **mat** by putting it back into the definition of representation :

```
Expand[basis /. rule2] == Expand[basis.mat]
```

```
True
```

The procedure above will become **MakeMatPoly[]**, which makes the rep mat for one rule. Then another operator **MakeRepPoly[]** will map it onto all the rules, producing the whole representation Click into the package to see the details. Here is its thumbnail syntax statement :

- **MakeRepPoly[basisList,gp(,opt)]** returns a matrix representation of group **gp** using the Cartesian polynomials in **basisList** as the basis functions. Option **Report→True** causes each matrix to be printed out as it is calculated. See Chapter 27 (MakeReps).

Click into the End Notes to see it tested. Perhaps you would like to stop here and go to the next chapter. The rest of this chapter is similar to what you have just seen. It details three different methods of solving Eq. 27.1 (the master representation equation); namely, for **f** objects that are matrices, atomic orbitals, or molecular orbitals.

27.3 Matrix basis (Ix, Iy, Iz)

Closed; please read from CD

27.4 Atomic orbital basis

Closed; please read from CD

27.5 Molecular orbital basis

Closed; please read from CD

27.6 Check the characters of the reps

Matrix representations like those that we have constructed here will find their main use later on in the projection operators. However, the simplest thing we can do with them is just to check that their class characters agree with those in the character table. Below, we use **MakeRep** operators to get concrete representations, and then we take the traces :

```
TabulatedBases["D3h"][[3]]
```

$$\left\{E',\ \left\{\{x,\ y\},\ \left\{\frac{1}{2}\left(x^2-y^2\right),\ x\,y\right\}\right\}\right\}$$

```
repxzyzD3h = MakeRepPoly[{x z , y z }, "D3h"];
MatrixList[%] // Size[6]
```

$$\left\{\begin{pmatrix}1&0\\0&1\end{pmatrix},\ \begin{pmatrix}-\frac{1}{2}&-\frac{\sqrt{3}}{2}\\\frac{\sqrt{3}}{2}&-\frac{1}{2}\end{pmatrix},\ \begin{pmatrix}-\frac{1}{2}&\frac{\sqrt{3}}{2}\\-\frac{\sqrt{3}}{2}&-\frac{1}{2}\end{pmatrix},\ \begin{pmatrix}-1&0\\0&1\end{pmatrix},\ \begin{pmatrix}\frac{1}{2}&-\frac{\sqrt{3}}{2}\\-\frac{\sqrt{3}}{2}&-\frac{1}{2}\end{pmatrix},\ \begin{pmatrix}\frac{1}{2}&\frac{\sqrt{3}}{2}\\\frac{\sqrt{3}}{2}&-\frac{1}{2}\end{pmatrix},\right.$$

$$\left.\begin{pmatrix}-1&0\\0&-1\end{pmatrix},\ \begin{pmatrix}\frac{1}{2}&\frac{\sqrt{3}}{2}\\-\frac{\sqrt{3}}{2}&\frac{1}{2}\end{pmatrix},\ \begin{pmatrix}\frac{1}{2}&-\frac{\sqrt{3}}{2}\\\frac{\sqrt{3}}{2}&\frac{1}{2}\end{pmatrix},\ \begin{pmatrix}1&0\\0&-1\end{pmatrix},\ \begin{pmatrix}-\frac{1}{2}&\frac{\sqrt{3}}{2}\\\frac{\sqrt{3}}{2}&\frac{1}{2}\end{pmatrix},\ \begin{pmatrix}-\frac{1}{2}&-\frac{\sqrt{3}}{2}\\-\frac{\sqrt{3}}{2}&\frac{1}{2}\end{pmatrix}\right\}$$

Apply the **Spur** operator (which takes the trace, or diagonal sum):

```
longVecD3h = Map[Spur, repxzyzD3h]
```

```
{2, -1, -1, 0, 0, 0, -2, 1, 1, 0, 0, 0}
```

This gives a list of characters as long as the group itself. Check the diagonal sum by eyeball examination. As advertised, every element within a class has the same **Spur** (diagonal sum). Take out the redundancy :

```
ChVecE′ = OnePerClass[longVecD3h, "D3h"]
```

```
{2, -1, 0, -2, 1, 0}
```

This checks with the character table. But the table says that species **E″** can also be constructed using matrices **{Ix, Iy}**. So do it :

```
repIxIyD3h = MakeRepRot[{Ix, Iy}, "D3h"];
MatrixList[%] // Size[6]
```

$$\left\{\begin{pmatrix}1 & 0\\ 0 & 1\end{pmatrix}, \begin{pmatrix}-\frac{1}{2} & -\frac{\sqrt{3}}{2}\\ \frac{\sqrt{3}}{2} & -\frac{1}{2}\end{pmatrix}, \begin{pmatrix}-\frac{1}{2} & \frac{\sqrt{3}}{2}\\ -\frac{\sqrt{3}}{2} & -\frac{1}{2}\end{pmatrix}, \begin{pmatrix}1 & 0\\ 0 & -1\end{pmatrix}, \begin{pmatrix}-\frac{1}{2} & \frac{\sqrt{3}}{2}\\ \frac{\sqrt{3}}{2} & \frac{1}{2}\end{pmatrix}, \begin{pmatrix}-\frac{1}{2} & -\frac{\sqrt{3}}{2}\\ -\frac{\sqrt{3}}{2} & \frac{1}{2}\end{pmatrix},\right.$$

$$\left.\begin{pmatrix}-1 & 0\\ 0 & -1\end{pmatrix}, \begin{pmatrix}\frac{1}{2} & \frac{\sqrt{3}}{2}\\ -\frac{\sqrt{3}}{2} & \frac{1}{2}\end{pmatrix}, \begin{pmatrix}\frac{1}{2} & -\frac{\sqrt{3}}{2}\\ \frac{\sqrt{3}}{2} & \frac{1}{2}\end{pmatrix}, \begin{pmatrix}-1 & 0\\ 0 & 1\end{pmatrix}, \begin{pmatrix}\frac{1}{2} & -\frac{\sqrt{3}}{2}\\ -\frac{\sqrt{3}}{2} & -\frac{1}{2}\end{pmatrix}, \begin{pmatrix}\frac{1}{2} & \frac{\sqrt{3}}{2}\\ \frac{\sqrt{3}}{2} & -\frac{1}{2}\end{pmatrix}\right\}$$

Are the two representations, **{x z,y z}** and **{Ix,Iy}**, the same?

```
repxzyzD3h == repIxIyD3h
```

```
False
```

Nothing said they had to be, and they are not. But the characters had better be the same :

```
OnePerClass[Map[Spur, repIxIyD3h], "D3h"] == ChVecE′
```

```
True
```

They are.

> **On your own**
> Call up the **BoxedCharacterTable** and compare the characters.
>
> Then do some examples of you own on other groups.

> The leftmost character, the character of class **E**, is the dimension of the representation .

Click back to Chapter 14 to be reminded why.

27.7 End Notes

27.7.1 A complex numerical rep ($E_{u,a}$ of T_h)

The 1-D numerical rep algorithm is simple, but when it produces complex numbers it needs some extra attention. Group $\mathbf{T_h}$ has four complex-valued one-dimensional reps. We will work with the **Eua** species.

```
tabulatedEuaRow = CharacterTable["Th"][[8]];
% // Size[8]
```

$$\left\{\text{Eua},\ \left\{1,\ -\frac{1}{2}-\frac{i\sqrt{3}}{2},\ -\frac{1}{2}+\frac{i\sqrt{3}}{2},\ 1,\ -1,\ \frac{1}{2}+\frac{i\sqrt{3}}{2},\ \frac{1}{2}-\frac{i\sqrt{3}}{2},\ -1\right\},\ \left\{\left\{x\,y\,z\left(x^2+y^2-i\sqrt{3}\,(x^2-y^2)-2\,z^2\right)\right\}\right\}\right\}$$

```
EuaFn = tabulatedEuaRow[[3, 1, 1]]
```

$$x\,y\,z\left(x^2+y^2-i\sqrt{3}\,(x^2-y^2)-2\,z^2\right)$$

Try the one-function algorithm on it :

```
repRaw = (EuaFn /. SymmetryRulesInverse["Th"]) / EuaFn;
Short[%, 6]
```

$$\left\{1,\ \frac{-2\,x^2+y^2+z^2-i\sqrt{3}\,(y^2-z^2)}{x^2+y^2-i\sqrt{3}\,(x^2-y^2)-2\,z^2},\ \frac{-2\,x^2+y^2+z^2-i\sqrt{3}\,(y^2-z^2)}{x^2+y^2-i\sqrt{3}\,(x^2-y^2)-2\,z^2},\ \ll 18\gg,\ -1,\ -1,\ -1\right\}$$

This looks like a total failure. Each result is supposed to be just a number, but many are not. But don't despair; it just needs more simplification. Below, we do it, and also shorten the result by using **OnePerClass** on it:

```
repEuaOfTh = repRaw // FullSimplify // ComplexExpand;
chsEuaOfTh = OnePerClass[repEuaOfTh, "Th"]
```

$$\left\{1,\ -\frac{1}{2}-\frac{i\sqrt{3}}{2},\ -\frac{1}{2}+\frac{i\sqrt{3}}{2},\ 1,\ -1,\ \frac{1}{2}+\frac{i\sqrt{3}}{2},\ \frac{1}{2}-\frac{i\sqrt{3}}{2},\ -1\right\}$$

This agrees with the tabulated characters:

```
chsEuaOfTh == (tabulatedEuaRow[[2]] // ComplexExpand)
```

```
True
```

27.7.2 The smart rules for complex symbols

In the **Boxed** version of the character table the complex numbers are represented by ϵ and several of its variants that humans are used to seeing; namely, **-ϵ, ϵ*, -ϵ*, i ϵ, -i ϵ, -i ϵ*, i ϵ***. Translating complex numbers into these symbols is highly prone to human error, so it has been automated by a tricky little operator **SmartRules** tucked away somewhere in the **Symmetry`** package :

```
ϵvalue = Exp[-2 π ⅈ / 3] // ComplexExpand
```

$-\frac{1}{2}-\frac{i\sqrt{3}}{2}$

```
Clear[ϵ];
ϵRules = SmartRules[ϵvalue, ϵ]
```

$\left\{-\frac{1}{2}-\frac{i\sqrt{3}}{2}\to\epsilon,\ \frac{1}{2}+\frac{i\sqrt{3}}{2}\to-\epsilon,\ -\frac{1}{2}+\frac{i\sqrt{3}}{2}\to\epsilon^*,\ \frac{1}{2}-\frac{i\sqrt{3}}{2}\to-\epsilon^*,\ -\frac{i}{2}+\frac{\sqrt{3}}{2}\to i\,\epsilon,\ \frac{i}{2}-\frac{\sqrt{3}}{2}\to-i\,\epsilon,\ \frac{i}{2}+\frac{\sqrt{3}}{2}\to-i\,\epsilon^*,\ -\frac{i}{2}-\frac{\sqrt{3}}{2}\to i\,\epsilon^*\right\}$

Here is the character row presented in the **BoxedCharacterTable** :

```
chsEuaOfTh /. ϵRules
```

$\{1, \epsilon, \epsilon^*, 1, -1, -\epsilon, -\epsilon^*, -1\}$

On your own
All cyclic groups with more than two elements have complex characters. Do some further examples using the model above.

27.7.3 A few tests of MakeRepPoly

The step-by-step development was done in the main text. Here we test the package version.

A one-dimensional representation A2u of D3d

```
TabulatedBases["D3d"][[5]]
```

$\{A2u, \{\{z\}, \{3x^2 y - y^3\}\}\}$

```
repA2uofD3d = MakeRepPoly[{3 x^2 y - y^3}, "D3d"]
```

```
{{{1}}, {{1}}, {{1}}, {{-1}}, {{-1}}, {{-1}},
 {{-1}}, {{-1}}, {{-1}}, {{1}}, {{1}}, {{1}}}
```

```
MorphTest[repA2uofD3d, "D3d"]
```

```
slot#1 represents slot#2 unfaithfully (homomorphism)
```

A two-dimensional representation Eu of D3d

```
TabulatedBases["D3d"][[6]]
```

$$\left\{\text{Eu}, \left\{\{x, y\}, \left\{\frac{1}{2}\left(x^2 - y^2\right) z, x y z\right\}\right\}\right\}$$

```
repEuofD3d =
  MakeRepPoly[{1/2 (x^2 - y^2) z, x y z}, "D3d"];
% // MatrixList // Size[6]
```

$$\left\{\begin{pmatrix} 1 & 0 \\ 0 & 1 \end{pmatrix}, \begin{pmatrix} -\frac{1}{2} & \frac{\sqrt{3}}{2} \\ -\frac{\sqrt{3}}{2} & -\frac{1}{2} \end{pmatrix}, \begin{pmatrix} -\frac{1}{2} & -\frac{\sqrt{3}}{2} \\ \frac{\sqrt{3}}{2} & -\frac{1}{2} \end{pmatrix}, \begin{pmatrix} \frac{1}{2} & \frac{\sqrt{3}}{2} \\ \frac{\sqrt{3}}{2} & -\frac{1}{2} \end{pmatrix}, \begin{pmatrix} \frac{1}{2} & -\frac{\sqrt{3}}{2} \\ -\frac{\sqrt{3}}{2} & -\frac{1}{2} \end{pmatrix}, \begin{pmatrix} -1 & 0 \\ 0 & 1 \end{pmatrix},\right.$$

$$\left.\begin{pmatrix} -1 & 0 \\ 0 & -1 \end{pmatrix}, \begin{pmatrix} \frac{1}{2} & \frac{\sqrt{3}}{2} \\ -\frac{\sqrt{3}}{2} & \frac{1}{2} \end{pmatrix}, \begin{pmatrix} \frac{1}{2} & -\frac{\sqrt{3}}{2} \\ \frac{\sqrt{3}}{2} & \frac{1}{2} \end{pmatrix}, \begin{pmatrix} -\frac{1}{2} & \frac{\sqrt{3}}{2} \\ \frac{\sqrt{3}}{2} & \frac{1}{2} \end{pmatrix}, \begin{pmatrix} 1 & 0 \\ 0 & -1 \end{pmatrix}, \begin{pmatrix} -\frac{1}{2} & -\frac{\sqrt{3}}{2} \\ -\frac{\sqrt{3}}{2} & \frac{1}{2} \end{pmatrix}\right\}$$

```
MorphTest[repEuofD3d, "D3d"]
```

Faithful, or Isomorphic

```
Map[UnitaryQ, repEuofD3d] // Union
```

{True}

A one-dimensional complex representation Ea of T

```
TabulatedBases["T"][[2]]
```

$$\left\{\text{Ea}, \left\{\left\{-x^2 - y^2 - i\sqrt{3}\left(x^2 - y^2\right) + 2 z^2\right\}\right\}\right\}$$

```
repEaofT =
 MakeRepPoly[{-x^2 - y^2 - i Sqrt[3] (x^2 - y^2) + 2 z^2}, "T"] //
   Expand // OnePerClass[#, "T"] &
```

$$\left\{\{\{1\}\}, \left\{\left\{-\frac{1}{2} - \frac{i\sqrt{3}}{2}\right\}\right\}, \left\{\left\{-\frac{1}{2} + \frac{i\sqrt{3}}{2}\right\}\right\}, \{\{1\}\}\right\}$$

This agrees with the tabulated characters :

```
CharacterTable["T"][[4, {1, 2}]]
```

$$\left\{\text{Ea}, \left\{1, -\frac{1}{2} - \frac{i\sqrt{3}}{2}, -\frac{1}{2} + \frac{i\sqrt{3}}{2}, 1\right\}\right\}$$

A three-dimensional representation, T of T

```
TabulatedBases["T"][[4]]
```

```
{T, {{x, y, z}, {x y, y z, x z}, {Ix, Iy, Iz}}}
```

```
repTofT = MakeRepPoly[{x y, y z, x z}, "T"];
% // MatrixList // Size[6]
```

$$\left\{\begin{pmatrix}1&0&0\\0&1&0\\0&0&1\end{pmatrix}, \begin{pmatrix}0&1&0\\0&0&1\\1&0&0\end{pmatrix}, \begin{pmatrix}0&-1&0\\0&0&-1\\1&0&0\end{pmatrix}, \begin{pmatrix}0&1&0\\0&0&-1\\-1&0&0\end{pmatrix}, \begin{pmatrix}0&-1&0\\0&0&1\\-1&0&0\end{pmatrix}, \begin{pmatrix}0&0&-1\\-1&0&0\\0&1&0\end{pmatrix},\right.$$
$$\left.\begin{pmatrix}0&0&-1\\1&0&0\\0&-1&0\end{pmatrix}, \begin{pmatrix}0&0&1\\-1&0&0\\0&-1&0\end{pmatrix}, \begin{pmatrix}0&0&1\\1&0&0\\0&1&0\end{pmatrix}, \begin{pmatrix}-1&0&0\\0&1&0\\0&0&-1\end{pmatrix}, \begin{pmatrix}-1&0&0\\0&-1&0\\0&0&1\end{pmatrix}, \begin{pmatrix}1&0&0\\0&-1&0\\0&0&-1\end{pmatrix}\right\}$$

```
MorphTest[repTofT, "T"]
```

```
Faithful, or Isomorphic
```

```
Map[UnitaryQ, repTofT] // Union
```

```
{True}
```

A five-dimensional representation H of I

This is the longest and hardest representation in our tabulated groups, 60 matrices of size 5×5. The polynomial basis for rep **H** of group **I** is

```
TabulatedBases["I"][[5]];
% // Size[8]
```

$$\left\{H, \left\{\left\{\frac{-x^2 - y^2 + 2 z^2}{2\sqrt{3}}, \frac{1}{2}\left(x^2 - y^2\right), x y, y z, x z\right\}\right\}\right\}$$

Construction of rep **H** takes about 40 seconds on an 867 MHz G4, so we block automatic evaluation below. If you want to see it run, select the cell bracket, go to **Cell(menu) > CellProperties** and click **Evaluatable**. Then run it.

```
(*NOT EVALUATABLE*)
{timeHofI, repHofI} = MakeRepPoly[{(-x^2 - y^2 + 2 z^2)/(2 Sqrt[3]),
    1/2 (x^2 - y^2), x y, y z, x z}, "I"] // Timing;
timeHofI
Short[repHofI, 8]
```

Here is one matrix from rep **H**:

```
repHofI[[59]]//MatrixForm//Size[5]
```

$$\begin{pmatrix} 1 & 0 & 0 & 0 & 0 \\ 0 & \frac{1}{4}\left(-1+\sqrt{5}\right) & \frac{1}{2}\sqrt{\frac{1}{2}\left(5+\sqrt{5}\right)} & 0 & 0 \\ 0 & \frac{1}{2}\sqrt{\frac{1}{2}\left(5+\sqrt{5}\right)} & \frac{1}{4}\left(1-\sqrt{5}\right) & 0 & 0 \\ 0 & 0 & 0 & \frac{1}{4}\left(1+\sqrt{5}\right) & \frac{1}{8}\left(\sqrt{2\left(5+\sqrt{5}\right)}-\sqrt{10\left(5+\sqrt{5}\right)}\right) \\ 0 & 0 & 0 & \frac{1}{8}\left(\sqrt{2\left(5+\sqrt{5}\right)}-\sqrt{10\left(5+\sqrt{5}\right)}\right) & \frac{1}{4}\left(-1-\sqrt{5}\right) \end{pmatrix}$$

The radicals of radicals are not dealt with very effectively by **Simplify**, and this limits what can be done in infinite precision. But often you can speed things up by handling it numerically. But even that takes 9 Second :

```
(*NOT EVALUATABLE; JUST READ THE RESULT*)
MorphTest[repHofI, "I", Numeric → True] // Timing
```

```
Faithful, or Isomorphic

{8.9924 Second, Null}
```

```
(*NOT EVALUATABLE; JUST READ THE RESULT*)
Map[UnitaryQ,repHofI//N]//Union
```

```
{True}
```

27.7.4 Development of MakeRepRot

Closed; please read from CD

27.7.5 Tests of MakeRepAtomic

Closed; please read from CD

28. Reducible representations

Preliminaries

28.1 The Cartesian representations

The geometrical transforms embodied in the Cartesian representations (or **CartReps**) of each group take point **{x,y,z}** and turn it into a geometrically related point. This is just a special case of a functional transform in which the transformed functions **{f$_1$, f$_2$, f$_3$}** are the axes themselves **{x,y,z}**.

Here is a feature of the Cartesian representations that is common to all the axial groups, but not to the Platonic groups. Look at one of the axial groups :

```
Map[MatrixForm, CartRep["D3"]] // Size[5]
```

$$\left\{\begin{pmatrix}1&0&0\\0&1&0\\0&0&1\end{pmatrix},\begin{pmatrix}-\frac{1}{2}&-\frac{\sqrt{3}}{2}&0\\\frac{\sqrt{3}}{2}&-\frac{1}{2}&0\\0&0&1\end{pmatrix},\begin{pmatrix}-\frac{1}{2}&\frac{\sqrt{3}}{2}&0\\-\frac{\sqrt{3}}{2}&-\frac{1}{2}&0\\0&0&1\end{pmatrix},\begin{pmatrix}1&0&0\\0&-1&0\\0&0&-1\end{pmatrix},\begin{pmatrix}-\frac{1}{2}&\frac{\sqrt{3}}{2}&0\\\frac{\sqrt{3}}{2}&\frac{1}{2}&0\\0&0&-1\end{pmatrix},\begin{pmatrix}-\frac{1}{2}&-\frac{\sqrt{3}}{2}&0\\-\frac{\sqrt{3}}{2}&\frac{1}{2}&0\\0&0&-1\end{pmatrix}\right\}$$

All these matrices are in block diagonal form, $\begin{pmatrix}☺&☺&\mathbf{0}\\☺&☺&\mathbf{0}\\\mathbf{0}&\mathbf{0}&☺\end{pmatrix}$. This occurs because both proper and improper **z**-axis rotations mix **x** and **y** together. They may change the sign of **z**, but they never mix it with anything. The same may be said of twofold rotations about axes perpendicular to **z**, and of reflections parallel or perpendicular to **z**. This covers all the symmetry elements not in the Platonic groups.

> **On your own**
> Pull up some other **CartReps** of non-Platonic groups and note their block diagonal form.

Now we will show that the upper and lower blocks are separate representations of the group. Pull out the **D$_3$** table and note that **z** is a basis for species **A$_2$**.

```
repz = MakeRepPoly[{z}, "D3"];
% // Flatten
```

```
{1, 1, 1, -1, -1, -1}
```

Never assume it is a rep without testing :

W.M. McClain, *Symmetry Theory in Molecular Physics with Mathematica*,
DOI 10.1007/b13137_28, © Springer Science+Business Media, LLC 2009

```
MorphTest[repz, "D3"]
```

```
slot#1 represents slot#2 unfaithfully (homomorphism)
```

Compare **repz** to the **3,3** elements of the tabulated **CartRep** for $\mathbf{D_3}$:

```
Map[#[[3, 3]] &, CartRep["D3"]]
```

```
{1, 1, 1, -1, -1, -1}
```

So the lower right block (**1 × 1**) is a rep by itself. Now make the rep based on **{x,y}** :

```
repxy = MakeRepPoly[{x, y}, "D3"];
% // MatrixList // Size[6]
```

$$\left\{\begin{pmatrix}1&0\\0&1\end{pmatrix},\begin{pmatrix}-\frac{1}{2}&-\frac{\sqrt{3}}{2}\\\frac{\sqrt{3}}{2}&-\frac{1}{2}\end{pmatrix},\begin{pmatrix}-\frac{1}{2}&\frac{\sqrt{3}}{2}\\-\frac{\sqrt{3}}{2}&-\frac{1}{2}\end{pmatrix},\begin{pmatrix}1&0\\0&-1\end{pmatrix},\begin{pmatrix}-\frac{1}{2}&\frac{\sqrt{3}}{2}\\\frac{\sqrt{3}}{2}&\frac{1}{2}\end{pmatrix},\begin{pmatrix}-\frac{1}{2}&-\frac{\sqrt{3}}{2}\\-\frac{\sqrt{3}}{2}&\frac{1}{2}\end{pmatrix}\right\}$$

```
MorphTest[repxy, "D3"]
```

```
Faithful, or Isomorphic
```

Compare this with the upper left **SubMatrix** of **CartRep["D3"]**:

```
repxy == Map[Take[#, 2, 2] &, CartRep["D3"]]
```

```
True
```

So **CartRep["D3"]** is of **{2 × 2, 1 × 1}** block diagonal form.

28.2 Block diagonal multiplication

The block diagonal matrices have zeroes everywhere except in square blocks located on the diagonal. They have an interesting property: when you multiply two block diagonal matrices that have the same pattern, the product has the same pattern as its parents. You could say that block structure comes down in the bloodline of matrices. Here are two simple matrix products :

$$\left\{\begin{pmatrix}1&1\\1&1\end{pmatrix}.\begin{pmatrix}2&2\\2&2\end{pmatrix}==\begin{pmatrix}4&4\\4&4\end{pmatrix},\quad\begin{pmatrix}3&3&3\\3&3&3\\3&3&3\end{pmatrix}.\begin{pmatrix}1&1&1\\1&1&1\\1&1&1\end{pmatrix}==\begin{pmatrix}9&9&9\\9&9&9\\9&9&9\end{pmatrix}\right\}$$

```
{True, True}
```

Put them into a 5×5 matrices with a 2×2 block in the upper left and a 3×3 block in the lower right; then multiply the two big matrices together :

$$\begin{pmatrix} 1 & 1 & 0 & 0 & 0 \\ 1 & 1 & 0 & 0 & 0 \\ 0 & 0 & 3 & 3 & 3 \\ 0 & 0 & 3 & 3 & 3 \\ 0 & 0 & 3 & 3 & 3 \end{pmatrix} . \begin{pmatrix} 2 & 2 & 0 & 0 & 0 \\ 2 & 2 & 0 & 0 & 0 \\ 0 & 0 & 1 & 1 & 1 \\ 0 & 0 & 1 & 1 & 1 \\ 0 & 0 & 1 & 1 & 1 \end{pmatrix} == \begin{pmatrix} 4 & 4 & 0 & 0 & 0 \\ 4 & 4 & 0 & 0 & 0 \\ 0 & 0 & 9 & 9 & 9 \\ 0 & 0 & 9 & 9 & 9 \\ 0 & 0 & 9 & 9 & 9 \end{pmatrix}$$

```
True
```

As you see, the diagonal blocks multiply independently. Do a few row-by-column sums outside the diagonal blocks, and you will see why the block pattern has to reproduce itself. Or, if you want a symbolic argument, here it is:

Let a diagonal block start at $\{\lambda,\lambda\}$ and end at $\{\mu,\mu\}$. Then any row-by-column product that intersects within the block can be decomposed into three summands: **before + inside + after**, Symbolically,

$$(\mathbb{A}.\mathbb{B})_{i,k} = \sum_{j=1}^{\lambda-1} A_{i,j}\, B_{j,k} + \sum_{j=\lambda}^{\mu} A_{i,j}\, B_{j,k} + \sum_{j=\mu+1}^{N} A_{i,j}\, B_{j,k}$$

But in the first sum all the $A_{i,j}$ are zero (they are all left of the block), and in the third sum all the $B_{j,k}$ are zero (being above the block). So we are left with

$$(\mathbb{A}.\mathbb{B})_{i,k} = 0 + \sum_{j=\lambda}^{\mu} A_{i,j}\, B_{j,k} + 0 \quad \text{(for } i \text{ and } k \text{ in the diagonal block)}$$

In words, each product element within the block is given by the matrix product of the blocks. This proves it when **i,k** is in the diagonal block.

Now consider a case when **i,k** is outside any of the diagonal blocks. For instance, the blue row times the blue column below gives the outside blue zero on the right :

$$\begin{pmatrix} 1 & 1 & 0 & 0 & 0 \\ 1 & 1 & 0 & 0 & 0 \\ 0 & 0 & 3 & 3 & 3 \\ 0 & 0 & 3 & 3 & 3 \\ 0 & 0 & 3 & 3 & 3 \end{pmatrix} . \begin{pmatrix} 2 & 2 & 0 & 0 & 0 \\ 2 & 2 & 0 & 0 & 0 \\ 0 & 0 & 1 & 1 & 1 \\ 0 & 0 & 1 & 1 & 1 \\ 0 & 0 & 1 & 1 & 1 \end{pmatrix} == \begin{pmatrix} 4 & 4 & 0 & 0 & 0 \\ 4 & 4 & 0 & 0 & 0 \\ 0 & 0 & 9 & 9 & 9 \\ 0 & 0 & 9 & 9 & 9 \\ 0 & 0 & 9 & 9 & 9 \end{pmatrix}$$

It is zero because one of the factors in the **Dot** product is always zero:

```
{1,1,0,0,0}.
{0,0,1,1,1} == 0
```

This could be made into a proof, but it would be too boring. I think you have the idea. Here is the theorem:

If $\mathbf{A}_i$ and $\mathbf{B}_j$ are square blocks on the diagonal, and if the 0's are appropriate blocks of zeroes, then the block diagonal product is

$$\begin{pmatrix} A_1 & 0 & 0 \\ 0 & A_2 & 0 \\ 0 & 0 & A_3 \ldots \\ & & \vdots \end{pmatrix} . \begin{pmatrix} B_1 & 0 & 0 \\ 0 & B_2 & 0 \\ 0 & 0 & B_3 \ldots \\ & & \vdots \end{pmatrix} == \begin{pmatrix} A_1 . B_1 & 0 & 0 \\ 0 & A_2 . B_2 & 0 \\ 0 & 0 & A_3 . B_3 \ldots \\ & & \vdots \end{pmatrix}$$

So q.e.d., in a sloppy way. There is more on this in the Matrix Review.

28.3 Reducible and irreducible representations

If a block diagonal representation is subjected to a similarity transform, the block diagonal pattern is generally destroyed, but the matrices continue to represent the group quite nicely (click for proof).

Usually, we are much more interested in the inverse of this process. In Chapter 30 (Schur'sReduction) (click for preview), we will show how to take a tangled non-block representation and transform it to block diagonal form, using only information from within the tangled representation itself. It's almost like a command to make shattered pottery come back together.

Discussion of these matters inevitably involves the following words

Reducible representation
A representation is **reducible** if there exists a similarity transform that puts it in block diagonal form.

But this is not always possible :

Irreducible representation
A representation for which no block-diagonalizing similarity transform exists is called an **irreducible** representation.

28.4 Mulliken names of irreducible representations

The symmetry species, or irreducible representations, are so important that they have been given standard names, called Mulliken names, after the great chemical theorist R. S. Mulliken. These names are used uniformly throughout the spectroscopic literature. They appear down the left column of every **Boxed**\ **CharacterTable**. Here are the rules for these names:

1. One dimensional reps are called **A** or **B**; two-dimensional, **E**; three dimensional **T** or **F**, four dimensional, **G**; five-dimensional, **H**. Exception: One dimensional reps that come in conjugate pairs act spectroscopically like **E** reps, and they are called $\mathbf{E_a}$ and $\mathbf{E_b}$.

2. One-dimensional reps have a name anchored on **A** or **B**. If the character for rotation about the principal axis by the smallest angle is **1**, the anchor is **A**; if **-1**, the anchor is **B**. Exception: In group $\mathbf{D_2}$, there is no unique principal axis, all the nontrivial reps are called **B**.

3. The anchor gets a subscript **1** if rotation about a two-fold axis in the **x,y** plane has character **1**, or lacking that, if reflection in a vertical plane has character **1**. The subscript is **2** if the character is **-1**. (Again, group $\mathbf{D_2}$ has to be exceptional.)

4. The anchor gets a single prime if the rep has character **1** for reflection in the **x,y** plane. If the character is **-1**, a double prime.

5. The anchor takes a subscript **g** (German *gerade*) if the character for inversion is **1**; it takes a **u** (German *ungerade*) if **-1**.

6. Several groups have multiple **E** or **T** reps, some with adornments as above. These are subscripted by **1**, **2**, ... just to distinguish them.

The *irreducible representation* names are also called *species* names, a much shorter and more descriptive way of speaking. We use "irreducible representations" and "species" quite interchangeably in this book. The word "species" should bring to mind a characteristic nodal pattern seen in the **ContourPlot** of the basis functions.

You have already seen many examples of irreducible representations. All the representation bases listed in the character tables produce irreducible representations. (That's why they are there.) Proving them irreducible is another matter; but curiously, there is a simple algorithm invented by I. Schur that performs the reduction very easily. We will study it in Chapter 30 (SchursReduction). But first, as motivation, you should see an example of a big reducible representation, and see what you learn by reducing it.

28.5 Example: the rep of the benzene orbitals

28.5.1 Construct the representation

If you need to remind yourself how to make a representation on the basis of atom permutations of a symmetric molecule, click back to the step-by-step construction in Chapter 27 (MakeReps). Here we do it by simply invoking the operator **MakeRepAtomic**, developed in that chapter :

```
{repBzπ, AOsToCart} = MakeRepAtomic["benzene",
   "D6h", atomKind → "C", preFactor → z];
repBzπ[[{2, 3, 4}]] // MatrixList // Size[7]
```

$$\left\{\begin{pmatrix} 0&0&0&0&0&1\\ 1&0&0&0&0&0\\ 0&1&0&0&0&0\\ 0&0&1&0&0&0\\ 0&0&0&1&0&0\\ 0&0&0&0&1&0 \end{pmatrix}, \begin{pmatrix} 0&1&0&0&0&0\\ 0&0&1&0&0&0\\ 0&0&0&1&0&0\\ 0&0&0&0&1&0\\ 0&0&0&0&0&1\\ 1&0&0&0&0&0 \end{pmatrix}, \begin{pmatrix} 0&0&0&0&1&0\\ 0&0&0&0&0&1\\ 1&0&0&0&0&0\\ 0&1&0&0&0&0\\ 0&0&1&0&0&0\\ 0&0&0&1&0&0 \end{pmatrix}\right\}$$

Even though you have only rep mats 2, 3, and 4 above you can see that the $\mathbf{p_z}$ representation is NOT in block diagonal form. The $\mathbf{D_{6\,h}}$ character table says that the largest irreducible representation is of size 2×2, so this is certainly not an irreducible representation. By linguistic reasoning then, it must be "reducible". Check immediately that it is a representation of $\mathbf{D_{6\,h}}$:

```
MorphTest[repBzπ, "D6h"]
```

```
Faithful, or Isomorphic
```

Is it unitary?

```
Map[UnitaryQ, repBzπ] // Union
```

```
{True}
```

Make its character vector :

```
χ⃗[repBzπ] = OnePerClass[Map[Spur, repBzπ], "D6h"]
```

```
{6, 0, 0, 0, 0, -2, 0, 0, 0, -6, 2, 0}
```

This example presents the remaining questions of representation theory in a very concrete form. These 24 representation matrices contain every bit of information about how these six atomic $\mathbf{p_z}$ orbitals transform under $\mathbf{D_{6\,h}}$. But these orbitals span a 6-dimensional function space that contains all the LCAO π molecular orbitals of benzene, so when we study this function space, we are studying the benzene π MO's. How can we characterize this six-dimensional space in a simple and understandable way that tells us something about benzene?

28.6 Reduction demo

We leap ahead to use the operator **ReduceRep**, not yet derived. We do not disclose the secret construction method until Chapter 30 (SchursReduction); here we just use it. First, make an arbitrary 6×6 matrix **Qmat**:

```
Qmat = {{9, 5, 1, 1, 5, 7}, {5, 2, 7, 5, 1, 2},
   {10, 5, 7, 2, 1, 4}, {5, 3, 9, 3, 6, 4},
   {8, 9, 3, 1, 6, 4}, {9, 6, 7, 5, 8, 7}};
```

You may think it surpassing strange that the **ReduceRep** operator requires an arbitrary matrix as part of its input. But that is the fact. We apply it and look at two of the resulting matrices:

```
blockRepBz = ReduceRep[repBzπ, Qmat];
%[[{2, 3}]] // GridList // Size[6]
```

$$\left\{\begin{pmatrix} 1 & 0 & 0 & 0 & 0 & 0 \\ 0 & \frac{1}{2} & \frac{\sqrt{3}}{2} & 0 & 0 & 0 \\ 0 & -\frac{\sqrt{3}}{2} & \frac{1}{2} & 0 & 0 & 0 \\ 0 & 0 & 0 & -1 & 0 & 0 \\ 0 & 0 & 0 & 0 & -\frac{1}{2} & \frac{\sqrt{3}}{2} \\ 0 & 0 & 0 & 0 & -\frac{\sqrt{3}}{2} & -\frac{1}{2} \end{pmatrix}, \begin{pmatrix} 1 & 0 & 0 & 0 & 0 & 0 \\ 0 & \frac{1}{2} & -\frac{\sqrt{3}}{2} & 0 & 0 & 0 \\ 0 & \frac{\sqrt{3}}{2} & \frac{1}{2} & 0 & 0 & 0 \\ 0 & 0 & 0 & -1 & 0 & 0 \\ 0 & 0 & 0 & 0 & -\frac{1}{2} & -\frac{\sqrt{3}}{2} \\ 0 & 0 & 0 & 0 & \frac{\sqrt{3}}{2} & -\frac{1}{2} \end{pmatrix}\right\}$$

Something rather marvellous has happened. They have been subjected to a clever similarity transform, that turned them all into block diagonal form; blocks of size **1**, **2**, **1**, and **2**. Interactive readers can see all **24** rep matrices. As we have said elsewhere, similarity transforms change the costumes in which a representation appears, but underneath it is the same old troupe of clowns, performing the same old show. Check it :

```
MorphTest[repBzπ, blockRepBz]
```

```
Faithful, or Isomorphic
```

The character of each clown is always the same, regardless of his costume :

```
OnePerClass[Map[Spur, blockRepBz], "D6h"] ==
  χ[repBzπ]
```

```
True
```

28.7 What is the point of rep reduction?

28.7.1 Identification of the blocks

We can find out a lot about the six π orbitals of benzene by finding the names the four irreducible representations in the blocks. If we know the names, we can look up the supporting basis functions in the character table, and these will give us important facts about the benzene orbitals. **L**ook at the character table to see how many irreducible reps are possessed by $\mathbf{D_{6\,h}}$:

```
BoxedCharacterTable["D6h"]
```

D6h	1	2	2	1	3	3	1	2	2	1	3	3	← Class populations
	E	C6	C3	C_2	C_2'	C_2''	inv	S3	S6	σh	σd	σv	↓ Basis functions
A1g	1	1	1	1	1	1	1	1	1	1	1	1	$\{(x^2+y^2),\ (z^2)\}$
A2g	1	1	1	1	-1	-1	1	1	1	1	-1	-1	$\{(x\,y\,(x^2-3\,y^2)\,(-3\,x^2+y^2)),\ ($ I
B1g	1	-1	1	-1	1	-1	1	-1	1	-1	1	-1	$\{(x\,(x^2-3\,y^2)\,z)\}$
B2g	1	-1	1	-1	-1	1	1	-1	1	-1	-1	1	$\{(y\,(-3\,x^2+y^2)\,z)\}$
E1g	2	1	-1	-2	0	0	2	1	-1	-2	0	0	$\left\{\begin{pmatrix} x\,z \\ y\,z \end{pmatrix},\ \begin{pmatrix} Ix \\ Iy \end{pmatrix}\right\}$
E2g	2	-1	-1	2	0	0	2	-1	-1	2	0	0	$\left\{\begin{pmatrix} \frac{1}{2}(x^2-y^2) \\ x\,y \end{pmatrix}\right\}$
A1u	1	1	1	1	1	1	-1	-1	-1	-1	-1	-1	$\{(x\,y\,(x^2-3\,y^2)\,(-3\,x^2+y^2)\,z)$
A2u	1	1	1	1	-1	-1	-1	-1	-1	-1	1	1	$\{(z)\}$
B1u	1	-1	1	-1	1	-1	-1	1	-1	1	-1	1	$\{(y\,(3\,x^2-y^2))\}$
B2u	1	-1	1	-1	-1	1	-1	1	-1	1	1	-1	$\{(x\,(-x^2+3\,y^2))\}$
E1u	2	1	-1	-2	0	0	-2	-1	1	2	0	0	$\left\{\begin{pmatrix} x \\ y \end{pmatrix}\right\}$
E2u	2	-1	-1	2	0	0	-2	1	1	-2	0	0	$\left\{\begin{pmatrix} \frac{1}{2}(x^2-y^2)\,z \\ x\,y\,z \end{pmatrix}\right\}$

Group $\mathbf{D_{6\,h}}$ has **12** irreducible reps, but we already know from the block diagonal form that only four contribute to the benzene $\mathbf{p_z}$ rep. Which ones are they? We are going to lift out the four blocks and test them one at a time to see if they are irreducible reps given in the character table.

This is a conceptually simple method for identifying the irreducible reps contained in the reducible representation, but not the one used in practice. But that quick and easy method depends on a long chain of theorems, and we cannot logically present it yet. To peek ahead, click to Chapter 35 (RepAnalysis).

Block {1,1} in blockRepBz

To test the blocks, we need a way to lift them out. Fortunately, the operator **`Take`** does exactly what we need, if supplied with the right options. Click to the End notes for a demo. Use it on the **1 × 1** rep in the upper left :

```
rep11 = Map[Take[#, 1, 1] &, blockRepBz] // Flatten
```

```
{1, 1, 1, 1, 1, 1, -1, -1, -1, -1, -1,
 -1, -1, -1, -1, -1, -1, -1, 1, 1, 1, 1, 1, 1}
```

This is the whole 24-element representation, and since the rep is one-dimensional, it is also a list of all the characters. Since all elements in the same class have the same character, we lift out just the first character in each class :

```
χ11 = rep11 // OnePerClass[#, "D6h"] &
```

```
{1, 1, 1, 1, -1, -1, -1, -1, -1, -1, 1, 1}
```

Does this match any character in the table for D_{6h}?

```
Select[CharacterTable["D6h"], (#[[2]] == χ11) &]
```

```
{{A2u, {1, 1, 1, 1, -1, -1, -1, -1, -1, -1, 1, 1}, {{z}}}}
```

Yes; it is species A_{2u}, which transforms like **z**. The symmetry of function **z** is exactly the symmetry of the lowest energy p_z orbital, in the sense that they both have a node at $z = 0$.

On your own
We have now shown that block **1** is A_{2u}. Show in a similar way that block **2** is E_{1g}, **2** is B_{1g}, and **4** is E_{2u}.

$$\begin{pmatrix} A_{2u} & 0 & 0 & 0 \\ 0 & E_{1g} & 0 & 0 \\ 0 & 0 & B_{1g} & 0 \\ 0 & 0 & 0 & E_{2u} \end{pmatrix}$$

Click for help.

So we have the orbital labels in Fig. 23. 5. In order of increasing energy, the species are A_{2u}(one nodal plane), E_{1g} (two nodes), E_{2u} (three), and B_{1g} (four).

Contour plot

The B_{1g} basis function is given in the character table as $x\left(x^2 - 3y^2\right) z$.

Here is its **x,y** contour plot at height $z = 1$:

```
grph1 = ContourPlot[x^3 - 3 x y^2, {x, -1, 1},
  {y, -1, 1}, FrameTicks → None, ImageSize → 120,
  ColorFunction → BlueShades, PlotLabel → "B1 g"]
```

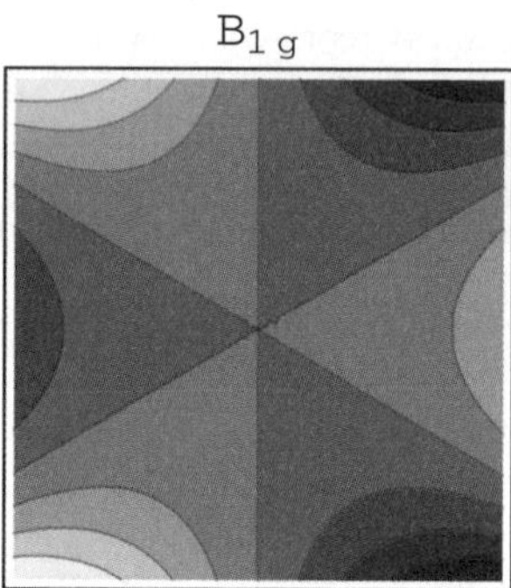

Fig. 28.1 The tabulated polynomial for B_{1g} of D_{6h}

Click back to the contour plots of the benzene orbitals to compare. You may be quite astonished. You may begin to see the true significance of the polynomial basis functions given in the character tables.

28.7.2 Implications for spectroscopy

These results explain the labels A_{2u}, E_{1g}, E_{2u}, and B_{1g} in Fig. 23.1, the contour plots of the benzene π orbitals (found in principle by an accurate solution of one-electron Schrödinger equation). We could not justify those names in Chapter 23, but now we see that the accurate eigenstates are basis functions for these four irreducible representations of D_{6h}.

> The basis of all electronic state notation in spectroscopy is the use of irreducible representation names for Schrödinger eigenstates that transform like the representation.

For instance, if a molecule has two observed electronic states that transform like the irreducible representation A_u, they will be named $1\ A_u$ (the lower), $2\ A_u$ (the higher).

Note that in this chapter we never found the molecular orbitals explicitly or exactly. We worked only with the atomic orbitals that span a space of approximate molecular orbitals. But nevertheless, we reached definite, detailed conclusions about the "exact" benzene orbitals.

Group theory cannot give quantitative energies, but it does give the energy ordering of the orbitals. Benzene's π orbitals all have a $\mathbf{p_z}$ node in the molecular plane, but the number of additional nodes (perpendicular to the $\mathbf{z = 0}$ plane) is $\mathbf{0}$ for $\mathbf{A_{2\,u}}$, $\mathbf{1}$ for $\mathbf{E_{1\,g}}$, $\mathbf{2}$ for $\mathbf{E_{2\,u}}$, and $\mathbf{3}$ for $\mathbf{B_{1\,g}}$, just as it is in the simple polynomial basis functions for these reps. Each additional node costs energy, so the increasing energy order must be $\mathbf{A_{2\,u}}$, $\mathbf{E_{1\,g}}$, $\mathbf{E_{2\,u}}$, $\mathbf{B_{1\,g}}$.

Further, group theory predicts that the corresponding degeneracy pattern is 1, 2, 2, 1. With six π electrons and two electrons per orbital, the filled orbitals in the ground state can only be the $\mathbf{A_{2\,u}}$ and the two $\mathbf{E_{1\,g}}$. The lowest spectroscopic transitions must therefore be from somewhere in twofold degenerate level $\mathbf{E_{1\,g}}$ to somewhere else in twofold degenerate level $\mathbf{E_{2\,u}}$

$$\left(\alpha\, \psi_{\mathbf{E1gA}} + \beta\, \psi_{\mathbf{E1gB}}\right) \to \left(\gamma\, \psi_{\mathbf{E2uA}} + \delta\, \psi_{\mathbf{E2uB}}\right)$$

where the coefficients $\alpha, \beta, \gamma, \delta$ would be picked by an intrinsic quantum uncertainty process.

But this refers only to one electron hovering around six fixed attractive centers. In real benzene, six π electrons are present and we must add electron-electron repulsion terms to the Hamiltonian. The degeneracies then vanish and we must consider four nondegenerate transitions $\psi_{\mathbf{E1gA}}\to\psi_{\mathbf{E2uA}}$, $\psi_{\mathbf{E1gA}}\to\psi_{\mathbf{E2uB}}$, $\psi_{\mathbf{E1gB}}\to\psi_{\mathbf{E2uA}}$, and $\psi_{\mathbf{E1gB}}\to\psi_{\mathbf{E2uB}}$. Three of these have been definitely identified in the spectroscopy of benzene. The fourth and highest was a puzzle for many years, but it is now known to be mixed into a broad continuum of the same symmetry above ionization, and is therefore unobservable as an isolated vibronic band. (The continuum does not appear in the π electron treatment, and for years people did not take it into account properly.)

As we pursue the benzene example through this book, we will see that this little π orbital treatment is a very reliable framework for understanding the vibronic spectroscopy of the three resolvable states. The polarization behavior of these states are all predicted as exact results from geometrical symmetry alone, without a single quantitative quantum chemistry calculation.

28.8 End Notes

28.8.1 Lifting out submatrices with **Take**

The operator **Take** has the ability to lift submatrices out of larger matrices.

$$\mathbb{G} = \begin{pmatrix} \mathbf{a} & \mathbf{b} & \mathbf{c} & \mathbf{d} \\ \mathbf{e} & \mathbf{f} & \mathbf{g} & \mathbf{h} \\ \mathbf{i} & \mathbf{j} & \mathbf{k} & \mathbf{l} \\ \mathbf{m} & \mathbf{n} & \mathbf{o} & \mathbf{p} \end{pmatrix};$$

```
Take[G, 2, 3] // GridForm
```

a	b	c
e	f	g

The **2** told it to take the first two rows; the **3** said to take the first three columns. Now tell it to take the upper left 3×3 :

```
Take[G, 3, 3] // GridForm
```

a	b	c
e	f	g
i	j	k

To lift out from the interior of the matrix you must apply **Take** twice:

```
Take[Take[G, 3, 3], -2, -2] // GridForm
```

f	g
j	k

28.8.2 Species of the four blocks in blockRepBz

We condense the three operations from the main text into one cell. For all the matrices in the group, we **(1)** lift out the block, **(2)** calculate the character vector, and **(3)** **Select** a line from the character table that contains a matching vector.

```
Map[Spur, Map[Take[#, 1, 1] &, blockRepBz]];
OnePerClass[%, "D6h"];
Select[CharacterTable["D6h"], (#[[2]] == %) &]
```

```
{{A2u, {1, 1, 1, 1, -1, -1, -1, -1, -1, -1, 1, 1}, {{z}}}}
```

So block **1** is $\mathbf{A_{2\,u}}$. Do the next :

```
Map[Take[Take[#, 3, 3], -2, -2] &, blockRepBz];
Map[Spur, %];
OnePerClass[%, "D6h"];
Select[CharacterTable["D6h"], (#[[2]] == %) &]
```

```
{{E1g, {2, 1, -1, -2, 0, 0, 2, 1, -1, -2, 0, 0},
  {{x z, y z}, {Ix, Iy}}}}
```

So block **2** is $\mathbf{E_{1\,g}}$.

```
Map[Take[Take[#, 4, 4], -1, -1] &, blockRepBz];
Map[Spur, %];
OnePerClass[%, "D6h"];
Select[CharacterTable["D6h"], (#[[2]] == %) &]
```

```
{{B1g, {1, -1, 1, -1, 1, -1, 1, -1, 1, -1, 1, -1},
  {{x (x^2 - 3 y^2) z}}}}
```

So block **3** is $\mathbf{B_{1\,g}}$.

```
Map[Take[#, -2, -2] &, blockRepBz];
Map[Spur, %];
OnePerClass[%, "D6h"];
Select[CharacterTable["D6h"], (#[[2]] == %) &]
```

```
{{E2u, {2, -1, -1, 2, 0, 0, -2, 1, 1, -2, 0, 0},
  {{1/2 (x^2 - y^2) z, x y z}}}}
```

So block **4** is $\mathbf{E_{2\,u}}$.

29. The MakeUnitary operator

Preliminaries

29.1 Unitary and non-unitary reps

All our tabulated **CartReps** are made of unitary matrices. Click on the blue if this term is a little hazy to you. As a basic reminder, unitary means that the inverse is equal to the transpose.

Several theorems in the chapters ahead use the assumption that the rep in question is unitary. This shortens the theorem significantly, but does it restrict the validity of the theorem? This chapter says no; if you have a non-unitary rep it can always be made unitary by a similarity transform, and then you can work with it using the assumption that it is unitary.

- Linguistic note: When we say "then you can work with it" , we have slipped into language where two reps related by a similarity transform are "the same" (and can both be referenced by the same "it") even though they are not literally mathematically equal. Mathematicians talk this way all the time. Click into the MatrixReview for a discussion of why they do this.

29.1.1 Make a rep non-unitary by a similarity transform

In this chapter you will see some non-unitary representations. It's not that they are useful; in fact we try to avoid them. The whole point of this chapter is to show how to get rid of them. But first we have to show how non-unitary reps can arise. We begin with a nice little unitary representation :

```
xyRepD3 = MakeRepPoly[{x, y}, "D4"];
Map[MatrixForm, %] // Size[7]
```

$$\left\{\begin{pmatrix}1&0\\0&1\end{pmatrix},\begin{pmatrix}0&-1\\1&0\end{pmatrix},\begin{pmatrix}0&1\\-1&0\end{pmatrix},\begin{pmatrix}-1&0\\0&-1\end{pmatrix},\begin{pmatrix}1&0\\0&-1\end{pmatrix},\begin{pmatrix}-1&0\\0&1\end{pmatrix},\begin{pmatrix}0&1\\1&0\end{pmatrix},\begin{pmatrix}0&-1\\-1&0\end{pmatrix}\right\}$$

These matrices are all either symmetric or antisymmetric. It is easy to test a matrix for the unitary property. There is an operator called **UnitaryQ[mat]** in the **Symmetry`** package. It simply take the **Inverse** and the **Transpose** of the given **mat**, and tests them for equality. Test **xyRepD3** :

```
Map[UnitaryQ, xyRepD3]
```


DOI 10.1007/b13137_29,

```
{True, True, True, True, True, True, True, True}
```

Now we make a similarity transform of this rep, using a random integer transforming matrix **Q**:

```
Q = 1/10^2 Table[RandomInteger[{1, 10^2}], {i, 2}, {j, 2}]
```

$$\left\{\left\{\frac{12}{25}, \frac{7}{100}\right\}, \left\{\frac{73}{100}, \frac{13}{20}\right\}\right\}$$

```
Qinv = Inverse[Q]
```

$$\left\{\left\{\frac{6500}{2609}, -\frac{700}{2609}\right\}, \left\{-\frac{7300}{2609}, \frac{4800}{2609}\right\}\right\}$$

```
Map[UnitaryQ, {Q, Qinv}]
```

```
{False, False}
```

To preserve the unitary property of a representation, it must be similarity transformed by unitary matrices (click for proof), But here, perversely, we destroy the unitary property by transforming with nonunitary matrices :

```
NUrep = Map[Q.#.Qinv &, xyRepD3];
Map[MatrixForm, NUrep]
```

$$\left\{\begin{pmatrix} 1 & 0 \\ 0 & 1 \end{pmatrix}, \begin{pmatrix} \frac{3959}{2609} & -\frac{2353}{2609} \\ \frac{9554}{2609} & -\frac{3959}{2609} \end{pmatrix}, \begin{pmatrix} -\frac{3959}{2609} & \frac{2353}{2609} \\ -\frac{9554}{2609} & \frac{3959}{2609} \end{pmatrix}, \begin{pmatrix} -1 & 0 \\ 0 & -1 \end{pmatrix}, \begin{pmatrix} \frac{3631}{2609} & -\frac{672}{2609} \\ \frac{9490}{2609} & -\frac{3631}{2609} \end{pmatrix}, \begin{pmatrix} -\frac{3631}{2609} & \frac{672}{2609} \\ -\frac{9490}{2609} & \frac{3631}{2609} \end{pmatrix}, \begin{pmatrix} -\frac{3049}{2609} & \frac{2255}{2609} \\ -\frac{1104}{2609} & \frac{3049}{2609} \end{pmatrix}, \begin{pmatrix} \frac{3049}{2609} & -\frac{2255}{2609} \\ \frac{1104}{2609} & -\frac{3049}{2609} \end{pmatrix}\right\}$$

The symmetry (or antisymmetry) of most of the matrices has been lost. No way could a person recognize this as a representation of $\mathbf{D}_3$. Yet it is :

```
MorphTest[NUrep, "D4"]
```

```
Faithful, or Isomorphic
```

But is it unitary?

```
Map[UnitaryQ, NUrep]
```

```
{True, False, False, True, False, False, False, False}
```

No; the only unitary matrices are the unit element and its negative. Below, in this chapter, we will develop an operator that will restore the unitary property to this representation, using only information contained in the representation itself.

29.1.2 Non-unitary rep from unbalanced bases

Another way to make a nonunitary rep is to construct it using "unbalanced" basis functions. For instance, the functions **{x,y}** support a unitary rep of group $\mathbf{D_3}$. But suppose we "unbalance" the basis by multiplying the **y** by **2** :

```
x2yRepD3 = MakeRepPoly[{x, 2 y}, "D3"];
Map[MatrixForm, %] // Size[8]
```

$$\left\{\begin{pmatrix}1 & 0\\ 0 & 1\end{pmatrix}, \begin{pmatrix}-\frac{1}{2} & -\sqrt{3}\\ \frac{\sqrt{3}}{4} & -\frac{1}{2}\end{pmatrix}, \begin{pmatrix}-\frac{1}{2} & \sqrt{3}\\ -\frac{\sqrt{3}}{4} & -\frac{1}{2}\end{pmatrix}, \begin{pmatrix}1 & 0\\ 0 & -1\end{pmatrix}, \begin{pmatrix}-\frac{1}{2} & \sqrt{3}\\ \frac{\sqrt{3}}{4} & \frac{1}{2}\end{pmatrix}, \begin{pmatrix}-\frac{1}{2} & -\sqrt{3}\\ -\frac{\sqrt{3}}{4} & \frac{1}{2}\end{pmatrix}\right\}$$

MakeRepPoly did indeed make a representation :

```
MorphTest[x2yRepD3, "D3"]
```

```
Faithful, or Isomorphic
```

but its matrices are not unitary :

```
Map[UnitaryQ, x2yRepD3]
```

```
{True, False, False, True, False, False}
```

Projection operators can easily give basis sets for representations, but they will usually come out unbalanced. The balancing operation described in the next section can always cure this problem.

29.1.3 Balancing a basis set

In Chapter 27 (MakeReps) we constructed an operator **MakeRepMolec** that makes representations on the basis of complete sets of molecular orbitals. We used it with a simple set of molecular orbitals, and it produced a representation, but the rep was not unitary. Click back to review.

Here we show a simple scheme for balancing the basis functions, in order to make a unitary representation. We use the same basis as before, except that we multiply one of the basis functions by an unknown multiplier **k** :

```
MObasis = {k (2 AO[1] - AO[2] - AO[3]), AO[2] - AO[3]};
AOsToCart = {
   AO[1] → z f[1 - 2 x + x^2 + y^2 + z^2],
```

```
    AO[2] → z f[1 + x + x^2 - √3 y + y^2 + z^2],
    AO[3] → z f[1 + x + x^2 + √3 y + y^2 + z^2]};
group = "D3h";
```

Now make the representation:

```
testRepk = MakeRepMolec[MObasis, AOsToCart, group]
```

$$\left\{\{\{1,0\},\{0,1\}\},\left\{\left\{-\frac{1}{2},-\frac{1}{2k}\right\},\left\{\frac{3k}{2},-\frac{1}{2}\right\}\right\},\right.$$
$$\left\{\left\{-\frac{1}{2},\frac{1}{2k}\right\},\left\{-\frac{3k}{2},-\frac{1}{2}\right\}\right\},\{\{-1,0\},\{0,1\}\},$$
$$\left\{\left\{\frac{1}{2},-\frac{1}{2k}\right\},\left\{-\frac{3k}{2},-\frac{1}{2}\right\}\right\},\left\{\left\{\frac{1}{2},\frac{1}{2k}\right\},\left\{\frac{3k}{2},-\frac{1}{2}\right\}\right\},$$
$$\{\{-1,0\},\{0,-1\}\},\left\{\left\{\frac{1}{2},\frac{1}{2k}\right\},\left\{-\frac{3k}{2},\frac{1}{2}\right\}\right\},$$
$$\left\{\left\{\frac{1}{2},-\frac{1}{2k}\right\},\left\{\frac{3k}{2},\frac{1}{2}\right\}\right\},\{\{1,0\},\{0,-1\}\},$$
$$\left.\left\{\left\{-\frac{1}{2},\frac{1}{2k}\right\},\left\{\frac{3k}{2},\frac{1}{2}\right\}\right\},\left\{\left\{-\frac{1}{2},-\frac{1}{2k}\right\},\left\{-\frac{3k}{2},\frac{1}{2}\right\}\right\}\right\}$$

For some value of **k**, this rep will be unitary. Apply the unitary criterion to each matrix :

```
remainders =
 Map[(Transpose[#] - Inverse[#]) &, testRepk]
```

$$\left\{\{\{0,0\},\{0,0\}\},\left\{\left\{0,-\frac{1}{2k}+\frac{3k}{2}\right\},\left\{-\frac{1}{2k}+\frac{3k}{2},0\right\}\right\},\right.$$
$$\left\{\left\{0,\frac{1}{2k}-\frac{3k}{2}\right\},\left\{\frac{1}{2k}-\frac{3k}{2},0\right\}\right\},\{\{0,0\},\{0,0\}\},$$
$$\left\{\left\{0,\frac{1}{2k}-\frac{3k}{2}\right\},\left\{-\frac{1}{2k}+\frac{3k}{2},0\right\}\right\},$$
$$\left\{\left\{0,-\frac{1}{2k}+\frac{3k}{2}\right\},\left\{\frac{1}{2k}-\frac{3k}{2},0\right\}\right\},$$
$$\{\{0,0\},\{0,0\}\},\left\{\left\{0,\frac{1}{2k}-\frac{3k}{2}\right\},\left\{\frac{1}{2k}-\frac{3k}{2},0\right\}\right\},$$
$$\left\{\left\{0,-\frac{1}{2k}+\frac{3k}{2}\right\},\left\{-\frac{1}{2k}+\frac{3k}{2},0\right\}\right\},$$
$$\{\{0,0\},\{0,0\}\},\left\{\left\{0,-\frac{1}{2k}+\frac{3k}{2}\right\},\left\{\frac{1}{2k}-\frac{3k}{2},0\right\}\right\},$$
$$\left.\left\{\left\{0,\frac{1}{2k}-\frac{3k}{2}\right\},\left\{-\frac{1}{2k}+\frac{3k}{2},0\right\}\right\}\right\}$$

All these remainders would be zero if the rep were unitary. So `Select` out all the nonzero entries, and **Unionize** them to remove redundancies :

```
kExprs =
 Select[Flatten[remainders], # =!= 0 &] // Union
```

$$\left\{\frac{1}{2\,k}-\frac{3\,k}{2},\ -\frac{1}{2\,k}+\frac{3\,k}{2}\right\}$$

If you get inconsistent expressions here, you have made a mistake, and you must find it by prayer and fasting. But these are fine, so we forge ahead:

```
Reduce[kExprs == 0, k]
```

$$k == -\frac{1}{\sqrt{3}}\ ||\ k == \frac{1}{\sqrt{3}}$$

Either solution will balance it. Take the positive balancing factor :

```
balancedBasis =
  {1/Sqrt[3] (2 AO[1] - AO[2] - AO[3]), AO[2] - AO[3]};

testRepBalanced =
  MakeRepMolec[balancedBasis, AOsToCart, group];
% // MatrixList // Size[8]
```

$$\left\{\begin{pmatrix}1&0\\0&1\end{pmatrix},\begin{pmatrix}-\frac{1}{2}&-\frac{\sqrt{3}}{2}\\\frac{\sqrt{3}}{2}&-\frac{1}{2}\end{pmatrix},\begin{pmatrix}-\frac{1}{2}&\frac{\sqrt{3}}{2}\\-\frac{\sqrt{3}}{2}&-\frac{1}{2}\end{pmatrix},\begin{pmatrix}-1&0\\0&1\end{pmatrix},\right.$$

$$\begin{pmatrix}\frac{1}{2}&-\frac{\sqrt{3}}{2}\\-\frac{\sqrt{3}}{2}&-\frac{1}{2}\end{pmatrix},\begin{pmatrix}\frac{1}{2}&\frac{\sqrt{3}}{2}\\\frac{\sqrt{3}}{2}&-\frac{1}{2}\end{pmatrix},\begin{pmatrix}-1&0\\0&-1\end{pmatrix},\begin{pmatrix}\frac{1}{2}&\frac{\sqrt{3}}{2}\\-\frac{\sqrt{3}}{2}&\frac{1}{2}\end{pmatrix},$$

$$\left.\begin{pmatrix}\frac{1}{2}&-\frac{\sqrt{3}}{2}\\\frac{\sqrt{3}}{2}&\frac{1}{2}\end{pmatrix},\begin{pmatrix}1&0\\0&-1\end{pmatrix},\begin{pmatrix}-\frac{1}{2}&\frac{\sqrt{3}}{2}\\\frac{\sqrt{3}}{2}&\frac{1}{2}\end{pmatrix},\begin{pmatrix}-\frac{1}{2}&-\frac{\sqrt{3}}{2}\\-\frac{\sqrt{3}}{2}&\frac{1}{2}\end{pmatrix}\right\}$$

```
MorphTest[testRepBalanced, group]
```

```
Faithful, or Isomorphic
```

```
Map[UnitaryQ, testRepBalanced] // Union
```

```
{True}
```

On your own

Put in the other balancing solution $\left(-1/\sqrt{3}\right)$ and test with `MorphTest` and `UnitaryQ`. Then compare the balanced reps to `CartRep["D3h"]`.

29.2 The MakeUnitary theorem

29.2.1 Statement and commentary

Theorem

Every matrix representation of a group is *similar* to a unitary matrix representation.

This will be particularly important in proving and using Schur's Lemmas. This chapter ends with a *Mathematica* operator that changes any rep into a similar unitary rep. Rarely will we really have to do this, but it is nice to have it handy for demonstrations. Also, it encloses all the blather of the theorem itself in a very tiny nutshell.

29.2.2 Proof

The first similarity transform

The proof is constructive, and requires two similarity transforms. Here we work on the first one. Let the $\mathbb{A}_i$ be a set of nonunitary matrices representing finite group $\mathcal{A}$,of order $\mathbf{h}$, and let

$$\mathbb{H} = \sum_{i=1}^{h} \mathbb{A}_i . \mathbb{A}_i^{\dagger}$$

Then by a Lemma proved in the Matrix Review (click here), each summand of $\mathbb{H}$ is hermitian, and therefore, so is $\mathbb{H}$. Therefore (click here) matrix $\mathbb{H}$ has real eigenvalues and a set of eigenvectors that are mutually orthonormalizable in the complex sense. Take it that this has been done, and let $\mathbb{U}$ be the matrix of orthonormal column eigenvectors. Orthonormality in the complex sense means that $\mathbb{U}^{\dagger}.\mathbb{U}$ is the unit matrix, or that $\mathbb{U}^{\dagger} = \mathbb{U}^{-1}$, the definition of unitary. We let

$\mathbb{d}$ be the diagonal matrix of the eigenvalues of $\mathbb{H}$. The eigenequation is

$$\mathbb{H}.\mathbb{U} = \mathbb{U}.\mathbb{d} \quad \text{or} \quad \mathbb{d} = \mathbb{U}^{-1}.\mathbb{H}.\mathbb{U}$$

How does the diagonal matrix $\mathbb{d}$ relate to the original $\mathbb{A}$ matrices?

$$\mathbb{d} = \mathbb{U}^{\dagger}.\mathbb{H}.\mathbb{U} = \mathbb{U}^{\dagger}.\left(\sum_{i=1}^{h}\mathbb{A}_i.\mathbb{A}_i^{\dagger}\right).\mathbb{U}$$

$$\mathbb{d} = \sum_{i=1}^{h}\mathbb{U}^{\dagger}.\mathbb{A}_i.\mathbb{A}_i^{\dagger}.\mathbb{U} = \sum_{i=1}^{h}\mathbb{U}^{\dagger}.\mathbb{A}_i.\left(\mathbb{U}.\mathbb{U}^{\dagger}\right).\mathbb{A}_i^{\dagger}.\mathbb{U}$$

$$\mathbb{d} = \sum_{i=1}^{h}\left(\mathbb{U}^{\dagger}.\mathbb{A}_i.\mathbb{U}\right).\left(\mathbb{U}^{\dagger}.\mathbb{A}_i^{\dagger}.\mathbb{U}\right)$$

We transform the parenthesis on the right using the identity $\mathbb{L}^{\dagger}.\mathbb{M}^{\dagger}.\mathbb{N}^{\dagger} == (\mathbb{N}.\mathbb{M}.\mathbb{L})^{\dagger}$, giving

$$\mathbb{d} = \sum_{i=1}^{h}\left(\mathbb{U}^{\dagger}.\mathbb{A}_i.\mathbb{U}\right).\left((\mathbb{U})^{\dagger}.\left(\mathbb{A}_i^{\dagger}\right)^{\dagger}.\left(\mathbb{U}^{\dagger}\right)^{\dagger}\right)^{\dagger}$$

or since the double dagger is no dagger

$$\mathbb{d} = \sum_{i=1}^{h}\left(\mathbb{U}^{\dagger}.\mathbb{A}_i.\mathbb{U}\right).\left(\mathbb{U}^{\dagger}.\mathbb{A}_i.\mathbb{U}\right)^{\dagger}$$

But because $\mathbb{U}$ is unitary, $\mathbb{U}$ is $\left(\mathbb{U}^{\dagger}\right)^{-1}$, and $\mathbb{U}^{\dagger}.\mathbb{A}_i.\mathbb{U}$ is the similarity transform of $\mathbb{A}_i$ by $\mathbb{U}^{\dagger}$. We call it $\overline{\mathbb{A}}_i$:

$$\boxed{\overline{\mathbb{A}}_i = \left(\mathbb{U}^{\dagger}.\mathbb{A}_i.\mathbb{U}\right)}$$

so that

$$\mathbb{d} = \sum_{i=1}^{h}\overline{\mathbb{A}}_i.\overline{\mathbb{A}}_i^{\dagger}$$

The transformed matrices $\overline{\mathbb{A}}$ represent the same group that the $\mathbb{A}$ matrices do, because they are just similarity transforms, which preserve the multiplication table of any group. The costumes have changed, but the clowns are the same.

Lemma: The $\mathbb{d}$ matrices are real and positive

For this maneuver we move the element index **i** up into a square bracket, leaving the subscript area free for Greek row and column indices like α and β. In other words, we replace $\overline{\mathbb{A}}_i$ by $\bar{\mathbf{A}}[\mathbf{i}]_{\alpha,\beta}$. (Individual matrix elements, with subscripts, are denoted by plain letters; it is the whole matrices that are denoted by double struck letters.) The individual elements of $\mathbb{d}$ are $\mathbf{d}_{\alpha,\gamma}$:

$$\mathbf{d}_{\alpha,\gamma} = \sum_{i=1}^{h}\sum_{\beta=1}^{\lambda}\bar{\mathbf{A}}[\mathbf{i}]_{\alpha,\beta}\,\bar{\mathbf{A}}^{\dagger}[\mathbf{i}]_{\beta,\gamma}$$

The off-diagonal elements are zero, so we consider just the diagonal elements

$$\mathbf{d}_{\alpha,\alpha} = \sum_{i=1}^{h}\sum_{\beta=1}^{\lambda}\bar{\mathbf{A}}[\mathbf{i}]_{\alpha,\beta}\,\bar{\mathbf{A}}^{\dagger}[\mathbf{i}]_{\beta,\alpha} = \sum_{i=1}^{h}\sum_{\beta=1}^{\lambda}\bar{\mathbf{A}}[\mathbf{i}]_{\alpha,\beta}\,\bar{\mathbf{A}}^{*}[\mathbf{i}]_{\alpha,\beta}$$

In the rightmost member above we performed the † operation explicitly. We transposed by reversing the row and column indices, and left a ***** to indicate complex conjugation. But in this form each summand $\bar{\mathbf{A}}[\mathbf{i}]_{\alpha,\beta}\,\bar{\mathbf{A}}^{*}[\mathbf{i}]_{\alpha,\beta}$ is a quantity $\bar{\mathbf{A}}[\mathbf{i}]_{\alpha,\beta}$ times its own conjugate, and is therefore real and positive, or at least non-negative. So $\mathbb{d}$ is a diagonal matrix with real, non-negative entries down the diagonal, as was to be shown.

The second transform

Diagonal matrix $\mathbb{d}$ has a square root $\sqrt{\mathbb{d}}$ found by applying the square root operator down its diagonal. Matrix $\sqrt{\mathbb{d}}$ is unique up to a sign ambiguity on each diagonal element. Also, the inverse of a diagonal matrix is found by taking numerical inverses down the diagonal.

> **On your own**
> If this is a new idea to you, try it out on a numerical example with negatives and complex numbers on the diagonal. **ComplexExpand** may be useful. Compute $(\sqrt{\mathbb{d}}).(1/\sqrt{\mathbb{d}})$.

Now comes a clever manipulation that moves us toward a second and final similarity transform of the matrices we are trying to make unitary.

$$\mathbb{d} == \sqrt{\mathbb{d}}\,.\,\sqrt{\mathbb{d}} == \sum_{i=1}^{h}\overline{\mathbb{A}}_i\,.\,\overline{\mathbb{A}}_i^{\dagger}$$

Multiplying this left and right by $1/\sqrt{\mathbb{d}}$ we have

$$1 == \left(1/\sqrt{\mathbb{d}}\right).\left(\sum_{i=1}^{h}\overline{\mathbb{A}}_i.\overline{\mathbb{A}}_i^\dagger\right).\left(1/\sqrt{\mathbb{d}}\right)$$

Remember the rearrangement theorem ? It says that inside a sum over all group elements **i** you may pick any **j** and replace $\overline{\mathbb{A}}_i$ by $\left(\overline{\mathbb{A}}_j.\overline{\mathbb{A}}_i\right)$. This merely rearranges the sum over **i**, but does not alter its value. We do so:

$$1 == \sum_{i=1}^{h}\left(\mathbb{d}^{-1/2}.\left(\overline{\mathbb{A}}_j.\overline{\mathbb{A}}_i\right).\left(\overline{\mathbb{A}}_j.\overline{\mathbb{A}}_i\right)^\dagger.\mathbb{d}^{-1/2}\right)$$

and since $\left(\overline{\mathbb{A}}_j.\overline{\mathbb{A}}_i\right)^\dagger = \left(\overline{\mathbb{A}}_i^\dagger.\overline{\mathbb{A}}_j^\dagger\right)$,

$$1 == \sum_{i=1}^{h}\mathbb{d}^{-1/2}.\left(\overline{\mathbb{A}}_j.\overline{\mathbb{A}}_i\right).\left(\overline{\mathbb{A}}_i^\dagger.\overline{\mathbb{A}}_j^\dagger\right).\mathbb{d}^{-1/2}$$

and moving the sum as far inward as possible

$$1 == \mathbb{d}^{-1/2}.\overline{\mathbb{A}}_j.\left(\sum_{i=1}^{h}\overline{\mathbb{A}}_i.\overline{\mathbb{A}}_i^\dagger\right).\overline{\mathbb{A}}_j^\dagger.\mathbb{d}^{-1/2}$$

But this sum over **i** is by definition just the matrix $\mathbb{d}$ itself, so

$$1 == \mathbb{d}^{-1/2}.\overline{\mathbb{A}}_j.\mathbb{d}.\overline{\mathbb{A}}_j^\dagger.\mathbb{d}^{-1/2}$$

Applying the square root trick again

$$1 == \left(\mathbb{d}^{-1/2}.\overline{\mathbb{A}}_j.\mathbb{d}^{1/2}\right).\left(\mathbb{d}^{1/2}.\overline{\mathbb{A}}_j^\dagger.\mathbb{d}^{-1/2}\right)$$

The right member of the dot product above may be rewritten as

$$1 == \left(\mathbb{d}^{-1/2}.\overline{\mathbb{A}}_j.\mathbb{d}^{1/2}\right).\left(\left(\mathbb{d}^{-1/2}\right)^\dagger.\overline{\mathbb{A}}_j.\left(\mathbb{d}^{1/2}\right)^\dagger\right)^\dagger$$

But both $\mathbb{d}^{1/2}$ and $\mathbb{d}^{-1/2}$ are symmetric and real, so the dagger operation leaves them the same. Therefore

$$1 == \left(\mathbb{d}^{-1/2}.\overline{\mathbb{A}}_j.\mathbb{d}^{1/2}\right).\left(\mathbb{d}^{-1/2}.\overline{\mathbb{A}}_j.\mathbb{d}^{1/2}\right)^\dagger$$

Now we have it. The parentheses contain a second similarity transform of the group, so we let it be a double-barred $\mathbb{A}$ with element index **j** :

$$\overline{\overline{\mathbb{A}}}_j = \mathbb{d}^{-1/2}.\overline{\mathbb{A}}_j.\mathbb{d}^{1/2}$$

Thus we have $\mathbf{1} == \overline{\overline{\mathbb{A}}}_j . \overline{\overline{\mathbb{A}}}_j^{\dagger}$ for every element **j** in the group. This is the defining property of a unitary matrix, and the $\overline{\overline{\mathbb{A}}}$ matrices still represent the group because they are nothing but a double similarity transform of the original representation. So every representation is similar to a unitary representation, as was to be proved.

29.3 The operator MakeUnitary

We have collected all the essential steps from the proof above into one operator, called **MakeUnitary** , kept in **Symmetry`**. The **"?"** returns

- **MakeUnitary[rep_]** returns a list of two objects :
  (1) the left matrix of a similarity transform, and
  (2) a unitary rep equivalent to the given rep, made by the returned left matrix

Read the full text of it in **UrSymmetry.nb**. Test it on **NUrep**, constructed above and already proved non-unitary :

```
NUrep // Dimensions
```

```
{8, 2, 2}
```

```
Map[UnitaryQ, NUrep]
```

```
{True, False, False, True, False, False, False, False}
```

We take the second part below, for the reason just above.

```
Urep2 = MakeUnitary[N[NUrep]][[2]];
% // MatrixList // Size[8]
```

$$\left\{\begin{pmatrix}1. & 0\\ 0 & 1.\end{pmatrix},\begin{pmatrix}0 & 1.\\ -1. & 0\end{pmatrix},\begin{pmatrix}0 & -1.\\ 1. & 0\end{pmatrix},\begin{pmatrix}-1. & 0\\ 0 & -1.\end{pmatrix},\right.$$
$$\begin{pmatrix}0.313844 & 0.949474\\ 0.949474 & -0.313844\end{pmatrix},\begin{pmatrix}-0.313844 & -0.949474\\ -0.949474 & 0.313844\end{pmatrix},$$
$$\left.\begin{pmatrix}0.949474 & -0.313844\\ -0.313844 & -0.949474\end{pmatrix},\begin{pmatrix}-0.949474 & 0.313844\\ 0.313844 & 0.949474\end{pmatrix}\right\}$$

Is it unitary now?

```
Map[UnitaryQ, Urep2] // Union
```

```
{True}
```

The **MakeUnitary** operator can be really useful when making representations on the basis of physically meaningful basis functions, like orbitals. Instead of balancing the functions so that they give a unitary representation, just use them as they are, and then call **MakeUnitary** on the rep they produce.

> **OnYour Own**
> Take any unitary representation, tangle it up with a random similarity transform, and then make it unitary again with the **MakeUnitary** operator. Does this restore the original representation?

29.4 End Notes

29.4.1 A side effect

In the interest of full disclosure, we must admit that **MakeUnitary** often does something you may not like. The **CartRep** of $\mathbf{D_3}$ is block diagonal

```
CartRep["D3"];
% // MatrixList // Size[7]
```

$$\left\{\begin{pmatrix}1&0&0\\0&1&0\\0&0&1\end{pmatrix},\begin{pmatrix}-\frac{1}{2}&-\frac{\sqrt{3}}{2}&0\\\frac{\sqrt{3}}{2}&-\frac{1}{2}&0\\0&0&1\end{pmatrix},\begin{pmatrix}-\frac{1}{2}&\frac{\sqrt{3}}{2}&0\\-\frac{\sqrt{3}}{2}&-\frac{1}{2}&0\\0&0&1\end{pmatrix},\right.$$
$$\left.\begin{pmatrix}1&0&0\\0&-1&0\\0&0&-1\end{pmatrix},\begin{pmatrix}-\frac{1}{2}&\frac{\sqrt{3}}{2}&0\\\frac{\sqrt{3}}{2}&\frac{1}{2}&0\\0&0&-1\end{pmatrix},\begin{pmatrix}-\frac{1}{2}&-\frac{\sqrt{3}}{2}&0\\-\frac{\sqrt{3}}{2}&\frac{1}{2}&0\\0&0&-1\end{pmatrix}\right\}$$

We make it non-unitary by similarity transforming it by a non-unitary **sMat** :

```
sMat = {{2, 3, 0}, {1, 2, 0}, {0, 0, 1}};
sInv = Inverse[sMat];
ℬ = Map[sMat.#.sInv &, CartRep["D3"]] // N;
% // MatrixList // Size[7]
```

$$\left\{\begin{pmatrix}1.&0.&0.\\0.&1.&0.\\0.&0.&1.\end{pmatrix},\begin{pmatrix}6.4282&-11.2583&0.\\4.33013&-7.4282&0.\\0.&0.&1.\end{pmatrix},\begin{pmatrix}-7.4282&11.2583&0.\\-4.33013&6.4282&0.\\0.&0.&1.\end{pmatrix},\right.$$
$$\left.\begin{pmatrix}7.&-12.&0.\\4.&-7.&0.\\0.&0.&-1.\end{pmatrix},\begin{pmatrix}-0.0358984&1.66987&0.\\0.598076&0.0358984&0.\\0.&0.&-1.\end{pmatrix},\begin{pmatrix}-6.9641&10.3301&0.\\-4.59808&6.9641&0.\\0.&0.&-1.\end{pmatrix}\right\}$$

Rep $\mathcal{B}$ is block diagonal, but nonunitary:

```
Map[UnitaryQ, ℬ]
```

```
{True, False, False, False, False, False}
```

Apply **MakeUnitary**, and look at the result:

```
ℬU = MakeUnitary[ℬ] [[2]];
Map[GridForm, %] // Size[6]
```

{

1.	0	0
0	1.	0
0	0	1.

,

-0.5	0	-0.866025
0	1.	0
0.866025	0	-0.5

,

-0.5	0	0.866025
0	1.	0
-0.866025	0	-0.5

,

-0.447214	0	-0.894427
0	-1.	0
-0.894427	0	0.447214

,

0.998203	0	0.0599153
0	-1.	0
0.0599153	0	-0.998203

,

-0.55099	0	0.834512
0	-1.	0
0.834512	0	0.55099

}

Is it unitary?

```
Map[UnitaryQ, ℬU]
```

```
{True, True, True, True, True, True}
```

Yes, but the block diagonal form has been lost. The moral of this story is simple. If you need to analyze some wretched nonunitary representation, make it unitary first, then block diagonalize it.

30. Schur's reduction

Preliminaries

30.1 Schur's idea

In Chapter 28 (ReducibleReps) we introduced the concept of reducible and irreducible representations, and demonstrated concretely the important things you learn by reducing a representation to block diagonal form. However, we had to leave the block reduction method unexplained, because at that time we did not yet have the concept of the universal pseudo-commuter matrix. or that of the universal true commuter matrix. That was developed in Chapter 26 (TestForSimilarity), and in this chapter we will show how Shur may have used it to reduce a representation. We say "may have" because mathematicians are very good at covering their tracks, and no one knows exactly how he arrived at his famous lemmas. This is just one attempt to find the trail that led him to glory.

Suppose all the matrices of a representation are reducible to the pattern $\begin{pmatrix} u & 0 & 0 \\ 0 & v & w \\ 0 & x & y \end{pmatrix}$. Then it is easy to see that the diagonal matrix $\begin{pmatrix} \alpha & 0 & 0 \\ 0 & \beta & 0 \\ 0 & 0 & \beta \end{pmatrix}$ commutes with all of them, multiplying the upper left block by α and the lower right block by β. In detail,

$$\begin{pmatrix} u & 0 & 0 \\ 0 & v & w \\ 0 & x & y \end{pmatrix} . \begin{pmatrix} \alpha & 0 & 0 \\ 0 & \beta & 0 \\ 0 & 0 & \beta \end{pmatrix} == \begin{pmatrix} \alpha & 0 & 0 \\ 0 & \beta & 0 \\ 0 & 0 & \beta \end{pmatrix} . \begin{pmatrix} u & 0 & 0 \\ 0 & v & w \\ 0 & x & y \end{pmatrix} == \begin{pmatrix} \alpha u & 0 & 0 \\ 0 & \beta v & \beta w \\ 0 & \beta x & \beta y \end{pmatrix}$$

```
True
```

or in more general symbols, for all $\mathbb{B}_i$ in the group

$$\mathbb{B}_i . \mathbb{D} = \mathbb{D} . \mathbb{B}_i \qquad (30.1)$$

where $\mathbb{D}$ is diagonal and the $\mathbb{B}_i$ are block diagonal. But of course, with a new, unfamiliar rep one does not know what the block diagonal form is, or if it even exists. Let the matrices $\mathbb{T}_i$ constitute such a tangled representation. We now know how to find a universal commuter for them, obeying

$$\mathbb{T}_i . \mathbb{C} = \mathbb{C} . \mathbb{T}_i \qquad (30.2)$$

The formula for $\mathbb{C}$ was given as Eq. 27.15

$$\mathbb{C} = \sum_i \mathbb{T}_i^{-1} . \mathbb{Q} . \mathbb{T}_i \qquad (27.15)$$

W.M. McClain, *Symmetry Theory in Molecular Physics with Mathematica*,
DOI 10.1007/b13137_30, © Springer Science+Business Media, LLC 2009

Strangely, $\mathbb{Q}$ can be "almost" any matrix of the right size (except the unit matrix, which leads to a trivial result). Click if you need to review that discussion. Also, the word "almost" above needs a little expansion, but it will be better to do it farther down.

Now just fantasize a little, the way Schur did (or seems to have done). Suppose you had a matrix $\mathbb{S}$ that diagonalizes matrix $\mathbb{C}$ by a similarity transform. Then you could similarity transform the whole of Eq. 30.2 as

$$\left(\mathbb{S}.\mathbb{T}_i.\mathbb{S}^{-1}\right).\left(\mathbb{S}.\mathbb{C}.\mathbb{S}^{-1}\right) = \left(\mathbb{S}.\mathbb{C}.\mathbb{S}^{-1}\right).\left(\mathbb{S}.\mathbb{T}_i.\mathbb{S}^{-1}\right) \qquad (30.3)$$

If $\mathbb{S}$ transforms $\mathbb{C}$ into a diagonal matrix $\mathbb{D}$ according to $\mathbb{D} = \mathbb{S}.\mathbb{C}.\mathbb{S}^{-1}$, then

$$\left(\mathbb{S}.\mathbb{T}_i.\mathbb{S}^{-1}\right).\mathbb{D} = \mathbb{D}.\left(\mathbb{S}.\mathbb{T}_i.\mathbb{S}^{-1}\right) \qquad (30.4)$$

Comparing this to Eq. 30.1, we see that

$$\mathbb{B}_i = \mathbb{S}.\mathbb{T}_i.\mathbb{S}^{-1} \quad \text{for every i} \qquad (30.5)$$

So $\mathbb{S}$ is the elusive matrix that transforms tangled matrices $\mathbb{T}_i$ into block diagonal matrices $\mathbb{B}_i$. But how can you construct this fantastic matrix $\mathbb{S}$? The diagonalization of matrices like $\mathbb{C}$ by means of a similarity transform was not unknown in the nineteenth century. However, it was a rather heroic undertaking, for all but the smallest matrices. Today, of course, finding the $\mathbb{S}$ that diagonalizes a numerical $\mathbb{C}$ is just a routine task, given the Eigensystem operator. So using this advantage to the full, let's work out a concrete example to make sure that it all goes smoothly.

30.2 Two experimental rep reductions

30.2.1 Pure real example

To automate this process, we need to do two examples by hand, one with a pure real eigensystem and one with a complex eigensystem. Both cases occur in the point groups.

Below we construct a reducible rep by transforming the **CartRep** for group **D3**. The tangling operator is in the preliminaries; click to see it. It is unitary because it uses an Euler rotation matrix with randomly chosen angles, so every time you call it, it returns a different result. We show only two matrices:

```
repD3 = TangleU33[CartRep["D3"]];
%[[{4, 5}]] // MatrixList // Size[6]
```

```
{( -0.749384  0.225501  -0.622554 )   ( -0.0764828  -0.739458  -0.668844 )}
 ( 0.225501  -0.797097  -0.560165 ) , ( -0.739458   -0.407917   0.535542 )
 ( -0.622554 -0.560165   0.546481 )   ( -0.668844    0.535542  -0.5156   )
```

Use **MorphTest** and **UnitaryQ** to verify that **repD3** is what we want:

```
MorphTest[repD3, "D3"]
```

```
Faithful, or Isomorphic
```

```
Map[UnitaryQ, repD3]
```

```
{True, True, True, True, True, True}
```

In the preliminaries there is also an operator that constructs the universal commuter of a rep (á la Eq. 25.13); you may click here to read it.. Its syntax is

UniversalCommuter[$\mathcal{R}$, $\mathbb{Q}$]

where $\mathcal{R}$ is a representation and $\mathbb{Q}$ is that strange arbitrary matrix. So use it on representation **repD3** :

```
ℚ33 = {{9, 4, 1}, {7, 9, 9}, {4, 5, 8}};
ℂD3 = UniversalCommuter[repD3, ℚ33];
% // MatrixForm // Size[8]
```

```
( 68.6098  29.1006  28.1188 )
( 29.1006  44.3384  18.7396 )
( 28.1188  18.7396  43.0518 )
```

Verify that $\mathbb{C}_{D3}$ is a universal commuter:

```
Map[(ℂD3.# - #.ℂD3) &, repD3] // Chop // Union
```

```
{{{0, 0, 0}, {0, 0, 0}, {0, 0, 0}}}
```

We must find the similarity transform that diagonalizes $\mathbb{C}_{D3}$. We find its **Eigensystem**, noting that its eigenvalues come out in increasing order, so degenerate eigenvectors come out as neighbors. This helps with the block diagonality.

```
{eVals, eVecRows} = Eigensystem[ℂD3];
eVals // Size[7]
eVecRows // GridForm // Size[7]
```

```
{106.111, 24.9445, 24.9445}
```

0.733465	0.488814	0.472323
0.622634	-0.761914	-0.178366
0.679727	-0.52746	-0.509664

Do the standard check that **Eigensystem** performed correctly :

```
eVecCols = Transpose[eVecRows];
𝔻D3 = DiagonalMatrix[eVals];
(ℂD3.eVecCols - eVecCols.𝔻D3) // Chop
```

```
{{0, 0, 0}, {0, 0, 0}, {0, 0, 0}}
```

Is the matrix **eVecCols** a unitary matrix?

```
eVecCols = Transpose[eVecRows];
eVecRows.eVecCols // ChopInteger // GridForm
```

1	0	0
0	1	0.916007
0	0.916007	1

Eigensystem does not necessarily orthogonalize degenerate eigenvectors. If you want them orthogonal (and usually you do) you have to **Orthogonalize** them :

```
eVecRowsOrth = Orthogonalize[eVecRows];
% // GridForm // Size[7]
```

0.733465	0.488814	0.472323
0.622634	-0.761914	-0.178366
0.272681	0.42491	-0.86319

```
eVecColsOrth = Transpose[eVecRowsOrth];
```

Now they have to be orthonormal:

```
eVecRowsOrth.eVecColsOrth // ChopInteger // GridForm
```

1	0	0
0	1	0
0	0	1

Does **eVecRowsOrth** diagonalize $\mathbb{C}_{D3}$?

```
eVecRowsOrth.ℂD3.eVecColsOrth;
% // ChopInteger // MatrixForm
```

$$\begin{pmatrix} 106.111 & 0 & 0 \\ 0 & 24.9445 & 0 \\ 0 & 0 & 24.9445 \end{pmatrix}$$

So **eVecRowsOrth** is the fantastic matrix $\mathbb{S}$ that Schur imagined. Does it really block-diagonalize the whole rep?

```
repD3Block =
  Map[eVecRowsOrth.#.eVecColsOrth &, repD3] //
   ChopInteger;
% // GridList // Size[7]
```

$$\left\{\begin{pmatrix} 1 & 0 & 0 \\ 0 & 1 & 0 \\ 0 & 0 & 1 \end{pmatrix}, \begin{pmatrix} 1 & 0 & 0 \\ 0 & -0.5 & 0.866025 \\ 0 & -0.866025 & -0.5 \end{pmatrix}, \begin{pmatrix} 1 & 0 & 0 \\ 0 & -0.5 & -0.866025 \\ 0 & 0.866025 & -0.5 \end{pmatrix}, \begin{pmatrix} -1 & 0 & 0 \\ 0 & -0.963782 & 0.266691 \\ 0 & 0.266691 & 0.963782 \end{pmatrix}, \begin{pmatrix} -1 & 0 & 0 \\ 0 & 0.712852 & 0.701314 \\ 0 & 0.701314 & -0.712852 \end{pmatrix}, \begin{pmatrix} -1 & 0 & 0 \\ 0 & 0.25093 & -0.968005 \\ 0 & -0.968005 & -0.25093 \end{pmatrix}\right\}$$

It does. Schur was right.

On your own

1. Use **`MorphTest`** to check that all is well.

2. Add up the characters of the two blocks and compare with the characters in the **`BoxedCharacterTable`** for $\mathbf{D_3}$. In some examples the 2×2 block is in the upper left; in others, the lower right. What determines this in general? (Think first)

30.2.2 Complex example

Some representations produce a complex **`Eigensystem`**, which puts a little different twist on things. If the functions **`f + i g`** and **`f - i g`** support conjugate complex one-dimensional reps, then **`{f,g}`** ought to support a two-dimensional real rep that will be reducible to two complex 1-D reps. We find such a situation in reps $\mathbf{E_{u,a}}$ and $\mathbf{E_{u,b}}$ of group $\mathbf{T_h}$.

```
BoxedCharacterTable["Th", 7]
```

Th	1	4	4	3	1	4	4	3	← Class populations Note : ϵ = Exp[-2πi/3]
	E	C3	C32	C2	inv	S6	S65	σh	↓ Basis functions
Ag	1	1	1	1	1	1	1	1	$\{(x^2+y^2+z^2)\}$
Ega	1	ϵ	ϵ^*	1	1	ϵ	ϵ^*	1	$\{(x^2+\sqrt{3}\;{-i}\,x^2+y^2+\sqrt{3}\;i\,y^2-2z^2)\}$
Egb	1	ϵ^*	ϵ	1	1	ϵ^*	ϵ	1	$\{(x^2+\sqrt{3}\;i\,x^2+y^2+\sqrt{3}\;{-i}\,y^2-2z^2)\}$
Tg	3	0	0	-1	3	0	0	-1	$\left\{\begin{pmatrix} xy \\ yz \\ xz \end{pmatrix}, \begin{pmatrix} Ix \\ Iy \\ Iz \end{pmatrix}\right\}$
Au	1	1	1	1	-1	-1	-1	-1	$\{(xyz)\}$
Eua	1	ϵ	ϵ^*	1	-1	$-\epsilon$	$-\epsilon^*$	-1	$\{(x^3yz+\sqrt{3}\;{-i}\,x^3yz+xy^3z+\sqrt{3}\;i\,xy^3z-2xyz^3)\}$
Eub	1	ϵ^*	ϵ	1	-1	$-\epsilon^*$	$-\epsilon$	-1	$\{(x^3yz+\sqrt{3}\;i\,x^3yz+xy^3z+\sqrt{3}\;{-i}\,xy^3z-2xyz^3)\}$
Tu	3	0	0	-1	-3	0	0	1	$\left\{\begin{pmatrix} x \\ y \\ z \end{pmatrix}\right\}$

```
basesThEu = TabulatedBases["Th"][[{6, 7}]]
```

$\{\{\text{Eua}, \{\{x\,y\,z\,(x^2+y^2-i\sqrt{3}\,(x^2-y^2)-2\,z^2)\}\}\}, \{\text{Eub}, \{\{x\,y\,z\,(x^2+y^2+i\sqrt{3}\,(x^2-y^2)-2\,z^2)\}\}\}\}$

```
fnEua = basesThEu[[1, 2, 1, 1]]
```

$x\,y\,z\,(x^2+y^2-i\sqrt{3}\,(x^2-y^2)-2\,z^2)$

```
fFn = fnEua /. -i → 0
```

$x\,y\,z\,(x^2+y^2-2\,z^2)$

$$\text{gFn} = \frac{(\text{fnEua} - \text{fFn})}{-i} \text{ // Collect[\#, z] \&}$$

$\sqrt{3}\,x\,y\,(x^2-y^2)\,z$

```
repTh = MakeRepPoly[
   {fFn, gFn}, "Th"];
Short[% // GridList] // Size[7]
```

$\left\{\begin{pmatrix}1 & 0\\ 0 & 1\end{pmatrix}, \begin{pmatrix}-\frac{1}{2} & \frac{\sqrt{3}}{2}\\ -\frac{\sqrt{3}}{2} & -\frac{1}{2}\end{pmatrix}, \ll 20 \gg, \begin{pmatrix}-1 & 0\\ 0 & -1\end{pmatrix}, \begin{pmatrix}-1 & 0\\ 0 & -1\end{pmatrix}\right\}$

Some are diagonal, some are not. We want them all diagonal, so start by making a universal commuter

```
Q22 = {{5, 7}, {2, 11}};
CTh = UniversalCommuter[repTh, Q22] // Simplify
```

{{192, 60}, {-60, 192}}

Check that it commutes:

```
Map[(#.CTh == CTh.#) &, repTh] // Union
```

{True}

We find the **Eigensystem** of $\mathbb{C}_{Th}$:

```
{eVals, eVecRows} = Eigensystem[CTh] // Chop;
eVals // Size[8]
eVecRows // MatrixForm // Size[8]
```

$\{192+60\,i, 192-60\,i\}$

$\begin{pmatrix}-i & 1\\ i & 1\end{pmatrix}$

Both eigenvectors and eigenvalues are complex. Make the standard check :

```
𝕌Th = Transpose[eVecRows];
𝔻Th = DiagonalMatrix[eVals];
(ℂTh.𝕌Th - 𝕌Th.𝔻Th) // Chop
```

```
{{0, 0}, {0, 0}}
```

Check that $\mathbb{U}_{Th}$ and its adjoint are orthonormal :

```
𝕌†Th = Conjugate[Transpose[𝕌Th]];
𝕌Th.𝕌†Th // ChopInteger // FullSimplify
```

```
{{2, 0}, {0, 2}}
```

They are orthogonal, but not normalized. We could just divide them by $\sqrt{2}$, but it is more general to call **Orthogonalize** on them. (Click here for details of this operator.) By default, it uses **Dot** to take scalar products, so it does not do the right job on complex vectors. This is easily fixed. We prepare a scalar product function **CxDot** for use with complex vectors. It conjugates the right factor before using **Dot** :

```
Clear[CxDot];
CxDot[a_, b_] := Dot[a, Conjugate[b]]
```

GramSchmidt is a bit of overkill in this case, but here it is :

```
eVecRowsOrth = 𝕍Th = Orthogonalize[eVecRows, CxDot]
```

$$\left\{\left\{-\frac{i}{\sqrt{2}}, \frac{1}{\sqrt{2}}\right\}, \left\{\frac{i}{\sqrt{2}}, \frac{1}{\sqrt{2}}\right\}\right\}$$

```
𝕍†Th = Transpose[Conjugate[𝕍Th]];
𝕍Th.𝕍†Th
```

```
{{1, 0}, {0, 1}}
```

Now the eigenvectors are orthonormal in the complex sense. Do they diagonalize the universal commuter?

```
𝕍Th.ℂTh.𝕍†Th // ChopInteger // MatrixForm
```

$$\begin{pmatrix} 192 - 60\,i & 0 \\ 0 & 192 + 60\,i \end{pmatrix}$$

Yes. So let them block-diagonalize the whole rep :

```
repThBlock =
  Map[𝕍Th.#.𝕍†Th &, repTh] // Simplify // ExpandAll;
Short[% // GridList, 3] // Size[7]
```

$$\left\{\begin{array}{|c|c|}\hline 1 & 0 \\ \hline 0 & 1 \\ \hline\end{array}, \begin{array}{|c|c|}\hline -\frac{1}{2}-\frac{i\sqrt{3}}{2} & 0 \\ \hline 0 & -\frac{1}{2}+\frac{i\sqrt{3}}{2} \\ \hline\end{array}, \begin{array}{|c|c|}\hline -\frac{1}{2}-\frac{i\sqrt{3}}{2} & 0 \\ \hline 0 & -\frac{1}{2}+\frac{i\sqrt{3}}{2} \\ \hline\end{array},\right.$$

$$\left.\ll 18 \gg, \begin{array}{|c|c|}\hline -1 & 0 \\ \hline 0 & -1 \\ \hline\end{array}, \begin{array}{|c|c|}\hline -1 & 0 \\ \hline 0 & -1 \\ \hline\end{array}, \begin{array}{|c|c|}\hline -1 & 0 \\ \hline 0 & -1 \\ \hline\end{array}\right\}$$

Display **repTh$_{\text{Block}}$** with the number $-\frac{1}{2}-\frac{i\sqrt{3}}{2}$ symbolized as ϵ :

```
ϵRules = SmartRules[-1/2 - (ⅈ √3)/2, ϵ];
(repThBlock /. ϵRules) // MatrixList // Size[7]
```

$$\left\{\begin{pmatrix} 1 & 0 \\ 0 & 1 \end{pmatrix}, \begin{pmatrix} \epsilon & 0 \\ 0 & \epsilon^* \end{pmatrix}, \begin{pmatrix} \epsilon & 0 \\ 0 & \epsilon^* \end{pmatrix}, \begin{pmatrix} \epsilon & 0 \\ 0 & \epsilon^* \end{pmatrix}, \begin{pmatrix} \epsilon & 0 \\ 0 & \epsilon^* \end{pmatrix}, \begin{pmatrix} \epsilon^* & 0 \\ 0 & \epsilon \end{pmatrix},\right.$$

$$\begin{pmatrix} \epsilon^* & 0 \\ 0 & \epsilon \end{pmatrix}, \begin{pmatrix} \epsilon^* & 0 \\ 0 & \epsilon \end{pmatrix}, \begin{pmatrix} \epsilon^* & 0 \\ 0 & \epsilon \end{pmatrix}, \begin{pmatrix} 1 & 0 \\ 0 & 1 \end{pmatrix}, \begin{pmatrix} 1 & 0 \\ 0 & 1 \end{pmatrix}, \begin{pmatrix} 1 & 0 \\ 0 & 1 \end{pmatrix}, \begin{pmatrix} -1 & 0 \\ 0 & -1 \end{pmatrix},$$

$$\begin{pmatrix} -\epsilon & 0 \\ 0 & -\epsilon^* \end{pmatrix}, \begin{pmatrix} -\epsilon & 0 \\ 0 & -\epsilon^* \end{pmatrix}, \begin{pmatrix} -\epsilon & 0 \\ 0 & -\epsilon^* \end{pmatrix}, \begin{pmatrix} -\epsilon & 0 \\ 0 & -\epsilon^* \end{pmatrix}, \begin{pmatrix} -\epsilon^* & 0 \\ 0 & -\epsilon \end{pmatrix},$$

$$\left.\begin{pmatrix} -\epsilon^* & 0 \\ 0 & -\epsilon \end{pmatrix}, \begin{pmatrix} -\epsilon^* & 0 \\ 0 & -\epsilon \end{pmatrix}, \begin{pmatrix} -\epsilon^* & 0 \\ 0 & -\epsilon \end{pmatrix}, \begin{pmatrix} -1 & 0 \\ 0 & -1 \end{pmatrix}, \begin{pmatrix} -1 & 0 \\ 0 & -1 \end{pmatrix}, \begin{pmatrix} -1 & 0 \\ 0 & -1 \end{pmatrix}\right\}$$

Yes; they are all made of two 1×1 blocks. If you have not seen **SmartRules** before, here is its thumbnail description:

- **SmartRules[cmplxExpr, s]** returns eight rules. The first turns **cmplexExpr** into symbol **s**. The other seven recognize variations of **cmplxExpr**, returning $-s$, s^*, $-s^*$, $i\,s$, $-i\,s$, $(is)^*$, or $(-is)^*$, as appropriate.

30.3 The ReduceRep operator

We are now ready to define an operator that will block diagonalize any representation. Drawing together the steps in the real and complex examples,

```
Clear[ReduceRep];
ReduceRep[rep_] := Module[{a, b, c, U},
  (*Body*)
  {a, b, c} = Dimensions[rep];
  If[b =!= c,
   Return["rep is not square"], Null, Null];
  U = IdentityMatrix[b];
  ReduceRep[rep, U]];
```

```
ReduceRep[rep_, Qmat_] :=
 Module[{Cmat, esysC, VrowsRaw,
   CxDot, a, b, Vrows, Vcols},
  (*Body*)
  Cmat =
   Simplify[Expand[UniversalCommuter[rep, Qmat]]];
  esysC = Eigensystem[Cmat]; VrowsRaw = esysC[[2]];
  CxDot[a_, b_] := a.Conjugate[b];
  Vrows = Chop[Orthogonalize[VrowsRaw,
     CxDot, Method → "GramSchmidt"]];
  Vcols = Conjugate[Transpose[Vrows]];
  Chop[ExpandAll[
    Simplify[(Vrows.#1.Vcols &) /@ rep]]]]
```

Remember that you have already seen an application of this, back in Chapter 28 (ReducibleReps). This is the secret construction method we used there.

30.4 Tests of ReduceRep

30.4.1 Basic tests of ReduceRep

Make sure it works on **repD3** and **repTh**, the examples we did by hand. We do them again, using the same $\mathbb{Q}$ matrices as before.

```
(*Real example*)
repD3Block - ReduceRep[repD3, Q33] // Chop // Union
```

```
{{{0, 0, 0}, {0, 0, 0}, {0, 0, 0}}}
```

```
(*Complex example*)
(repThBlock - ReduceRep[repTh, Q22]) // Chop // Union
```

```
{{{0, 0}, {0, 0}}}
```

30.4.2 Benzene orbitals again, by ReduceRep

As a little tougher test, we let **ReduceRep** do the $\mathbf{p_z}$ rep of $\mathbf{D_{6h}}$, which supports the π orbitals of benzene.

```
{pzRepD6h, bzRules} = MakeRepAtomic["benzene",
    "D6h", atomKind → "C", preFactor → z];
Dimensions[pzRepD6h]
```

```
{24, 6, 6}
```

```
Q66 = {{-6, -2, -5, 2, 6, -6}, {-5, -3, 3, -6, -9, 2},
    {3, 2, 5, -3, -1, -9}, {5, -3, -10, 6, 7, 0},
    {8, 5, -1, 3, 9, 6}, {6, -5, -1, 2, -3, 0}};
```

Use **ReduceRep** on it, displaying only the first 3 elements :

```
pzRepReduced = ReduceRep[pzRepD6h, Q66];
Take[pzRepReduced, 3] // GridList // Size[6]
```

{

1	0	0	0	0	0
0	1	0	0	0	0
0	0	1	0	0	0
0	0	0	1	0	0
0	0	0	0	1	0
0	0	0	0	0	1

,

-1	0	0	0	0	0
0	$\frac{1}{2}$	$\frac{\sqrt{3}}{2}$	0	0	0
0	$-\frac{\sqrt{3}}{2}$	$\frac{1}{2}$	0	0	0
0	0	0	$-\frac{1}{2}$	$\frac{\sqrt{3}}{2}$	0
0	0	0	$-\frac{\sqrt{3}}{2}$	$-\frac{1}{2}$	0
0	0	0	0	0	1

,

-1	0	0	0	0	0
0	$\frac{1}{2}$	$-\frac{\sqrt{3}}{2}$	0	0	0
0	$\frac{\sqrt{3}}{2}$	$\frac{1}{2}$	0	0	0
0	0	0	$-\frac{1}{2}$	$-\frac{\sqrt{3}}{2}$	0
0	0	0	$\frac{\sqrt{3}}{2}$	$-\frac{1}{2}$	0
0	0	0	0	0	1

}

The block sizes are $\mathbf{1} \oplus \mathbf{2} \oplus \mathbf{2} \oplus \mathbf{1}$. The logical next step, already carried out earlier, is to find the characters of the blocks and use the characters to identify the names of the irreducible representations. A better and quicker method lies ahead in Chapter 34 (Rep Reduction), but it depends on the Great Orthogonality, which we are still developing. Be patient.

30.5 What have we learned?

30.5.1 A whiff of Schur's First Lemma

In the last chapter we asked the pseudo-commuter algorithm to do the impossible; namely, to find the similarity transform matrix for two reps that were not related by a symmetry transform. It responded surprisingly but reasonably, returning a trivial but formally correct result, an all-zero matrix. Now let's set the computer on another impossible task, just to see what we get.

What happens if we try to reduce an irreducible rep? The character table of group $\mathbf{T_d}$ says that **{x, y, z}** is the basis of irreducible rep $\mathbf{T_2}$; that means the **CartRep** of $\mathbf{T_d}$ is irreducible. We pull it up :

```
repTd = CartRep["Td"];
Map[MatrixForm, repTd] // Size[5]
```

$$\left\{\begin{pmatrix}1&0&0\\0&1&0\\0&0&1\end{pmatrix}, \begin{pmatrix}0&0&-1\\-1&0&0\\0&1&0\end{pmatrix}, \begin{pmatrix}0&-1&0\\0&0&1\\-1&0&0\end{pmatrix}, \begin{pmatrix}0&-1&0\\0&0&-1\\1&0&0\end{pmatrix}, \begin{pmatrix}0&0&1\\-1&0&0\\0&-1&0\end{pmatrix}, \begin{pmatrix}0&0&-1\\1&0&0\\0&-1&0\end{pmatrix}, \begin{pmatrix}0&1&0\\0&0&-1\\-1&0&0\end{pmatrix}, \begin{pmatrix}0&1&0\\0&0&1\\1&0&0\end{pmatrix},\right.$$

$$\begin{pmatrix}0&0&1\\1&0&0\\0&1&0\end{pmatrix}, \begin{pmatrix}-1&0&0\\0&-1&0\\0&0&1\end{pmatrix}, \begin{pmatrix}1&0&0\\0&-1&0\\0&0&-1\end{pmatrix}, \begin{pmatrix}-1&0&0\\0&1&0\\0&0&-1\end{pmatrix}, \begin{pmatrix}0&1&0\\-1&0&0\\0&0&-1\end{pmatrix}, \begin{pmatrix}0&-1&0\\1&0&0\\0&0&-1\end{pmatrix}, \begin{pmatrix}-1&0&0\\0&0&1\\0&-1&0\end{pmatrix}, \begin{pmatrix}-1&0&0\\0&0&-1\\0&1&0\end{pmatrix},$$

$$\left.\begin{pmatrix}0&0&1\\0&-1&0\\-1&0&0\end{pmatrix}, \begin{pmatrix}0&0&-1\\0&-1&0\\1&0&0\end{pmatrix}, \begin{pmatrix}0&0&1\\0&1&0\\1&0&0\end{pmatrix}, \begin{pmatrix}0&1&0\\1&0&0\\0&0&1\end{pmatrix}, \begin{pmatrix}1&0&0\\0&0&1\\0&1&0\end{pmatrix}, \begin{pmatrix}0&0&-1\\0&1&0\\-1&0&0\end{pmatrix}, \begin{pmatrix}0&-1&0\\-1&0&0\\0&0&1\end{pmatrix}, \begin{pmatrix}1&0&0\\0&0&-1\\0&-1&0\end{pmatrix}\right\}$$

There; it is nice and simple and guaranteed irreducible. So let's try to reduce it. The first step is to make the universal commuter, for which we have the command already written :

```
ℭTd = UniversalCommuter[repTd, 𝔔33] // MatrixForm //
  Size[7]
```

$$\begin{pmatrix}208&0&0\\0&208&0\\0&0&208\end{pmatrix}$$

The next step is to diagonalize $\mathfrak{C}_{\mathbf{Td}}$, and then ... But wait! It is already diagonal! Not only that, it is a constant matrix (a number times the unit matrix). Constant matrices commute with all matrices, so this is indeed a correct, but trivial, universal commuter. We remark again that the universal commuter algorithm is very hardy. Even when asked for the impossible, it returns something that is formally correct.

> **On your own**
> Run that `UniversalCommuter` command several times. Remember, it uses a random matrix, so every time you run it the answer is different. You will see different values of the constant, but it is always a constant matrix. Try it.

30.5.2 Generalizing from examples

Examples can help one to sniff out a theorem, but unless you can prove that your examples exhaust all possibilities, they do not constitute a proof. The real work is in figuring out why the regularities seen in the examples work as they do, and in creating a statement that encapsulates exactly what is seen in all the examples, including the "failures".

For instance, here is a bad initial attempt to state a theorem consistent with the observations above. "If rep $\mathcal{A}$ has a constant universal commuter matrix, then $\mathcal{A}$ is irreducible."

This is consistent with what you will see nearly all the time if you choose the arbitrary $\mathbb{Q}$ matrix of the algorithm by a random process, to an accuracy of fourteen or sixteen digits. But it will not do as a theorem statement, because there might be certain special $\mathbb{Q}$ matrices that produce a constant universal commuter, even in reducible cases. We need a way to state it so that this possible exception does not ruin it. In the next chapter you will see how Schur did this, in his famous theorem known today as Schur's First Lemma.

Click into the End notes to see an actual example of a $\mathbb{Q}$ matrix (originally randomly chosen) which shows quite peculiar behavior.

> **On your own**
> Try to formulate a proper statement of Schur's First Lemma before you see how Schur did it. Don't be dreamy about it; actually write it out.

30.6 End notes

30.6.1 Characters of the blocks

If you do not **RoundSort** the eigensystem, the arbitrary matrix **Q** used in **MakeUnitary** determines the order of the blocks. If it is generated randomly on each call, the order of the blocks may change. The 2×2 block is either upper left or lower right, so one of the following diagrams will be correct:

$$\begin{pmatrix} \square & 0 & 0 \\ 0 & \square & \square \\ 0 & \square & \square \end{pmatrix} \qquad \begin{pmatrix} \square & \square & 0 \\ \square & \square & 0 \\ 0 & 0 & \square \end{pmatrix}$$

But if the eigensystem is **Transposed** and **RoundSorted**, then the **{{eval,evec}}** pairs come out with **evals** in ascending order. Then if **Transposed** again to make **{{evals},{evecs}}** the **evecs** will appear in the order of their **evals**, and this determines the order of the blocks. So the upper left block has the lowest **eval**, etc.

30.6.2 Orthogonalization

Orthogonalize is a core operator in Version 6. It can use any of several algorithms, but by default it uses the Gram-Schmidt procedure. For a clear, detailed explanation of how it works, click here to go to Eric Weisstein's MathWorld encyclopedia article on it.

30.6.3 Don't trust those Q matrices

The **RepReduce** algorithm is not perfect. Here we redo the benzene orbital reduction using a **Q66** matrix that causes a block diagonal transform with three twofold degeneracies. This is not totally wrong, but it is incomplete. One of the twofold degeneracies could have been reduced further to two 1×1 blocks. The problem appears when you use the following **Q** matrix :

```
Q66BAD = {{4, 5, -3, -9, 2, 6}, {-5, 2, 8, 1, -3, 7},
    {-10, 6, 2, 10, -8, 2}, {-10, -7, -1, 7, 1, 4},
    {7, 9, -7, 10, 4, -6}, {-10, -5, -7, -2, -6, 6}};
```

```
pzRepReduced = ReduceRep[pzRepD6h, Q66BAD];
Take[pzRepReduced, 3] // MatrixList // Size[6]
```

$$\left\{\begin{pmatrix}1&0&0&0&0&0\\0&1&0&0&0&0\\0&0&1&0&0&0\\0&0&0&1&0&0\\0&0&0&0&1&0\\0&0&0&0&0&1\end{pmatrix},\begin{pmatrix}\frac{1}{2}&\frac{\sqrt{3}}{2}&0&0&0&0\\-\frac{\sqrt{3}}{2}&\frac{1}{2}&0&0&0&0\\0&0&-\frac{1}{2}&\frac{\sqrt{3}}{2}&0&0\\0&0&-\frac{\sqrt{3}}{2}&-\frac{1}{2}&0&0\\0&0&0&0&0&1\\0&0&0&0&1&0\end{pmatrix},\begin{pmatrix}\frac{1}{2}&-\frac{\sqrt{3}}{2}&0&0&0&0\\\frac{\sqrt{3}}{2}&\frac{1}{2}&0&0&0&0\\0&0&-\frac{1}{2}&-\frac{\sqrt{3}}{2}&0&0\\0&0&\frac{\sqrt{3}}{2}&-\frac{1}{2}&0&0\\0&0&0&0&0&1\\0&0&0&0&1&0\end{pmatrix}\right\}$$

Interactive readers can look at them all. They all consistent with three blocks of size 2×2. **MorphTest** says the rep is faithful, and it passes the unitary test. Very seldom does a randomly chosen **Q** matrix give such a result, but it did happen once and we have preserved it as a curiosity. Below we try to find the root of this behavior:

```
C = UniversalCommuter[pzRepD6h, Q66BAD] // Simplify;
% // MatrixForm // Size[6]
```

$$\begin{pmatrix}100&36&-42&-72&-42&36\\36&100&36&-42&-72&-42\\-42&36&100&36&-42&-72\\-72&-42&36&100&36&-42\\-42&-72&-42&36&100&36\\36&-42&-72&-42&36&100\end{pmatrix}$$

Matrix **C** passes the universal commuter test. But there is something strange about its eigenvalues :

```
{eVals, eVecRows} = Eigensystem[C];
eVals
```

```
{250, 250, 34, 34, 16, 16}
```

There is an extra degeneracy, caused by something about **Q66BAD**. Follow it out step-by-step if you like; this is the only strange feature that appears. The eigenvalues are indeed a function of the **Q** matrix, so we seem to have hit on one that just happens to induce an extra degeneracy. It does not seem very profound.

A cleverly chosen **Q** matrix plays an essential role in Schur's Lemmas, but you can't trust random examples in the reduction algorithm. The rep reduction method that everyone uses lies ahead in Chapter 35 (RepAnalysis). It does not involve any arbitrary **Q** matrix, and is utterly reliable.

31. Schur's First Lemma

Preliminaries

31.1 Statement of the Lemma

We read Schur's First Lemma remembering that a constant matrix (a number times a unit matrix) commutes with all matrices.

Schur's First Lemma
1. If a rep has a non-constant universal commuter, it is reducible.
2. If a rep is irreducible, its only universal commuters are constant matrices.

Note that the two parts of this lemma are not simple alternatives "assume X" and then "assume not-X". Let X be the statement " The rep has a non-constant universal commuter", and let Y be the statement "The rep is reducible". The first part is

1. If X, then Y.

The second part has the form

2. If not-Y, then not-X.

These statements are mutually contrapositive, and Boolean logic says that if a statement is True, then its contrapositive is automatically True also.

31.2 Proof

31.2.1 Introduction to the proof

We begin with any general representation $\mathcal{A}$. Schur's reduction algorithm shows us how to compute a universal commuter $\mathbb{C}$ that obeys

$$\mathbb{A}_i . \mathbb{C} == \mathbb{C} . \mathbb{A}_i \tag{31.1}$$

for all i in the group. It is then always possible to find (by eigensystem techniques) a similarity transform that diagonalizes $\mathbb{C}$, converting Eq. 31.1 into a form like

$$\mathbb{B}_i . \mathbb{D} = \mathbb{D} . \mathbb{B}_i \tag{31.2}$$

W.M. McClain, *Symmetry Theory in Molecular Physics with Mathematica*,
DOI 10.1007/b13137_31, © Springer Science+Business Media, LLC 2009

where $\mathbb{D}$ is diagonal and $\mathbb{B}_i$ (possibly block diagonal) are the transforms of the $\mathbb{A}_i$. To be concrete we take a 4×4 example :

$$\begin{pmatrix} \blacklozenge & \blacklozenge & \blacklozenge & \blacklozenge \\ \blacklozenge & \blacklozenge & \blacklozenge & \blacklozenge \\ \blacklozenge & \blacklozenge & \blacklozenge & \blacklozenge \\ \blacklozenge & \blacklozenge & \blacklozenge & \blacklozenge \end{pmatrix} \cdot \begin{pmatrix} a & 0 & 0 & 0 \\ 0 & b & 0 & 0 \\ 0 & 0 & c & 0 \\ 0 & 0 & 0 & d \end{pmatrix} == \begin{pmatrix} a & 0 & 0 & 0 \\ 0 & b & 0 & 0 \\ 0 & 0 & c & 0 \\ 0 & 0 & 0 & d \end{pmatrix} \cdot \begin{pmatrix} \blacklozenge & \blacklozenge & \blacklozenge & \blacklozenge \\ \blacklozenge & \blacklozenge & \blacklozenge & \blacklozenge \\ \blacklozenge & \blacklozenge & \blacklozenge & \blacklozenge \\ \blacklozenge & \blacklozenge & \blacklozenge & \blacklozenge \end{pmatrix} \tag{31.3}$$

where the numbers {a,b,c,d} are the eigenvalues of matrix $\mathbb{C}$. The $\mathbb{B}_i$ matrices are pure diamonds, each diamond ◆ generally standing for a different number.

31.2.2 Part I: Assume a non-constant universal commuter

To have the minimal case of a non-constant commuter, let all the eigenvalues have value **a** except the last, which has a different value **b**. Then Eq. 31.3 multiplies out to give

$$\begin{pmatrix} a\blacklozenge & a\blacklozenge & a\blacklozenge & b\blacklozenge \\ a\blacklozenge & a\blacklozenge & a\blacklozenge & b\blacklozenge \\ a\blacklozenge & a\blacklozenge & a\blacklozenge & b\blacklozenge \\ a\blacklozenge & a\blacklozenge & a\blacklozenge & b\blacklozenge \end{pmatrix} == \begin{pmatrix} a\blacklozenge & a\blacklozenge & a\blacklozenge & a\blacklozenge \\ a\blacklozenge & a\blacklozenge & a\blacklozenge & a\blacklozenge \\ a\blacklozenge & a\blacklozenge & a\blacklozenge & a\blacklozenge \\ b\blacklozenge & b\blacklozenge & b\blacklozenge & b\blacklozenge \end{pmatrix} \tag{31.4}$$

Subtracting the right side from both sides, we get

$$\begin{pmatrix} 0 & 0 & 0 & -(a-b)\blacklozenge \\ 0 & 0 & 0 & -(a-b)\blacklozenge \\ 0 & 0 & 0 & -(a-b)\blacklozenge \\ (a-b)\blacklozenge & (a-b)\blacklozenge & (a-b)\blacklozenge & 0 \end{pmatrix} == \begin{pmatrix} 0 & 0 & 0 & 0 \\ 0 & 0 & 0 & 0 \\ 0 & 0 & 0 & 0 \\ 0 & 0 & 0 & 0 \end{pmatrix} \tag{31.5}$$

Since (a-b) is nonzero by assumption, the values of the ◆ that multiply (a-b) are now known: they must be zero. So every matrix of the representation must have the pattern

$$\mathbb{A}_i = \begin{pmatrix} \blacklozenge & \blacklozenge & \blacklozenge & 0 \\ \blacklozenge & \blacklozenge & \blacklozenge & 0 \\ \blacklozenge & \blacklozenge & \blacklozenge & 0 \\ 0 & 0 & 0 & \blacklozenge \end{pmatrix} \tag{31.6}$$

Therefore, if even one eigenvalue is different from the others (making the diagonalized commuter non-constant), then the rep is reducible, which was to be shown.

A corollary

Above, we assumed the eigenvalues were {a,a,a,b} and we found blocks of size 3×3 and 1×1. By a similar argument, if the eigenvalues are {a,a,b,b} then all the rep matrices have two blocks of size 2×2 :

$$\mathbb{A}_i = \begin{pmatrix} \blacklozenge & \blacklozenge & 0 & 0 \\ \blacklozenge & \blacklozenge & 0 & 0 \\ 0 & 0 & \blacklozenge & \blacklozenge \\ 0 & 0 & \blacklozenge & \blacklozenge \end{pmatrix}$$

Do the row-by-column multiplications on paper, and you will soon see that every set of **N** equal eigenvalues yields a diagonal block of size **N×N**. The automatic sort done by **Eigensystem** insures that the equal eigenvalues are listed together, forming the square blocks that we desire.

> **On your own**
> Carry out the similar argument explicitly.
> Also, look at a case where equal eigenvalues are scattered around, instead of being listed all together.

31.2.3 Part II: Assume the rep is irreducible

Once a statement has been proved true, its contrapositive does not really need a separate proof. But this may seem like skating on thin ice, so we prove it anyway. It is short and sweet:

Eq. 31.5 was

$$\begin{pmatrix} 0 & 0 & 0 & -(a-b)\,\blacklozenge \\ 0 & 0 & 0 & -(a-b)\,\blacklozenge \\ 0 & 0 & 0 & -(a-b)\,\blacklozenge \\ (a-b)\,\blacklozenge & (a-b)\,\blacklozenge & (a-b)\,\blacklozenge & 0 \end{pmatrix} == \begin{pmatrix} 0 & 0 & 0 & 0 \\ 0 & 0 & 0 & 0 \\ 0 & 0 & 0 & 0 \\ 0 & 0 & 0 & 0 \end{pmatrix}$$

This time assume that the rep is not reducible, and the values of **a** and **b** are unknown. Then the ◆ symbols are not all zero in all the rep matrices, because under this hypothesis there does not exist any transform that could change them all into zeroes. (This is the meaning of *irreducible*.) This means that in at least one element of one rep matrix we have **(a - b) ◆ = 0**, where the ◆ is *not* zero. From this, we must conclude that **a = b**, so that in the diagonalized universal commuter all the elements are equal. So the only universal commuter of an irreducible rep is a constant matrix, which was to be shown.

Now we have direct proofs of both parts of Schur's First Lemma.

31.3 End Notes

31.3.1 Converse and Contrapositive, in English

In the statements below, we accept that nothing but water today will save the garden. But we do not accept that diamond is the only stone that scratches glass.

Statement:

1. "If the garden gets water today, it will live."
   `(Water today) ⇒ (Garden lives)` **True**
2. "If this is a diamond, it will scratch glass"
   `(Diamond) ⇒(Scratch)` **True**

The converse of a true statement may be true or false.

Converse: (swap the premises)

1. "If the garden lives, it (will have had water) today."
   `(Garden lives) ⇒ (Water today)` **True**
2. "If this stone scratches glass, it is a diamond."
   `(Scratch) ⇒ (Diamond)` **False**

The contrapositive of a true statement is always true.

Contrapositive: (swap the premises and negate both)

1. "If the garden (dies), it (will not have had water) today."
   `Not[Garden lives] ⇒ Not[Water today]` **True**
2. "If this will not scratch glass, it is not diamond."
   `Not[Scratch] ⇒ Not[Diamond]` **True**

32. Schur's Second Lemma

Preliminaries

32.1 Introduction

Schur's First and Second Lemmas are lemmas for the Great Orthogonality. They are generalized statements about the results of two closely related algorithms; the First generalizes results seen using ReduceRep, presented in Chapter 30 (SchursReduction); the Second generalizes a result seen in Chapter 26 (TestForSimilarity).

32.2 Statement and discussion

32.2.1 Statement

Schur's Second Lemma
If a group has representations $\mathcal{A}$ and $\mathcal{B}$ that are irreducible and not similar, then the equations

$$\mathbb{A}_i.\mathbb{P} = \mathbb{P}.\mathbb{B}_i \text{, for all } i \text{ in the group,}$$

have only the trivial solution $\mathbb{P} = 0$.

32.2.2 Comment

This lemma is a statement about the way the similarity algorithm fails when there is no similarity transform that links two representations $\mathcal{A}$ and $\mathcal{B}$. The algorithm finds $\mathbb{P}$ in the equation $\mathbb{A}_i.\mathbb{P} = \mathbb{P}.\mathbb{B}_i$ and you might think that failure would be a matrix $\mathbb{P}$ that has no inverse, so that it cannot be used in the transform equation $\mathbb{A}_i = \mathbb{P}.\mathbb{B}_i.\mathbb{P}^{-1}$. But Schur's Second Lemma says that failure has to be much more drastic than this. When no similarity transform exists, the only possible $\mathbb{P}$ is one that is all zeroes.

These zeroes are central to the structure of representation theory. The next chapter will show them to be the orthogonality zeroes of the Great

W.M. McClain, *Symmetry Theory in Molecular Physics with Mathematica*,
DOI 10.1007/b13137_32,

Orthogonality.

32.3 Proof, using the unitary property

32.3.1 An equation to which Schur's First Lemma applies

We start with the definition of the pseudocommuter, and make a few simple maneuvers

$$\begin{gathered} \mathbb{A}_i.\mathbb{P} = \mathbb{P}.\mathbb{B}_i \\ (\mathbb{A}_i.\mathbb{P})^\dagger = (\mathbb{P}.\mathbb{B}_i)^\dagger \\ \mathbb{P}^\dagger.\mathbb{A}_i^\dagger = \mathbb{B}_i^\dagger.\mathbb{P}^\dagger \end{gathered} \tag{32.1}$$

If both $\mathcal{A}$ and $\mathcal{B}$ are unitary, we can replace the dagger by an inverse:

$$\mathbb{P}^\dagger.\mathbb{A}_i^{-1} = \mathbb{B}_i^{-1}.\mathbb{P}^\dagger \tag{32.2}$$

or multiplying from the left by $\mathbb{P}$,

$$\left(\mathbb{P}.\mathbb{P}^\dagger\right).\mathbb{A}_i^{-1} = \mathbb{P}.\mathbb{B}_i^{-1}.\mathbb{P}^\dagger \tag{32.3}$$

Don't set $\left(\mathbb{P}.\mathbb{P}^\dagger\right)$ to $\mathbf{1}$ just yet. It could be 1, but it does not have to be, as you will see. If element $\mathbf{k}$ is inverse to element $\mathbf{i}$, this is

$$\left(\mathbb{P}.\mathbb{P}^\dagger\right).\mathbb{A}_k = \mathbb{P}.\mathbb{B}_k.\mathbb{P}^\dagger \tag{32.4}$$

But in the first line of this proof we assumed that $\mathbb{P}.\mathbb{B}_k$ was the same as $\mathbb{A}_k.\mathbb{P}$ so making this replacement on the right, we have

$$\left(\mathbb{P}.\mathbb{P}^\dagger\right).\mathbb{A}_k = \mathbb{A}_k.\left(\mathbb{P}.\mathbb{P}^\dagger\right) \tag{32.5}$$

Look at this with squinty eyes. It says that $\mathbb{P}.\mathbb{P}^\dagger$ is a universal commuter for rep $\mathcal{A}$. But $\mathcal{A}$ was assumed irreducible, and Schur's First Lemma says any universal commuter for an irreducible rep must be a constant matrix; i. e.,

$$\mathbb{P}.\mathbb{P}^\dagger = \mathbf{c}\,\mathbb{U} \tag{32.6}$$

where $\mathbf{c}$ is a number and $\mathbb{U}$ is the unit matrix.

To proceed farther, we must divide our thinking into two cases. So far, reps $\mathcal{A}$ and $\mathcal{B}$ might have been either the same size or different sizes, and $\mathbb{P}$ either square or rectangular. In either case, of course, $\mathbb{P}.\mathbb{P}^\dagger$ is square and the same size as $\mathcal{A}$. But now we need to do something that is possible only with square matrices, so we specialize to the Square Case and the Rectangular Case.

32.3.2 Square Case

Taking the determinant of both sides of Eq. 32.6, we have

$$\mathrm{Det}\left[\mathbb{P}\,.\,\mathbb{P}^{\dagger}\right] = \mathrm{Det}[\mathrm{c}\,\mathbb{I}] = \mathrm{c}^{\mathrm{dim}\mathcal{B}} \tag{32.7}$$

where $\mathbf{dim}\mathcal{B}$ is the dimension of rep $\mathcal{B}$ But if $\mathbb{P}$ is square

$$\mathrm{Det}\left[\mathbb{P}\,.\,\mathbb{P}^{\dagger}\right] = \mathrm{Det}[\mathbb{P}]\;\mathrm{Det}\left[\mathbb{P}^{\dagger}\right] \tag{32.8}$$

Now the $\mathbf{c}$'s down the diagonal of $\mathbb{P}\,.\,\mathbb{P}^{\dagger}$ must be either zero or nonzero. If nonzero, the inverse of $\mathbb{P}$ exists and we can write from our original assumption that

$$\mathbb{A}_{\mathrm{i}} = \mathbb{P}\,.\,\mathbb{B}_{\mathrm{i}}\,.\,\mathbb{P}^{-1} \tag{32.9}$$

This says that $\mathcal{A}$ and $\mathcal{B}$ are similar. But by hypothesis $\mathcal{A}$ and $\mathcal{B}$ are not similar, so it must be true that $\mathbf{c} = \mathbf{0}$, since this is the only thing that can prevent the conclusion that they are similar. Therefore

$$\mathbb{P}\,.\,\mathbb{P}^{\dagger} = \mathrm{c}\,\mathbb{I} = 0 \tag{32.10}$$

We cannot conclude from this that $\mathbb{P}$ is a zero matrix, because it is easy to find two nonzero matrices that multiply to give a zero matrix. However, look at the subscripted form of the last equation. It is

$$\sum_{k} \mathbf{P}_{\mathrm{i,k}}\,\mathbf{P}^{\dagger}_{\mathrm{k,m}} = \mathbf{0} \tag{32.11}$$

and by the definition of the dagger

$$\sum_{k} \mathbf{P}_{\mathrm{i,k}}\,\mathbf{P}^{*}_{\mathrm{m,k}} = \mathbf{0} \tag{32.12}$$

The case $\mathbf{i} = \mathbf{m}$ tells us everything we want to know:

$$\sum_{k} \mathbf{P}_{\mathrm{i,k}}\,\mathbf{P}^{*}_{\mathrm{i,k}} = \sum_{k} |\mathbf{P}_{\mathrm{i,k}}|^{2} = \mathbf{0} \tag{32.13}$$

The summands are absolute squares and cannot be negative, so if the sum is zero, the summands must all be zero individually. Thus

$$\mathbf{P}_{\mathrm{i,k}} \overset{\mathrm{i,k}}{\equiv} \mathbf{0} \tag{32.14}$$

This is the end of the square case. We conclude that if $\mathcal{A}$ and $\mathcal{B}$ are of the same size, and irreducible, and not similar, then $\mathbb{A}_{\mathrm{R}}\,.\,\mathbb{P} = \mathbb{P}\,.\,\mathbb{B}_{\mathrm{R}}$ can be true for all $\mathbf{R}$ in the group only if $\mathbb{P}$ is composed entirely of zeroes.

32.3.3 Rectangular case

This case is now very easy. Take the equation with rectangular $\mathbb{P}$ and pad the smaller rep with rows and columns of zeroes until it is the same size as the larger rep. Now pad $\mathbb{P}$ appropriately, and you have, for example,

$$\begin{matrix} \mathbb{A}_i & . & \mathbb{P} & = & \mathbb{P} & . & \mathbb{B}_i \\ \begin{pmatrix} \square & \square & \square \\ \square & \square & \square \\ \square & \square & \square \end{pmatrix} & . & \begin{pmatrix} \square & \square & \mathbf{0} \\ \square & \square & \mathbf{0} \\ \square & \square & \mathbf{0} \end{pmatrix} & = & \begin{pmatrix} \square & \square & \mathbf{0} \\ \square & \square & \mathbf{0} \\ \square & \square & \mathbf{0} \end{pmatrix} & . & \begin{pmatrix} \square & \square & \mathbf{0} \\ \square & \square & \mathbf{0} \\ \mathbf{0} & \mathbf{0} & \mathbf{0} \end{pmatrix} \end{matrix} \tag{32.15}$$

This is now the square case, and by the reasoning above, the padded $\mathbb{P}$ must be all zeroes. But the rectangular $\mathbb{P}$ is just a subset of the square padded $\mathbb{P}$, so it is all zeroes too.

32.3.4 Coda

The two cases are obviously exhaustive (square or not square), so the theorem is proved.

32.4 A whiff of the great orthogonality

If the arbitrary matrix $\mathbb{Q}$ is the unit matrix, the corresponding pseudocommuter is the special case $\sum_{\mathbb{R}} \mathbb{A}_{\mathbb{R}}^{-1} . \mathbb{B}_{\mathbb{R}}$. This is a nice simple formula for investigating the cancellations that give those all-zero matrices.

Let's carry this sum out without an automated **Dot** operator, just as Schur might have done by hand. Group $\mathbf{D_6}$ will make a good choice because it has two 2-D irreducible reps :

```
BoxedCharacterTable["D6"]
```

D6	1	2	2	1	3	3	← Class populations
	E	C6	C3	C_2	C_2'	C_2''	↓ Basis functions
A1	1	1	1	1	1	1	$\{(x^2+y^2), (z^2)\}$
A2	1	1	1	1	-1	-1	$\{(z), (Iz)\}$
B1	1	-1	1	-1	1	-1	$\{(x^3-3xy^2)\}$
B2	1	-1	1	-1	-1	1	$\{(-3x^2y+y^3)\}$
E1	2	1	-1	-2	0	0	$\left\{\begin{pmatrix}x\\y\end{pmatrix}, \begin{pmatrix}xz\\yz\end{pmatrix}, \begin{pmatrix}Ix\\Iy\end{pmatrix}\right\}$
E2	2	-1	-1	2	0	0	$\left\{\begin{pmatrix}\frac{x^2}{2}-\frac{y^2}{2}\\xy\end{pmatrix}\right\}$

We take $\mathcal{A}$ to be rep $\mathbf{E_1}$ of group $\mathbf{D_6}$, and we take $\mathcal{B}$ to be rep $\mathbf{E_2}$. They are easy to construct, and we simultaneously make the inverse rep for $\mathcal{A}$:

```
𝒜 = MakeRepPoly[{x, y}, "D6"];
𝒜^-1 = Map[Inverse, 𝒜];
% // GridList // Size[6]
```

$$\left\{\begin{pmatrix}1&0\\0&1\end{pmatrix}, \begin{pmatrix}\frac{1}{2}&\frac{\sqrt{3}}{2}\\-\frac{\sqrt{3}}{2}&\frac{1}{2}\end{pmatrix}, \begin{pmatrix}\frac{1}{2}&-\frac{\sqrt{3}}{2}\\\frac{\sqrt{3}}{2}&\frac{1}{2}\end{pmatrix}, \begin{pmatrix}-\frac{1}{2}&\frac{\sqrt{3}}{2}\\-\frac{\sqrt{3}}{2}&-\frac{1}{2}\end{pmatrix},\right.$$

$$\begin{pmatrix}-\frac{1}{2}&-\frac{\sqrt{3}}{2}\\\frac{\sqrt{3}}{2}&-\frac{1}{2}\end{pmatrix}, \begin{pmatrix}-1&0\\0&-1\end{pmatrix}, \begin{pmatrix}1&0\\0&-1\end{pmatrix}, \begin{pmatrix}-\frac{1}{2}&\frac{\sqrt{3}}{2}\\\frac{\sqrt{3}}{2}&\frac{1}{2}\end{pmatrix},$$

$$\left.\begin{pmatrix}-\frac{1}{2}&-\frac{\sqrt{3}}{2}\\-\frac{\sqrt{3}}{2}&\frac{1}{2}\end{pmatrix}, \begin{pmatrix}\frac{1}{2}&\frac{\sqrt{3}}{2}\\\frac{\sqrt{3}}{2}&-\frac{1}{2}\end{pmatrix}, \begin{pmatrix}-1&0\\0&1\end{pmatrix}, \begin{pmatrix}\frac{1}{2}&-\frac{\sqrt{3}}{2}\\-\frac{\sqrt{3}}{2}&-\frac{1}{2}\end{pmatrix}\right\}$$

```
ℬ = MakeRepPoly[{x^2 - y^2, x y}, "D6"];
% // GridList // Size[6]
```

$$\left\{\begin{pmatrix}1&0\\0&1\end{pmatrix}, \begin{pmatrix}-\frac{1}{2}&-\frac{\sqrt{3}}{4}\\\sqrt{3}&-\frac{1}{2}\end{pmatrix}, \begin{pmatrix}-\frac{1}{2}&\frac{\sqrt{3}}{4}\\-\sqrt{3}&-\frac{1}{2}\end{pmatrix}, \begin{pmatrix}-\frac{1}{2}&\frac{\sqrt{3}}{4}\\-\sqrt{3}&-\frac{1}{2}\end{pmatrix},\right.$$

$$\begin{pmatrix}-\frac{1}{2}&-\frac{\sqrt{3}}{4}\\\sqrt{3}&-\frac{1}{2}\end{pmatrix}, \begin{pmatrix}1&0\\0&1\end{pmatrix}, \begin{pmatrix}1&0\\0&-1\end{pmatrix}, \begin{pmatrix}-\frac{1}{2}&-\frac{\sqrt{3}}{4}\\-\sqrt{3}&\frac{1}{2}\end{pmatrix},$$

$$\left.\begin{pmatrix}-\frac{1}{2}&\frac{\sqrt{3}}{4}\\\sqrt{3}&\frac{1}{2}\end{pmatrix}, \begin{pmatrix}-\frac{1}{2}&\frac{\sqrt{3}}{4}\\\sqrt{3}&\frac{1}{2}\end{pmatrix}, \begin{pmatrix}1&0\\0&-1\end{pmatrix}, \begin{pmatrix}-\frac{1}{2}&-\frac{\sqrt{3}}{4}\\-\sqrt{3}&\frac{1}{2}\end{pmatrix}\right\}$$

The Second Lemma says we should get all zeroes if we sum over both **R** and **k** :

```
Table[Σ_{R=1}^{12} Σ_{k=1}^{2} 𝒜^-1[[R, m, k]] ℬ[[R, k, i]], {i, 2}, {m, 2}]
```

```
{{0, 0}, {0, 0}}
```

and we do. But just out of curiosity, let us delay the sum over **k** so that we can see the two matrices that add together to give the zero matrix. For the summand with **k = 1** :

```
k = 1;
Table[Σ_{R=1}^{12} 𝒜^-1[[R, m, k]] ℬ[[R, k, i]], {i, 2}, {m, 2}]
```

```
{{0, 0}, {0, 0}}
```

For the summand with **k = 2** :

```
k = 2;
Table[Σ_{r=1}^{12} 𝒜^-1[[r, m, k]] ℬ[[r, k, i]], {i, 2}, {m, 2}]
```

```
{{0, 0}, {0, 0}}
```

```
Clear[k]
```

Very surprising! The sum over **k** has nothing to do with the vanishing. If you sum over the group first, the matrices that you later add together are already pure zeroes.

If you see something like this once, you will probably say it is a coincidence. You were looking for a nontrivial sum like

$$\begin{pmatrix} a & 0 \\ 0 & b \end{pmatrix} + \begin{pmatrix} -a & 0 \\ 0 & -b \end{pmatrix}$$

```
{{0, 0}, {0, 0}}
```

but maybe you were cheated out of it by the simplicity of group $\mathbf{D_3}$.

On your own
Do more examples, looking for those elusive nonzero matrices that add together to give all zeroes. The 3-D reps $\mathbf{T_1}$ and $\mathbf{T_2}$ of $\mathbf{T_d}$ would be good.

Finally, you may begin to sense the dim outlines of a Great Orthogonality, and give up this fruitless search.

33. The Great Orthogonality

Preliminaries

33.1 Skewers

We define a new term which helps greatly in stating and in understanding the Great Orthogonality. Consider the irreducible representations of some group. In the figure below, we take the octahedral group **O**, with reps named $\mathbf{A_1}$, $\mathbf{A_2}$, **E**, $\mathbf{T_1}$, and $\mathbf{T_2}$. Each representation is a **List** of **h** matrices, where **h** is the group order. In each rep, think of the **h** matrices as stacked up like a deck of cards. (For the octahedral group, **h** is **24**, but we show only the first **6** below.) For the two **T** reps the matrices are 3×3; for **E**, 2×2, for the two **A** reps, 1×1. Now thrust a skewer through the deck at the **i,j** position, making a shishkebab of all the **i,j** elements in the stack of matrices. This list we call the "skewer of the matrix list at the **i,j** position".

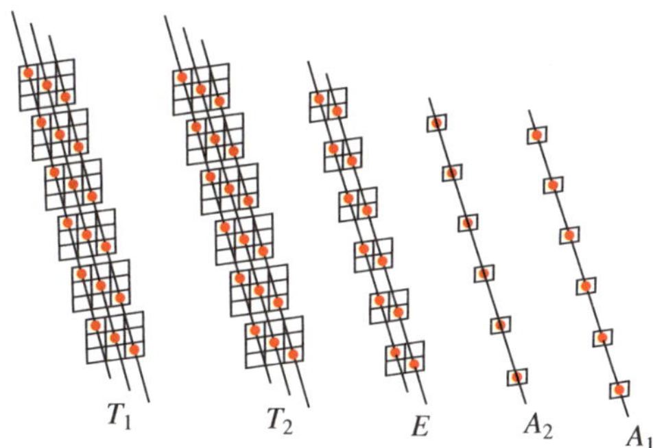

Fig.33.1 For visual clarity, we show only the first six matrices and only the diagonal skewers. But a skewer goes through every element position of every matrix. The octahedral group O has 24 skewers (count 'em), each one being a 24D vector.

W.M. McClain, *Symmetry Theory in Molecular Physics with Mathematica*, DOI 10.1007/b13137_33, © Springer Science+Business Media, LLC 2009

The skewer is therefore just a vector (sometimes complex), and the number of elements in a skewer vector is equal to the order of the group. In other words, if there are **h** elements in the group, the skewers are vectors in an **h**-dimensional space.

33.2 Two statements of the Great Orthogonality

33.2.1 In skewer language

The Great Orthogonality

1. All the skewers of all the irreducible representations of a group are orthogonal.

2. The scalar product of each skewer with itself is **h/d**, where **h** is the order of the group and **d** is the dimension of the skewer's representation.

1. The phrase " all the irreducible representations of a group" must be clarified. There are infinitely many similarity transforms of any representation, and obviously we not counting them separately. Perhaps the word " nonsimilar" or " inequivalent" should be inserted somewhere in the statement, but usually it is left unsaid.

2. If the skewers are complex, orthogonality is understood in the complex sense, $\underline{\vec{u}} \cdot (\underline{\vec{v}}^*) = 0$.

3. Orthogonality of skewers holds both within irreducible representations and between different irreducible representations.

Important comment

The Great Orthogonality sets a limit on the number of irreducible representations that can exist. In an **h** dimensional space, it is impossible to have more than **h** linearly independent orthogonal vectors (or skewers). This limits the number of independent representations that can exist. We turn this into a quantitative statement :

All representations are square , so if species number **s** is of size $\mathbf{d_s} \times \mathbf{d_s}$ the number of skewers it possesses is $\mathbf{d_s^2}$. But because of their orthogonality in an **h**-dimensional space, the total number of such skewers cannot be more than **h**. Therefore we have the inequality

$$\sum_{s=1}^{Ns} d_s^2 \le h$$

where the sum limit **Ns** is the number of species. Previously, one might have thought that **Ns** might be very large, or infinite, and that our failure to find more than a few irreducible reps merely meant that the missing ones were very complicated or very large. But now, this thought is forbidden.

Later, we will show that the equality actually does hold in all cases.

33.2.2 In detailed subscript language

Here is a more traditional statement of the Great Orthogonality :
Let every element of every irreducible representation be symbolized by a **Γ** with four indices. There cannot be less that four, because we must specify which deck of cards (a lower case label **i** for *irreducible representation*), which card in the deck (an upper case Greek label (like **Λ**)for element number), and which position on the card (two lower case Greek indices like **α, β**). So each individual matrix element in the complete set of irreducible representations has its own $\Gamma^{(i)}_{\mu,\nu}[\Lambda]$. We could also attach a capital Script index $\mathcal{G}$ to indicate the group, but there is not any room left, so we suppress it.

To state the Great Orthogonality we need only these **Γ** symbols, plus **h** (the order of the group, and **d[i]**, the dimension of each representation. The statement is fairly compact :

The Great Orthogonality

$$\sum_{\Lambda=1}^{h} \Gamma^{(i)}_{\mu,\nu}[\Lambda] \left(\Gamma^{(j)}_{\alpha,\beta}[\Lambda]\right)^* = \frac{h}{d[i]} \delta_{i,j}\, \delta_{\mu,\alpha}\, \delta_{\nu,\beta} \tag{33.1}$$

The δ symbols are Kronecker deltas; **0** when the subscripts are not equal and **1** when they are. In other words, they are unit matrices. This is not an abstraction. It is real. Set **nKD** to any dimension you like :

```
nKD = 3;
Table[δi,j, {i, nKD}, {j, nKD}];
% // MatrixForm
Null
```

$$\begin{pmatrix} 1 & 0 & 0 \\ 0 & 1 & 0 \\ 0 & 0 & 1 \end{pmatrix}$$

The Great Orthogonality will not be hard to prove now that we have Schur's

Lemmas. But first, to make friends with this monster, we do two numerical examples.

33.3 Two Great Orthogonality demos

33.3.1 With all reps one-dimensional

First we do an example that has only 1-D reps, for simplicity. But we choose one that has complex character vectors :

```
grp = "C6";
BoxedCharacterTable[grp]
```

C6	1	1	1	1	1	1	← Class populations Note : ϵ = Exp[- 2πi / 3]
	E	C6	C62	C63	C64	C65	↓ Basis functions
A	1	1	1	1	1	1	{(z), (x^2+y^2), (z^2), (Iz)}
B	1	-1	1	-1	1	-1	{($(x \pm i\ y)^3$)}
E1a	1	-ϵ	ϵ*	-1	ϵ	-ϵ*	{(x - i y)}
E1b	1	-ϵ*	ϵ	-1	ϵ*	-ϵ	{(x + i y)}
E2a	1	ϵ*	ϵ	1	ϵ*	ϵ	{($(x - i\ y)^2$)}
E2b	1	ϵ	ϵ*	1	ϵ	ϵ*	{($(x + i\ y)^2$)}

Usually, you must use **Extend** on the character vectors to make them into skewers. But here all classes contain only a single element, so the skewers are identical to the character vectors, and you can read them right from the table. Note that ϵ = **Exp[2π i/6]** $=\frac{1}{2}+\frac{i\sqrt{3}}{2}$.

Make a multiplication table of all the complex **Dot** products of the skewers. It uses **Conjugate** on the second member of each **Dot**, without which not :

	A	B	E1a	E1b	E2a	E2b
A	6	0	0	0	0	0
B	0	6	0	0	0	0
E1a	0	0	6	0	0	0
E1b	0	0	0	6	0	0
E2a	0	0	0	0	6	0
E2b	0	0	0	0	0	6

Fig. 33.2 The diagonal elements should be group order (6) divided by species dimension (always 1 in this group). Indeed, they are.

The Great Orthogonality is verified for group $\mathbf{C_6}$.

33.3.2 With a two-dimensional rep

The real power of the Great Orthogonality is in groups with multidimensional representations. Group $\mathbf{C_{3\,v}}$ is one of the simplest :

```
gp = "C3v";
BoxedCharacterTable[gp ]
```

C3v	1	2	3	← Class populations
	E	C3	σv	↓ Basis functions
A1	1	1	1	$\{(z), (x^2+y^2), (z^2)\}$
A2	1	1	-1	$\{(Iz)\}$
E	2	-1	0	$\left\{\begin{pmatrix} x \\ y \end{pmatrix}, \begin{pmatrix} \frac{1}{2}(x^2-y^2) \\ xy \end{pmatrix}, \begin{pmatrix} xz \\ yz \end{pmatrix}, \begin{pmatrix} Ix \\ Iy \end{pmatrix}\right\}$

It will have six skewers: One for $\mathbf{A_1}$, one for $\mathbf{A_2}$, and four for **E**. But class **C3** has two members and class **σv** has three members, so to turn the character vectors into full skewers, the second character element must be repeated twice; the third, three times. This is done by the `Extend` operator :

```
skewerA1 = ChVec["A1", gp] // Extend[#, gp] &
{1, 1, 1, 1, 1, 1}
```

```
skewerA2 = ChVec["A2", gp] // Extend[#, gp] &
```

```
{1, 1, 1, -1, -1, -1}
```

To make the four skewers of rep **E**, we must first make a representation. The character table says basis polynomials **{x,y}** should do the job:

```
repExy = MakeRepPoly[{x, y}, "C3v"];
% // GridList // Size[6]
```

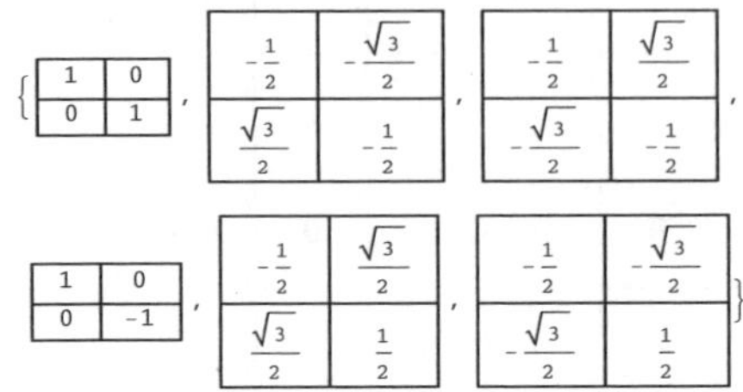

It never hurts to make sure:

```
MorphTest[repExy, "C3v"]
```

```
Faithful, or Isomorphic
```

We make all four **E** skewers at once, then Join them to the A_1 and A_2 skewers:

```
skewersE =
  Table[Skewer[repExy, i, j], {i, 1, 2}, {j, 1, 2}];
allSkewersC3v =
  Flatten[{skewerA1, skewerA2, skewersE}] //
   Partition[#, 6] &;
% // GridForm
```

1	1	1	1	1	1
1	1	1	-1	-1	-1
1	$-\frac{1}{2}$	$-\frac{1}{2}$	1	$-\frac{1}{2}$	$-\frac{1}{2}$
0	$-\frac{\sqrt{3}}{2}$	$\frac{\sqrt{3}}{2}$	0	$\frac{\sqrt{3}}{2}$	$-\frac{\sqrt{3}}{2}$
0	$\frac{\sqrt{3}}{2}$	$-\frac{\sqrt{3}}{2}$	0	$\frac{\sqrt{3}}{2}$	$-\frac{\sqrt{3}}{2}$
1	$-\frac{1}{2}$	$-\frac{1}{2}$	-1	$\frac{1}{2}$	$\frac{1}{2}$

Make a vector multiplication table. We got a little carried away and colored it:

	A_1	A_2	$E_{1,1}$	$E_{1,2}$	$E_{2,1}$	$E_{2,2}$
A_1	6	0	0	0	0	0
A_2	0	6	0	0	0	0
$E_{1,1}$	0	0	3	0	0	0
$E_{1,2}$	0	0	0	3	0	0
$E_{2,1}$	0	0	0	0	3	0
$E_{2,2}$	0	0	0	0	0	3

Fig. 33.3 Green indicates the zeroes that come from a mismatch in species indices, while blue marks the zeroes that come from a mismatch of matrix indices. The diagonal elements should be group order (6) divided by species dimension (1 or 2). Indeed, they are.

The Great Orthogonality is verified for group C_{3v}.

The Great Orthogonality is not an ancient truth, handed down from time immemorial. It is a jewel in the crown of modern Western Civilization, your own heritage. So buckle down and follow the general proof. It is not that hard.

33.4 Proof of the Great Orthogonality

33.4.1 How Schur's Lemmas help

The proof is remarkably easy, after the results of the Chapter 30 ((SchursFirstLemma) and Chapter 31 (SchursSecondLemma). The proof is divided into two parts; the first uses Schur's First; the second, Schur's Second. The first says certain universal commuters can only be constant matrices; the second says that certain others can only be zero. Both contain the famous arbitrary $\mathbb{Q}$ matrix, and a clever choice of $\mathbb{Q}$ in both cases leads directly to the Great Orthogonality.

We assume in the proof below that the representations of the group are unitary. But the theorem of Chapter 29 (MakeUnitary) says that every rep is similar to a unitary rep. If your rep of interest is nonunitary, run the package operator **`MakeUnitary`** on it, and in that guise it will behave as required below.

33.4.2 Orthogonality within reps

Let $\mathcal{A}$ be an irreducible rep. There is always a universal commuter matrix $\mathbb{C}$ for this rep, and Schur's First Lemma says that it can only be $\mathbf{c}\,\mathbb{U}$, a constant matrix. Therefore

$$\mathbf{c}\,\mathbb{U} = \sum_{\Lambda} \mathbb{A}_{\Lambda} . \mathbb{Q} . \mathbb{A}_{\Lambda}^{-1} \tag{33.2}$$

or in index notation

$$\mathbf{c}\,\delta_{\mu,\alpha} = \sum_{\Lambda,\mathbf{k},\mathbf{m}} (\mathbf{A}_{\Lambda})_{\mu,\mathbf{k}}\, \mathbb{Q}_{\mathbf{k},\mathbf{m}} \left(\mathbf{A}_{\Lambda}^{-1}\right)_{\mathbf{m},\alpha} \tag{33.3}$$

Choose $\mathbb{Q}$ to be all zeroes except for a **1** in some one element. If it is the **2,3** element we would write it as $\delta_{\mathbf{k},2}\,\delta_{\mathbf{m},3}$ because

```
Table[δk,2 δm,3, {k, 3}, {m, 3}] // MatrixForm
```

$$\begin{pmatrix} 0 & 0 & 0 \\ 0 & 0 & 1 \\ 0 & 0 & 0 \end{pmatrix}$$

More generally, let the position of the unique **1** be chosen as ν, β :

$$\mathbf{c}\,\delta_{\mu,\alpha} = \sum_{\Lambda,\mathbf{k},\mathbf{m}} (\mathbf{A}_{\Lambda})_{\mu,\mathbf{k}}\, \delta_{\mathbf{k},\nu}\, \delta_{\mathbf{m},\beta} \left(\mathbf{A}_{\Lambda}^{-1}\right)_{\mathbf{m},\alpha} \tag{33.4}$$

The δ functions allow us to perform the sums over **k** and **m** symbolically:

$$\mathbf{c}\,\delta_{\mu,\alpha} = \sum_{\Lambda} \left(\mathbf{A}_{\Lambda}\right)_{\mu,\nu} \left(\mathbf{A}_{\Lambda}^{-1}\right)_{\beta,\alpha} \tag{33.5}$$

Symbols μ and α appear on both sides, so they are completely free. Let $\mu \to \alpha$, and sum on α from **1** to **d**, the dimension of the rep

$$\sum_{\alpha=1}^{\mathbf{d}} \mathbf{c}\,\delta_{\alpha,\alpha} = \sum_{\alpha=1}^{\mathbf{d}} \sum_{\Lambda} \left(\mathbf{A}_{\Lambda}\right)_{\alpha,\nu} \left(\mathbf{A}_{\Lambda}^{-1}\right)_{\beta,\alpha} \tag{33.6}$$

On the left, $\delta_{\alpha,\alpha}$ is always **1**, and the sum then gives $\mathbf{c\,d}$. On the right we swap the two **A** factors to put their α indices adjacent (because a sum over adjacent repeated indices is a dot product). Also we swap the order of summation.

$$\mathbf{c\,d} = \sum_{\Lambda} \left(\sum_{\alpha=1}^{\mathbf{d}} \left(\mathbf{A}_{\Lambda}^{-1}\right)_{\beta,\alpha} \left(\mathbf{A}_{\Lambda}\right)_{\alpha,\nu} \right) \tag{33.7}$$

On the right the parenthesis is the dot product of mutually inverse matrices, so it gives the unit matrix δ_{β_0,ν_0} :

$$\mathbf{c\,d} = \sum_\Lambda \delta_{\beta,\nu}$$

The sum over Λ is just $\mathbf{h}$ repeats of the unit matrix $\delta_{\beta,\nu}$, so

$$\mathbf{c} = (\mathbf{h}/\mathbf{d})\ \delta_{\beta,\nu} \tag{33.9}$$

Now go back to Eq. 33.5, insert this expression for $\mathbf{c}$

$$(\mathbf{h}/\mathbf{d})\ \delta_{\beta,\nu_0}\ \delta_{\mu,\alpha} = \sum_\Lambda (\mathbf{A}_\Lambda)_{\mu,\nu} \left(\mathbf{A}_\Lambda^{-1}\right)_{\beta,\alpha} \tag{33.10}$$

and use the unitarity of $\mathbb{A}_\Lambda$:

$$(\mathbf{h}/\mathbf{d})\ \delta_{\beta,\nu}\ \delta_{\mu,\alpha} = \sum_\Lambda (\mathbf{A}_\Lambda)_{\mu,\nu} \left(\mathbf{A}_\Lambda^{T*}\right)_{\beta,\alpha} \tag{33.11}$$

or finally

$$(\mathbf{h}/\mathbf{d})\ \delta_{\mu,\alpha}\,\delta_{\nu,\beta} = \sum_\Lambda (\mathbf{A}_\Lambda)_{\mu,\nu}\ (\mathbf{A}_\Lambda^{*})_{\alpha,\beta} \tag{33.12}$$

On the right we have the $\mathbf{h}$-dimensional complex dot product of two skewers from within rep $\mathcal{A}$; the left side says that dot product is zero unless $\mu = \alpha$ and $\nu = \beta$; i.e., unless it is a skewer dotted with itself.

We also get the length squared of each skewer: it is $\mathbf{h}/\mathbf{d}$. So the first part of the Great Orthogonality is proved.

We have written this for a rep called $\mathcal{A}$. But $\mathcal{A}$ is some rep number $\mathbf{i}$, so to include the $\mathbf{i}$ in the notation, we let $\mathbf{A}_\Lambda \to \Gamma^{(i)}[\Lambda]$ and $\mathbf{A}_\Lambda^* \to \left(\Gamma^{(i)}[\Lambda]\right)^*$:

$$\sum_{\Lambda=1}^{\mathbf{h}} \Gamma_{\mu,\nu}^{(i)}[\Lambda] \left(\Gamma_{\alpha,\beta}^{(i)}[\Lambda]\right)^* = \frac{\mathbf{h}}{\mathbf{d[i]}}\ \delta_{\mu,\alpha}\,\delta_{\nu,\beta} \tag{33.13}$$

33.4.3 Orthogonality between reps

Now for the second part. Let reps $\mathcal{A}$ and $\mathcal{B}$ be inequivalent irreducible reps. There is always a pseudocommuter matrix $\mathbb{P}$ for these reps, but because they are inequivalent and irreducible, Schur's Second Lemma says that the pseudocommuter can only be zero. Thus for any matrix $\mathbb{Q}$ we have the equation

$$\sum_{\Lambda=1}^{\mathbf{h}} \mathbb{A}_\Lambda . \mathbb{Q} . \mathbb{B}_\Lambda^{-1} = 0 \tag{33.14}$$

or with explicit sums instead of `Dot` products

$$\sum_{\Lambda,\xi,\eta} (\mathbf{A}_\Lambda)_{\mu,\xi}\, \mathbf{Q}_{\xi,\eta} \left(\mathbf{B}_\Lambda^{-1}\right)_{\eta,\alpha} = 0 \tag{33.15}$$

Just as in the first part, we are free to pick $\mathbb{Q}$ to be all zeroes except for the ν,β element, which is $\mathbf{1}$. This means that $Q_{\xi,\eta} = \delta_{\xi,\nu}\,\delta_{\eta,\beta}$. Also, by unitarity, $\mathbf{B}_\Lambda^{-1} = \mathbf{B}_\Lambda^{\mathbf{T}*}$. Therefore

$$\sum_{\Lambda,\xi,\eta} (\mathbf{A}_\Lambda)_{\mu,\xi}\,\delta_{\xi,\nu}\,\delta_{\eta,\beta}\left(\mathbf{B}_\Lambda^{\mathbf{T}*}\right)_{\eta,\alpha} = 0 \tag{33.16}$$

or carrying out the sum over ξ and η and transposing $\mathbf{B}_\Lambda^{\mathbf{T}*}$

$$\sum_{\Lambda} (\mathbf{A}_\Lambda)_{\mu,\nu}\,(\mathbf{B}_\Lambda^*)_{\alpha,\beta} = 0 \tag{33.17}$$

Indices μ,ν and α,β can be whatever you like; for as long as $\mathcal{A}$ and $\mathcal{B}$ are inequivalent and irreducible, the value is that zero on the right. Thus all skewers from different irreducible reps are orthogonal. If we let $\mathbf{A}_\Lambda \to \Gamma^{(i)}[\Lambda]$ and $\mathbf{B}_\Lambda^* \to \left(\Gamma^{(j)}[\Lambda]\right)^*$ we have the second part of the Great Orthogonality in standard notation :

$$\sum_{\Lambda=1}^{h} \Gamma^{(i)}_{\mu,\nu}[\Lambda]\left(\Gamma^{(j)}_{\alpha,\beta}[\Lambda]\right)^* = 0, \quad \text{for } i \neq j \tag{33.18}$$

33.4.4 Grand conclusion

We can write the first part and the second part as a single equation if we insert $\delta_{i,j}$ on the right side of Eq. 33.13, the first part conclusion. If $\mathbf{i}\quad\mathbf{j}$, you have the second part case, and the $\delta_{i,j} = 0$ rules. If $\mathbf{i} = \mathbf{j}$, you have the first part case, and the other two δ functions rule. So we have in both cases

$$\sum_{\Lambda=1}^{h} \Gamma^{(i)}_{\mu,\nu}[\Lambda]\left(\Gamma^{(j)}_{\alpha,\beta}[\Lambda]\right)^* = \frac{h}{d[i]}\,\delta_{i,j}\,\delta_{\mu,\alpha}\,\delta_{\nu,\beta} \tag{33.19}$$

which was to be shown.

34. Character orthogonalities

Preliminaries

34.1 Introduction and demonstrations

34.1.1 Introduction

The character orthogonalities are special cases of the Great Orthogonality. They can be seen directly in all character tables. This is also the chapter where we prove that the character table is square, and that it has an inverse. So perk up.

The symbol $\mathbb{X}$ is a DoubleStruckCapitalX. We will write $\mathbb{X}$ when we mean a character matrix without explicit indices, or $\chi_{\mathtt{species},\mathtt{CLASS}}$ when indices are explicit. The **species** index (or irreducible representation index) will be a lower case letter, the **CLASS** index will be upper case. Thus character tables are symbolized as $\mathbb{X}$ or $\chi_{\mathtt{s},\mathtt{C}}$. Or, transposed and conjugated, as $\mathbb{X}^\dagger$ or $\chi^*_{\mathtt{C},\mathtt{s}}$.

Here is a fine point of the notation in this chapter. We will never write both a dagger and explicit indices on the same anchor symbol. If $\mathbb{X} = \chi_{\mathtt{s},\mathtt{C}}$ then its conjugate-transpose (or adjoint) is $\mathbb{X}^\dagger = \chi^*_{\mathtt{C},\mathtt{s}}$. Since the dagger already implies transpose, $\chi^\dagger_{\mathtt{C},\mathtt{s}}$ would have to mean $\chi^*_{\mathtt{s},\mathtt{C}}$. That is dangerously subject to misunderstanding, so we stay away from it by simply making a rule that we never put a dagger and explicit indices on the same anchor.

We will also need symbols $\overleftrightarrow{\mathbb{X}}$ or $\chi_{\mathtt{s},\Lambda}$ (and their adjoints $\overleftrightarrow{\mathbb{X}}^\dagger$ or $\chi^*_{\Lambda,\mathtt{s}}$) for the *extended* characters, where a capital Greek index (like Λ) runs over all the elements of the group. The double-headed overarrow is the reminder that when $\overleftrightarrow{\mathbb{X}}$ is translated to subscript notation, its second index must be capital Greek. So $\overleftrightarrow{\mathbb{X}}$ is generally a wide rectangular matrix; its adjoint $\overleftrightarrow{\mathbb{X}}^\dagger$ is tall.

There are two important orthogonalities that exist within the character table of every group. We introduce them by numerical example.

W.M. McClain, *Symmetry Theory in Molecular Physics with Mathematica*,
DOI 10.1007/b13137_34, © Springer Science+Business Media, LLC 2009

34.1.2 Column orthogonality demonstration

Set up to work with group **T**, all the rotations of a tetrahedron:

```
grp = "T";
```

$$\text{to}\epsilon = \left\{-\frac{1}{2}-\frac{\mathrm{i}\sqrt{3}}{2} \to \epsilon,\ -\frac{1}{2}+\frac{\mathrm{i}\sqrt{3}}{2} \to \epsilon^*\right\};$$

Call up its character vectors :

```
X = ChVecs[grp];
% /. toϵ // GridForm
```

1	1	1	1
1	ϵ	ϵ^*	1
1	ϵ^*	ϵ	1
3	0	0	-1

The character vectors are the rows of this matrix. We will need its **Adjoint** :

```
X† = X // Transpose // Conjugate;
% /. toϵ // GridForm
```

1	1	1	3
1	ϵ^*	ϵ	0
1	ϵ	ϵ^*	0
1	1	1	-1

We **Dot** each column of **X** with every other column by putting the **Adjoint** on the left of the **Dot** product :

```
X†.X // Simplify // GridForm
```

12	0	0	0
0	3	0	0
0	0	3	0
0	0	0	4

The column orthogonality has been demonstrated.

On your own

1. Call up the **BoxedCharacterTable** to see that **X** is the "center of the table".

2. Change **grp** and rerun these orthogonalities. Use a group with complex-valued characters ($\mathbf{S_4}$ or $\mathbf{S_8}$, for example).

34.1.3 Row orthogonality demonstration

The character vectors have one element for every *class* in the group. The whole table contains one vector for every *species* in the group. Now we define the *extended* character vector which has one element for every *element* in the group. They make a rectangular matrix (# species) × (# elements) :

```
X⃡ = AllExtendedChVecs[grp];
% /. toϵ // GridForm // Size[7]
```

1	1	1	1	1	1	1	1	1	1	1	1
1	ϵ	ϵ	ϵ	ϵ	ϵ*	ϵ*	ϵ*	ϵ*	1	1	1
1	ϵ*	ϵ*	ϵ*	ϵ*	ϵ	ϵ	ϵ	ϵ	1	1	1
3	0	0	0	0	0	0	0	0	-1	-1	-1

Note the repeating columns: the class populations are **{1,4,4,3}**.

```
X⃡† = X⃡ // Transpose // Conjugate;
% /. toϵ // GridForm // Size[7]
```

1	1	1	3
1	ϵ*	ϵ	0
1	ϵ*	ϵ	0
1	ϵ*	ϵ	0
1	ϵ*	ϵ	0
1	ϵ	ϵ*	0
1	ϵ	ϵ*	0
1	ϵ	ϵ*	0
1	ϵ	ϵ*	0
1	1	1	-1
1	1	1	-1
1	1	1	-1

```
X⃡.X⃡† // Expand;
% // GridForm // Size[7]
```

12	0	0	0
0	12	0	0
0	0	12	0
0	0	0	12

The row orthogonality has been demonstrated. Are you surprised that two rectangular matrices can **Dot** to give a unit matrix ? (Well, a multiple thereof)

Two rectangular matrices of size **long** × **short** and **short** × **long** are said to be *pseudoinverse* if their **short** × **short** product is a unit matrix. There is a discussion of this, and of the core operator **PseudoInverse**, in the MatrixReview. Unlike true inverses, pseudoinverses do not give a unit matrix in the other order. Try it :

```
X↔† . X↔ // Expand // GridForm // Size[7]
```

12	0	0	0	0	0	0	0	0	0	0	0
0	3	3	3	3	0	0	0	0	0	0	0
0	3	3	3	3	0	0	0	0	0	0	0
0	3	3	3	3	0	0	0	0	0	0	0
0	3	3	3	3	0	0	0	0	0	0	0
0	0	0	0	0	3	3	3	3	0	0	0
0	0	0	0	0	3	3	3	3	0	0	0
0	0	0	0	0	3	3	3	3	0	0	0
0	0	0	0	0	3	3	3	3	0	0	0
0	0	0	0	0	0	0	0	0	4	4	4
0	0	0	0	0	0	0	0	0	4	4	4
0	0	0	0	0	0	0	0	0	4	4	4

It is certainly not a unit matrix. Here it gave a block diagonal matrix, but the MatrixReview shows that in random numerical cases, it just gives a mess.

On your own
Go back to the top of this section, reset **grp** to some other group name. If it is complex, you may have to change the definition of ϵ and ϵ^* also. Then evaluate down to here. Do this for several groups.

34.2 Row orthogonality proof

Now that you have seen numerical examples of the character orthogonalities, you must be eager to see how they relate to the Great Orthogonality, which is

$$\sum_{\Lambda=1}^{h} \Gamma^{(i)}_{\mu,\nu}[\Lambda] \left(\Gamma^{(j)}_{\alpha,\beta}[\Lambda]\right)^* = \frac{h}{d[i]} \delta_{i,j}\, \delta_{\mu,\alpha}\, \delta_{\nu,\beta} \qquad (33.19)$$

We take the diagonal case by letting $\mu \to \nu$ and $\alpha \to \beta$. On the left, this specifies the diagonal elements of the Γ matrices. On the right, it causes $\delta_{\mu,\alpha}\,\delta_{\nu,\beta} \to \delta_{\nu,\beta}\,\delta_{\nu,\beta} = \delta_{\nu,\beta}$. Then sum both sides on ν and β from **1** to the dimension of the representation. This gives

$$\sum_{\Lambda=1}^{h} \left(\sum_{\nu=1}^{d[i]} \Gamma^{(i)}_{\nu,\nu}[\Lambda]\right) \left(\sum_{\beta=1}^{d[j]} \Gamma^{(j)}_{\beta,\beta}[\Lambda]\right)^* = \frac{h}{d[i]} \delta_{i,j} \sum_{\nu=1}^{d[i]} \sum_{\beta=1}^{d[j]} \delta_{\nu,\beta}$$

On the left each inner sum produces the character of the representation matrix, $\chi_{i,\Lambda}$ and $(\chi_{j,\Lambda})^*$, respectively. On the right every summand vanishes except when $\nu = \beta$, so the sum over β gives **1**. Then summing that **1** over ν from **1** to **d[i]** gives **d[i]**, which cancels with the **d[i]** in the denominator of the prefactor, leaving

The long row orthogonality

$$\sum_{\Lambda=1}^{h} \chi_{i,\Lambda} (\chi_{j,\Lambda})^* = h\, \delta_{i,j} \quad (34.1)$$

i and **j** index species; Λ indexes *elements*; **h** = group order

Now $(\chi_{j,\Lambda})$ can also be written as $(\chi_{\Lambda,j})^T$ (because the transpose **T** plus the explicit index reversal undo each other). This makes $\chi_{i,\Lambda} (\chi_{j,\Lambda})^*$ into $\chi_{i,\Lambda} (\chi_{\Lambda,j})^{T*}$ which has the summed indices Λ adjacent.

In $\chi_{i,\Lambda}$ index **i** runs over the species of the group (**NS** of them), while Λ runs over all elements of the group (**NE** of them). So $\chi_{i,\Lambda}$ is rectangular, of size **NS** $\times$ **NE**, We let it be matrix $\overleftrightarrow{\mathbb{X}}$ in subscript-free notation. Its adjoint $\overleftrightarrow{\mathbb{X}}^{\dagger}$ is therefore of size **NE** $\times$ **NS**, and the dot product **(NS** $\times$ **NE).(NE** $\times$ **NS)** is a matrix of size **NS** $\times$ **NS**. So the long row orthogonality can be written very neatly as

The long row orthogonality as a Dot product

$$\overleftrightarrow{\mathbb{X}} \,.\, \overleftrightarrow{\mathbb{X}}^{\dagger} = h\, \mathbb{U}_{NS\times NS} \quad (34.2)$$

where $\mathbb{U}_{NS\times NS}$ is a unit matrix of size **NS** $\times$ **NS**.

There is a lot of redundancy in the long character vector $\chi_{i,\Lambda}$ because its value for every element in a given class is the same. Therefore as Λ runs over a class, all the summands $\chi_{i,\Lambda}\, \chi^*_{j,\Lambda}$ in that class are identical, and may be denoted by $\chi_{i,c}\, \chi^*_{j,c}$, where **c** is a class index. If the population of the class is P_c, then the contribution of the whole class is $P_c\, \chi_{i,c}\, \chi^*_{j,c}$, and we may rewrite the sum over elements Λ as a sum over classes **c**

The short, weighted row orthogonality

$$\sum_{C=1}^{NC} P_C \left(\chi_{i,c}\, \chi^*_{j,c}\right) = h\, \delta_{i,j} \tag{34.3}$$

i and j index species; c indexes *classes*; NC = number of classes

To rewrite this symmetrically, divide both sides by h and let $\left(\sqrt{\frac{P_C}{h}}\ \chi_{i,c}\right) \to$ $\mathbb{Y}$ (size $NS \times NC$) and let $\left(\sqrt{\frac{P_C}{h}}\ \chi^{T*}_{c,j}\right) \to \mathbb{Y}^\dagger$(size $NC \times NS$) . The number of columns of $\mathbb{Y}$ is the same as the number of rows of $\mathbb{Y}^\dagger$, so the sum over them can be written as a dot product. Letting $\delta_{i,j}$ be the unit matrix $\mathbb{U}$ of size $NS \times NS$ we can write

The short row orthogonality as a Dot product

$$\mathbb{Y}.\mathbb{Y}^\dagger = \mathbb{U}_{NS\times NS} \tag{34.4}$$

We will prove that $\mathbb{Y}$ is square, but we did not have to assume it to reach this conclusion.

34.3 Column orthogonality proof

The proof of column orthogonality cannot assume that the character table, viewed as a matrix $\mathbb{X}$, is square. It is true, but we have not proved it yet. It is not hard to find books with bogus "proofs" of column orthogonality that assume the square shape without proving it. There is another class of bogus proofs in which $\mathbb{X}$ is assumed to have an inverse without proving it. Here is a real proof that does not assume that $\mathbb{X}$ is square, nor that it has an inverse. It is necessarily a bit lengthy, so get ready.

Let $\chi_{s,c}$ be the character table, where s labels the rows of irreducible representations (also called symmetry species) and c labels the classes. The rows of the table (known by the previous theorem to be orthogonal) span a vector space, in which the general vector, constructed with arbitrary coefficients a_s, is

$$\underset{\rightharpoondown}{\mathbf{g}} = \mathbf{g}_C = \sum_{s=1}^{NS} \mathbf{a}_s \, \chi_{s,C}$$

The strategy of this proof is a little unusual, and you should be forewarned about it. We are going to construct a matrix $\mathbf{M}$ and show that $\underset{\rightharpoondown}{\mathbf{g}}.\mathbf{M} \equiv \underset{\rightharpoondown}{\mathbf{g}}$, no matter what $\underset{\rightharpoondown}{\mathbf{g}}$ is. From this we will conclude that $\mathbf{M}$ must be the unit matrix. The column orthogonality is then written as $\mathbf{M}_{C,D} == \delta_{C,D}$.

The species index $\mathbf{s}$ runs from $\mathbf{1}$ to $\mathbf{NS}$, the (unknown) number of species. Now make a scalar product of this arbitrary vector with the adjoint of the character table. We write it as a sum over classes, so the population factor $\mathbf{P_C}$must be included.

$$\sum_{C=1}^{NC} \mathbf{g}_C \, \mathbf{P}_C \, \chi^*_{C,r} = \sum_{C=1}^{NC} \sum_{s=1}^{NS} \mathbf{a}_s \, \mathbf{P}_C \, \chi_{s,C} \, \chi^*_{C,r} \tag{34.6}$$

Switch the order of summation on the right

$$\sum_{C=1}^{NC} \mathbf{g}_C \, \mathbf{P}_C \, \chi^*_{C,r} = \sum_{s=1}^{NS} \mathbf{a}_s \sum_{C=1}^{NC} \mathbf{P}_C \left(\chi_{s,C} \, \chi^*_{C,r}\right) \tag{34.7}$$

But the sum over $\mathbf{C}$ on the right is the weighted row orthogonality, so

$$\sum_{C=1}^{NC} \mathbf{g}_C \, \mathbf{P}_C \, \chi^*_{C,r} = \sum_{s=1}^{NS} \mathbf{a}_s \left(\mathbf{h} \, \delta_{s,r}\right) \tag{34.8}$$

Do the sum over $\mathbf{s}$:

$$\sum_{C=1}^{NC} \mathbf{g}_C \, \mathbf{P}_C \, \chi^*_{C,r} = \mathbf{h} \, \mathbf{a}_r \tag{34.9}$$

Divide both sides by $\mathbf{h}$, multiply by $\chi_{r,D}$, and sum over $\mathbf{r}$:

$$\frac{1}{\mathbf{h}} \sum_{r=1}^{NC} \sum_{C=1}^{NC} \mathbf{g}_C \, \mathbf{P}_C \;\; \chi^*_{C,r} \, \chi_{r,D} = \sum_{r=1}^{NS} \sum_{s=1}^{NS} \mathbf{a}_s \, \delta_{s,r} \, \chi_{r,D} \tag{34.10}$$

On the right, the sum over $\mathbf{r}$ can be done explicitly. On the left, move the sum over $\mathbf{r}$ as far inward as possible :

$$\frac{1}{\mathbf{h}} \sum_{C=1}^{NC} \mathbf{g}_C \, \mathbf{P}_C \sum_{r=1}^{NC} \;\chi^*_{C,r} \, \chi_{r,D} = \sum_{s=1}^{NS} \mathbf{a}_s \, \chi_{s,D} \tag{34.11}$$

But the sum over $\mathbf{s}$ is the general vector $\underset{\rightharpoondown}{\mathbf{g}}$ that we first wrote in Eq. 34.5, so

$$\sum_{C=1}^{NS} \mathbf{g}_C \left(\frac{\mathbf{P}_C}{\mathbf{h}} \sum_{r=1}^{NS} \chi^*_{C,r} \, \chi_{r,D} \right) = \mathbf{g}_D \tag{34.12}$$

But $\underset{\rightharpoondown}{\mathbf{g}}$ was absolutely arbitrary, so the parenthesis cannot contain anything other than $\delta_{C,D}$. Therefore we can conclude

The column orthogonality

$$\sum_{s=1}^{NS} \frac{P_C}{h} \chi^*_{C,s} \chi_{s,D} = \delta_{C,D} \qquad (34.13)$$

C and D index classes, s indexes species.

As before, let $\left(\sqrt{\frac{P_C}{h}}\ \chi_{s,D}\right) \to \mathbb{Y}$ (size **NS** × **NC**) and let $\left(\sqrt{\frac{P_C}{h}}\ \chi^*_{C,s}\right) \to$ $\mathbb{Y}^\dagger$(size **NC** × **NS**) . The sum is over **s**, so the dot product is of size **(NC × NS) . (NS × NC)**. Therefore $\delta_{C,D}$ is a unit matrix of size **NC** × **NC** :

The column orthogonality as a Dot product

$$\mathbb{Y}^\dagger . \mathbb{Y} = \mathbb{U}_{NC\times NC} \qquad (34.14)$$

34.4 Character tables are square

Now we are ready to draw one of the most important conclusions in representation theory. It will be quick. We have deduced the orthogonalities

$$\mathbb{Y}^\dagger . \mathbb{Y} = \mathbb{U}_{NC\times NC} \quad \text{and} \quad \mathbb{Y} . \mathbb{Y}^\dagger = \mathbb{U}_{NS\times NS} \qquad (34.15)$$

But we do not know whether **NC** (the number of classes) is the same as **NS** (the number of species), or not. If they are not the same, the $\mathbb{Y}^\dagger . \mathbb{Y}$ product is a rectangular multiplication of size (**NC** × **L**) . (**L** × **NC**), where **L** is unknown. And the other orthogonality says the $\mathbb{Y} . \mathbb{Y}^\dagger$ multiplication is of size (**NS** × **M**) . (**M** × **NS**), with **M** unknown.

So by one reasoning $\mathbb{Y}$ must be of size (**L** × **NC**), and by the other it must be of size (**NS** × **M**). But it is the same matrix, so the unknown **L** must be **NS**, and the unknown **M** must be **NC**. Now the size of $\mathbb{Y}$ can be expressed both as (**NS** × **NC**) and as (**NC** × **NS**). But both of these ways can be true only if **NS** = **NC**, so :

> The weighted character table $\mathbb{Y}$ of every group is square. But its size is the same as the unweighted character table $\mathbb{X}$, which must also be square.
>
> That is, the number of classes is identical to the number of irreducible representations (or species).

We have also proved another important fact:

Instead of making symmetric definitions $\left(\sqrt{\frac{\mathbf{P_C}}{\mathbf{h}}}\ \chi_{\mathbf{s,D}}\right) \to \mathbb{Y}$ and $\left(\sqrt{\frac{\mathbf{P_C}}{\mathbf{h}}}\ \chi^{*}_{\mathbf{C,s}}\right) \to \mathbb{Y}^{\dagger}$, let $\left(\chi_{\mathbf{s,D}}\right) \to \mathbb{X}$ and $\left(\frac{\mathbf{P_C}}{\mathbf{h}}\ \chi^{*}_{\mathbf{C,s}}\right) \to \tilde{\mathbb{X}}$. We showed that $\mathbb{X}.\tilde{\mathbb{X}} = \tilde{\mathbb{X}}.\mathbb{X} = \mathbb{U}$, so matrices $\mathbb{X}$ and $\tilde{\mathbb{X}}$ meet the complete definition of mutually inverse matrices.

> So the character table $\mathbb{X}$ of every group has an inverse, given by $\tilde{\mathbb{X}}$.

34.5 Check the twiddle matrix numerically

Too much index manipulation makes some people uneasy, and well it should; it is so easy to make a mistake. Fortunately, you can now check it numerically. Try the following with any group. Just change the **grp** name at the top.

```
grp = "T";
h = GroupOrder[grp];
X = ChVecs[grp];
% // GridForm // Size[6];
```

1	1	1	1
1	$-\frac{1}{2}-\frac{i\sqrt{3}}{2}$	$-\frac{1}{2}+\frac{i\sqrt{3}}{2}$	1
1	$-\frac{1}{2}+\frac{i\sqrt{3}}{2}$	$-\frac{1}{2}-\frac{i\sqrt{3}}{2}$	1
3	0	0	-1

```
P = ClassPopulations[grp]
```

```
{1, 4, 4, 3}
```

Here is the construction we are checking, the twiddle matrix :

```
X̃ = P * Transpose[Conjugate[X]] / h // Expand;
% // GridForm
```

$\frac{1}{12}$	$\frac{1}{12}$	$\frac{1}{12}$	$\frac{1}{4}$
$\frac{1}{3}$	$-\frac{1}{6}+\frac{i}{2\sqrt{3}}$	$-\frac{1}{6}-\frac{i}{2\sqrt{3}}$	0
$\frac{1}{3}$	$-\frac{1}{6}-\frac{i}{2\sqrt{3}}$	$-\frac{1}{6}+\frac{i}{2\sqrt{3}}$	0
$\frac{1}{4}$	$\frac{1}{4}$	$\frac{1}{4}$	$-\frac{1}{4}$

Test both the left and right orthonormalities :

```
{X.X̃, X̃.X} // Simplify;
% // GridList // Size[7]
```

$$\left\{\begin{array}{|c|c|c|c|}\hline 1&0&0&0\\\hline 0&1&0&0\\\hline 0&0&1&0\\\hline 0&0&0&1\\\hline\end{array},\ \begin{array}{|c|c|c|c|}\hline 1&0&0&0\\\hline 0&1&0&0\\\hline 0&0&1&0\\\hline 0&0&0&1\\\hline\end{array}\right\}$$

34.6 Unique rows and columns

We now know that the character table **X** of every group possesses an inverse, given by $\tilde{\mathbb{X}}$. This implies that the determinant of **X** is nonzero. But if two rows are equal, or if two columns are equal, (or even proportional) the determinant must vanish. So **X** has unique, linearly independent columns and unique, linearly independent rows. This is a major result:

> Each row of the character table identifies its irreducible representation uniquely, and each column identifies its class uniquely.

34.7 A theorem of your very own

The following exercises have more meat than usual :

1. Let $\underset{\rightharpoondown}{\mathbf{d}}$ be a list of the dimensions of all the irreducible reps of a group, and let **h** be its order. Write a **Module** that you can **Map** onto **GroupCatalog**, returning a list

$$\{\text{groupname}, h, \sum_{i=1}^{h} \underset{\rightharpoondown}{d}[[i]], \sum_{i=1}^{h} \underset{\rightharpoondown}{d}[[i]]^2\}.$$

You will make a surprising numerical discovery. Module construction is a basic *Mathematica* skill, but if you still need help with it, click here.

2. Firm up your discovery with a symbolic proof. Think outside the lines a little. What other quantity do you know that is always equal to **h**? For the answer, click here.

34.8 End Notes

34.8.1 A surprising regularity

Look at the list you will **Map** onto :

```
GroupCatalog
```

```
{C1, C2, C2h, C2v, C3, C3h, C3v, C4, C4h, C4v, C5, C6, C6h,
 C6v, Ch, Ci, D2, D2d, D2h, D3, D3d, D3h, D4, D4d, D4h, D5,
 D5d, D5h, D6, D6d, D6h, I, O, Oh, S4, S6, S8, T, Td, Th}
```

Remember that the first element of every character vector is the dimension of the representation.

```
expt[g_String] := Module[{dvec, h},
  dvec = Transpose[ChVecs[g]][[1]];
  h = GroupOrder[g];
  {g, h, Apply[Plus, dvec], dvec.dvec}]
```

Interactive readers should look at the whole thing.

```
Map[expt, GroupCatalog];
%[[{3, 12, 25, 36}]] // ColumnForm
```

```
{C2h, 4, 4, 4}
{C6, 6, 6, 6}
{D4h, 16, 12, 16}
{S6, 6, 6, 6}
```

What are you supposed to see? The first number is the group order, the second is the sum of the species dimensions, and the third is the sum of the dimensions squared. Is it obvious that two of these should *always* be the same?

34.8.2 The $\vec{d}.\vec{d}$ theorem

Let $\vec{\mathbf{d}}$ be a list of the dimensions of all the irreducible reps of a group. In every group, $\vec{\mathbf{d}}.\vec{\mathbf{d}} = \mathbf{h}$. Why does this always hold?

Consider a representation in block diagonal form, in which each block is an irreducible representation, and with each rep present once and only once. Each block **i** is of size **d[i]** × **d[i]**. There is a unique skewer for each element, so the number of skewers in block **i** is $\mathbf{d[i]}^2$. But we know that the skewers span a space of dimension **h**, so the total number of skewers is **h**. Therefore $\sum_{i=1}^{NR} \mathbf{d[i]}^2 = \vec{\mathbf{d}}.\vec{\mathbf{d}} = \mathbf{h}$, which was to be proved.

You might think there would be some nice simple formula for the value of $\sum_{i=1}^{NR} \mathbf{d[i]}$, the dimension of the block diagonal representation considered above. But there isn't. If you can find one, let somebody know.

35. Reducible rep analysis

Preliminaries

35.1 The point of rep analysis

This is the "quick and easy" way to analyze a rep, promised long ago in Chapter 28 (ReducibleRepresentations). We could not present it there because we did not yet have Chapter 33 (TheGreatOrthogonality) and its offspring, CharacterOrthogonalities of Chapter 34. Now we have them, and we can go forward.

But first, remember why we need rep analysis. Here is the standard procedure for symmetry analysis, used over and over in applications :

1. Identify **N** symmetric basis objects $\underset{\rightharpoondown}{\mathbf{b}}$ = $\{\mathbf{b}_1, \mathbf{b}_2, .., \mathbf{b}_N\}$ whose symmetry you want to analyze. These could be orbitals, or spectroscopic transition operators, or many other things. These are your basis objects.

2. Let group $\mathcal{G}$ be the group of a molecule of interest.

3. Make a representation $\mathcal{B}$ of $\mathcal{G}$ using $\underset{\rightharpoondown}{\mathbf{b}}$ as the basis objects. The command is

$\mathcal{B}$ = **MakeRepX[{$\mathbf{b}_1, \mathbf{b}_2, .., \mathbf{b}_N$}, $\mathcal{G}$]**,

as developed in Chapter 27 (MakeReps). $\mathcal{B}$ is a "big" rep, because its **N×N** matrices are generally bigger than any of the irreducible rep matrices of $\mathcal{G}$.

4. Compute $\underset{\rightharpoondown}{\chi}^{(\mathcal{B})}$, the short character vector of rep $\mathcal{B}$.

5. Call **Analyze[**$\underset{\rightharpoondown}{\chi}^{(\mathcal{B})}$**,** $\mathcal{G}$**]**, developed below. This returns all the symmetry information that there is about the list of basis objects, as they relate to your molecule.

Now you are ready for the quick and easy analysis method.

W.M. McClain, *Symmetry Theory in Molecular Physics with Mathematica*,
DOI 10.1007/b13137_35,

35.2 Symbolic development of `Analyze`

Consider a big, tangled representation $\mathcal{B}$ of some group $\mathcal{G}$, and consider also a similarity transform of it, $\mathcal{B}_{\mathbf{block}}$, in block diagonal form. The transform leaves diagonal sums invariant, so the diagonal sum, or character, of each $\mathcal{B}$ matrix is the same as that of the corresponding $\mathcal{B}_{\mathbf{block}}$ matrix. But the blocks are irreducible representations, and their characters are listed in the character table of group $\mathcal{G}$. An irreducible rep may appear in $\mathcal{B}_{\mathbf{block}}$ one time, several times, or not at all. Let the number of times that rep $\mathbf{r}$ appears in $\mathcal{B}_{\mathbf{block}}$ be $\mathbf{n}_{\mathbf{r}}^{(\mathcal{B})}$. Then the character of each class $\mathbf{C}$ of $\mathcal{B}$ can be expressed as a weighted sum down the species column of class $\mathbf{C}$ in the character table:

This may be better understood from an example diagram. On the left we have a big rep matrix belonging to class C. Its trace, or spur, is $\mathbf{a+b+\ldots+m}$. On the right, this matrix has been block diagonalized.

A matrix from rep $\mathcal{B}$ Same matrix from rep $\mathcal{B}_{\text{block}}$

$$\begin{pmatrix}
a & x & x & x & x & x & x & x & x & x & x & x \\
x & b & x & x & x & x & x & x & x & x & x & x \\
x & x & c & x & x & x & x & x & x & x & x & x \\
x & x & x & d & x & x & x & x & x & x & x & x \\
x & x & x & x & e & x & x & x & x & x & x & x \\
x & x & x & x & x & f & x & x & x & x & x & x \\
x & x & x & x & x & x & g & x & x & x & x & x \\
x & x & x & x & x & x & x & h & x & x & x & x \\
x & x & x & x & x & x & x & x & i & x & x & x \\
x & x & x & x & x & x & x & x & x & j & x & x \\
x & x & x & x & x & x & x & x & x & x & k & x \\
x & x & x & x & x & x & x & x & x & x & x & m
\end{pmatrix}
\rightarrow
\begin{pmatrix}
A & \blacklozenge & \blacklozenge & 0 & 0 & 0 & 0 & 0 & 0 & 0 & 0 & 0 \\
\blacklozenge & B & \blacklozenge & 0 & 0 & 0 & 0 & 0 & 0 & 0 & 0 & 0 \\
\blacklozenge & \blacklozenge & C & 0 & 0 & 0 & 0 & 0 & 0 & 0 & 0 & 0 \\
0 & 0 & 0 & D & \bullet & 0 & 0 & 0 & 0 & 0 & 0 & 0 \\
0 & 0 & 0 & \bullet & E & 0 & 0 & 0 & 0 & 0 & 0 & 0 \\
0 & 0 & 0 & 0 & 0 & A & \blacklozenge & \blacklozenge & 0 & 0 & 0 & 0 \\
0 & 0 & 0 & 0 & 0 & \blacklozenge & B & \blacklozenge & 0 & 0 & 0 & 0 \\
0 & 0 & 0 & 0 & 0 & \blacklozenge & \blacklozenge & C & 0 & 0 & 0 & 0 \\
0 & 0 & 0 & 0 & 0 & 0 & 0 & 0 & D & \bullet & 0 & 0 \\
0 & 0 & 0 & 0 & 0 & 0 & 0 & 0 & \bullet & E & 0 & 0 \\
0 & 0 & 0 & 0 & 0 & 0 & 0 & 0 & 0 & 0 & F & 0 \\
0 & 0 & 0 & 0 & 0 & 0 & 0 & 0 & 0 & 0 & 0 & G
\end{pmatrix}$$

By similarity we know that the two traces are equal

$$(\mathbf{a} + \mathbf{b} + \ldots + \mathbf{m}) = (\mathbf{A} + \mathbf{B} + \mathbf{C}) + (\mathbf{D} + \mathbf{E}) + (\mathbf{A} + \mathbf{B} + \mathbf{C}) + (\mathbf{D} + \mathbf{E}) + \mathbf{F} + \mathbf{G}$$

Suppose that you have gone through the clumsy analysis procedure of Chapter 28 (ReducibleRepresentations), and you know that $\begin{pmatrix} A & \blacklozenge & \blacklozenge \\ \blacklozenge & B & \blacklozenge \\ \blacklozenge & \blacklozenge & C \end{pmatrix}$ is from species **6**, while $\begin{pmatrix} D & \blacklozenge \\ \blacklozenge & E \end{pmatrix}$ is from species **5**, and **E**, **F**, and **G** are from species **1**, **2**, and **3**, respectively. Then you can rewrite the diagonal sums as

$$(\mathbf{a} + \mathbf{b} + \ldots + \mathbf{m}) = \left(1 * \chi_1^{(\mathcal{G})}\right) + \left(1 * \chi_2^{(\mathcal{G})}\right) + \left(1 * \chi_3^{(\mathcal{G})}\right) + \left(0 * \chi_4^{(\mathcal{G})}\right) + \left(2 * \chi_5^{(\mathcal{G})}\right) + \left(2 * \chi_6^{(\mathcal{G})}\right)$$

These matrices belong to some class **C** of the group, so $\left(\mathbf{a}+\mathbf{b}+\ldots+\mathbf{m}\right)$ is the character of the whole class **C** of rep $\mathcal{B}$, symbolized as $\chi_{\mathrm{C}}^{(\mathcal{B})}$. Similarly, the χ's on the right side also belong to the whole class **C**, so we write it s a **Dot** product with class **C** specified on both sides :

$$\chi_{\mathrm{C}}^{(\mathcal{B})} = \{1, 1, 1, 0, 2, 2\}.\{\chi_{1,\mathrm{C}}, \chi_{2,\mathrm{C}}, \chi_{3,\mathrm{C}}, \chi_{4,\mathrm{C}}, \chi_{5,\mathrm{C}}, \chi_{6,\mathrm{C}}\}$$

Now let **{1,1,1,0,2,2}** be the species count vector $\underset{\rightharpoondown}{\mathbf{n}}$. It is specific to rep $\mathcal{B}$ of group $\mathcal{G}$, so we give it the superscript $(\mathcal{B},\mathcal{G})$. Also the χ vector above is just a row from the character table of group $\mathcal{G}$, with standard symbol $\underset{\rightharpoondown \mathrm{C}}{\chi}^{(\mathcal{G})}$. Now the right side becomes a **Dot** product of two vectors :

$$\chi_{\mathrm{C}}^{(\mathcal{B})} == \underset{\rightharpoondown}{\mathbf{n}}^{(\mathcal{B},\mathcal{G})} . \underset{\rightharpoondown \mathrm{C}}{\chi}^{(\mathcal{G})} \qquad (35.1)$$

The class index **C** appears on both sides above, meaning that this equation holds for all classes. Therefore we can replace it by a vector underscript on the left, (the vector running over classes) while on the right when **C** runs over all classes in $\underset{\rightharpoondown \mathrm{C}}{\chi}^{(\mathcal{G})}$ you have the character matrix $\mathbb{X}^{(\mathcal{G})}$. So this equation, holding for all classes and all species at once, is

$$\underset{\rightharpoondown}{\chi}^{(\mathcal{B})} == \underset{\rightharpoondown}{\mathbf{n}}^{(\mathcal{B},\mathcal{G})} . \mathbb{X}^{(\mathcal{G})} \qquad (35.2)$$

We now know that the number of classes is the same as the number of irreducible representations, so $\underset{\rightharpoondown}{\chi}^{(\mathcal{B})}$ and $\underset{\rightharpoondown}{\mathbf{n}}^{(\mathcal{B},\mathcal{G})}$ are vectors of same length **N**, and $\mathbb{X}^{(\mathcal{G})}$ is a matrix of size $\mathbf{N}\times\mathbf{N}$,

In easy cases you can solve this by mental arithmetic. Write the character vector $\underset{\rightharpoondown}{\chi}^{(\mathcal{B})}$ on a piece of paper and slip it up against the bottom of the character table $\mathbb{X}^{(\mathcal{G})}$. For each class, run your finger down the column while thinking what coefficient vector $\underset{\rightharpoondown}{\mathbf{n}}^{(\mathcal{B},\mathcal{G})}$ will make the sum shown on the paper for $\underset{\rightharpoondown}{\chi}^{(\mathcal{B})}$.

When you have one set of coefficients that makes all the characters correct, you have solved the problem. But if this takes more than a few seconds, feed it to **LinearSolve** and get the answer without human thought :

Quick and easy rep analysis

$$\underset{\rightharpoondown}{\mathbf{n}}^{(\mathcal{B},\mathcal{G})} == \mathbf{LinearSolve}\left[\mathbf{Transpose}\left[\mathbb{X}^{(\mathcal{G})}\right], \underset{\rightharpoondown}{\chi}^{(\mathcal{B})}\right] \qquad (35.3)$$

That's all there is to it.

> **On your own**
> Pull up the syntax of **LinearSolve** and see why we had to **Transpose** $\mathbb{X}^{(\mathcal{G})}$ before we used it.

Of course, if you insist, it can also be solved symbolically. The character matrix $\mathbb{X}^{(\mathcal{G})}$ has an inverse $\tilde{\mathbb{X}}^{(\mathcal{G})}$ given by

$$\tilde{\mathbb{X}}^{(\mathcal{G})} = \frac{1}{h} \underset{\rightharpoondown}{\mathbf{P}}^{(\mathcal{G})} * \left(\mathbb{X}^{(\mathcal{G})\,\dagger}\right)$$

where $\underset{\rightharpoondown}{\mathbf{P}}^{(\mathcal{G})}$ is the class population vector for group $\mathcal{G}$. So dotting $\tilde{\mathbb{X}}^{(\mathcal{G})}$ in on the right of Eq. 35.2 we isolate the unknown $\underset{\rightharpoondown}{\mathbf{n}}$ vector:

Not quite so quick and easy rep analysis

$$\underset{\rightharpoondown}{\mathbf{n}}^{(\mathcal{B},\mathcal{G})} == \underset{\rightharpoondown}{\chi}^{(\mathcal{B})} \cdot \left(\frac{1}{h} \underset{\rightharpoondown}{\mathbf{P}} * \left(\mathbb{X}^{(\mathcal{G})\,\dagger}\right)\right) \quad (35.4)$$

The **Dot** runs over classes

If you think that $\underset{\rightharpoondown}{\mathbf{P}}$ vector makes it too complicated, you can change the short characters to long characters, and change the sum over classes into a sum over elements. Then

Another pretty quick and easy rep analysis

$$\underset{\rightharpoondown}{\mathbf{n}}^{(\mathcal{B},\mathcal{G})} == \frac{1}{h} \underset{\rightharpoondown}{\overleftrightarrow{\chi}}^{(\mathcal{B})} \cdot \overleftrightarrow{\mathbb{X}}^{(\mathcal{G})\,\dagger} \quad (35.5)$$

The **Dot** runs over elements

35.3 Numerical example of rep analysis

We continue with the benzene example we have used throughout the book. We have a handy operator that cuts through the Gordian knot of rep construction. Use it to construct a rep on the basis of the six π orbitals of benzene :

```
grp = "D6h";
B = MakeRepAtomic["benzene",
```

```
    grp, atomKind → "C", preFactor → z] [[1]];
MorphTest[%, grp]
```

```
Faithful, or Isomorphic
```

Sum up the diagonals of all 24 matrices, and select the first from each class :

```
χ(B) = Map[Spur, B] // OnePerClass[#, grp] &
```

```
{6, 0, 0, 0, 0, -2, 0, 0, 0, -6, 2, 0}
```

This is the character vector to be analyzed. We pull up the square essentials from the character table :

```
X(G) = ChVecs[grp];
```

The next line solves the problem:

```
n(B,G) = LinearSolve[Transpose[X(G)], χ(B)] // Simplify
```

```
{0, 0, 1, 0, 1, 0, 0, 1, 0, 0, 0, 1}
```

To make it more easily readable by humans, pull the species names and bracket them with the number of occurrences :

```
Transpose[{SpeciesNames[grp], n(B,G)}]
```

```
{{A1g, 0}, {A2g, 0}, {B1g, 1}, {B2g, 0}, {E1g, 1}, {E2g, 0},
 {A1u, 0}, {A2u, 1}, {B1u, 0}, {B2u, 0}, {E1u, 0}, {E2u, 1}}
```

This says that in the block diagonal reduction of rep $\mathcal{B}$, four species appeared one time each; namely, $\mathbf{B_{1\,g}}$, $\mathbf{E_{1\,g}}$, $\mathbf{A_{2\,u}}$, and $\mathbf{E_{2\,u}}$. These are exactly the species labels used in Fig. 23. 5, to which you may click back.

35.4 The `Analyze` operator

The procedure of Eq. 35.3 has been immortalized in the **`Symmetry`` ** package under the name **`Analyze`**.

- **Analyze[chVec,"g"]** returns the composition of **chVec** in terms of the irreducible character vectors of group **"g"**.

In Chapter 28 (ReducibleReps) we promised a "quick and easy" way to find the block diagonal components of a big reducible representation. Our example there was a 6×6 representation of group $\mathbf{D_{6\,h}}$, using as basis functions the six π orbitals of benzene. We called the rep $\mathcal{B}$, and found its character vector. Here it is again :

$$\underset{\rightharpoondown}{\chi}^{\mathcal{B}} = \{6,\ 0,\ 0,\ 0,\ 0,\ 2,\ 0,\ 0,\ 0,\ 6,\ 2,\ 0\};$$

Perhaps you recall how much trouble it was to reduce that representation by methods based plainly on the definitions. Compare all that to the use of the **Analyze** operator :

```
Analyze[χ^B, "D6h"] // Size[8]
```

```
{{A1g, 1}, {A2g, 0}, {B1g, 0}, {B2g, 0}, {E1g, 0}, {E2g, 1},
 {A1u, 0}, {A2u, 0}, {B1u, 0}, {B2u, 1}, {E1u, 1}, {E2u, 0}}
```

That's the complete answer. This problem, rather formidable if done by hand, was solved by three group theoretic operators: First, **MakeRepAtomic**, then a mapping of **Spur,** and finally **Analyze**. You can study the orbitals of any tabulated molecule in any tabulated group the same way.

Things are building up.

36. The regular representation

Preliminaries

36.1 What is the regular representation?

Every group has a regular representation that can be constructed solely from knowledge of the group multiplication table. It has two miraculous properties. First, it allows us to determine the total number of irreducible representations that the group possesses. Second, it allows us to construct the character table of the group. We will prove and demonstrate the first property in this chapter, saving the second property for Chapter 48 (MakeCharacterTable).

But first, the definition and construction of the regular representation. In Chapter 25 (Representations) we gave the formula for constructing any representation. For element Λ it was

$$\mathcal{T}_\Lambda[\underset{\neg}{\mathtt{f}}] = \underset{\neg}{\mathtt{f}}.\mathbb{T}^{(\Lambda)} \qquad \text{(Eq. 25.1 again)}$$

where the vector $\underset{\neg}{\mathtt{f}}$ was a list of unspecified objects and $\mathbb{T}^{(\Lambda)}$ is the desired (unknown) rep matrix for element Λ of the group. The only requirement on the $\underset{\neg}{\mathtt{f}}$ objects is that they be symmetric (or invariant) under the transform $\mathcal{T}_\Lambda[\underset{\neg}{\mathtt{f}}]$ for every Λ in the group.

Now we make an ingenious interpretation of the undefined elements of vector $\underset{\neg}{\mathtt{f}}$. We take them to be the abstract group elements themselves. Also, we take the transform rule $\mathcal{T}_\Lambda[\underset{\neg}{\mathtt{f}}]$ to be multiplication of the element list $\underset{\neg}{\mathtt{f}}$ by element Λ. Then the general rep mat definition, Eq. 25.1 becomes concrete as

> **The regular representation**
>
> $$\mathbf{L} \otimes \{\mathbf{E}, \mathbf{A}, .., \mathbf{Z}\} = \{\mathbb{E}, \mathbb{A}, .., \mathbb{Z}\} . \mathbb{L} \qquad (36.1)$$

where the abstract group elements are $\{\mathbf{E},\mathbf{A},\ldots,\mathbf{Z}\}$ ($\mathbf{L}$ being some one of them and the DoubleStruck capitals $\{\mathbb{E},\mathbb{A},\ldots,\mathbb{Z}\}$ are the representation matrices $\mathbb{L}$ being the matrix for $\mathbf{L}$) Since the left side is just the $\mathbf{L}$th-row of the group's multiplication table, all you need is knowledge of that table. By the rearrangement theorem the left side is just a rearrangement of all the group elements, and therefore matrix $\mathbb{L}$ on the right has to be a permutation matrix, all zeroes and

DOI 10.1007/b13137_36, © Springer Science+Business Media, LLC 2009

ones. We can solve this using the package operator **MatrixOfCoefficients**. Here is its thumbnail sketch:

- **MatrixOfCoefficients[exprs,vars]** returns matrix **MC**, where **MC.vars == exprs**. It may fail if the **vars** are not simple symbols, and it will fail if the **exprs** are not linear in the **vars**.

Note that **MatrixOfCoefficients** returns a rightward matrix, whereas representation theory requires a leftward matrix. The switch is easy: Write the equation for a rightward matrix, solve it using **MatrixOfCoefficients**, and then **Transpose** the result to make it into a leftward matrix.

Below, we set **grp** to be our old war horse, group $\mathbf{C}_{3\,\mathbf{v}}$. You can either use it as is, or reset it to another group. Everything below should run perfectly for any tabulated group. Begin with the multiplication table for the group :

```
grp = "C3v";
MT = MultiplicationTable[grp];
BoxUp[MT, MT[[1]], "C3 v", .8]
```

$\mathbf{C}_{3\,\mathbf{v}}$	E	C3a	C3b	σva	σvb	σvc
E	E	C3a	C3b	σva	σvb	σvc
C3a	C3a	C3b	E	σvb	σvc	σva
C3b	C3b	E	C3a	σvc	σva	σvb
σva	σva	σvc	σvb	E	C3b	C3a
σvb	σvb	σva	σvc	C3a	E	C3b
σvc	σvc	σvb	σva	C3b	C3a	E

Define an operator that works on any row of multiplication table **MT**, returning the regular rep matrix of the corresponding element :

```
RRofMT[row_] :=
 Transpose[MatrixOfCoefficients[row, MT[[1]]]]
```

Map it over the whole matrix:

```
RegRepC3v = Map[RRofMT, MT];
% // GridList // Size[7]
```

$$\left\{\begin{matrix}1&0&0&0&0&0\\0&1&0&0&0&0\\0&0&1&0&0&0\\0&0&0&1&0&0\\0&0&0&0&1&0\\0&0&0&0&0&1\end{matrix},\ \begin{matrix}0&0&1&0&0&0\\1&0&0&0&0&0\\0&1&0&0&0&0\\0&0&0&0&0&1\\0&0&0&1&0&0\\0&0&0&0&1&0\end{matrix},\right.$$

$$\begin{matrix}0&1&0&0&0&0\\0&0&1&0&0&0\\1&0&0&0&0&0\\0&0&0&0&1&0\\0&0&0&0&0&1\\0&0&0&1&0&0\end{matrix},\ \begin{matrix}0&0&0&1&0&0\\0&0&0&0&0&1\\0&0&0&0&1&0\\1&0&0&0&0&0\\0&0&1&0&0&0\\0&1&0&0&0&0\end{matrix},$$

$$\left.\begin{matrix}0&0&0&0&1&0\\0&0&0&1&0&0\\0&0&0&0&0&1\\0&1&0&0&0&0\\1&0&0&0&0&0\\0&0&1&0&0&0\end{matrix},\ \begin{matrix}0&0&0&0&0&1\\0&0&0&0&1&0\\0&0&0&1&0&0\\0&0&1&0&0&0\\0&1&0&0&0&0\\1&0&0&0&0&0\end{matrix}\right\}$$

That is indeed a fine list of matrices, but does it represent the group ?

```
MorphTest[RegRepC3v, grp]
```

```
Faithful, or Isomorphic
```

It does. The representation theorem really works, even with this rather weird interpretation. Condense all this into an operator called **MakeRepRegular**:

```
MakeRepRegular[grp_] := Module[{MT},
   MT = MultiplicationTable[grp];
   Map[Transpose[MatrixOfCoefficients[#, MT[[1]]]] &,
    MT]];
```

Try it out:

```
MakeRepRegular["C3v"] == RegRepC3v
```

```
True
```

Most applied group theory books give a different definition of the regular representation, much better adapted to hand calculation than this one. If you want to see it, click to the End notes of this chapter. It contains a small surprise, not mentioned in any other group theory book we have seen.

36.2 Characters of the regular representation

Look at the diagonals of the example regular rep computed in the last section. Except in the first matrix (the unit matrix), the diagonals are all zeroes. If there is a **1** on the diagonal, the corresponding element did not move when the transform was applied. That always happens when the transform is multiplication of the group list by **E**, but it never happens for any other element.

But by the rearrangement theorem, when the multiplier is anything else, all the elements move, and the diagonal of the rep matrix is pure zeroes. So the character of the unit matrix is **h**, the order the group; but for every other matrix, the character is **0**. Check it :

```
χvecLong = Map[Spur, RegRepC3v]
```

```
{6, 0, 0, 0, 0, 0}
```

36.3 The golden property

In Chapter 35 (RepAnalysis) the second form of the algebraic formula for rep analysis was

$$\underset{\rightharpoondown}{\mathbf{n}}^{(\mathcal{B},\mathcal{G})} == \frac{1}{\mathbf{h}} \overset{\leftrightarrow}{\underset{\rightharpoondown}{\chi}}^{(\mathcal{B})} \cdot \overset{\leftrightarrow}{\mathbb{X}}^{(\mathcal{G})\,\dagger} \qquad \text{Dot over elements}$$

Replace the $\mathcal{B}$ (the $\mathcal{B}$ig rep to be analyzed) by a curly $\mathcal{R}$ (the $\mathcal{R}$egular rep), and rewrite in index notation. The symbol $\overset{\leftrightarrow}{\mathbb{X}}^{(\mathcal{G})\,\dagger}$ is $\left(\chi_{\mathbf{s},\Lambda}^{(\mathcal{G})}\right)^{\dagger} = \left(\chi_{\Lambda,\mathbf{s}}^{(\mathcal{G})}\right)^{*}$:

$$\mathbf{n}_{\mathbf{s}}^{(\mathcal{R},\mathcal{G})} = \frac{1}{\mathbf{h}} \sum_{\Lambda=1}^{\mathbf{h}} \chi_{\Lambda}^{(\mathcal{R})}\, \chi_{\Lambda,\mathbf{s}}^{(\mathcal{G})\,*}$$

Remember, $\mathbf{n}_{\mathbf{s}}^{(\mathcal{R},\mathcal{G})}$ is the number of times that species **s** appears in the regular representation $\mathcal{R}$ of group $\mathcal{G}$. For the regular representation, $\chi_{\Lambda}^{(\mathcal{R})}$ is zero except when Λ references the unit element **E**, so in the sum over Λ , only the term for element **E** survives :

$$\mathbf{n}_{\mathbf{s}}^{(\mathcal{R},\mathcal{G})} = \frac{1}{\mathbf{h}} \chi_{\mathbf{E}}^{(\mathcal{R})}\, \chi_{\mathbf{E},\mathbf{s}}^{(\mathcal{G})\,*}$$

But $\chi_{\mathbf{E}}^{(\mathcal{R})}$ is **h**, and $\chi_{\mathbf{E},\mathbf{s}}^{(\mathcal{G})\,*}$ is the real number $\mathbf{d}_{\mathbf{s}}^{(\mathcal{G})}$, the dimension of species **s** of group $\mathcal{G}$. Therefore,

$$n_s^{(\mathcal{R},\mathcal{G})} = d_s^{(\mathcal{G})} \tag{36.2}$$

Species **s** appears in the regular representation $\mathcal{R}$ of group $\mathcal{G}$ the same number of times as its dimension.

This is a hugely significant result. First of all, every irreducible rep has a dimension of at least **1**, so ALL the irreducible representations appear in the regular rep. This means that if we reduce the regular representation we are guaranteed to find every irreducible representation. Nothing before now has guaranteed a way to find them all, much less all at once.

36.4 Demo of the golden property

In order to use the golden property, you do not really have to make the whole regular representation at all. In Chapter 48 (MakeCharacterTable) we will need the whole regular representation, but to use the golden property all you need is its character vector, which is predictable: It consists of **h**, followed by **NC-1** zeroes, where **NC** is the Number of Classes. Here we cook it up for group D_{6h}:

```
grp = "D6h";
h = GroupOrder[grp];
NC = Length[ClassNames[grp]];
chVecRegRep = Join[{h}, Table[0, {i, NC - 1}]]
```

```
{24, 0, 0, 0, 0, 0, 0, 0, 0, 0, 0, 0}
```

Now apply **Analyze** to it :

```
Analyze[chVecRegRep, grp]
```

```
{{A1g, 1}, {A2g, 1}, {B1g, 1}, {B2g, 1}, {E1g, 2}, {E2g, 2},
 {A1u, 1}, {A2u, 1}, {B1u, 1}, {B2u, 1}, {E1u, 2}, {E2u, 2}}
```

As advertised, all 12 species are present at least once. All the one-dimensional species (anchored on **A** or **B**) appear one time, while all the two-dimensional species (anchored on **E**) appear two times. If this were an untabulated group, you would have to supply generic species names like Γ_i. The return would give them all back, each bracketed with its dimension. The dimensions would likely be new information for you.

36.5 The d^2 corollary

Now for a little corollary. The dimension of the regular representation is **h**, but if it is reduced, its dimension has to be the same as the sum of the dimensions of all the diagonal blocks. Species **s** contributes $\mathbf{d_s}$ to this sum, but it does so $\mathbf{n_s^{(\mathcal{R})}}$ times. Therefore $\mathbf{h} = \sum_{\mathbf{s=1}}^{\mathbf{NS}} \mathbf{n_s^{(\mathcal{R})}}\ \mathbf{d_s}$, where **NS** is the number of species. But we just proved that $\mathbf{n_s^{(\mathcal{R})}}$ is the same as $\mathbf{d_s}$, so

$$\mathbf{h} == \sum_{\mathbf{s=1}}^{\mathbf{NS}} \mathbf{d_s^2} \qquad (36.3)$$

This is consistent with, but stronger than, the **NS** inequality that we deduced earlier from the Great Orthogonality.

36.6 End Notes

36.6.1 Alternative construction of the regular rep

The alternative algorithm is easy to state in words, and nearly every group theory book does it :

(1) Reorder the rows of the group multiplication table so that the unit element falls on the principal diagonal.

(2) To make the regular rep matrix for element **X**, replace every **X** in the table by `1`. Then replace everything else by `0`.

This is easy for a human with pencil and paper, but a little clumsy to specify for a computer. Besides, this recipe needs a proof that it produces a representation. This was automatically assured for the **RegularRep** algorithm above, because it starts with the definition of representation. But for the traditionalists out there, here is a demonstration that *Mathematica* can carry out the traditional construction:

```
grp = "C3v";
h = GroupOrder[grp]
```

```
6
```

The traditional construction is a little easier if we use the integer name trick. Construct a multiplication table for the group, using integers **1** through **h** as the names of the elements.

```
MTstd =
  MultiplicationTable[Dot, CartRep[grp], Range[h]];
% // MatrixForm
```

$$\begin{pmatrix} 1 & 2 & 3 & 4 & 5 & 6 \\ 2 & 3 & 1 & 5 & 6 & 4 \\ 3 & 1 & 2 & 6 & 4 & 5 \\ 4 & 6 & 5 & 1 & 3 & 2 \\ 5 & 4 & 6 & 2 & 1 & 3 \\ 6 & 5 & 4 & 3 & 2 & 1 \end{pmatrix}$$

Now the product **i⊗j** lies as position **[[i,j]]** in this table. We assume the unit element is the first element, named **1**. Use the **Position** operator to where the **1** is in every row :

```
diagPermStd =
 Table[Position[MTstd[[i]], 1], {i, 1, h}] // Flatten
```

```
{1, 3, 2, 4, 5, 6}
```

Permute the rows so that the unit element falls on the principal diagonal of the multiplication table :

```
MT1 = MTstd[[diagPermStd]];
% // MatrixForm
```

$$\begin{pmatrix} 1 & 2 & 3 & 4 & 5 & 6 \\ 3 & 1 & 2 & 6 & 4 & 5 \\ 2 & 3 & 1 & 5 & 6 & 4 \\ 4 & 6 & 5 & 1 & 3 & 2 \\ 5 & 4 & 6 & 2 & 1 & 3 \\ 6 & 5 & 4 & 3 & 2 & 1 \end{pmatrix}$$

Suppose we want the matrix for element number **3** of this group. Replace every **3** by the letter **U**:

```
MT2 = MT1 /. {3 → "U"};
% // MatrixForm
```

$$\begin{pmatrix} 1 & 2 & \mathrm{U} & 4 & 5 & 6 \\ \mathrm{U} & 1 & 2 & 6 & 4 & 5 \\ 2 & \mathrm{U} & 1 & 5 & 6 & 4 \\ 4 & 6 & 5 & 1 & \mathrm{U} & 2 \\ 5 & 4 & 6 & 2 & 1 & \mathrm{U} \\ 6 & 5 & 4 & \mathrm{U} & 2 & 1 \end{pmatrix}$$

Replace every remaining integer by a zero :

```
MT3 = MT2 /. {_Integer → 0};
% // MatrixForm
```

$$\begin{pmatrix} 0 & 0 & U & 0 & 0 & 0 \\ U & 0 & 0 & 0 & 0 & 0 \\ 0 & U & 0 & 0 & 0 & 0 \\ 0 & 0 & 0 & 0 & U & 0 \\ 0 & 0 & 0 & 0 & 0 & U \\ 0 & 0 & 0 & U & 0 & 0 \end{pmatrix}$$

Re-replace every **U** by a **1** :

```
MT4 = MT3 /. {"U" → 1};
% // MatrixForm
```

$$\begin{pmatrix} 0 & 0 & 1 & 0 & 0 & 0 \\ 1 & 0 & 0 & 0 & 0 & 0 \\ 0 & 1 & 0 & 0 & 0 & 0 \\ 0 & 0 & 0 & 0 & 1 & 0 \\ 0 & 0 & 0 & 0 & 0 & 1 \\ 0 & 0 & 0 & 1 & 0 & 0 \end{pmatrix}$$

and that is the regular rep matrix for element #3. Put these operations together into a new alternate operator **MakeRepRegularRepAlt** :

```
MakeRepRegularAlt[grp_] :=
 Module[{h, MTstd, diagPermStd,
   MTreordered, MakeRRmat, n},
  h = GroupOrder[grp];
  MTstd = MultiplicationTable[
    Dot, CartRep[grp], Range[h]];
  diagPermStd = Table[Position[MTstd〚i〛, 1],
     {i, 1, h}] // Flatten;
  MTreordered = MTstd〚diagPermStd〛;
  MakeRRmat[n_] := MTreordered /. n → "U" /.
     el_Integer → 0 /. "U" → 1;
  Map[MakeRRmat, Range[h]]]
```

Try it out:

```
RegRepC3vAlt = MakeRepRegularAlt["C3v"];
% // GridList // Size[6]
```

{

1	0	0	0	0	0
0	1	0	0	0	0
0	0	1	0	0	0
0	0	0	1	0	0
0	0	0	0	1	0
0	0	0	0	0	1

,

0	1	0	0	0	0
0	0	1	0	0	0
1	0	0	0	0	0
0	0	0	0	0	1
0	0	0	1	0	0
0	0	0	0	1	0

,

0	0	1	0	0	0
1	0	0	0	0	0
0	1	0	0	0	0
0	0	0	0	1	0
0	0	0	0	0	1
0	0	0	1	0	0

,

0	0	0	1	0	0
0	0	0	0	1	0
0	0	0	0	0	1
1	0	0	0	0	0
0	1	0	0	0	0
0	0	1	0	0	0

,

0	0	0	0	1	0
0	0	0	0	0	1
0	0	0	1	0	0
0	0	1	0	0	0
1	0	0	0	0	0
0	1	0	0	0	0

,

0	0	0	0	0	1
0	0	0	1	0	0
0	0	0	0	1	0
0	1	0	0	0	0
0	0	1	0	0	0
1	0	0	0	0	0

}

The two algorithms do not have to produce identical reps. But did they?

```
RegRepC3vAlt == RegRepC3v
```

```
False
```

On your own

They are not the same, but you can prove that they are similar. You have all the tools.

37. Projection operators

Preliminaries

37.1 Projection operators in geometry

In geometry, a projection matrix can project out one component of a vector:

$$\begin{pmatrix} 1 & 0 & 0 \\ 0 & 0 & 0 \\ 0 & 0 & 0 \end{pmatrix} . \begin{pmatrix} x \\ y \\ z \end{pmatrix} == \begin{pmatrix} x \\ 0 \\ 0 \end{pmatrix}$$

The projection matrix is *idempotent*: any power is simply the matrix itself :

$$\begin{pmatrix} 1 & 0 & 0 \\ 0 & 0 & 0 \\ 0 & 0 & 0 \end{pmatrix} . \begin{pmatrix} 1 & 0 & 0 \\ 0 & 0 & 0 \\ 0 & 0 & 0 \end{pmatrix} . \begin{pmatrix} 1 & 0 & 0 \\ 0 & 0 & 0 \\ 0 & 0 & 0 \end{pmatrix} == \begin{pmatrix} 1 & 0 & 0 \\ 0 & 0 & 0 \\ 0 & 0 & 0 \end{pmatrix}$$

```
True
```

Therefore, no matter how many times you apply the **x**-projection matrix to **{x,y,z}**, the answer is always **{x,0,0}**. The projection operators of group theory are similar, but they operate in function spaces rather than in Cartesian space.

The determinant of projection matrices is zero, and they have no inverse. Therefore, projections cannot be undone. This makes perfect sense, because projections entail an irreversible loss of information in the dimensions that are wiped out. However, if you had a complete set of projections (like the **x**, **y**, and **z** projections) their sum would be the original.

W.M. McClain, *Symmetry Theory in Molecular Physics with Mathematica*,
DOI 10.1007/b13137_37,

37.2 Projection operators in algebra

37.2.1 The basic idea

The simplest example of function space projection is no doubt already familiar to you; it is the splitting of any function of one variable into parts of positive and negative "parity". If **f[z]** is any function, then

(f[z]+f[-z])/2

is invariant when **z** changes sign. On the other hand, the function

(f[z]-f[-z])/2

changes its sign, but not its absolute value, when **z** changes sign. The first is said to be of "even parity"; the second, of "odd parity". Note that they add to make the original **f[z]** again.

This extends easily to three dimensions. The inversion operator changes the sign of all three axes, and any function of **x**,**y**,**z** can be split into parts that are even and odd with respect to inversion. In the Symmetry` package there are two kinds of group theoretic projection operators, and they will be derived in detail in this chapter. But as an introduction, here is your a first example:

```
ProjectET[f[x, y, z], "Ci"] // Expand
```

$$\left\{\left\{\text{Ag}, \frac{1}{2}\,\text{f}[-\text{x}, -\text{y}, -\text{z}] + \frac{1}{2}\,\text{f}[\text{x}, \text{y}, \text{z}]\right\}, \left\{\text{Au}, -\frac{1}{2}\,\text{f}[-\text{x}, -\text{y}, -\text{z}] + \frac{1}{2}\,\text{f}[\text{x}, \text{y}, \text{z}]\right\}\right\}$$

We specified group C_i above. It consists of only the identity and the inversion, and the projector returned two functions, one even and one odd with respect to inversion.

What are those symbols **Ag** and **Au**? They are the names of the irreducible representations (or "symmetry species") of group C_i :

```
BoxedCharacterTable["Ci"]
```

Ci	1	1	← Class populations
	E	inv	↓ Basis functions
Ag	1	1	{x^2}, {y^2}, {z^2}, {x*y}, {x*z}, {y*z}, {Ix}, {Iy}, {Iz}
Au	1	-1	{(x), (y), (z)}

The **g** stands for the German word *gerade* (even) and **u** stands for *ungerade* (odd). The projection operator for group C_i splits the general function

f[x,y,z] into two parts; one is a basis for rep $\mathbf{A_g}$; the other, for $\mathbf{A_u}$. Here is a concrete example, with graphics :

```
f[x_] := 1/2 + 2 x^2 + x^3;
ProjfCi = ProjectET[f[x], "Ci"] // ExpandAll
```

$$\left\{\left\{\text{Ag}, \frac{1}{2} + 2\,x^2\right\}, \left\{\text{Au}, x^3\right\}\right\}$$

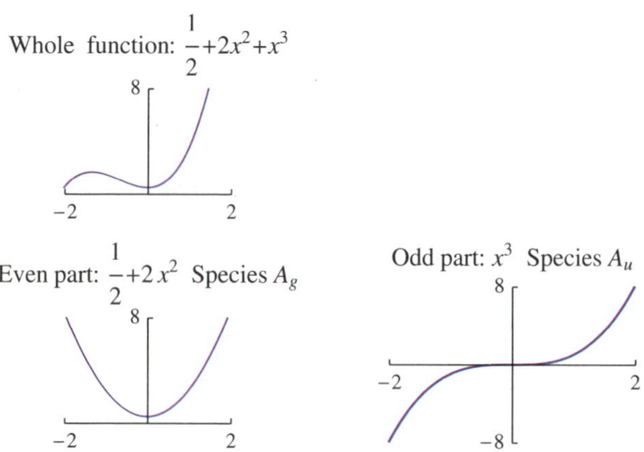

Fig.37.1 An asymmetric function, f[x], is split into two parts, odd and even with respect to reflection at the origin.

The concept of splitting a function into symmetric parts is marvelously generalized in group theory. The next section shows a quite simple example that would be very tedious to think through from first principles.

37.2.2 Introductory example (group D4)

Group $\mathbf{D_4}$ consists of fourfold and twofold rotations around **z**, plus twofold rotations around **x** and **y**. It has four 1-D species and one 2-D species:

```
BoxedCharacterTable["D4", 9]
```

D4	1	2	1	2	2	← Class populations
	E	C4	C_2	C'_2	C''_2	↓ Basis functions
A1	1	1	1	1	1	$\{(x^2+y^2), (z^2)\}$
A2	1	1	1	-1	-1	{(z), (Iz)}
B1	1	-1	1	1	-1	$\{(x^2-y^2)\}$
B2	1	-1	1	-1	1	{(xy)}
E	2	0	-2	0	0	$\left\{\begin{pmatrix}x\\y\end{pmatrix}, \begin{pmatrix}yz\\xz\end{pmatrix}, \begin{pmatrix}Ix\\Iy\end{pmatrix}\right\}$

The two projection operators defined in the **Symmetry`** package are **ProjectET** (**ProjectE**xpression**T**race) and **ProjectED** (**ProjectE**xpression**De**tailed) We leap ahead here and use **ProjectET** just to demonstrate its power. We use it on an asymmetric function **q[x,y]**:

```
Clear[q];
q[x_, y_] :=
  1/2 e^(-x^2-y^2) (1 + 2 x^2 + 2 x^3) (36 + 28 y + 84 y^2 + 12 y^3 + y^4);
```

$$q[x\_, y\_] := \frac{1}{2} e^{-x^2-y^2} \left(1 + 2x^2 + 2x^3\right) \left(36 + 28y + 84y^2 + 12y^3 + y^4\right);$$

ProjectET[_,"D4"] returns formulas for five functions, one for each species of group $\mathbf{D_4}$:

```
qProjections = Simplify[ProjectET[q[x, y], "D4"]];
% // Column // Size[8]
```

$\left\{A1, \frac{1}{4} e^{-x^2-y^2} \left(72 + 156 y^2 + y^4 + x^4 \left(1 + 2 y^2\right) + 2 x^2 \left(78 + 168 y^2 + y^4\right)\right)\right\}$

$\left\{A2, 14 e^{-x^2-y^2} x y \left(x^2 - y^2\right)\right\}$

$\left\{B1, \frac{1}{4} e^{-x^2-y^2} \left(y^2 \left(12 + y^2\right) - x^4 \left(1 + 2 y^2\right) + 2 x^2 \left(-6 + y^4\right)\right)\right\}$

$\left\{B2, 2 e^{-x^2-y^2} x y \left(7 y^2 + x^2 \left(7 + 6 y^2\right)\right)\right\}$

$\left\{E, e^{-x^2-y^2} \left(2 y \left(7 + 3 y^2\right) + 4 x^2 y \left(7 + 3 y^2\right) + x^3 \left(36 + 84 y^2 + y^4\right)\right)\right\}$

Look immediately at the contour plots :

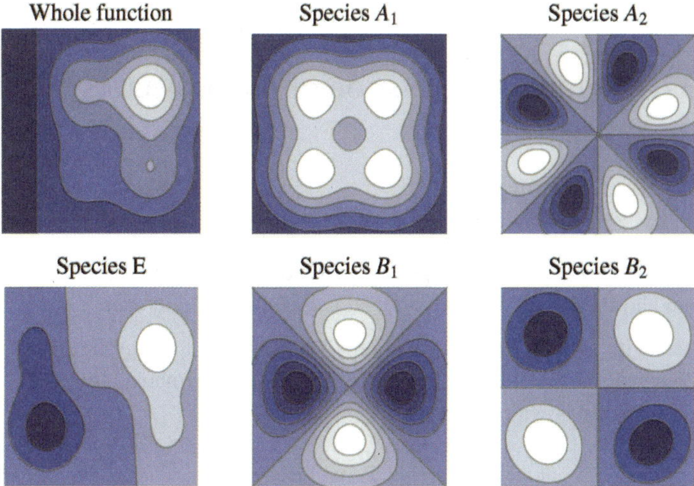

Fig.37.2 An asymmetric function, `q[x, y]`, is split into a five parts, each belonging to one of the species of group D_4.

Four of them are beautifully symmetric. Species **E** is a problem child; its irregular behavior is the reason we need that second operator **ProjectED**.

> **On your own**
> This is a good place to click back to Mulliken's species name rules in Chapter 28 (ReducibleReps), and see how they apply to these functions.

> **On your own**
> (1) Add up the projections and see if their sum is the original **q[x,y]**.
>
> (2) Construct the parity operator
>
> $$\texttt{Simplify}\left[\frac{\texttt{fn/.SymmetryRules[D4]}}{\texttt{fn}}\right]$$
>
> and apply it to the four 1-D projected functions. Then apply it to the 2-D projection **E**.

37.3 Symbolic derivation of the projectors

37.3.1 Strategy

After seeing what projection operators do, you are ready to understand exactly why and how they work. Here is our strategy (click for reference) :

1. Start with a given, explicit function **ψ[x,y,z]** and any group $\mathcal{G}$ for which you know the complete character table. Express ψ symbolically as an expansion in some unknown functions **u**:

$$\psi[x, y, z] = \sum_{i=1}^{N} u_i[x, y, z] \tag{37.1}$$

The number **N** is also an unknown at this point.

2. Let $\mathcal{T}_\Lambda$ be the transform rule for element Λ of group $\mathcal{G}$. Apply $\mathcal{T}_\Lambda$ to both sides of the expansion above :

$$\mathcal{T}_\Lambda[\psi[x, y, z]] = \sum_{k=1}^{N} \mathcal{T}_\Lambda\left[u_k[x, y, z]\right] \tag{37.2}$$

On the left, $\mathcal{T}_\Lambda[\psi]$ can be carried out using the **SymmetryRules** for group $\mathcal{G}$, but on the expansion side the $\mathcal{T}_\Lambda[u_k]$ are only symbolic, because the **u** functions are unknown.

3. To make further progress we must assert that the function space spanned by the **u** functions is invariant to all the transforms $\mathcal{T}_\Lambda$. Then we can use the definition of the representation to replace the $\mathcal{T}_\Lambda[u_k]$ by suitable linear combinations of the **u** functions.

4. Operate so as to create, on the right side, the left side of the Great Orthogonality. Then replace it by the right side, a numerical factor times three δ functions.

5. Do a sum enabled by one of the δ functions, plus other maneuvers that bring the equation into the form of an eigenequation. The eigenfunctions are functions of pure symmetry under group $\mathcal{G}$, and the eigenvalues are only **0** and **1**. Therefore, the operator is a projection operator.

37.3.2 Derivation of `ProjectED`

On the right side of Eq. 37.2 above, the summand is the left side of the rep ma formula from Chapter 25. We replace it by the right side :

$$\mathcal{T}_\Lambda[\psi] = \sum_{m=1}^{N} \mathcal{T}_\Lambda[u_m] == \sum_{m=1}^{N}\sum_{k=1}^{N} u_k\, T_{k,m}[\Lambda]$$

where the **u**'s are general basis objects that transform invariantly under the transforms of the group. But now we know about species and irreducible representations, and $T_{k,m}[\Lambda]$ can be the irreducible representation matrix for {species σ, element Λ} if we let the u_k functions be basis functions for it. We mark this change by switching from **u** to **b**, and introducing a species index σ :

$$\mathcal{T}_\Lambda\left[b_m^{(\sigma)}\right] = \sum_{k=1}^{d[\sigma]} b_k^{(\sigma)}\, T_{k,m}^{(\sigma)}[\Lambda] \tag{37.4}$$

where summation limit **N** has become **d[σ]**, the dimension of species σ; the $T_{k,m}^{(\sigma)}[\Lambda]$ are the matrices of rep σ; and the basis objects $b_k^{(\sigma)}$ are now specifically basis functions for species σ.

We will call the functions $b_k^{(\sigma)}$ "functions of pure $\mathcal{G}$ symmetry". Simple examples of them may be found on the right side of the character table, but the point here is that many much more complicated functions also have pure $\mathcal{G}$ symmetry. Once we have the projection operator, we will be able to generate them with a single **Enter** command.

Multiply on both sides by $T_{p,q}^{(\tau)*}[\Lambda]$ (note the conjugation) and sum over Λ (all elements of the group) :

$$\sum_{\Lambda=1}^{h} \mathcal{T}_\Lambda\left[b_m^{(\sigma)}\right] T_{p,q}^{(\tau)*}[\Lambda] = \sum_{\Lambda=1}^{h}\sum_{k=1}^{d[\sigma]} b_k^{(\sigma)}\, T_{k,m}^{(\sigma)}[\Lambda]\, T_{p,q}^{(\tau)*}[\Lambda] \tag{37.5}$$

These are finite sums and can be rearranged however you like :

$$\sum_{\Lambda=1}^{h} \mathcal{T}_\Lambda\left[b_m^{(\sigma)}\right] T_{p,q}^{(\tau)*}[\Lambda] = \sum_{k=1}^{d[\sigma]} b_k^{(\sigma)} \sum_{\Lambda=1}^{h} T_{k,m}^{(\sigma)}[\Lambda]\, T_{p,q}^{(\tau)*}[\Lambda] \tag{37.6}$$

Click back to the Great Orthogonality and pull a copy down here. You will see that the sum over Λ on the right above is identical to the left side of the Great Orthogonality. Therefore (this is huge) we replace it by the right side:

$$\sum_{\Lambda=1}^{h} \mathcal{T}_\Lambda\left[b_m^{(\sigma)}\right] T_{p,q}^{(\tau)*}[\Lambda] = \sum_{k=1}^{d[\sigma]} b_k^{(\sigma)} \frac{h}{d[\tau]}\, \delta_{\sigma,\tau}\, \delta_{k,p}\, \delta_{m,q} \tag{37.7}$$

Move the sum over **k** as far right as possible :

$$\sum_{\Lambda=1}^{h} \mathcal{T}_\Lambda\left[b_m^{(\sigma)}\right] T_{p,q}^{(\tau)*}[\Lambda] = \frac{h}{d[\tau]}\, \delta_{\sigma,\tau}\, \delta_{m,q} \sum_{k=1}^{d[\sigma]} b_k^{(\sigma)}\, \delta_{k,p} \tag{37.8}$$

Now perform the sum over **k**. Every summand is zero except when $k = p$, so

$$\sum_{\Lambda=1}^{h} \mathcal{T}_\Lambda\left[\mathbf{b}_m^{(\sigma)}\right] \mathbf{T}_{p,q}^{(\tau)*}[\Lambda] = \frac{h}{d[\tau]} \mathbf{b}_p^{(\sigma)} \delta_{\sigma,\tau} \delta_{m,q}$$

Rearrange a bit :

$$\frac{d[\tau]}{h} \sum_{\Lambda=1}^{h} \mathcal{T}_\Lambda\left[\mathbf{b}_m^{(\sigma)}\right] \mathbf{T}_{p,q}^{(\tau)*}[\Lambda] = \mathbf{b}_p^{(\sigma)} \delta_{\sigma,\tau} \delta_{m,q} \tag{37.10}$$

This equation has five free indices; σ, τ, $\mathbf{m}$, $\mathbf{p}$ and $\mathbf{q}$. So we can turn this into an eigenequation by simply setting the free index $\mathbf{m}$ equal to $\mathbf{p}$. It becomes an eigenequation because the same function will then appear on both sides:

$$\frac{d[\tau]}{h} \sum_{\Lambda=1}^{h} \mathcal{T}_\Lambda\left[\mathbf{b}_p^{(\sigma)}\right] \mathbf{T}_{p,q}^{(\tau)*}[\Lambda] = \mathbf{b}_p^{(\sigma)} \delta_{\sigma,\tau} \delta_{p,q} \tag{37.11}$$

The fog will lift a little if we make a simple symbol for the operator of this eigenequation. Letting $\mathbf{g}$ be a dummy operand, the operator definition is

$$\mathcal{P}_{p,q}^{(\tau)}[g\_] := \frac{d[\tau]}{h} \sum_{\Lambda=1}^{h} \mathcal{T}_\Lambda[g]\, \mathbf{T}_{p,q}^{(\tau)*}[\Lambda] \tag{37.12}$$

enabling us to write

$$\mathcal{P}_{p,q}^{(\tau)}\left[\mathbf{b}_p^{(\sigma)}\right] = \mathbf{b}_p^{(\sigma)} \delta_{\sigma,\tau} \delta_{p,q} \tag{37.13}$$

If we change $\mathcal{P}_{p,q}^{(\tau)}$ on the left into $\mathcal{P}_{q,q}^{(\tau)}$ the equation is really the same, because there is still a $\mathbf{p}$ and a $\mathbf{q}$ on the left and a $\delta_{p,q}$ on the right. But since the two lower indices on $\mathcal{P}$ will now always be the same, we can write only one. So

$$\mathcal{P}_{q}^{(\tau)}\left[\mathbf{b}_p^{(\sigma)}\right] = \mathbf{b}_p^{(\sigma)} \delta_{\sigma,\tau} \delta_{p,q} \tag{37.14}$$

This is the simple eigenequation we wanted. It says that whenever $\mathcal{P}_q^{(\tau)}$ operates on a function of pure $\mathcal{G}$ symmetry, it either gives the function back unchanged (if both index pairs match) or else it gives zero (if one index pair doesn't match). This is what is meant by a symmetry projection operator. This $\mathcal{P}_q^{(\tau)}$ becomes **ProjectED** in the **Symmetry`** package . So box it up:

$$\boxed{\mathcal{P}_{q}^{(\tau)}[g\_] := \frac{d[\tau]}{h} \sum_{\Lambda=1}^{h} \mathcal{T}_\Lambda[g]\, \mathbf{T}_{q,q}^{(\tau)*}[\Lambda]} \tag{37.15}$$

Think about applying $\mathcal{P}_q^{(\tau)}$ to some function from a space that is spanned by all the functions that have pure $\mathcal{G}$ symmetry. This is an enormous space, containing every function that a chemist will ever be interested in. To be concrete, let the group $\mathcal{G}$ be $\mathbf{D_4}$, as in the introductory numerical example. Then any function Ψ can be written as a sum of parts having pure $\mathcal{G}$ symmetry

$$\Psi = \psi_1^{A_1} + \psi_1^{A_2} + \psi_1^{B_1} + \psi_1^{B_2} + \psi_1^{E} + \psi_2^{E}$$

The subscripts **1** on the **A** and **B** species are redundant because there is only one function in those cases, but we need it because it appears in the operator. The particular rep **E** used in the construction of $\mathcal{P}$ (namely, the matrices $\mathbf{T}_{a,b}^{(\tau)*}$) requires two basis functions, and there they are, ψ_1^{E} and ψ_2^{E}. Now apply a particular $\mathcal{P}$ operator to Ψ; for example, $\mathcal{P}_2^{(E)}$. Since the eigenvalue is zero except when both index pairs match

$$\mathcal{P}_2^{(E)}[\Psi] = 0 + 0 + 0 + 0 + 0 + \psi_2^{E} \tag{37.17}$$

This is what is meant by *group theoretic projection*. In words, operator $\mathcal{P}_2^{(E)}$ projects out from any function Ψ a part that has pure symmetry **E,2**. If Ψ contains no such part, then it will return zero. But "usually", it will not return zero. There is no limit to how complicated Ψ can be, and therefore no limit on how complicated ψ_2^{E} can be.

We will hold the numerical examples until after the next section.

On your own
Call up the syntax statement for **ProjectED** and read it. Click into the Symmetry` package to read it. Then use it.

37.3.3 Derivation of `ProjectET`

But wait! There's more! If you buy all this, we will give you, absolutely FREE, a SECOND projection operator that's EVEN NEATER than the first!

Start with the general projection operator, Eq. 37.15 and sum over **q**. We denote the summed projector by the same $\mathcal{P}$, just leaving off the subscript **q** :

$$\mathcal{P}^{(\tau)}[g\_] = \sum_{q=1}^{d[\tau]} \mathcal{P}_q^{(\tau)}[g] = \sum_{\Lambda=1}^{h} \mathcal{T}_\Lambda[g] \sum_{q=1}^{d[\tau]} \mathbf{T}_{q,q}^{(\tau)*}[\Lambda] \tag{37.18}$$

Do you see it ? The summing $\mathbf{T}_{q,q}^{(\tau)*}$ over **q** creates $\chi^{(\tau)*}$, the conjugated character of the matrix Λ.

$$\mathcal{P}^{(\tau)}[g\_] := \sum_{\Lambda=1}^{h} \mathcal{T}_\Lambda[g]\, \chi^{(\tau)*}[\Lambda] \tag{37.19}$$

This is called the trace projection operator, **ProjectET**. Its construction no longer requires a whole representation; it needs only the characters, and these can be read straight from character tables. To write its eigenequation, bring down the general eigenequation, Eq. 37.17 , and look at it again:

$$\mathcal{P}_{\mathbf{q}}^{(\tau)}\left[\mathbf{b}_{\mathbf{p}}^{(\sigma)}\right] = \mathbf{b}_{\mathbf{p}}^{(\sigma)}\ \delta_{\sigma,\tau}\,\delta_{\mathbf{p},\mathbf{q}} \qquad \text{(Eq. 37.17 again)}$$

Take the case $\mathbf{p}\to\mathbf{q}$, and sum over $\mathbf{q}$:

$$\sum_{\mathbf{q}=1}^{\mathbf{d}[\tau]} \mathcal{P}_{\mathbf{q}}^{(\tau)}\left[\mathbf{b}_{\mathbf{q}}^{(\sigma)}\right] = \left(\sum_{\mathbf{q}=1}^{\mathbf{d}[\tau]} \mathbf{b}_{\mathbf{q}}^{(\sigma)}\right)\delta_{\sigma,\tau} = \mathbf{b}_{\mathbf{sum}}^{(\sigma)}\,\delta_{\sigma,\tau} \qquad (37.20)$$

On the right, $\mathbf{b}_{\mathbf{sum}}^{(\sigma)}$belongs to species σ, but the components are all mixed together in it. On the left, we do not have a simple sum of $\mathcal{P}$ operators, and we must examine it more closely. Write it out explicitly:

$$\left(\mathcal{P}_{1}^{(\tau)}\left[\mathbf{b}_{1}^{(\sigma)}\right] + \ldots + \mathcal{P}_{\mathbf{d}}^{(\tau)}\left[\mathbf{b}_{\mathbf{d}}^{(\sigma)}\right]\right) = \mathbf{b}_{\mathbf{sum}}^{(\sigma)}\,\delta_{\sigma,\tau} \qquad (37.21)$$

But you can add as many things as you like to the operands of the $\mathcal{P}$'s, provided they produce zero when $\mathcal{P}$ operates. Zero is exactly what all the other $\mathbf{b}$'s produce, even when $\sigma=\tau$, so add them in. Then each $\mathcal{P}$ operator acts on $\mathbf{b}_{\mathbf{sum}}^{(\sigma)}$, and the eigenequation becomes

$$\mathcal{P}_{1}^{(\tau)}\left[\mathbf{b}_{\mathbf{sum}}^{(\sigma)}\right] + \ldots + \mathcal{P}_{\mathbf{d}}^{(\tau)}\left[\mathbf{b}_{\mathbf{sum}}^{(\sigma)}\right] = \mathbf{b}_{\mathbf{sum}}^{(\sigma)}\,\delta_{\sigma,\tau} \qquad (37.22)$$

Since each $\mathcal{P}_{\mathbf{q}}^{(\tau)}$ is linear

$$\left(\mathcal{P}_{1}^{(\tau)} + \ldots + \mathcal{P}_{\mathbf{d}}^{(\tau)}\right)\left[\mathbf{b}_{\mathbf{sum}}^{(\sigma)}\right] = \mathbf{b}_{\mathbf{sum}}^{(\sigma)}\,\delta_{\sigma,\tau} \qquad (37.23)$$

But that operator sum is exactly what we defined as the trace operator $\mathcal{P}^{(\tau)}$, so the eigenequation for the trace projector is

$$\boxed{\mathcal{P}^{(\tau)}\left[\mathbf{b}_{\mathbf{sum}}^{(\sigma)}\right] = \mathbf{b}_{\mathbf{sum}}^{(\sigma)}\,\delta_{\sigma,\tau}} \qquad (37.24)$$

For one-dimensional species, this is just as powerful as the detailed projector. But for multidimensional species it gives only the sum of all the component functions you would get by detailed projection. The usual approach is to use **`ProjectET`** first, giving the 1-D basis functions and all the various $\mathbf{b}_{\mathbf{sum}}^{(\tau)}$ functions. Then construct full representations **`rep`**$^{(\tau)}$**`[`**$\mathcal{G}$**`]`** needed by **`Proj-ectED`** and use that operator to crack the nuts that have not yet broken.

37.4 Your own examples of ProjectET

On your own, you can now do as many examples as you find amusing. The main bother is the contour plots of the results, so we automate that as **Display⋱ Projections**. in the preliminaries.

To do your own example, change the **urFn** and the **groupName** below.

```
urFn = q[x, y];
groupName = "D3";
prjs = ProjectET[urFn, groupName] // Simplify;
allD3Fns = Join[{{"Ur", urFn}, prjs}] //
   Partition[Flatten[#], 2] &;
```

Interactive readers can remove the semicolon and see all the algebra. But the contour plots will suffice for the book:

```
DisplayProjections[allD3Fns]
```

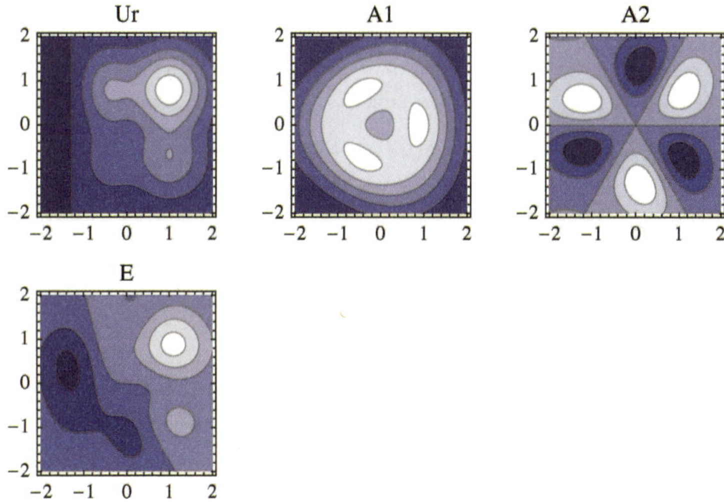

This is the same Ur function as before, but now the 1-D projections involve three-fold rotational symmetry. Before, under group D_4, it was four-fold. The E function is a sum of symmetries (E,1) and (E,2), which we split apart below.

37.5 Your own examples of ProjectED

To see **ProjectED** at work, you must supply all the inputs for **ProjectED[sumFn,rep,cmp,grp]**. We assume that **grp** will be the same:

```
grp = groupName
```

```
D3
```

and that **sumFnE** was returned by **ProjectET[urFn,grp]**. Pull it up :

```
{spLabel, sumFnE} = prjs[[3]]
```

$$\left\{E,\ \frac{1}{64}\, e^{-x^2-y^2} \left(-6 x^6 + 3 x^7 - 3 x^5 \left(-112 - 48 y + y^2\right) + 18 x^4 \left(-38 - 8 y + y^2\right) - 9 x y^2 \left(-192 + 112 y^2 - 16 y^3 + y^4\right) + 2 x^2 \left(-96 + 368 y + 2004 y^2 + 336 y^3 + 11 y^4\right) + x^3 \left(1728 + 1792 y + 4704 y^2 + 288 y^3 + 49 y^4\right) + 2 y \left(448 + 96 y + 368 y^2 - 326 y^3 + 24 y^4 - y^5\right)\right)\right\}$$

Its species is **E**, so to split it you must first make a **rep** of species **E**. Choose basis functions from the character table :

```
BoxedCharacterTable[groupName]
```

D3	1	2	3	← Class populations
	E	C3	C2	↓ Basis functions
A1	1	1	1	$\{(x^2+y^2),\ (z^2)\}$
A2	1	1	-1	$\{(z),\ (Iz)\}$
E	2	-1	0	$\left\{\begin{pmatrix} x \\ y \end{pmatrix},\ \begin{pmatrix} \frac{x^2}{2}-\frac{y^2}{2} \\ x y \end{pmatrix},\ \begin{pmatrix} x z \\ y z \end{pmatrix},\ \begin{pmatrix} Ix \\ Iy \end{pmatrix}\right\}$

```
basisFns = TabulatedBases[grp][[3, 2, 1]]
```

```
{x, y}
```

Make the rep:

```
repE = MakeRepPoly[basisFns, grp]
```

$$\left\{\{\{1,0\},\{0,1\}\},\left\{\left\{-\frac{1}{2},-\frac{\sqrt{3}}{2}\right\},\left\{\frac{\sqrt{3}}{2},-\frac{1}{2}\right\}\right\},\right.$$
$$\left\{\left\{-\frac{1}{2},\frac{\sqrt{3}}{2}\right\},\left\{-\frac{\sqrt{3}}{2},-\frac{1}{2}\right\}\right\},\{\{1,0\},\{0,-1\}\},$$
$$\left.\left\{\left\{-\frac{1}{2},\frac{\sqrt{3}}{2}\right\},\left\{\frac{\sqrt{3}}{2},\frac{1}{2}\right\}\right\},\left\{\left\{-\frac{1}{2},-\frac{\sqrt{3}}{2}\right\},\left\{-\frac{\sqrt{3}}{2},\frac{1}{2}\right\}\right\}\right\}$$

Do the detailed projection for component **1** . The next two operations are quite slow, but they do return. Be patient.

```
prj1 = ProjectED[sumFnE, repE, 1, grp] // Simplify
```

$$\frac{1}{64}e^{-x^2-y^2}\left(-6x^6+3x^7-3x^5\left(-112+y^2\right)+18x^4\left(-38+y^2\right)-\right.$$
$$9xy^2\left(-192+112y^2+y^4\right)-2y^2\left(-96+326y^2+y^4\right)+$$
$$\left.2x^2\left(-96+2004y^2+11y^4\right)+x^3\left(1728+4704y^2+49y^4\right)\right)$$

Now do the detailed projection for component **2**.

```
prj2 = ProjectED[sumFnE, repE, 2, grp] // Simplify
```

$$\frac{1}{4}e^{-x^2-y^2}y\left(56-9x^4+9x^5+46y^2+\right.$$
$$\left.3y^4+9xy^4+2x^3\left(56+9y^2\right)+x^2\left(46+42y^2\right)\right)$$

Put them all together and display the contour plots :

```
EFns = {{"Esum", sumFnE}, {"E1", prj1}, {"E2", prj2}};
DisplayProjections[EFns]
```

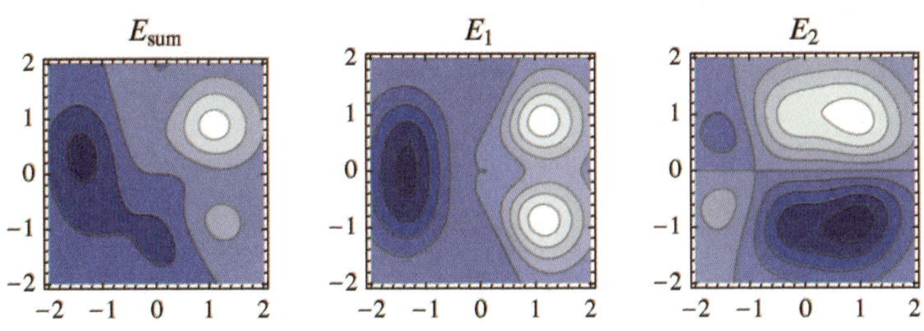

Make sure it all adds up:

```
Simplify[prj1 + prj2] == sumFnE
```

True

37.6 End notes

37.6.1 The "Totally symmetric factor " message

The simple incarnation of **ProjectED** can be very slow if it does not have a little help. Often the problem is in a totally symmetric convergence factor like **Exp**$\left[-\left(x^2+y^2\right)\right]$. *Mathematica* does not simplify powers like a human would do it, and sometimes they are like molasses. The cure we hit on seems clumsy, but it is quite fast when implemented.

1) Split **expr** into parts with head **Power**, times a **remainder**
2) Split each **Power** part into a base and an exponent
3) Run **TotallySymmetricQ** on each base and exponent.
(This is a private test that lives only in the package.)
4) If the powers are totally symmetric, project **remainder** and recombine.
5) If not, project the whole expression, however long it takes.

If it finds a totally symmetric factor, it issues an appropriate message.

38. Tabulated bases for representations

Preliminaries

38.1 Introduction

This chapter presents a straightforward application of the material from Chapter 37 (ProjectionOperators). All the character tables in this book have a basis function for every species of every group. Here we will show how they were found- by a simple application of the projection operators to the spherical harmonic functions. Chemists use these functions a lot because they are the angular part of the wave function for the electrons of any free atom. So the basis functions in our character tables should be especially good for chemical problems.

38.2 Basis sets from harmonic functions

38.2.1 Spherical harmonics as functions of spherical angles

Every spherical harmonic depends on two angles, the polar angle θ (latitude) and the equatorial angle ϕ (longitude) of the spherical coordinate system. They all have two parameters, **L** (the *rank*) and **m** (the *component*). They are best thought of in a ziggurat array:

```
Show[labelledZig]
```

			0, 0	s			
		1, −1	1, 0	1, +1	p		
	2, −2	2, −1	2, 0	2, +1	2, +2	d	
3, −3	3, −2	3, −1	3, 0	3, +1	3, +2	3, +3	f

The block labels are **L,m**. Each layer **L** of the ziggurat is a *rank*, and the individual blocks on that layer are its *components*. One rank spans the functions in one shell of an atom; s, p, d, f, …, as shown. The rank integers **L** begin with

W.M. McClain, *Symmetry Theory in Molecular Physics with Mathematica*,
DOI 10.1007/b13137_38, © Springer Science+Business Media, LLC 2009

zero at the top, and increase without limit going down. The component numbers **m** for rank **L** begin with **m = -L** on the left and run by integer increments to **m = +L** on the right, making **2L+1** components in each rank.

Any spherical harmonic can be called up in *Mathematica* by **SphericalHarmonicY[L,m,θ,ϕ]**. In the preliminaries we define an abbreviated notation $Y_{Ang}^{L,m}[\theta, \phi]$ closer to that seen in mathematics:

$$Y_{Ang}^{3,2}[\theta, \phi]$$

$$\frac{1}{4} e^{2 i \phi} \sqrt{\frac{105}{2 \pi}} \text{Cos}[\theta] \text{Sin}[\theta]^2$$

On your own
Try different values of **L** and **m** above; **m** is legal when **-L ≤ m ≤ +L**.
How does *Mathematica* deal with illegal **m**? Try one.

38.2.2 Cartesian spherical harmonics

Spherical to Cartesian

The spherical harmonic angular variables **θ** and **ϕ** designate a unique spot on the surface of a unit sphere. But of course so do the variables **x**, **y**, and **z** under the constraint $x^2 + y^2 + z^2 = 1$. But we must then deal with ambiguities like $x^2 + y^2 = 1 - z^2$. We will do so by a defining a *standard guise* for Cartesian harmonics.

We need the harmonics in Cartesian form in order to extract from them the tabulated basis functions for representations. In the preliminaries we define a function $Y_{Car}^{L,m}[x, y, z]$ that converts the angular form to Cartesian and returns them in a very compact form. If you are interested, click here and read how it is done. Here are the first rank functions:

$$\left\{Y_{Car}^{1,-1}[x, y, z], Y_{Car}^{1,0}[x, y, z], Y_{Car}^{1,+1}[x, y, z]\right\}$$

$$\left\{\frac{1}{2}\sqrt{\frac{3}{2\pi}} (x - i y), \frac{1}{2}\sqrt{\frac{3}{\pi}} z, -\frac{1}{2}\sqrt{\frac{3}{2\pi}} (x + i y)\right\}$$

The **L, ±m** pairs are always conjugate to each other.

The Cartesian harmonics look like they are functions of three independent variables **x,y,z** . But that is impossible, because they trace back to angular functions of only two independent angles. There can be only two *independent*

Cartesian variables. Indeed, the spherical-to-Cartesian rules in the definition of $Y_{Car}^{L,m}$ imply $x^2 + y^2 + z^2 = 1$. Thus, $Y_{Car}^{L,m}[x, y, z] = Y_{Ang}^{L,m}[\theta, \phi]$ provided that **x,y,z** and **θ,ϕ** refer to the same point on the unit sphere.

> **On your own**
> Verify this by numerical example. Generate two random angles in the right range, and convert them to a Cartesian point on the unit sphere. Then test whether $Y_{Ang}^{L,m}$ and $Y_{Car}^{L,m}$ return the same number. If you remember the spherical-to-Cartesian formula you will not need to click here.

Standard guise for Cartesian harmonics

Cartesian spherical harmonics are tougher to deal with than simple polynomials. Because of the unit sphere identity, the harmonics may appear in a number of different guises. For instance, z^2 is really the same as $1 - x^2 - y^2$. So we must settle on a standard guise for the spherical harmonics.

> We take our "standard guise" to be a form in which **z** appears to no power higher than **1**.

In the preliminaries we define an operator called **ToStandardGuise**. It uses **FixedPoint** to apply $z^2 \to 1 - x^2 - y^2$ repeatedly until the result stops changing. Try it on four arbitrarily chosen polynomials:

```
ToStandardGuise[
  {x^2 y^5 z^3, 5 x^2 y^5 z^2, 6 x^2 y^5 z, 7 x^4 y^3}];
% // Column
```

$x^2 y^5 z - x^4 y^5 z - x^2 y^7 z$

$5 x^2 y^5 - 5 x^4 y^5 - 5 x^2 y^7$

$6 x^2 y^5 z$

$7 x^4 y^3$

As advertised, the highest power of **z** in the output is **1**.

38.3 Example: basis functions for $D_{6\,d}$

38.3.1 Introduction

Here, as a single example, we find a complete set of basis functions for one group. We choose $\mathbf{D_{6\,d}}$ because the supporting polynomial for one of its species, $\mathbf{A_2}$, was particularly elusive. But as you will see, the systematic spherical harmonic method does finally provide it in the seventh rank. Here are the polynomials that are to be found :

```
TabulatedBases["D6d"] // MatrixForm // Size[8]
```

$$\begin{pmatrix} A1 & \{\{x^2+y^2\},\ \{z^2\}\} \\ A2 & \{\{x^6 z - 15 x^4 y^2 z + 15 x^2 y^4 z - y^6 z\},\ \{Iz\}\} \\ B1 & \{\{2 x y (x^2 - 3 y^2)(3 x^2 - y^2) z\}\} \\ B2 & \{\{z\},\ \{z^3\},\ \{6 x^5 y - 20 x^3 y^3 + 6 x y^5\}\} \\ E1 & \{\{x,\ y\}\} \\ E2 & \{\{x^2 - y^2,\ 2 x y\}\} \\ E3 & \{\{x^3 - 3 x y^2,\ 3 x^2 y - y^3\}\} \\ E4 & \{\{x^2 z - y^2 z,\ 2 x y z\}\} \\ E5 & \{\{x z,\ y z\},\ \{x^5 - 10 x^3 y^2 + 5 x y^4,\ 5 x^4 y - 10 x^2 y^3 + y^5\},\ \{Ix,\ Iy\}\} \end{pmatrix}$$

We will project the first few ranks explicitly. The rank **0** harmonic is just a constant, which always belongs to the totally symmetric species. But after that the first few ranks always provide very useful small polynomials.

38.3.2 Rank 0, the s shell (species A_1)

```
L = 0;
polys0 = Table[Y_Car^{L,m}[x, y, z], {m, 0, L}]
```

$$\left\{\frac{1}{2\sqrt{\pi}}\right\}$$

There is only a constant, which we interpret as $\mathbf{x^2 + y^2 + z^2}$, the constant radius of all the Cartesian harmonics. Since $\mathbf{D_{6\,d}}$ is an axial group, we suspect that this is the sum of two invariant polynomials. Try it :

```
ProjectET[{x^2 + y^2, z^2}, "D6d"]
```

```
{{A1, {x² + y², z²}}, {A2, {0, 0}}, {B1, {0, 0}},
 {B2, {0, 0}}, {E1, {0, 0}}, {E2, {0, 0}},
 {E3, {0, 0}}, {E4, {0, 0}}, {E5, {0, 0}}}
```

We list both of these under species $\mathbf{A_1}$.

38.3.3 Rank 1, the p shell (species B_2 and E_1)

```
L = 1;
polys1 = Table[Y_Car^{L,m}[x, y, z], {m, 0, L}]
```

$$\left\{\frac{1}{2}\sqrt{\frac{3}{\pi}}\, z,\ -\frac{1}{2}\sqrt{\frac{3}{2\pi}}\,(x + i\, y)\right\}$$

The harmonic amplitudes are of no importance to us here, so we renormalize each polynomial so that its leading term has no coefficient. The operator is **NeatPoly**, defined in the preliminaries of this chapter. Click to see it.

```
neatPolys1 = Map[NeatPoly, polys1]
```

```
{z, x + i y}
```

We project these simplified functions under group $\mathbf{D_{6d}}$:

```
prj1 = ProjectET[neatPolys1, "D6d"];
Map[Flatten, %] // MatrixForm // Size[6]
```

$$\begin{pmatrix} A1 & 0 & 0 \\ A2 & 0 & 0 \\ B1 & 0 & 0 \\ B2 & z & 0 \\ E1 & 0 & x + i\, y \\ E2 & 0 & 0 \\ E3 & 0 & 0 \\ E4 & 0 & 0 \\ E5 & 0 & 0 \end{pmatrix}$$

1. For species $\mathbf{B_2}$ we take the function **z**.

2. For species $\mathbf{E_1}$ we take **x + i y**. But when a complex expression appears in a two-dimensional species, the real and imaginary parts span a real 2-dimensional function space invariant under all the transforms of the group. Test this:

```
{x, y} /. SymmetryRules["D6h"];
Short[%, 6] // Size[7]
```

$$\left\{\{x,\ y\},\ \left\{\frac{x}{2}-\frac{\sqrt{3}\ y}{2},\ \frac{\sqrt{3}\ x}{2}+\frac{y}{2}\right\},\right.$$
$$\left\{\frac{x}{2}+\frac{\sqrt{3}\ y}{2},\ -\frac{\sqrt{3}\ x}{2}+\frac{y}{2}\right\},\ \left\{-\frac{x}{2}-\frac{\sqrt{3}\ y}{2},\ \frac{\sqrt{3}\ x}{2}-\frac{y}{2}\right\},\ \ll 17\gg,$$
$$\left.\{-x,\ y\},\ \left\{\frac{x}{2}-\frac{\sqrt{3}\ y}{2},\ -\frac{\sqrt{3}\ x}{2}-\frac{y}{2}\right\},\ \left\{\frac{x}{2}+\frac{\sqrt{3}\ y}{2},\ \frac{\sqrt{3}\ x}{2}-\frac{y}{2}\right\}\right\}$$

Indeed, each transform is a linear combination of **x** and **y**. So the table entry will be **{x,y}**.

> **On your own**
> Make a representation of $\mathbf{D_{6\,d}}$ on the basis **{x,y}**, and show that it is unitary. Click here if you can't think how to do this.

38.3.4 Rank 2, the d shell (species E_2 and E_5)

Do everything in one cell :

```
L = 2;
polys2 = Table[Y_Car^(L,m)[x, y, z], {m, 0, L}];
neatPolys2 = Map[NeatPoly, polys2];
prj2 = ProjectET[neatPolys2, "D6d"];
Map[Flatten, %] // MatrixForm // Size[6];
```

$$\begin{pmatrix} A1 & 1-3z^2 & 0 & 0 \\ A2 & 0 & 0 & 0 \\ B1 & 0 & 0 & 0 \\ B2 & 0 & 0 & 0 \\ E1 & 0 & 0 & 0 \\ E2 & 0 & 0 & x^2+2\,i\,x\,y-y^2 \\ E3 & 0 & 0 & 0 \\ E4 & 0 & 0 & 0 \\ E5 & 0 & x\,z+i\,y\,z & 0 \end{pmatrix}$$

3. In species **A1** the constant **1** and the polynomial $\mathbf{z^2}$ project as a weighted sum, but actually the parts belong separately to **A1**, because constants always project into the totally symmetric species. So just the $\mathbf{z^2}$ part is tabulated.

4. In species $\mathbf{E_2}$, the real and imaginary parts produce the pair $\{\mathbf{x^2-y^2,\ 2\,x\,y}\}$. We retain the coefficient **2** that goes with **x y**. This is quite important; we will comment on this further below.

5. In species **E5** we tabulate **{x z, y z}**.

> **On your own**
> Make a representation of D_{6d} on the basis $\{x^2 - y^2, xy\}$, (leaving off the factor of **2** from **2 x y**), and show that it is **not** unitary.

38.3.5 Rank 3, the f shell (species E_3 and E_4)

We take another step toward complete automation by removing the uninteresting zeroes from the projected results. The operator is **NoZeroes** ; click here to see it.

```
L = 3;
polys3 = Table[Y^{L,m}_Car[x, y, z], {m, 0, L}];
neatPolys3 = Map[NeatPoly, polys3];
prj3 = ProjectET[neatPolys3, "D6d"];
NoZeroes[prj3];
% // MatrixForm
```

$$\begin{pmatrix} B2 & \left\{z - \frac{5z^3}{3}\right\} \\ E1 & \{x + i\,y - 5\,x\,z^2 - 5\,i\,y\,z^2\} \\ E3 & \{x^3 + 3\,i\,x^2\,y - 3\,x\,y^2 - i\,y^3\} \\ E4 & \{x^2\,z + 2\,i\,x\,y\,z - y^2\,z\} \end{pmatrix}$$

6. In species B_2, we see **z** again. Clearly z^3 projects separately; we use it.

7. In species E_1, we already know $x + i\,y$, so subtracting it we are left with $-5\,z^2\,(x - i\,y)$. We could have made an entry $\{z^2\,x,\ z^2\,y\}$, but we elected not to.

8. From species E_3 we tabulate $\{x^3 - 3\,x\,y^2,\ 3\,x^2\,y - y^3\}$.

9. From species E_4,we tabulate $\{x^2\,z - y^2\,z,\ 2\,x\,y\,z\}$.

38.3.6 What about A_2 and B_1?

And so it goes. We are up to rank **3** and we have nothing yet for **A2** or **B1**. Ranks **4** and **5** will not produce them either, but finally an **A2** projection appears when **L = 7**, and a **B1** appears when **L = 8**. You can try them below, changing **L** as necessary:

```
L = 7; (*or any other rank *)
gp = "D6d"; (* or any other group *)
polys = Table[Y_Car^{L,m}[x, y, z], {m, 0, L}];
neatPolys = Map[NeatPoly, polys];
prj1 = ProjectET[neatPolys, gp];
{prj2a, prj2b} = Transpose[prj1];
prj2c = Map[Select[#, (# =!= 0) &] &, prj2b];
prj3 = Transpose[{prj2a, prj2c}];
prj4 = Select[prj3, #[[2]] =!= {} &];
prj5 = Map[Simplify, prj4]
```

$$\Big\{\{A1, \{2\,i\,x\,y\,(3x^4 - 10x^2y^2 + 3y^4)\,z\}\},$$
$$\{A2, \{(x^6 - 15x^4y^2 + 15x^2y^4 - y^6)\,z\}\},$$
$$\{B2, \{z - 9z^3 + \frac{99z^5}{5} - \frac{429z^7}{35}\}\},$$
$$\{E1, \{-\frac{1}{5}(x + i\,y)\,(-5 + 135z^2 - 495z^4 + 429z^6)\}\},$$
$$\{E2, \{-\frac{1}{3}(x + i\,y)^4 z\,(-3 + 13z^2)\}\},$$
$$\{E3, \{\frac{1}{3}(x + i\,y)^3\,(3 - 66z^2 + 143z^4)\}\},$$
$$\{E4, \{\frac{1}{15}(x + i\,y)^2 z\,(15 - 110z^2 + 143z^4)\}\},$$
$$\{E5, \{-(x + i\,y)^5\,(-1 + 13z^2),\ (x + i\,y)^7\}\}\Big\}$$

In the lower harmonics, each row (each species) has only one nonzero entry, i.e., each harmonic function belongs to some pure symmetry species. But higher harmonics become so complicated that they have to split into several parts. You can see an example of this in rank **7** under $\mathbf{D_{6\,d}}$, above. Component **6** in this rank is $\mathbf{(x + i\,y)^6\,z}$, but this does not appear among the projections. The real part is in $\mathbf{A_2}$, but the imaginary part is in $\mathbf{A_1}$.

38.4 End Notes

38.4.1 Ziggurats

The *ziggurat* is a Babylonian temple built as a series of recessed terraces. Click here for a web page on ziggurats. Click here for the earlier similar Egyptian structures.

38.4.2 Equivalence of Y_{Ang} and Y_{Car}

```
{α, β} = π { RandomReal[{0, 1}], RandomReal[{0, 2}]}
```

```
{0.857543, 1.3431}
```

On the unit sphere, `x = Cos[β] Sin[α]`, `y = Sin[β] Sin[α]`, and `z = Cos[α]`. Therefore we set

```
{ξ, η, ζ} = {Cos[β] Sin[α], Sin[β] Sin[α], Cos[α]}
```

```
{0.170708, 0.736718, 0.654298}
```

Pick arbitrary rank and component indices (below, 5 and 2). The acid test:

$$\left(Y^{5,2}_{Car}[\xi, \eta, \zeta] - Y^{5,2}_{Ang}[\alpha, \beta]\right) \text{ // Chop}$$

```
0
```

So the angular and Cartesian forms return the same output.

38.4.3 Checking unitarity

It's a one-liner:

```
MakeRepPoly[{x, y}, "D6d"];
Map[UnitaryQ, %] // Union
```

```
{True}
```

39. Quantum matrix elements

Preliminaries

39.1 Matrix elements defined

In quantum mechanics we often encounter expressions of the form

$$\int \psi_i^*[\underset{\rightharpoondown}{\tau}]\ Q\left[\psi_j[\underset{\rightharpoondown}{\tau}]\right]\ d\underset{\rightharpoondown}{\tau}$$

where ψ_i and ψ_j are the eigenfunctions of some Hamiltonian. Operator Q may be anything, but it is usually Hermitian. The integral is a multiple integral over the whole domain of variables $\underset{\rightharpoondown}{\tau}$, which might be, for instance, $\{x, y, z\}$, each running from $-\infty$ to $+\infty$. Explicitly,

$$\int_{-\infty}^{+\infty}\int_{-\infty}^{+\infty}\int_{-\infty}^{+\infty}\psi_i^*[x,\ y,\ z]\ Q_{x,y,z}[\psi_j[x,\ y,\ z]]\ dx\, dy\, dz$$

Two shorthand notations are commonly seen; namely,

$$\langle i|Q|j\rangle \quad \text{or} \quad Q_{i,j}.$$

The latter notation brings to mind a matrix, $Q_{i,j}$ being the element in row i, column j. In fact, these expressions are called *matrix elements*.

Group theory has a lot to say about matrix elements. It is especially helpful in identifying those that are zero, or nonzero matrix elements that are equal. It does not help with finding the numerical values, however.

The term **totally symmetric** comes up a lot in this chapter. The totally symmetric species of any group is the one whose character vector is a list of **1**'s. All groups have a totally symmetric species; it is always given as the top row of the character table; it is the “trivial” rep. In other words, a function that belongs to the totally symmetric species is unchanged under every transform in the group.

W.M. McClain, *Symmetry Theory in Molecular Physics with Mathematica*,
DOI 10.1007/b13137_39, © Springer Science+Business Media, LLC 2009

39.2 The Bedrock Theorem, and its corollary

39.2.1 The theorem

Here is the full formal statement, for the record. We will pull it apart into more digestible chunks below.

The Bedrock Theorem, general statement

Let the functions $\left\{\mathbf{p}_1^{\mathbf{U}}, \ldots, \mathbf{p}_{\mathbf{nU}}^{\mathbf{U}}\right\}$ be the basis for a unitary irreducible rep of species $\mathbf{U}$ of group $\mathcal{G}$, and let functions $\left\{\mathbf{q}_1^{\mathbf{V}}, \ldots, \mathbf{q}_{\mathbf{nV}}^{\mathbf{V}}\right\}$ be the basis for a unitary irreducible rep of species $\mathbf{V}$. Then (39.1)

$$\langle \mathbf{p}_i^{\mathbf{U}*}\, \mathbf{q}_j^{\mathbf{V}} \rangle = 0$$

if the **species** are different ($\mathbf{U}$ $\mathbf{V}$) or if the **components** are different ($\mathbf{i}$ $\mathbf{j}$).

Important comment:

If the functions $\left\{\mathbf{p}_1^{\mathbf{U}}, \ldots, \mathbf{p}_{\mathbf{nU}}^{\mathbf{U}}\right\}$ (or $\left\{\mathbf{q}_1^{\mathbf{U}}, \ldots, \mathbf{q}_{\mathbf{nU}}^{\mathbf{U}}\right\}$) are basis functions for a rep, they must span a space that is invariant under all the transforms of the group. Click here to review the term *invariant space.*

Specialize the general statement for functions of different species:

The Bedrock Theorem for different species

If function $\mathbf{p}^{\mathbf{U}}$ belongs to species $\mathbf{U}$ of group $\mathcal{G}$, and if $\mathbf{q}^{\mathbf{not}\,\mathbf{U}}$ belongs to a different species of the same group, then (39.2)

$$\langle \mathbf{p}^{\mathbf{U}*}\ \mathbf{q}^{\mathbf{not}\,\mathbf{U}} \rangle = 0$$

This is the way most people remember the Bedrock Theorem. But the general statement (involving that *invariant space* stuff) becomes important when both functions belong to the same multidimensional species (a species $\mathbf{E}$, or $\mathbf{T}$, say).

The Bedrock Theorem has an important corollary. To keep it simple, we state it below only for the two-dimensional species $\mathbf{E}$:

39.2.2 The corollary

Bedrock Corollary for species E

If $\{p_1^E, p_2^E\}$ and $\{q_1^E, q_2^E\}$) support the same unitary representation of the two dimensional species **E**, then

$$\langle p_1^{E*}\, q_1^E\rangle = \langle p_2^{E*}\, q_2^E\rangle \quad \text{and}$$

$$\langle p_1^{E*}\, q_2^E\rangle = \langle p_2^{E*}\, q_1^E\rangle = 0 \tag{39.3}$$

First comment

This often causes two matrix elements to have exactly the same size, even when it is not clear geometrically how symmetry is making it happen. We will do in detail the two perpendicular transition moments of benzene, which are equal by this corollary.

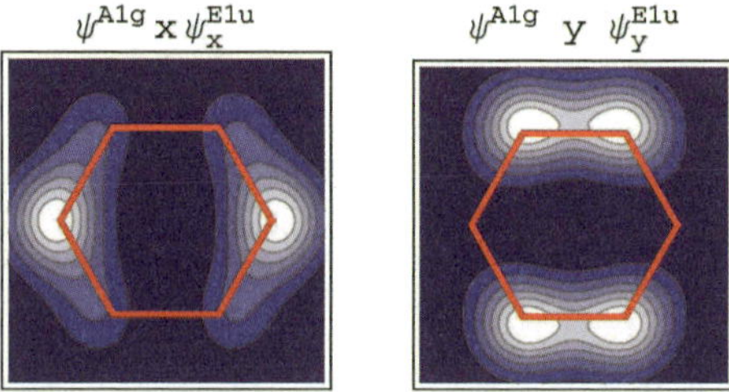

Fig. 39.1 **Two integrands that have exactly the same integral. Can you see why, by geometrical reasoning? (No way ...)**

Second comment

The word *unitary* above is important. The relative sizes of p_1^E and p_2^E (and of q_1^E and q_2^E) are set by the fact that they support a unitary representation. To see the difficulty that non-unitarity leads to, double the second member to make basis sets $\{p_1^E, 2\,p_2^E\}$ and $\{q_1^E, 2\,q_2^E\}$. These will still support a common representation, but it will not be unitary. And of course the **1,1** and **2,2** matrix elements can no longer be equal, since **2,2** will be four times as large as formerly.

39.3 Example using two polynomials

39.3.1 Define two polynomials and project them

Define polynomials **pFn** and **qFn**. Each one is a homogeneous polynomial plus a totally symmetric polynomial. Click to review the definition and basic properties. We use orders 5 and 7. But homogeneous polynomials of odd order (times a totally symmetric convergence factor) always integrate to zero, and this is undesirable here. So the totally symmetric polynomial is added on to keep that projection from integrating to zero. This does not affect the species **E** part of the projection, which is our focus below.

```
pFn[x_, y_] :=
  (x^7 - 5 x^6 y + 6 x^5 y^2 - 50 x^4 y^3 + 2 x^3 y^4 + 7 x^2 y^5 -
     9 x y^6 + 5 y^7) + 10 (x^2 + y^2)^3
```

```
qFn[x_, y_] :=
  (x^5 + 7 x^4 y + 60 x^3 y^2 + 7 x^2 y^3 + 3 x y^4 + y^5) - 3 (x^2 + y^2)^2;
```

```
DisplayCPs[
 {{"pFn", pFn[x, y]}, {"qFn", qFn[x, y]}}, 3, 220]
```

pFn

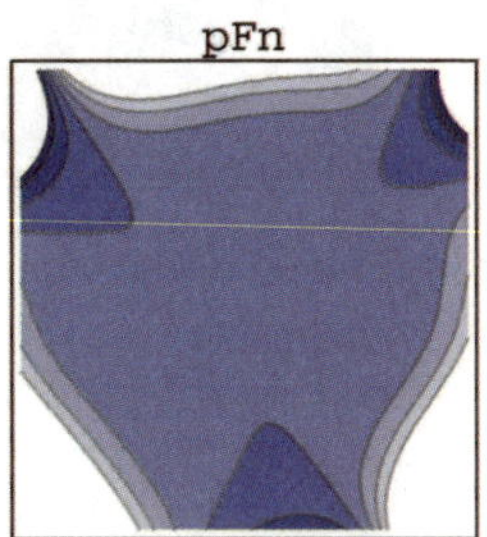

qFn

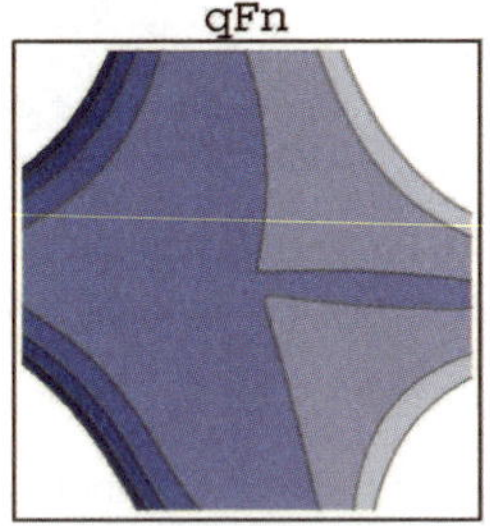

Fig. 39.2 Contour plots of two asymmetric polynomials in x and y

We will use **repE**, constructed below, to do the detailed projection.

```
repE = MakeRepPoly[{x, y}, "D3"];
% // GridList // Size[6]
```

$$\left\{\begin{array}{|c|c|}\hline 1 & 0 \\ \hline 0 & 1 \\ \hline\end{array}, \begin{array}{|c|c|}\hline -\frac{1}{2} & -\frac{\sqrt{3}}{2} \\ \hline \frac{\sqrt{3}}{2} & -\frac{1}{2} \\ \hline\end{array}, \begin{array}{|c|c|}\hline -\frac{1}{2} & \frac{\sqrt{3}}{2} \\ \hline -\frac{\sqrt{3}}{2} & -\frac{1}{2} \\ \hline\end{array}, \begin{array}{|c|c|}\hline 1 & 0 \\ \hline 0 & -1 \\ \hline\end{array}, \begin{array}{|c|c|}\hline -\frac{1}{2} & \frac{\sqrt{3}}{2} \\ \hline \frac{\sqrt{3}}{2} & \frac{1}{2} \\ \hline\end{array}, \begin{array}{|c|c|}\hline -\frac{1}{2} & -\frac{\sqrt{3}}{2} \\ \hline -\frac{\sqrt{3}}{2} & \frac{1}{2} \\ \hline\end{array}\right\}$$

On your own

The character table for $\mathbf{D_3}$ shows an alternative basis set for the **E** representation; namely, $\{(x^2 - y^2),\ 2\,x\,y\}$. Show that the representation supported by $\{(x^2 - y^2),\ -2\,x\,y\}$ (note the minus sign) is identical to that supported by **{x,y}**. That is why we call it **repE**, not **repExy**.

```
Sort[repE] ==
 Sort[MakeRepPoly[{(x^2 - y^2), -2 x y}, "D3"]]
```

```
True
```

In the closed cell below, we carry out the entire projection of both the **p** and **q** polynomials, using both **ProjectET** and **ProjectED**, and then plot the projected parts.

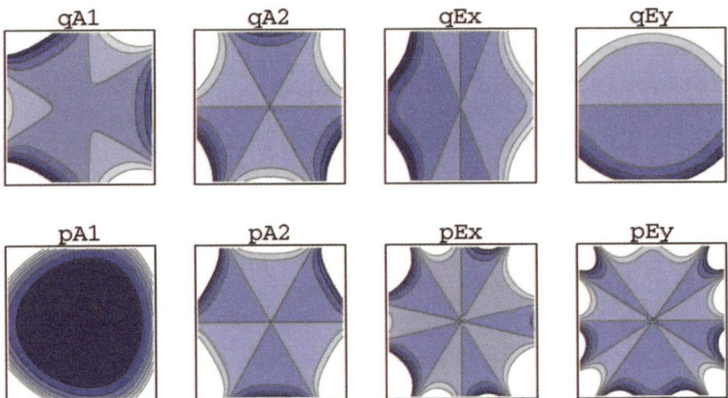

Fig. 39.3 Complete projections of the two asymmetric polynomials, group $\mathbf{D_3}$

Note that **pEx** looks odd in **x** and even in **y**, like **x** itself, while **pEy** looks odd in **y** and even in **x**, like **y** itself. Similarly, for **q**. Check this in the formulas:

```
{{32 pEx, 64 pEy}, {qEx, 16 qEy}} // Expand //
  GridForm // Size[9]
```

$-13x^7 + 237x^5y^2 + 289x^3y^4 - 153xy^6$	$559x^6y - 1735x^4y^3 + 741x^2y^5 + 27y^7$
$5x^5 + 52x^3y^2 - 9xy^4$	$43x^4y + 66x^2y^3 + 39y^5$

Do the projected parts add up correctly?

```
{Expand[pFn[x, y]] == Expand[pA1 + pA2 + pEx + pEy],
 Expand[qFn[x, y]] == Expand[qA1 + qA2 + qEx + qEy]}
```

```
{True, True}
```

39.3.2 Integrate the projected parts

There is a corollary to the Bedrock Theorem that is easy to demonstrate numerically right now. To see it, all you have to do is integrate the projected parts.

We define an operator **integral2D** containing a totally symmetric convergence factor. This is essentially a redefinition of **p** and **q** which gives them a new factor $\texttt{Exp}\left[-\left(x^2+y^2\right)/2\right]$. This does not alter their symmetries because the new factor is totally symmetric under all point groups.

Go to **Help > DocumentationCenter** and in the top bar type in **FunctionsThatRememberValuesTheyHaveFound** to read about the slightly strange definition below. The strangeness is a speedup trick that was useful in developing this chapter, to avoid computing the same integral twice.

```
Clear[integral2D];
integral2D[u_] := integral2D[u] =
```

$$\int_{-\infty}^{+\infty}\int_{-\infty}^{+\infty} \texttt{Evaluate}\left[\texttt{Exp}\left[-\left(x^2+y^2\right)\right]u\right] \,dx\,dy$$

Two-dimensional integrals can be slow, but the following four are not too bad:

```
Map[integral2D, {pA1, pA2, pEx, pEy}]
```

$\{60\pi, 0, 0, 0\}$

Everything except **pA1**, the totally symmetric part, integrated to zero. Try the same thing on the **q** polynomial :

```
Map[integral2D, {qA1, qA2, qEx, qEy}]
```

$\{-6\pi, 0, 0, 0\}$

Again, the same thing. This always happens. This is not the complete Bedrock Theorem, but this is an important consequence, and it is the way many people remember the theorem. So it is good to see it happening.

On your own
This does **not** mean that you never get zero when you integrate a totally symmetric function. Make a homogeneous polynomial of odd degree, project it, and use **integral2D** on its totally symmetric part. You will see.

39.3.3 The pair product integrals

Now we are ready to test the Bedrock Theorem. Make a 4×4 array of the integrands, **pPart × qPart** :

```
pqIntegrands = Outer[Times, pParts, qParts];
Dimensions[pqIntegrands]
```

```
{4, 4}
```

Mathematica can do all 16 of the **pqIntegrands** exactly, but the first time through it takes a little while. So be patient.

Integrals	qA1	qA2	qEx	qEy
pA1	-5625π	0	0	0
pA2	0	$-\frac{303255\pi}{128}$	0	0
pEx	0	0	2385π	0
pEy	0	0	0	$\frac{61605\pi}{128}$

We certainly see a nice diagonality, as predicted by the Bedrock Theorem. But didn't the Bedrock Corollary say that **⟨pEx ∗ qEx⟩ = ⟨pEy ∗ qEy⟩**? This is just not true. We see **⟨pEx*qEx⟩=2385**π, while **⟨pEy*qEy⟩=** $\frac{61605\pi}{128}$. Something is amiss.

To quote Groucho Marx (caught cheating in *Duck Soup*) "This is not what it looks like… Who are you going to believe? Me, or your own eyes? "

39.3.4 A very common error

Before we denounce and repudiate the Bedrock Corollary, read it again, very carefully :

> **Bedrock Corollary for species E**
>
> If $\{p_1^E, p_2^E\}$ and $\{q_1^E, q_2^E\}$) support the same unitary representation of the two dimensional species **E**, then ...
>
> HEY, HOLD IT RIGHT THERE

(39.4)

In the numerical example, do these function pairs support the same unitary representation? Neither pair can support any representation unless it spans an invariant space. We can test that very easily. Make an invariant space based on **pEx**, and see if **pEy** pops forth as the complementary invariant partner:

```
{MISx, MISy} = MakeInvariantSpace[pEx, "D3"]
```

$$\left\{x^7 - \frac{237\,x^5\,y^2}{13} - \frac{289\,x^3\,y^4}{13} + \frac{153\,x\,y^6}{13},\; x^6\,y - \frac{211\,x^4\,y^3}{179} - \frac{159\,x^2\,y^5}{179} + \frac{39\,y^7}{179}\right\}$$

We don't need equality, just proportionality:

$$\left\{\frac{\mathbf{MISx}}{\mathbf{pEx}}, \frac{\mathbf{MISy}}{\mathbf{pEy}}\right\} \text{ // FullSimplify}$$

$$\left\{-\frac{32}{13}, \frac{64\left(179\,x^6 - 211\,x^4\,y^2 - 159\,x^2\,y^4 + 39\,y^6\right)}{179\left(559\,x^6 - 1735\,x^4\,y^2 + 741\,x^2\,y^4 + 27\,y^6\right)}\right\}$$

Functions **MISx** and **pEx** are indeed proportional (that is the way the invariant space operator works) but **MISy** and **pEy** are not proportional. So the pair **{pEx, pEy}** does not support an invariant space, and therefore cannot support any representation at all. The Bedrock Corollary simply doesn't apply. And, as we saw by calculation, it didn't.

We made this blunder to illustrate one of the most common and deeply puzzling errors in the use of group theory. The Bedrock Corollary does produce surprising equalities between matrix elements, but you have to obey all the fine print. In other words, you must actually construct the two representations and verify that they are identical and unitary. Now we will do it the right way.

39.3.5 Correct the error

What do we have to do to see the Corollary obeyed? We need two pairs of functions that "support the same unitary representation of the two-dimensional species **E**". For one pair we can use **{x,y}**, lifted right from the $\mathbf{D_3}$ character table. That will be our **{qEx,qEy}**. For the other pair **{pEx,pEy}** we will use **invarX** and **invarY**, calculated just above.

We know by construction that **invarX** and **invarY** support a representation, but we are not guaranteed that it is unitary. For that, they must be have the right relative scaling. This is done by the usual "balancing" procedure, automated in an End note. The **BalancePolyBasis** operator yields two solutions

```
{solnMinus, solnPlus} =
 BalancePolyBasis[invarX, invarY, "D3"]
```

$$\left\{\left\{x^7 - \frac{237\, x^5 y^2}{13} - \frac{289\, x^3 y^4}{13} + \frac{153\, x\, y^6}{13},\right.\right.$$
$$\left. -\frac{179\, x^6 y}{13} + \frac{211\, x^4 y^3}{13} + \frac{159\, x^2 y^5}{13} - 3\, y^7\right\},$$
$$\left\{x^7 - \frac{237\, x^5 y^2}{13} - \frac{289\, x^3 y^4}{13} + \frac{153\, x\, y^6}{13},\right.$$
$$\left.\left.\frac{179\, x^6 y}{13} - \frac{211\, x^4 y^3}{13} - \frac{159\, x^2 y^5}{13} + 3\, y^7\right\}\right\}$$

The two solutions differ only in relative sign. After trying both we know that **solnMinus** is the one we want:

```
{baseFnX, baseFnY} = solnMinus;
```

Now we can answer the big question. Do **{baseFnX,baseFnY}** and **{x,y}** support the same representation? The test is simple; make the reps and ask if they are equal:

```
pRep = MakeRepPoly[{baseFnX, baseFnY}, "D3"];
qRep = MakeRepPoly[{x, y}, "D3"];
pRep == qRep
```

```
True
```

Then ask if they are unitary :

```
Map[UnitaryQ, pRep] // Union
```

```
{True}
```

We are on track, and ready to integrate. But look at the integrands first:

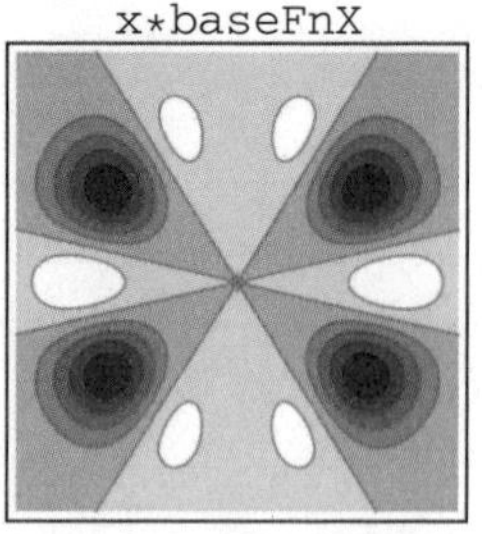

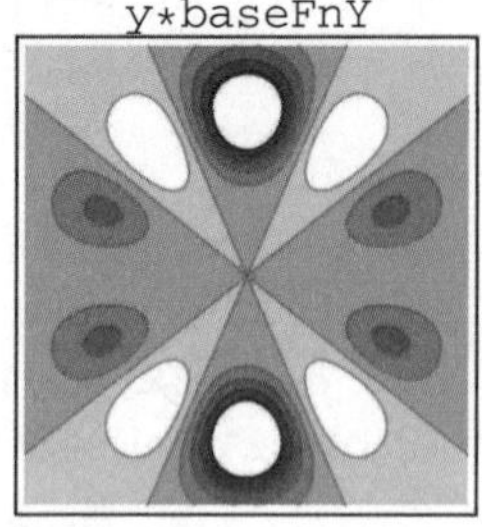

Fig. 39.4 Contour plots of two polynomials that integrate to the same value.

Can you see geometrically that these integrals should be the same? Personally. I will be amazed if they are equal. Calculate them:

```
{integral2D[x * baseFnX], integral2D[y * baseFnY]}
```

```
{-12 π, -12 π}
```

They are equal. I am amazed. Finally, the Bedrock Corollary is confirmed numerically.

> **On your own**
> We based all this on the **pEx** function. But all we needed from it was that it was a function of species **E**. We have three others: **pEy**, **qEx**, and **qEy**. You can go through the same thing using any of them. Try it.

39.4 Major example: benzene transition moments

39.4.1 Make the benzene orbitals

Invariant function spaces usually arise in molecular physics as degenerate molecular eigenfunctions. We will demonstrate it on a famous case, the strongest electronic transition of benzene in the near ultraviolet, named $\mathbf{A_{1\,g}} \rightarrow \mathbf{E_{1\,u}}$.

For this numerical demonstration we do not need accurate wave functions for benzene. They are known, but they are far too large for the small computer you are using to read this book. Instead we will use relatively simple functions that serve as an "artist's conception" of the accurate wave functions, having the same nodal structure and similarly shaped lobes. They have the great advantage that *Mathematica* can do their matrix element integrals exactly, in infinite precision, with computation times that are usually under a minute.

All the calculations may be read in the End Notes by interactive readers, but we close them up in the printed version. Here we just recap briefly what the calculations are, because they are almost like the polynomial case in Section 39.3, above.

Instead of postulating arbitrary polynomials, we construct linear combinations of carbon-centered atomic wave functions on a benzene framework. We show how to do this in Chapter 40, on SALC construction. We pick out the SALCs of symmetries $\mathbf{A_{1\,g}}$, $\mathbf{E_{1\,u,x}}$ and $\mathbf{E_{1\,u,y}}$, balance the two $\mathbf{E}$ functions, and normalize all three. Then we verify that $\mathbf{E_{1\,u,x}}$ and $\mathbf{E_{1\,u,y}}$ support the same representation that $\mathbf{x}$ and $\mathbf{y}$ do. The contour plots of these functions are

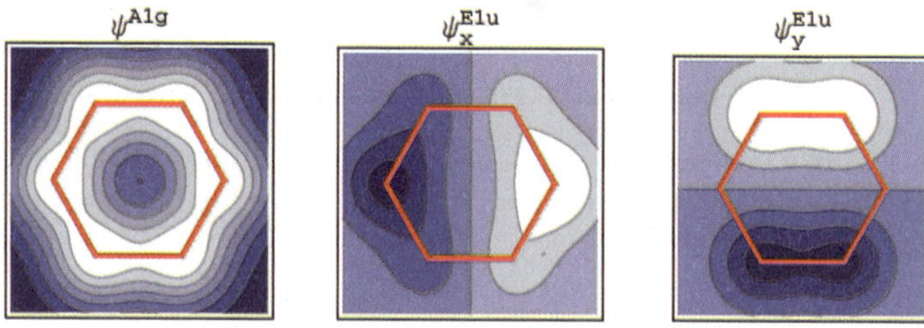

Fig. 39.5 Benzene orbitals used in the transition moment calculations

The detailed formulas are assigned to the symbols ψ^{A1g}, ψ_x^{E1u}, and ψ_y^{E1u}. Interactive readers, bring them up and look at them. You will see that they are just sums of gaussians.

39.4.2 Compute the transition moments

To verify the Bedrock corollary we need to compute four integrals:

$$\left\{\left\{\int_{-\infty}^{+\infty}\int_{-\infty}^{+\infty}\left(\psi^{A1g}\,\mathbf{x}\,\psi_x^{E1u}\right)\,d\mathbf{x}\,d\mathbf{y},\ \int_{-\infty}^{+\infty}\int_{-\infty}^{+\infty}\left(\psi^{A1g}\,\mathbf{x}\,\psi_y^{E1u}\right)\,d\mathbf{x}\,d\mathbf{y}\right\},\right.$$
$$\left.\left\{\int_{-\infty}^{+\infty}\int_{-\infty}^{+\infty}\left(\psi^{A1g}\,\mathbf{y}\,\psi_x^{E1u}\right)\,d\mathbf{x}\,d\mathbf{y},\ \int_{-\infty}^{+\infty}\int_{-\infty}^{+\infty}\left(\psi^{A1g}\,\mathbf{y}\,\psi_y^{E1u}\right)\,d\mathbf{x}\,d\mathbf{y}\right\}\right\}$$

The **ContourPlot**s of these integrands are very interesting:

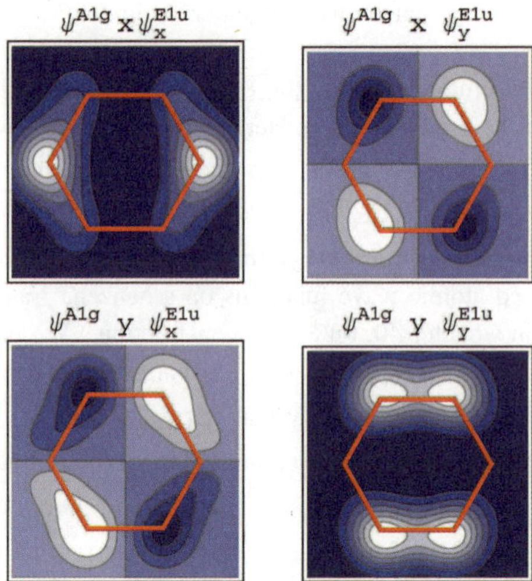

Fig. 39.6 Contour plots of the four integrands calculated below

Who would guess that those two diagonal contours have exactly the same integral? Personally, I won't believe it until I see it. So do the integrals and tabulate them. Here they are:

benzene	ψ_x^{E1u}	ψ_y^{E1u}
$\psi^{A1g}\, x$	$\frac{1}{2}\left(2 + \frac{1}{e^{9/2}} + \frac{3}{e^{3/2}}\right)\pi$	0
$\psi^{A1g}\, y$	0	$\frac{1}{2}\left(2 + \frac{1}{e^{9/2}} + \frac{3}{e^{3/2}}\right)\pi$

That is the Bedrock Corollary, obeyed. The function $\psi^{A1g}\, x\, \psi_x^{E1u}$ gives exactly the same integral as $\psi^{A1g}\, y\, \psi_y^{E1u}$.

This is why benzene has the same absorption coefficient for any linear polarization that vibrates in the plane of the molecule.

> **On your own**
> Take a polarization vector given as **x Cos[θ] + y Sin[θ]**, and prove that this is true.
>
> Remember, the rule for combining independent processes is to add *probabilities*.

After two numerical demonstrations, one for polynomials and one for molecular orbitals, you should now be ready for the complete proof of the Bedrock

Theorem and its Corollary.

39.5 Proof of the Bedrock Theorem

39.5.1 Main theorem

Our reasoning begins with a fairly simple Lemma :

> Matrix elements are invariant under point group transformations of the integrand.
>
> $$\langle \mathbf{f} \rangle = \left\langle \mathcal{T}_\Lambda[\mathbf{f}] \right\rangle \tag{39.5}$$

A point group transformation amounts to turning the contour plot of the integrand about the origin, or reflecting it in a plane that passes through the origin. But all point group transformations are unitary, so all point-to-point distances in the contour plot are preserved. This means that the *shape* of the contour plot does not change. If it is integrated over all space, the integral of the new function will have the same value as the integral of the old function. If you want a more formal argument, click here. If you want a numerical example, click here.

$$\left\langle \mathbf{p}_i^U, \mathbf{q}_j^V \right\rangle = \left\langle \left(\mathbf{p}_i^U\right)^* \mathbf{q}_j^V \right\rangle = \left\langle \left(\mathcal{T}_\Lambda\left[\mathbf{p}_i^U\right]\right)^* \mathcal{T}_\Lambda\left[\mathbf{q}_j^V\right] \right\rangle$$

In Chapter 25 (Representations) (click to review) we defined the representation matrix $\mathbb{T}$ as the solution to $\mathcal{T}_\Lambda[\mathbf{f}_k] = \sum_{h=1}^{nf} \mathbf{f}_h \, \mathbb{T}_{h,k}[\Lambda]$. In the derivation there, the functions $\mathbf{f}_h$ were basis functions for a particular irreducible representation which was understood throughout the derivation. But now we assume explicitly that the functions $\mathbf{p}_i^U$ are basis functions for rep $\mathbf{U}$, while the $\mathbf{q}_j^V$ are basis functions for rep $\mathbf{V}$. So reluctantly we include another index on the $\mathbb{T}$, to indicate its species. Also the sum limits may be different for the two species :

$$\left\langle \left(\mathbf{p}_i^U\right)^* \mathbf{q}_j^V \right\rangle = \left\langle \left(\sum_{h=1}^{nU} \mathbf{p}_h^U \, \mathbb{T}_{h,i}^U[\Lambda]\right)^* \sum_{k=1}^{nV} \mathbf{q}_k^V \, \mathbb{T}_{k,j}^V[\Lambda] \right\rangle$$

Since this is true for every transform in the group, we can sum over all the elements Λ of the group. If $\mathbf{g}$ is the order of the group, this gives us $\mathbf{g}$ copies of the same thing on the left :

$$\mathbf{g}\left\langle \left(\mathbf{p}_i^U\right)^* \mathbf{q}_j^V \right\rangle = \sum_\Lambda \left\langle \left(\sum_{h=1}^{nU} \mathbf{p}_h^U \,\mathbb{T}_{h,i}^U[\Lambda]\right)^* \sum_{k=1}^{nV} \mathbf{q}_k^V \,\mathbb{T}_{k,j}^V[\Lambda] \right\rangle$$

Reversing the order of summation, and factoring as much as possible out of the Λ sum

$$\left\langle \left(\mathbf{p}_i^U\right)^* \mathbf{q}_j^V \right\rangle = \left(\frac{1}{g}\right) \left\langle \sum_{h=1}^{nU} \sum_{k=1}^{nV} \left(\mathbf{p}_h^U\right)^* \mathbf{q}_k^V \right\rangle \sum_\Lambda \left(\mathbb{T}^U_{h,i}[\Lambda]\right)^* \mathbb{T}_{k,j}^V[\Lambda]$$

But the Λ sum is just the left side of the Great Orthogonality

$$\sum_\Lambda \left(\mathbb{T}^U_{h,i}[\Lambda]\right)^* \mathbb{T}_{k,j}^V[\Lambda] == \frac{g}{nV}\,\delta_{U,V}\,\delta_{h,k}\,\delta_{i,j}$$

so putting the right side in its place, canceling the $\mathbf{g}$'s, and rearranging

$$\left\langle \left(\mathbf{p}_i^U\right)^* \mathbf{q}_j^V \right\rangle = \left(\frac{1}{nV} \sum_{h=1}^{nU} \sum_{k=1}^{nV} \delta_{h,k} \left\langle \left(\mathbf{p}_h^U\right)^* \mathbf{q}_k^V \right\rangle\right) \delta_{U,V}\,\delta_{i,j}$$

The factor $\delta_{h,k}$ in the sum over $\mathbf{h}$ ensures that each summand is zero unless $\mathbf{h} = \mathbf{k}$, so we may perform that sum explicitly:

$$\left\langle \left(\mathbf{p}_i^U\right)^* \mathbf{q}_j^V \right\rangle = \left(\frac{1}{nV} \sum_{k=1}^{nV} \left\langle \left(\mathbf{p}_k^U\right)^* \mathbf{q}_k^V \right\rangle\right) \delta_{U,V}\,\delta_{i,j}$$

The sum depends on the species symbols $\mathbf{U}$ and $\mathbf{V}$, but the sum over $\mathbf{k}$ erases all component index dependence. Therefore we may denote it by $\mathbf{C}^{U,V}$.

The Bedrock Theorem, general statement

Let the functions $\left\{\mathbf{p}_1^U, \ldots, \mathbf{p}_{nU}^U\right\}$ be the basis for a unitary irreducible rep of species $\mathbf{U}$ of group $\mathcal{G}$, and let functions $\left\{\mathbf{q}_1^V, \ldots, \mathbf{q}_{nV}^V\right\}$ be the basis for a unitary irreducible rep of species $\mathbf{V}$. Then (39.6)

$$\left\langle \left(\mathbf{p}_i^U\right)^* \mathbf{q}_j^V \right\rangle = \mathbf{C}^{U,V}\,\delta_{U,V}\,\delta_{i,j}$$

This says that the integral vanishes if the **species** are different ($\mathbf{U}\ \ \mathbf{V}$) or if the **components** are the different ($\mathbf{i}\ \ \mathbf{j}$), which was to be shown.

39.5.2 Corollary

In the Bedrock Theorem, Eq. 39.6 , above, when $\mathbf{V} \to \mathbf{U}$ and $\mathbf{j} \to \mathbf{i}$, we have

$$\langle (p_i^U)^* q_i^U \rangle = C^U$$

Index **i** appears on the left, but no longer appears on the right. We did not drop it accidentally; it was honestly used up; after setting **i = j**, it appears on the right only in δ_{ii}, and that evaluated to **1**.

This equation is trying to tell us that all the integrals $\langle (p_1^U)^* q_1^U \rangle$, $\langle (p_2^U)^* q_2^U \rangle$, … **all have the same value** C^U, which depends only on the species label **U**. It does not give a useful formula for C^U, it just says that this constant exists.

Bedrock Corollary

If $\{p_1^S, \ldots, p_n^S\}$ and $\{q_1^S, \ldots, q_n^S\}$) support the same irreducible unitary representation of the multi-dimensional species **S**, then (39.7)

$$\langle p_i^{S*} q_i^S \rangle = C^S$$

where C^S depends only on species **S**, and not at all on index **i**.

Next, we make these careful, intricate statements into rules of thumb.

39.6 Integrals over functions of pure symmetry

39.6.1 Integrands of pure symmetry

Look again at Fig. 39.2, showing the projected parts of a function (click here) . We are going to show formally that the integrals of all these projected parts are zero, with the possible exception of the totally symmetric part. Visually, you can see it immediately, because of the equal-but-opposite lobes that characterize functions of pure symmetry. The formal proof is based on the Bedrock Theorem, using a simple constant **1** as one of its functions.

Mathematicians are careful to define " function" so that it include constants. Try projecting a constant, using any group you like :

```
ProjectET[1, "D6h"]
```

```
{{A1g, 1}, {A2g, 0}, {B1g, 0}, {B2g, 0}, {E1g, 0}, {E2g, 0},
 {A1u, 0}, {A2u, 0}, {B1u, 0}, {B2u, 0}, {E1u, 0}, {E2u, 0}}
```

The function **1** always comes out whole as the component of total symmetry.

All other species are always zero. This makes sense: no point group transform ever changes **1** to anything else, so it does indeed have total symmetry.

Now rethink the Bedrock Theorem Eq. 39.6 (click back) denoting a generic species of totally symmetry as using $\mathbf{A_{1\,g}}$:

$$\left\langle \left(\mathbf{p}_1^{\mathbf{A1g}}\right)^* \mathbf{q}_j^{\mathbf{V}} \right\rangle = \mathbf{C}^{\mathbf{U},\mathbf{V}}\, \delta_{\mathbf{A1g},\mathbf{V}}\, \delta_{\mathbf{1},\mathbf{j}}$$

Now taking a case where $\mathbf{p}_1^{\mathbf{A1g}} = \mathbf{1}$, we see that $\left\langle \mathbf{q}_j^{\mathbf{V}} \right\rangle = \mathbf{0}$ unless $\mathbf{V}$ is the totally symmetric species (in which case $\mathbf{j}$ can only be $\mathbf{1}$), it collapses to

$$\left\langle \mathbf{q}_j^{\text{not totally symmetric}} \right\rangle = \mathbf{0} \tag{39.8}$$

That is, pure symmetric functions other than totally symmetric functions always have symmetric lobes of opposite sign that make their integrals vanish. But don't get too carried away. This is different from saying $\left\langle \mathbf{q}^{\mathbf{A1g}} \right\rangle \neq \mathbf{0}$, which cannot be proved because it is not true. Often it is true, but exceptions are not hard to cook up.

39.6.2 Selection rules, electronic and vibronic

Now go one little step more. The projection theorem says that any function **q[x,y,z]** can be split into summands with pure symmetry under any group :

$$\mathbf{q}[\mathbf{x},\mathbf{y},\mathbf{z}] = \mathbf{q_A}[\mathbf{x},\mathbf{y},\mathbf{z}] + \mathbf{q_B}[\mathbf{x},\mathbf{y},\mathbf{z}] + \ldots + \mathbf{q_{E,1}}[\mathbf{x},\mathbf{y},\mathbf{z}] + \mathbf{q_{E,2}}[\mathbf{x},\mathbf{y},\mathbf{z}] + \ldots$$

where **A** is the species of total symmetry. Now integrate both sides. On the right, every integral vanishes except **A**. So we have shown that

$$\int_{-\infty}^{+\infty}\cdots\int_{-\infty}^{+\infty}\mathbf{q}\; d\tau = \int_{-\infty}^{+\infty}\cdots\int_{-\infty}^{+\infty}\mathbf{q_A}\; d\tau$$

In words,

The integral of any function over all space is the same as the integral of its totally symmetric component.

Or to put it another way,

If a function has no totally symmetric component, its integral will vanish.

This is of enormous use in quantum theory. The question is whether or not a matrix element of the form

$$\int \psi_i^*[\underline{\tau}]\; Q\left[\psi_j[\underline{\tau}]\right]\; d\underline{\tau}$$

vanishes or not. The answer is: Apply **ProjectET** to the integrand. If it has no totally symmetric component, the integral vanishes.

39.6.3 Laporte's rule

Here is another benzene transition moment: $\int \psi^{A1g}[\underline{\tau}]\; \mathbf{x}\; \psi^{E2g}[\underline{\tau}]\; d\underline{\tau}$. Is it necessarily zero, or not? If you know LaPorte's rule, you can say immediately that it is zero.

In systems with an inversion center, eigenfunctions of even inversion parity are labelled **g** (for *gerade*) and those of odd parity are labelled **u** (for *ungerade*).

Laporte's rule
Dipole transitions of type **g→u** and **u→g** are allowed; those of type **g→g** and **u→u** are forbidden.

The reason is that the dipole transition operator (**x**, **y**, or **z**) is a function of type **u**. So integrands that are either **g × u × u = g** or **u × u × g = g** do not generally integrate to zero, and are allowed. But integrands of type **g × u × g = u**, or of type **u × u × u = u** must always integrate to zero, and are forbidden.

39.7 End Notes

39.7.1 Homogeneous polynomials

A homogeneous polynomial is one in which the powers of each summand add to the same integer. It cannot have a constant term, because the powers of the constant term add to **0**.

A homogeneous polynomial of order **5** in **x** and **y**, for instance, has summands $c\, x^m\, y^n$, where $m + n = 5$. So either **m** or n must be odd. When the polynomial (times a totally symmetric convergence factor like $e^{-(x^2+y^2)}$), is integrated over all space the odd-power coordinate produces zero. So odd-order polynomials always integrate to 0.

Homogeneous polynomials multiply like integers: even×even = even, odd×odd = even, and odd×even = odd.

39.7.2 Balancing basis polynomials

As before, in Chapter 29 (MakeUnitary), we go step by step through the balancing procedure; but this time we will automate it at the end. Start with a pair of functions known to belong to species **E** of group D_3, and known to span an invariant space, but not yet "balanced" to support a unitary representation. They were projected out of **pE1** using a rep based on **{x,y}**, so one transforms like **x**, the other like **y**:

We copy the raw functions come from Subsection 39.3.4 (click back):

$$\text{pE1x} = x^7 - \frac{237\, x^5\, y^2}{13} - \frac{289\, x^3\, y^4}{13} + \frac{153\, x\, y^6}{13};$$

$$\text{pE1y} = x^6\, y - \frac{211\, x^4\, y^3}{179} - \frac{159\, x^2\, y^5}{179} + \frac{39\, y^7}{179};$$

Multiply one of the functions by an unknown **k**, and make the rep :

```
kRep = MakeRepPoly[{pE1x, k pE1y}, "D3"]
```

$$\Big\{\{\{1,\,0\},\,\{0,\,1\}\},\ \Big\{\Big\{-\frac{1}{2},\,\frac{13\sqrt{3}\,k}{358}\Big\},\,\Big\{-\frac{179\sqrt{3}}{26\,k},\,-\frac{1}{2}\Big\}\Big\},$$

$$\Big\{\Big\{-\frac{1}{2},\,-\frac{13\sqrt{3}\,k}{358}\Big\},\,\Big\{\frac{179\sqrt{3}}{26\,k},\,-\frac{1}{2}\Big\}\Big\},$$

$$\{\{1, 0\}, \{0, -1\}\}, \left\{\left\{-\frac{1}{2}, -\frac{13\sqrt{3}\,k}{358}\right\}, \left\{-\frac{179\sqrt{3}}{26\,k}, \frac{1}{2}\right\}\right\},$$
$$\left\{\left\{-\frac{1}{2}, \frac{13\sqrt{3}\,k}{358}\right\}, \left\{\frac{179\sqrt{3}}{26\,k}, \frac{1}{2}\right\}\right\}\}$$

```
remainders =
 Map[(Transpose[#] - Inverse[#]) &, kRep] //
   Flatten // Union
```

$$\left\{0, \frac{179\sqrt{3}}{26\,k} - \frac{13\sqrt{3}\,k}{358}, -\frac{179\sqrt{3}}{26\,k} + \frac{13\sqrt{3}\,k}{358}\right\}$$

```
kSoln = Solve[remainders[[2]] == 0, k]
```

$$\left\{\left\{k \to -\frac{179}{13}\right\}, \left\{k \to \frac{179}{13}\right\}\right\}$$

We do not know whether to use the positive or negative solution, so we pick the positive one (always the second one) arbitrarily, and make the balanced basis :

```
{xBasisBal, yBasisBal} =
 {pE1x, k pE1y} /. kSoln[[2]] // Expand
```

$$\left\{x^7 - \frac{237\,x^5\,y^2}{13} - \frac{289\,x^3\,y^4}{13} + \frac{153\,x\,y^6}{13},\right.$$
$$\left.\frac{179\,x^6\,y}{13} - \frac{211\,x^4\,y^3}{13} - \frac{159\,x^2\,y^5}{13} + 3\,y^7\right\}$$

This is the end of the construction. See if it worked :

```
repBB = MakeRepPoly[{xBasisBal, yBasisBal}, "D3"];
% // GridList // Size[6]
```

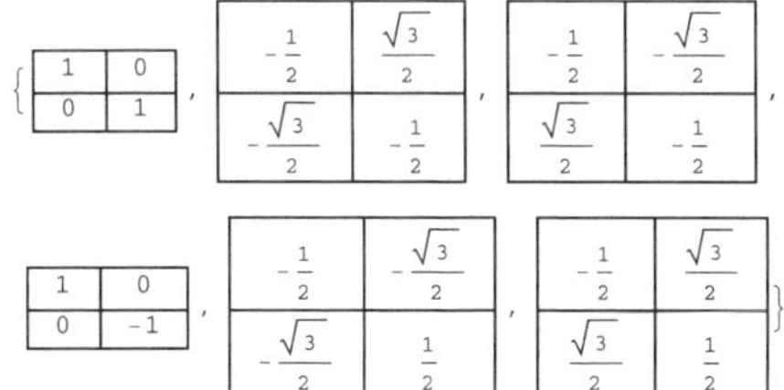

Is it unitary?

```
Map[UnitaryQ, repBB]
```

{True, True, True, True, True, True}

It worked. Click into the preliminaries to read the exact text of the **BalancePolyBasis** operator.

Test it:

```
{xBasisBal, yBasisBal} ==
 BalancePolyBasis[pE1x, pE1y, "D3"]
```

```
True
```

That was not much of a test, but at least the wheels are turning.

> **On your own**
> Show that **repBB** really is a rep of species **E** of group **D₃**.

It is unitary, but is it exactly the same as **repE** that was used in the original projection of **pE1x** and **pE1y** ? Make that rep again, and compare:

```
repE = MakeRepPoly[{x, y}, "D3"];
% // GridList // Size[6]
```

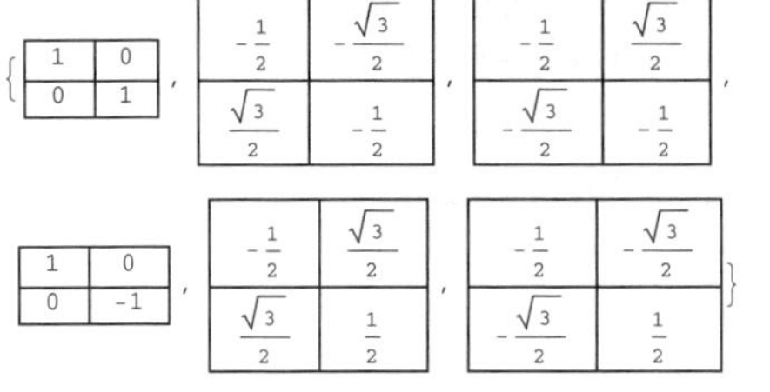

$$\left(\frac{\text{Select[repBB // Flatten, \# =!= 0 \&]}}{\text{Select[repE // Flatten, \# =!= 0 \&]}}\right) \text{ // Union}$$

```
{-1, 1}
```

There are some sign differences. We had a 50-50 chance of that when we arbitrarily picked the sign of **k** inside the module. So go with the other sign, switching to Mars and Venus as our complementary pair :

```
{pE1♂, pE1♀} = {xBasisBal, -yBasisBal};
MakeRepPoly[{pE1♂, pE1♀}, "D3"] == repE
```

```
True
```

Yes, that is what it needed.

39.7.3 Balancing $E_{1u,x}$ and $E_{1u,y}$

BalancePolyBasis works only when **MakeRepPoly** works. Here we need to use **MakeRepMolec**. As before, multiply one basis function by an unknown **k** and construct the rep :

```
kRep =
  MakeRepMolec[basisRaw * {1, k}, πsToCart, "D6h"];
% // GridList // Short[#, 3] & // Size[6]
```

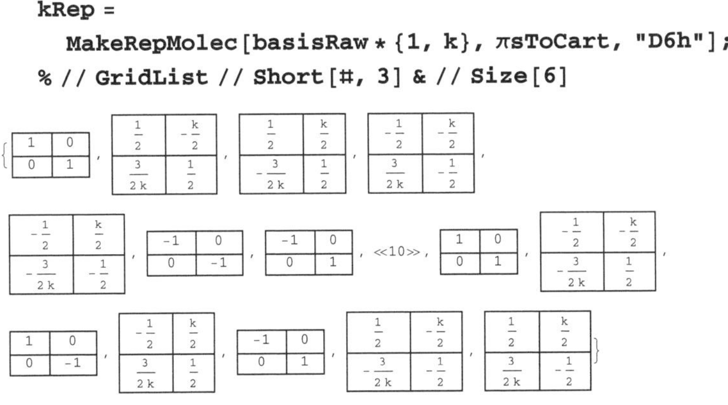

The rest is the same as for polynomials.

39.7.4 Unitary transform of an integrand

In the **MatrixReview** we show that all unitary transforms preserve point-to-point distances (click for proof). Think of an **x,y,z** space with a function defined in it, and imagine further that a diamond lattice is thrown over this same space. The cells of the diamond lattice are tetrahedra.

If the lattice is transformed by a unitary transformation, each tetrahedral unit cell will be rotated and/or reflected. But the lattice edge lengths will be faithfully carried over from old to new. Tetrahedra, like triangles, are completely determined by their edge lengths, so the new tetrahedra must have the same volume as the old ones. If the function value at the center of each tetrahedron is carried over from old to new at the same time, then the Riemann integral sum (tetrahedron volumes times central function values) will be invariant to the transform at every mesh size, and also in the limit of increasingly fine lattices (the integral) .

Example of a unitary integrand transform

A numerical demo makes it absolutely clear what we are talking about above. Take a quite arbitrary function in the **x,y** plane :

```
qFn[x_, y_] :=
  (1 + 2 x^2 + 2 x^3) (36 + 28 y + 84 y^2 + 12 y^3 + y^4);
```

Define an integral containing a rotationally symmetric convergence factor:

$$\text{int2D[u\_]} := \int_{-\infty}^{+\infty}\int_{-\infty}^{+\infty} \text{Exp}\left[-\left(x^2+y^2\right)\right] u \, dx \, dy$$

The integrals below are a little slow, but they are exact. Be patient.

```
before = int2D[qFn[x, y]]
```

$$\frac{315\pi}{2}$$

Transform #2 is rotation by a third of a turn. Its rules are

```
Tf2 = SymmetryRules["D3"][[2]]
```

$$\left\{x \to -\frac{x}{2} - \frac{\sqrt{3}\,y}{2},\ y \to \frac{\sqrt{3}\,x}{2} - \frac{y}{2},\ z \to z\right\}$$

Subject the function to this transform and integrate it again :

```
after = int2D[Evaluate[qFn[x, y] /. Tf2]]
```

$$\frac{315\pi}{2}$$

```
before == after
```

```
True
```

As advertised, they are the same. Just for fun, look directly at the algebraic forms of the integrands of **before** and **after** (use **Expand** on the forms **qFn[x,y]** and **qFn[x,y]/.Tf2**). Can you see why their integrals would be the same ? (No way…)

39.7.5 Preparation of the benzene functions

We can make invariant function spaces with a single hit on **MakeSALCs**. In the preliminaries we made up molecule "**unitBenzene**". It is exactly like benzene, but scaled to have unit **C-C** bond length. To make the Cartesian integrals a little quicker, we want functions that depend only on **x** and **y**. So we specify an angular factor of **1**, rather than the usual **z** :

```
{πMOs, πsToCartRaw} =
 MakeSALCs["unitBenzene", "C", π, 1]
```

We are being extremely scrupulous, so without delay we verify that these "π" atomic orbitals really are invariant under all the $\mathbf{D_{6\,h}}$ transforms. First, put them into Cartesian, abolishing their **z**-dependence :

```
πsToCart = πsToCartRaw /. z → 0;
AOlist = πsToCart /. (a_ → b_) → b
```

The invariance test is simply to carry out the definition of invariance: Transform all six orbitals under every transform in the group (making 24×6 transformed orbitals); then get rid of redundancies using **//Flatten//Union**. Since **Union** also sorts its output, we must also apply it to the originals, to put them in the same order :

```
(AOlist // Union) ==
  (AOlist /. SymmetryRules["D6h"] // ExpandAll //
     Flatten // Union)
```

```
True
```

Invariance is verified. Now we must look at the representation supported by the two functions $\mathbf{E_{1\,u,x}}$ and $E_{1\,u,y}$. It is very unlikely to be unitary as it stands, so we begin by "balancing" it.

```
{speciesName, basisRaw} = πMOs[[4, 1]]
```

> **On your own**
>
> Show that the balancing factor for this basis set is $\sqrt{3}$, applied to the second member. Click here for help.

```
basisE1u = basisRaw {1, √3}
```

Make the rep supported by this basis :

```
repE1u = MakeRepMolec[basisE1u, πsToCart, "D6h"];
% // GridList // Short[#, 3] & // Size[6]
```

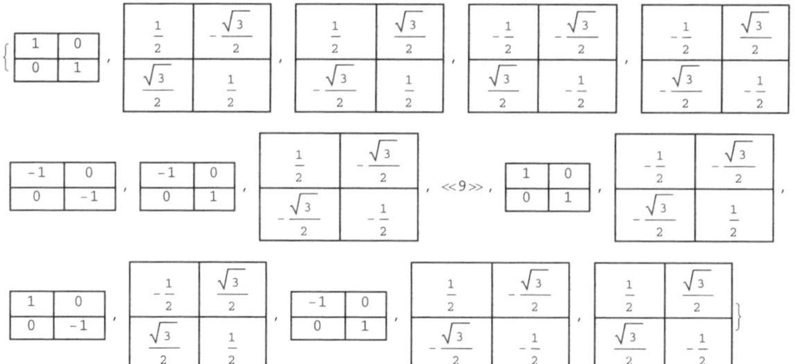

It ought to be unitary :

```
Map[UnitaryQ, repE1u] // Union
```

```
{True}
```

Click back to see that **ProjectED** used the **{x,y}** rep when it split up these $\mathbf{E_{1\,u}}$ functions. Our **p** functions are going to be $\{\mathbf{A_{1\,g}\,x},\ \mathbf{A_{1\,g}\,y}\}$. But since the

A1 g factor always transforms into itself, this rep will be identical to the **{x,y}** rep. We must compare the **xyRep** to **repE1u** :

```
xyRep = MakeRepPoly[{x, y}, "D6h"];
xyRep == repE1u
```

```
True
```

They had to be identical, and they are. The identity of these reps is the *sine qua non* of the Bedrock Corollary. Most apparent "failures" of this corollary founder on this very rock. In the past, there was at least the excuse that rep construction was too tedious to do, just for a check. No more.

Now we prepare pseudo-wave functions, taking **f** as a Gaussian shape around each carbon atom :

```
Clear[f];
f[uSq_] := Exp[-3 uSq];
ψ^A1g = πMOs[[1, 2, 1]] /. πsToCart;
ψ_x^E1u = basisE1u[[1]] /. πsToCart;
ψ_y^E1u = basisE1u[[2]] /. πsToCart;
Clear[f];
```

It is important Clear **f**. It is an external used by a module.

ψ^{A1g}

$$e^{-3\,(1-2\,x+x^2+y^2)}+e^{-3\,(1+2\,x+x^2+y^2)}+e^{-3\left(1-x+x^2-\sqrt{3}\,y+y^2\right)}+e^{-3\left(1+x+x^2-\sqrt{3}\,y+y^2\right)}+e^{-3\left(1-x+x^2+\sqrt{3}\,y+y^2\right)}+e^{-3\left(1+x+x^2+\sqrt{3}\,y+y^2\right)}$$

ψ_x^{E1u}

$$2\,e^{-3\,(1-2\,x+x^2+y^2)}-2\,e^{-3\,(1+2\,x+x^2+y^2)}+e^{-3\left(1-x+x^2-\sqrt{3}\,y+y^2\right)}-e^{-3\left(1+x+x^2-\sqrt{3}\,y+y^2\right)}+e^{-3\left(1-x+x^2+\sqrt{3}\,y+y^2\right)}-e^{-3\left(1+x+x^2+\sqrt{3}\,y+y^2\right)}$$

ψ_y^{E1u}

$$\sqrt{3}\left(e^{-3\left(1-x+x^2-\sqrt{3}\,y+y^2\right)}+e^{-3\left(1+x+x^2-\sqrt{3}\,y+y^2\right)}-e^{-3\left(1-x+x^2+\sqrt{3}\,y+y^2\right)}-e^{-3\left(1+x+x^2+\sqrt{3}\,y+y^2\right)}\right)$$

With these functions, you can verify the Bedrock Corollary by exact symbolic integration.

40. Constructing SALCs

Symmetry Adapted Linear Combinations of atomic orbitals

Preliminaries

40.1 What are SALCs ?

The acronym SALC stands for Symmetry Adapted Linear Combinations of atomic orbitals. In small molecules, SALCs can be identical to the one-electron orbitals found by Huckel calculations, but in larger molecules they must be linearly combined to make Huckel orbitals. But their utility goes far beyond the Huckel approximation. By using SALCs as basis functions, the most sophisticated of quantum calculations can greatly reduce the size of the matrices that need to be diagonalized. The reduction of matrix size goes hand in hand with a higher level of understanding of the problem. If you are interested in *understanding* molecules (as opposed to just calculating their properties) you can hardly understand them at all without SALCs.

The famous Schrödinger equation for one electron in an arbitrary potential can always be solved without making use of symmetry, and indeed that is how many modern calculation packages approach the problem. The central object in the finite basis method is the Hamiltonian matrix, which is extremely large in accurate solutions. It is written as

$$\begin{pmatrix} \int \chi_1 \hat{H}[\chi_1]\, d\tau & \int \chi_1 \hat{H}[\chi_2]\, d\tau & \cdots & \int \chi_1 \hat{H}[\chi_n]\, d\tau \\ \int \chi_2 \hat{H}[\chi_1]\, d\tau & \int \chi_2 \hat{H}[\chi_2]\, d\tau & \cdots & \int \chi_2 \hat{H}[\chi_n]\, d\tau \\ \vdots & \vdots & \cdots & \vdots \\ \int \chi_n \hat{H}[\chi_1]\, d\tau & \int \chi_n \hat{H}[\chi_2]\, d\tau & \cdots & \chi_n \hat{H}[\chi_n] \end{pmatrix}$$

where all the χ functions are atomic orbitals centered on atoms. Those on different atoms are not orthogonal, so this matrix has few, if any, zeroes in it.

In this chapter we show how to use the group theoretic projections operators to simplify this matrix before we try to solve it. We form linear combinations of the χ functions that belong to different species of the group of the Hamiltonian (i.e., SALCs). Taking these combinations as the basis functions, those that belong to different species have a vanishing Hamiltonian integral because of The Bedrock Theorem. Click back to it for a quick refresher.

The Bedrock Theorem, properly employed, produces huge regions of zeroes in the Hamiltonian matrix; in fact, with minimal cleverness it can become block diagonal. Then you can diagonalize the central blocks one by one, reducing the

W.M. McClain, *Symmetry Theory in Molecular Physics with Mathematica*,
DOI 10.1007/b13137_40,

calculational problem enormously. But even more important, the block diagonal structure of the problem is easily and naturally grasped by humans. Now you can say things like: "Oh, yes, of course. Even though they are in same region of space, the B_1 orbitals don't interact with the B_2 orbitals (i.e., by the Bedrock Theorem, they have $H_{i,k} = 0$) because they are of different species."

40.2 Make SALCs, step by step

40.2.1. The input

As the principal demonstration, we will make the SALCs for **cyclopropenyl Cation**, $C_3H_3^+$, going step by step. The neutral molecule is a radical called cyclopropenyl (15 electrons); but it is unstable in equilateral geometry. The loss of one electron settles it down into the simplest D_{3h} shape, an equilateral triangle.

If you haven't "had" Organic, the σ-bonded framework of the molecule is

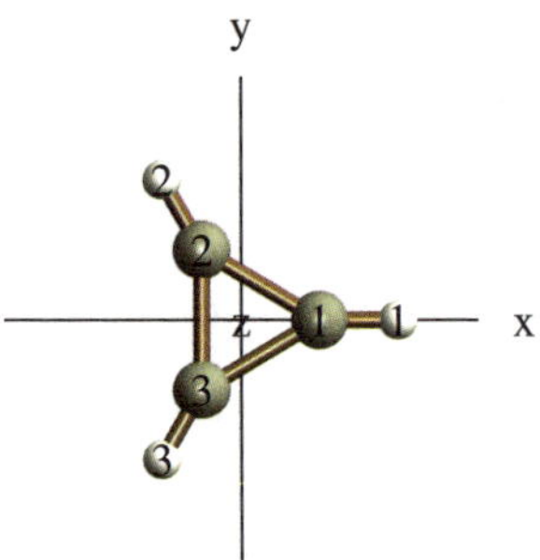

Fig. 40.1 The σ core of $C_3H_3^+$. The π bonds are not shown.

Finding the π bonding SALCs is the problem before us. They will be linear combinations of three p_z orbitals, each on one **C** atom. Step through this section and watch the SALCs being constructed. Then set **mol** to the name of another aromatic (flat, π-bonded) molecule and do it again. After you have grasped the process, you can **Select** the Section bracket to the right, and **Enter**. Computation will go straight down to the final answer, and a graphic will be drawn. A totally automated operator **MakeSALCs**, based on these operations, is described below in Subsection 40.3.1 .

40.2.2. Construct symbolic atomic orbitals

```
(*Input*)
mol = "cyclopropenylCation";
atomKind = "C";
orbName = π;
angFactor = z;
```

The input has been entered above. First, look at the molecule list:

```
molList = MoleculeToList[mol];
Size[8][Column[%]]
```

```
{C, {1, 0, 0}, {1, D3h}}
{C, {-1/2, √3/2, 0}, {2}}
{C, {-1/2, -√3/2, 0}, {3}}
{H, {2, 0, 0}, {1}}
{H, {-1, √3, 0}, {2}}
{H, {-1, -√3, 0}, {3}}
```

By convention, the group of the molecule is carried as the last item in the tag list of the first atom. Extract it :

```
gp = molList[[1, 3, -1]]
```

```
D3h
```

We want to make an orbital on each atom of the chosen **atomKind**. They are

```
atomList = Select[molList, (#[[1]] == atomKind) &]
```

```
{{C, {1, 0, 0}, {1, D3h}},
 {C, {-1/2, √3/2, 0}, {2}}, {C, {-1/2, -√3/2, 0}, {3}}}
```

This list will be empty if you specify an **atomKind** that is not in the molecule. But here, one atomic orbital is centered at each carbon atom point :

```
atomPoints = Transpose[atomList][[2]]
```

```
{{1, 0, 0}, {-1/2, √3/2, 0}, {-1/2, -√3/2, 0}}
```

Construct a list of vector differences between the electron position **{x,y,z}** and each **atomPoint** :

```
Δvecs = Map[({x, y, z} - #) &, atomPoints]
```

$$\left\{\{-1+x,\ y,\ z\},\ \left\{\frac{1}{2}+x,\ -\frac{\sqrt{3}}{2}+y,\ z\right\},\ \left\{\frac{1}{2}+x,\ \frac{\sqrt{3}}{2}+y,\ z\right\}\right\}$$

The lengths squared of the **Δvecs** are

```
args = Map[(#.#) &, Δvecs] // ExpandAll
```

$$\left\{1-2x+x^2+y^2+z^2,\ 1+x+x^2-\sqrt{3}\,y+y^2+z^2,\ 1+x+x^2+\sqrt{3}\,y+y^2+z^2\right\}$$

The **A**tomic **O**rbital basis set is then

```
Clear[f];
AOs = (angFactor ) * (Map[f, args])
```

$$\left\{z\,f\left[1-2x+x^2+y^2+z^2\right],\ z\,f\left[1+x+x^2-\sqrt{3}\,y+y^2+z^2\right],\ z\,f\left[1+x+x^2+\sqrt{3}\,y+y^2+z^2\right]\right\}$$

For an **s** orbital the angular prefactor is **1**; for $\mathbf{p_z}$ it is **z**; for $\mathbf{d_{xy}}$ it is **x y**; etc. The form of **f** is that function which minimizes the energy expectation for the whole molecule. But we do not need to know it to determine the SALCs. For graphics purposes, **f** can be any centrally peaked function.

The **AOs** need abbreviated arguments. In cyclopropenyl they are π orbitals, so we let **π[]** stand for **z*f[]**. The π names we will need are

```
orbitalSymbols = Map[orbName, Range[Length[args]]]
```

$$\{\pi[1],\ \pi[2],\ \pi[3]\}$$

It will be quite useful below to have recognition rules that turn the Cartesian formulas into π symbols, and vice-versa. Here is the vice

```
toSymbs = Thread[AOs → orbitalSymbols]
```

$$\left\{z\,f\left[1-2x+x^2+y^2+z^2\right]\to\pi[1],\ z\,f\left[1+x+x^2-\sqrt{3}\,y+y^2+z^2\right]\to\pi[2],\ z\,f\left[1+x+x^2+\sqrt{3}\,y+y^2+z^2\right]\to\pi[3]\right\}$$

and here is the versa

```
toCartMOs = Thread[orbitalSymbols → AOs]
```

$$\left\{\pi[1]\to z\,f\left[1-2x+x^2+y^2+z^2\right],\ \pi[2]\to z\,f\left[1+x+x^2-\sqrt{3}\,y+y^2+z^2\right],\ \pi[3]\to z\,f\left[1+x+x^2+\sqrt{3}\,y+y^2+z^2\right]\right\}$$

40.2.3. Use `ProjectET` to make the trace SALCs

Given all the machinery we have developed, making the SALCs is a piece of cake. All the heavy lifting is done by **ProjectET** (the trace projector) followed, when required, by **ProjectED** (the detailed projector). First, we apply **ProjectET** to *all three* of the π ligand orbitals constructed above. Why apply it to all three? This will produce a lot of redundancy. But we really do want all three; you will see why.

```
rawTraceSALCs = ProjectET[AOs, gp] /. toSymbs
```

$$\left\{\{A_1', \{0, 0, 0\}\}, \{A_2', \{0, 0, 0\}\}, \{E', \{0, 0, 0\}\}, \{A_1'', \{0, 0, 0\}\}, \left\{A_2'', \left\{\frac{\pi[1]}{3}+\frac{\pi[2]}{3}+\frac{\pi[3]}{3}, \frac{\pi[1]}{3}+\frac{\pi[2]}{3}+\frac{\pi[3]}{3}, \frac{\pi[1]}{3}+\frac{\pi[2]}{3}+\frac{\pi[3]}{3}\right\}\right\}, \left\{E'', \left\{\frac{2\pi[1]}{3}-\frac{\pi[2]}{3}-\frac{\pi[3]}{3}, -\frac{\pi[1]}{3}+\frac{2\pi[2]}{3}-\frac{\pi[3]}{3}, -\frac{\pi[1]}{3}-\frac{\pi[2]}{3}+\frac{2\pi[3]}{3}\right\}\right\}\right\}$$

That was wonderfully easy, but it needs clean-up. Many zeroes were generated, and several projected functions are identical, or differ only by a scale factor. This is cleaned up by an operator **NeatTraceSALCs** defined in the preliminaries. If you want to see it work step by step, click to the End Notes.

```
labeledTraceSALCs =
 NeatTraceSALCs[rawTraceSALCs, orbName]
```

$$\{\{A_2'', \{\pi[1]+\pi[2]+\pi[3]\}\}, \{E'', \{\pi[1]+\pi[2]-2\pi[3], 2\pi[1]-\pi[2]-\pi[3], \pi[1]-2\pi[2]+\pi[3]\}\}\}$$

Three $\mathbf{p_z}$ atomic orbitals projected into one function of species $\mathbf{A_2''}$ and three functions of species $\mathbf{E''}$. Three orbitals in and four out? That doesn't sound right (and it isn't). The multidimensional species need a detailed projection, which will reveal their redundancy and allow us to deal with it. So the next step is to separate the one-D SALCs (which are complete as they stand) from the multi-D SALCs (which need more work).

40.2.4. Separation of one-D cases from multi-D cases

Separation of the one-D SALCs from the multi-D SALCs is simple for humans, but clumsy for computers. Our goal is total automation, so we do it the clumsy way.

```
{labels, orbs} = Transpose[labeledTraceSALCs]
```

```
{{A2'', E''}, {{π[1] + π[2] + π[3]}, {π[1] + π[2] - 2 π[3],
   2 π[1] - π[2] - π[3], π[1] - 2 π[2] + π[3]}}}
```

The degeneracy of a species is the first element of its character vector :

```
speciesDims = Map[ChVec[#, gp]〚1〛 &, labels]
```

```
{1, 2}
```

Tag each species with its own degeneracy, just before the species name:

```
orbsTagged = Transpose[{speciesDims, labels, orbs}]
```

```
{{1, A2'', {π[1] + π[2] + π[3]}}, {2, E'', {π[1] + π[2] - 2 π[3],
   2 π[1] - π[2] - π[3], π[1] - 2 π[2] + π[3]}}}
```

Now we can pull out just the one-D cases :

```
oneDs1 = Select[orbsTagged, #〚1〛 == 1 &] /.
   {a_, b_, c__} → {b, c}
```

```
{{A2'', {π[1] + π[2] + π[3]}}}
```

In simple cases this is what we want, but when there are several one-D functions that belong to the same species, it is not. We want each one-D function to have its own species label. The little operation below does exactly that:

```
oneDs2 = Table[
   spNm = oneDs1〚i, 1〛;
   fns = oneDs1〚i, 2〛;
   Map[{spNm, {#}} &, fns],
   {i, 1, Length[oneDs1]}]
```

```
{{{A2'', {π[1] + π[2] + π[3]}}}}
```

In the initial example, nothing happened. In other examples, you will see its usefulness. But even here the bracketing is too complicated. Fix it:

```
oneDs = Partition[Flatten[oneDs2], 2] /.
   {nm_String, fn_} → {nm, {fn}}
```

```
{{A2'', {π[1] + π[2] + π[3]}}}
```

That is the desired format for the **oneD** output. Turn now to the **multiD** projections. Pull them out by selecting on multiplicity tags greater than **1**; then apply a rule that removes the tags:

```
multiDsums = Select[orbsTagged, #〚1〛 > 1 &] /.
   {a_, b_, c__} → {b, c}
```

```
{{E″, {π[1] + π[2] - 2 π[3],
   2 π[1] - π[2] - π[3], π[1] - 2 π[2] + π[3]}}}
```

As you probably know by now, the three π combinations above, all of type **E″**, belong to a 2-D function space. A basis set for it will be found by **ProjectED**.

40.2.5. Use ProjectED to make the detailed SALCs

Graphics uses redAtoms and the CP function

The next command, **PrjMultiD**, finishes the job. It doesn't do any real work; it is just an administrator. It is fairly complicated because it must deal with so many cases. Its operand **multiDsums** can be empty, or it can contain one or several species, and each species can contain one or several sums. Beyond that, it has to call **MakeRepPoly** to construct an entire representation for use by **ProjectED**. Finally, it removes redundancies and zeroes, and massages the result into the desired output form, **{spName,{orthogonalSALCs}}**. Click here to see **PrjMultiD** developed step by step. Click here to read the final module. Now watch it work:

```
multiDs = PrjMultiD[multiDsums, toCartMOs, gp]
```

```
Using E″ basis {x z, y z}
```

```
{{{E″, {2 π[1] - π[2] - π[3], π[2] - π[3]}}}}
```

The three cyclopropenyl **E″** orbital sums reported by **ProjectET** were redundant; they were spanned by the two orthogonal orbitals you see above. Finally, we join all the results together, stripping the extra outer brackets from **E″**:

```
allSALCs = Join[oneDs, multiDs[[1]]]
```

```
{{A2″, {π[1] + π[2] + π[3]}},
 {E″, {2 π[1] - π[2] - π[3], π[2] - π[3]}}}
```

That is the desired final bracketing, and the SALC problem is now solved. All that remains is the graphics. Automated graphics never look good in every case, but we offer a first rough draft, fairly satisfactory for flat planar molecules. For other cases you will want to take charge yourself. The red dots are the atoms of the **atomKind** you asked for; black dots are other atoms in the molecule.

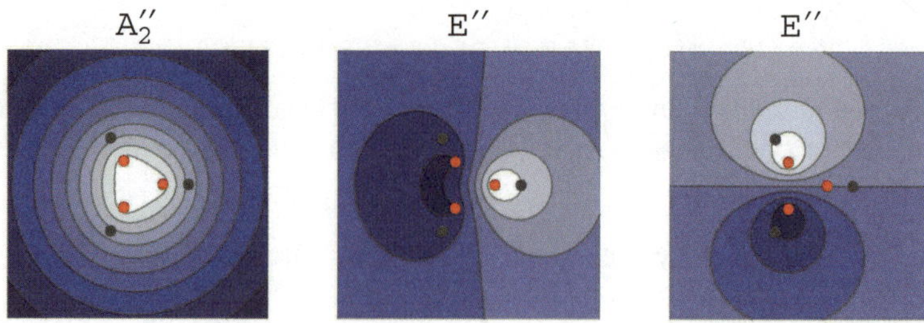

Fig. 40.2 The red dots are C atoms; the gray dots are H. The π orbitals are centered only on C atoms.

You can change the input lines way at the top and rerun with other molecules. Or, you can go on to the total automation section below, and do other cases with one click. We don't know of any molecule for which this algorithm fails. If you find one, let us know.

> **On your own**
> Try the molecules
> **{"cyclopropenylCation","benzene", "naphthaleneCatoms","cyclobutadiene","ethylene", "octaComplex","PCl5"}**

40.3 Automation

40.3.1. The complete MakeSALCs operator

In the preliminaries we define an operator

```
MakeSALCs[mol,atomKind,orbName,angFactor]
```

which makes the SALCs using the procedure of Section 40.2, above.
Try it out on a non-trivial example:

```
SALCs[UF9] = MakeSALCs["UF9", "F", σ, 1];
```

```
The group is D3h

Using E′ basis {x, y}

Using E″ basis {x z, y z}
```

The output has four parts: (1) the molecule name, (2) the atom kind, (3) a SALCs list, and (4) a set of rules for transforming to Cartesian. Humans are mainly interested in part 3; look at it:

```
SALCs[UF9][[3]];
Column[%] // Size[8]
```

```
{A'1, {σ[1] + σ[2] + σ[3] + σ[4] + σ[5] + σ[6]}}
{A'1, {σ[7] + σ[8] + σ[9]}}
{A''2, {σ[1] + σ[2] + σ[3] - σ[4] - σ[5] - σ[6]}}
{E', {2 σ[1] - σ[2] - σ[3] + 2 σ[4] - σ[5] - σ[6], 2 σ[7] - σ[8] - σ[9]}}
{E', {σ[2] - σ[3] + σ[5] - σ[6], σ[8] - σ[9]}}
{E'', {2 σ[1] - σ[2] - σ[3] - 2 σ[4] + σ[5] + σ[6], σ[2] - σ[3] - σ[5] + σ[6]}}
```

Each species name is bracketed with a SALC list of appropriate length (the detailed projections from one trace projection). If a species has more than one SALC set, each set has a separate entry. Above, there are two 1-D SALCs of species $\mathbf{A}_1'$, one of species $\mathbf{A}_2''$, two 2-D SALC pairs for $\mathbf{E}'$, and one pair for $\mathbf{E}''$.

Parts 1, 2, and 4 of the output are used mainly by graphics operators; but look if you want, at **SALCs[UF₉][[1]]**, ... **[[2]]**, and **..[[4]]**.

40.3.2. Automated two-dimensional graphics

Make the 2D eyeGuide

The first job is to make a standard backdrop for the SALC graphics. This is done by the operator **RedAtomsXY** , which makes a 2D projection of the molecule by removing the **z** coordinate of all atoms, rendering the SALC centers in red, all other atoms in grey. It will work on any tabulated molecule:

```
eyeGuideBz = RedAtomsXY["benzene", "C"];
Show[Graphics[eyeGuideBz], ImageSize → 100,
 PlotRange → All, AspectRatio → Automatic]
```

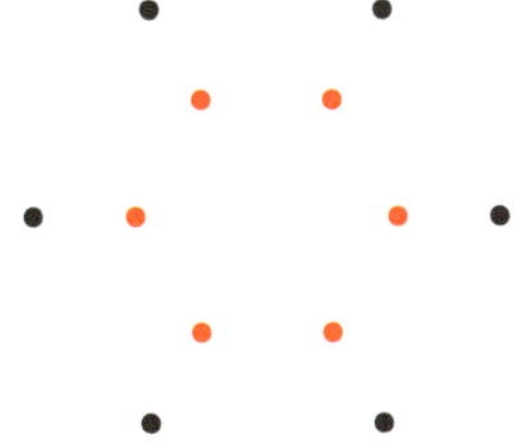

Fig. 40.3 The eye guide for benzene. Red is C, gray is H.

To see a step-by-step construction of **RedAtomsXY**, click here. The **z**-axis view, as above, is nearly always what you want for planar molecules. If you are doing one that needs a different view, click here.

The second job is to make a graphics cartouche of the **ContourPlot** of the desired SALC.

We will make the graphics using **SALCfigs2D**. To see that operator constructed step by step, click here.

Do the contour plots of the SALC functions

Two-dimensional graphics are easier than three-dimensional graphics, so we do them first. 2D is appropriate for all planar molecules.

We have defined some semi-automatic graphics operators that produce a decent first draft of two-dimensional SALC graphics. They were used above in the step-by-step construction of the SALCs of $C_3H_3^+$. Now we discuss how to use them in general, taking benzene as the example. First, run **MakeSALCs** :

```
SALCs[C6H6] = MakeSALCs["benzene", "C", π, 1];
```

```
The group is D6h
```

Using E2g basis $\left\{\frac{1}{2}\left(x^2 - y^2\right), x\,y\right\}$

```
Using E1u basis {x, y}
```

Check the important part of the output:

```
SALCs[C6H6]⟦3⟧;
Size[7][Column[%]]
```

```
{A1g, {π[1] + π[2] + π[3] + π[4] + π[5] + π[6]}}
{B2u, {π[1] - π[2] + π[3] - π[4] + π[5] - π[6]}}
{E2g, {2 π[1] - π[2] - π[3] + 2 π[4] - π[5] - π[6], π[2] - π[3] + π[5] - π[6]}}
{E1u, {2 π[1] + π[2] - π[3] - 2 π[4] - π[5] + π[6], π[2] + π[3] - π[5] - π[6]}}
```

The **SALCgraphic2D** operator has been written to make graphics of the output above. We call it, suppressing the output with a semicolon. The graphics are reasonably nice looking, but they are not in energy order. It takes a human to count nodes and order them correctly, as in the **GraphicsColumn** operator:

```
grTbl =
  Table[SALCgraphic2D[SALCs[C6H6], i], {i, 1, 4}];
GraphicsColumn[
 {grTbl⟦2⟧, grTbl⟦3⟧, grTbl⟦4⟧, grTbl⟦1⟧},
 ImageSize → {150, 300}]
```

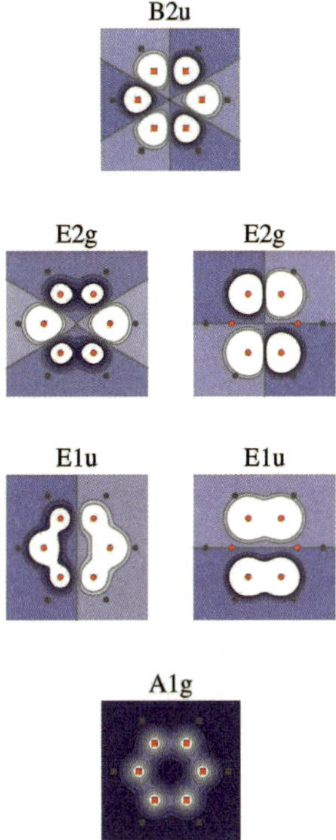

Fig. 40.4 Benzene SALCs in energy order. Each higher figure has one more node.

40.3.3. Semi-automated three-dimensional graphics

Make a 3D `eyeGuide` for $\mathbf{UF_9}$

Here is the idiosyncratic hand work. We must construct a framework to help the eye to see the structure of $\mathbf{UF_9}$ in three dimensions. Six of the **`F`** atoms are at the corners of a vertical triangular biprism; the other three are staggered in the equatorial plane. The **`eyeGuideUF9`** is made in the closed cell below. Open it to see the ghastly details.

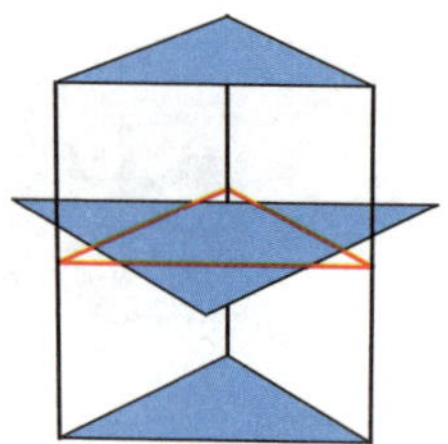

Fig. 40.5 An eyeguide for the SALCs of UF$_9$

The red lines help you to see how the larger middle triangle pokes out through the central prism. The **U** atom (not shown) is in the center of the red triangle. the **F** atoms are at the corners of the three triangles. The bond lengths are adjusted so that the nine **U-F** distances are equal (nearly true in nature).

Raw materials: SALCs[UF$_9$]

Three dimensional graphics can also be automated, except for the **eyeGuide**, which must be made by hand for each molecule. But the rest will run without too much individual attention.

Make sure that **MakeSALCs** has been run on **UF$_9$**:

```
Length[SALCs[UF9][[3]]]
```

```
6
```

If this does not return a **6**, click back and construct the **SALCs[UF$_9$]**.

The command that produces the 3D graphics is

```
SALCgraphic3D[MakeSALCsOutput,imSz,rowNbr, Δ]
```

where **MakeSALCsOutput** is the entire output of **MakeSALCs**, **imSz** goes into **ImageSize→imSz**, **rowNbr** is the row of **MakeSALCsOutput** to be used, and Δ goes into **PlotRange→{{-Δ,Δ},{-Δ,Δ},{-Δ, Δ}}**. The default for **imSz** is **{100,100}**, the default for Δ is **3**.

```
tblUF9 = Table[SALCgraphic3D[
   SALCs[UF9], {60, 60}, i, 3.5], {i, 1, 6}]
```

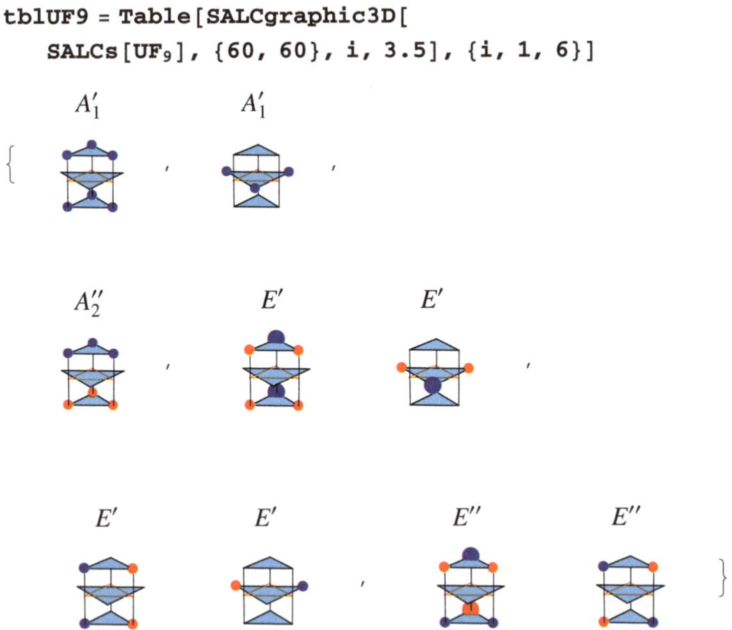

Fig. 40.6 The SALCs of UF_9. Blue and red dots indicate the algebraic sign of the atomic orbital; size indicates amplitude. No dot means 0 coefficient.

40.4 End notes

These End notes are quite lengthy and detailed, and may be read from the CD that accompanies the book.

41. Hybrid orbitals

Preliminaries

41.1 Hybrid orbitals in general

The tetrahedral nature of carbon bonding was worked out by pre-quantum chemists who had little more knowledge of atoms than Plato did. They reasoned like this: CH_2Cl_2 has only one isomer, not two. Therefore it cannot be square planar, but it can be tetrahedral. Think about it. It is not proof, but it makes the tetrahedral structure look plausible. Look it up in your old organic book.

However, the atom orbitals first worked out by Schrödinger in 1926 showed no tetrahedral proclivities. It was Linus Pauling (1931) who first combined them to make *directed valence bonds*. Starting with Schrodinger's atomic orbitals, he produced *all* of the common bonding geometries worked out by organic chemists in over a century of empirical investigations. After this, it was hard to doubt that quantum physics was going to explain chemistry. Today a qualitative description of Pauling's results is presented in freshman chemistry, or even in high school, under the name of *hybrid orbitals*.

Here is a typical presentation. First you see the four bonding orbitals of carbon, the Schrödinger orbitals of shell 2; a **2s** and three perpendicular **2p**'s .

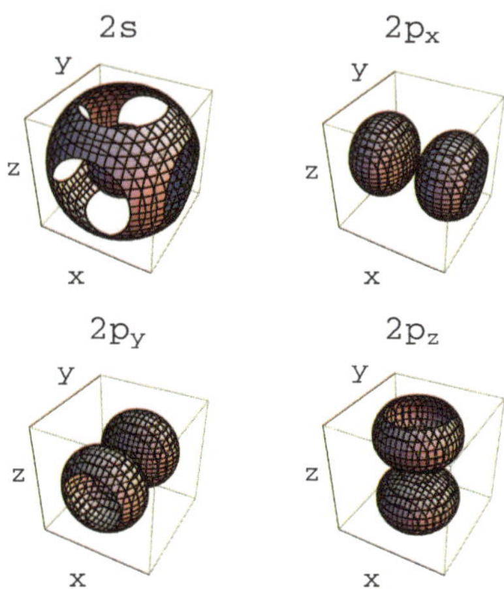

Fig. 41.1 Schrödinger's bonding orbitals of carbon.

(We plot them with holes so you can see inside. Note the inner and outer

W.M. McClain, *Symmetry Theory in Molecular Physics with Mathematica*,
DOI 10.1007/b13137_41, © Springer Science+Business Media, LLC 2009

surfaces of 2s. The two lobes of the 2p orbitals have opposite signs, not indicated.)

Then there is a magic arrow with the word *hybridization* written over it, and then you see

The closed graphics cell below is NOT Evaluatable

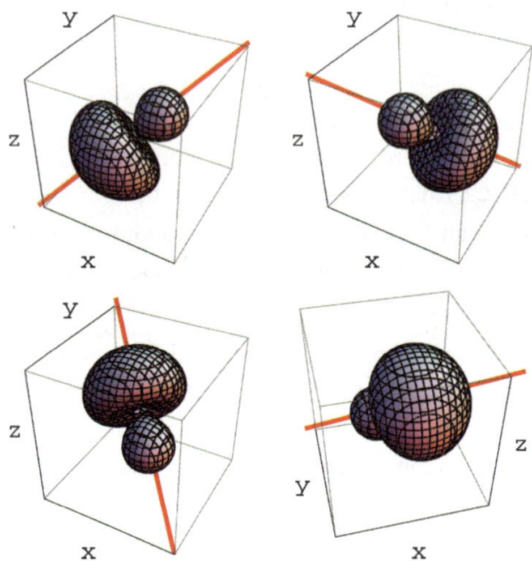

Fig. 41.2 Hybridized bonding orbitals of carbon, directed to the corners of a tetrahedron. The two lobes have opposite signs, not indicated.

Or often, you see only one of these "mushroom" orbitals. It is then said that the cap of the mushroom contains the major electron density that lets it bond tetrahedrally with other atoms.

It is an enduring surprise that linear combinations of the four functions of Fig. 41.1 can produce the four tetrahedrally directed functions of Fig. 41.2. Our purpose in this chapter is to finally give you a thorough, complete understanding of just what hybridization really is, and how Pauling found it.

41.2 Pauling's hybridization strategy

We will stick to the tetrahedral example, the most famous. We seek four linear combinations of the carbon orbitals **{2s, 2p$_x$, 2p$_y$, 2p$_z$}** skewered on the four diagonals of a cube, and therefore pointing to the four apices of a tetrahedron. Each must be cylindrically symmetric about its own diagonal, and they must be identical, except for orientation.

On your own

You should be able to follow this construction method to make any other hybrids your want; for instance, the $\mathbf{sp}^2$ hybrids of three-fold symmetry involved in the bonding of flat aromatic hydrocarbons. Or the **sp** hybrids of two-fold symmetry that make the bonds of linear acetylene chains. And of course there are many metal-based hybrids involved in coordination compounds. They all yield to the same method.

However, the method of 41.3 is didactic, not practical. Wait until you have read the quick and easy method of Section 41.4 before you try your own hybrids.

Each orbital will belong to group **T**; that is, the transforms of group **T** only swap them around from one tetrahedral apex to another. The problem is to find 16 coefficients $\mathbf{a}_{m,n}$ that produce the desired geometry. (If you set up a chemically impossible geometry, you will not be able to find solutions. Pauling was lucky that pre-quantum organic chemistry was exactly right about the geometries that were possible.) So all he had to do was find the **a** numbers in

$$\begin{pmatrix} \mathbf{hybrid}_1 \\ \mathbf{hybrid}_2 \\ \mathbf{hybrid}_3 \\ \mathbf{hybrid}_4 \end{pmatrix} == \begin{pmatrix} a_{11} & a_{12} & a_{13} & a_{14} \\ a_{21} & a_{22} & a_{23} & a_{24} \\ a_{31} & a_{32} & a_{33} & a_{34} \\ a_{41} & a_{42} & a_{43} & a_{44} \end{pmatrix} \cdot \begin{pmatrix} 2s \\ 2p_x \\ 2p_y \\ 2p_z \end{pmatrix} \quad (41.1)$$

Pauling's stroke of genius was to realize that the **a** coefficients do not depend on the exact shape of the orbitals. All that is required of the functions $\{2s, 2p_x, 2p_y, 2p_z\}$ is that they transform , respectively, like a sphere, and like the **x**-axis, **y**-axis and **z**-axis. All that is required of the functions $\{\mathbf{hybrid}_1, .., \mathbf{hybrid}_4\}$ is that they be cylindrically symmetric around the cube diagonals.

You can use a mock **2s** function, as long as it belongs to the totally symmetric species $\mathbf{A}_1$ of group **T**; along with a mock $2p_x$ of species **T** and component **x**; etc. for $2p_y$ and $2p_z$. The procedure is to cook up some mock hybrids that have the desired symmetry properties, project them to get mock orbitals of species $\{A_1, T_x, T_y, T_z\}$, and then put them into the inverse of Eq. 41.1, namely

$$\begin{pmatrix} \text{mock2s} \\ \text{mock2p}_x \\ \text{mock2p}_y \\ \text{mock2p}_z \end{pmatrix} == \begin{pmatrix} b_{11} & b_{12} & b_{13} & b_{14} \\ b_{21} & b_{22} & b_{23} & b_{24} \\ b_{31} & b_{32} & b_{33} & b_{34} \\ b_{41} & b_{42} & b_{43} & b_{44} \end{pmatrix} . \begin{pmatrix} \text{mockHybrid}_1 \\ \text{mockHybrid}_2 \\ \text{mockHybrid}_3 \\ \text{mockHybrid}_4 \end{pmatrix}$$

in which **bMat** becomes the only unknown. Solve for it (using **MatrixOfCo·efficients**). Its inverse is then the desired **aMat**. Take this **aMat** back to Eq. 41.1 and put in good carbon orbitals. Eq. 41.1 will then return good tetrahedral hybrid orbitals on the carbon atom, ready to make methane.

In the next section we carry this out literally; then in the last section we will show a shortcut that uses the **MakeSALCs** operator of Chapter 40 (ConstructingSALCs).

41.3 Implement Pauling's strategy

41.3.1 Make the mockHybrids

We need a symbolic **bondFn** that depends on electron position **{x,y,z}** through the distance from two fixed points, the carbon nucleus and a ligand nucleus (like the H atom in methane). Put the four tetrahedral vertices at

```
rps3D =
  {{1, 1, 1}, {-1, -1, 1}, {1, -1, -1}, {-1, 1, -1}};
```

The vectors from each vertex to the point **{x,y,z}** are

```
vecs = Map[{x, y, z} - # &, rps3D]
```

```
{{-1 + x, -1 + y, -1 + z}, {1 + x, 1 + y, -1 + z},
 {-1 + x, 1 + y, 1 + z}, {1 + x, -1 + y, 1 + z}}
```

and the squared lengths of these vectors are

```
rSqList = Map[#.# &, vecs] // ExpandAll
```

$\{3 - 2x + x^2 - 2y + y^2 - 2z + z^2,\ 3 + 2x + x^2 + 2y + y^2 - 2z + z^2,$
$3 - 2x + x^2 + 2y + y^2 + 2z + z^2,\ 3 + 2x + x^2 - 2y + y^2 + 2z + z^2\}$

Put the head ϕ on these square distances to make them into mock tetrahedral orbitals :

```
φList = Map[φ, rSqList]
```

$\{\phi[3 - 2x + x^2 - 2y + y^2 - 2z + z^2],$
$\phi[3 + 2x + x^2 + 2y + y^2 - 2z + z^2],$

```
ϕ[3 - 2 x + x^2 + 2 y + y^2 + 2 z + z^2],
ϕ[3 + 2 x + x^2 - 2 y + y^2 + 2 z + z^2]}
```

We want to refer to these four function very briefly as **ϕ[1]** through **ϕ[4]**:

```
mockHybrids = {ϕ[1], ϕ[2], ϕ[3], ϕ[4]};
CartesianToϕ = Thread[ϕList → mockHybrids]
```

```
{ϕ[3 - 2 x + x^2 - 2 y + y^2 - 2 z + z^2] → ϕ[1],
 ϕ[3 + 2 x + x^2 + 2 y + y^2 - 2 z + z^2] → ϕ[2],
 ϕ[3 - 2 x + x^2 + 2 y + y^2 + 2 z + z^2] → ϕ[3],
 ϕ[3 + 2 x + x^2 - 2 y + y^2 + 2 z + z^2] → ϕ[4]}
```

and the reverse transform is

```
ϕToCartesian = Map[Reverse, CartesianToϕ]
```

```
{ϕ[1] → ϕ[3 - 2 x + x^2 - 2 y + y^2 - 2 z + z^2],
 ϕ[2] → ϕ[3 + 2 x + x^2 + 2 y + y^2 - 2 z + z^2],
 ϕ[3] → ϕ[3 - 2 x + x^2 + 2 y + y^2 + 2 z + z^2],
 ϕ[4] → ϕ[3 + 2 x + x^2 - 2 y + y^2 + 2 z + z^2]}
```

41.3.2 Visual check of mockHybrid symmetry

We now make a visual check that we have done what we intended. Purely for graphic purposes, we define a very rough embodiment of the bond function that does about what a chemist thinks it ought to do: it puts electron density between the atoms. Each point in space is at distance $\sqrt{x^2 + y^2 + z^2}$ from the carbon nucleus and at distance $\sqrt{\mathtt{rSq[i]}}$ from ligand nucleus **i**. Consider an ellipsoidal shell with one focus at the carbon atom and the other on one of the ligand atoms. Everywhere on this shell, the sum of these two distances is a constant. Therefore we can define an elliptical mock bonding function as

$$\mathtt{ellipticalBond[rSq\_]} := e^{-\left(\sqrt{x^2+y^2+z^2}+\sqrt{\mathtt{rSq}}\right)}$$

The decaying exponential puts maximum density right between the foci.

The closed graphics cell below is NOT Evaluatable

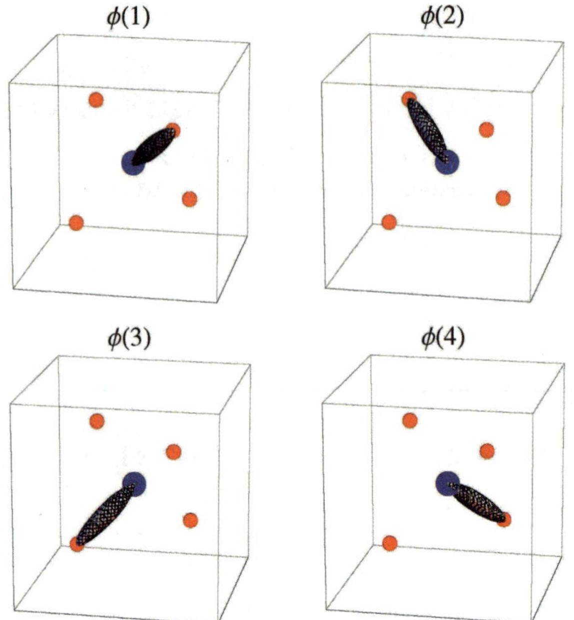

Fig. 41.3 The mock hybrids are four tetrahedrally oriented ellipsoids, as desired.

41.3.3 The trace projector, `ProjectET`

Now we are ready to extract the really important information. What are the symmetry species of the mock hybrids that we have so painfully created?

```
prjRaw =
  ExpandAll[ProjectET[ϕList, "T"]] /. CartesianToϕ;
prjϕList = NeatProjection[prjRaw, ϕ]
```

```
{{A1, {ϕ[1] + ϕ[2] + ϕ[3] + ϕ[4]}},
 {T, {ϕ[1] + ϕ[2] + ϕ[3] - 3 ϕ[4], 3 ϕ[1] - ϕ[2] - ϕ[3] - ϕ[4],
   ϕ[1] + ϕ[2] - 3 ϕ[3] + ϕ[4], ϕ[1] - 3 ϕ[2] + ϕ[3] + ϕ[4]}}}
```

That is the answer: In group **`T`**, four tetrahedrally pointed functions are spanned by one function of species **`A1`** and three functions of species **`T`**. For future use, we give these functions names :

```
mockA1g = prjϕList[[1, 2, 1]]
```

```
ϕ[1] + ϕ[2] + ϕ[3] + ϕ[4]
```

```
mockOrbitalsT = prjϕList[[2, 2]]
```

```
{ϕ[1] + ϕ[2] + ϕ[3] - 3 ϕ[4], 3 ϕ[1] - ϕ[2] - ϕ[3] - ϕ[4],
 ϕ[1] + ϕ[2] - 3 ϕ[3] + ϕ[4], ϕ[1] - 3 ϕ[2] + ϕ[3] + ϕ[4]}
```

There you see four functions that all live together in the same three-dimensional function space. Clearly, we need to find three nice orthogonal basis functions for that space. That is what the detailed projector is for.

41.3.4 The detailed projector, `ProjectED`

There are four **T** functions, but we rather suspect that only three of them are linearly independent. The detailed projection **ProjectED** will tell us.

We want the projected **mockHybrids** to transform like **x**, **y**, and **z**, just as Schrödinger's **2p** functions do. So in **ProjectED** we will specify a representation of group **T** made with basis functions **{x,y,z}** :

```
Trep = MakeRepPoly[{x, y, z}, "T"];
% // MatrixList // Size[6]
```

$$\left\{\begin{pmatrix}1&0&0\\0&1&0\\0&0&1\end{pmatrix},\begin{pmatrix}0&1&0\\0&0&1\\1&0&0\end{pmatrix},\begin{pmatrix}0&-1&0\\0&0&1\\-1&0&0\end{pmatrix},\begin{pmatrix}0&-1&0\\0&0&-1\\1&0&0\end{pmatrix},\begin{pmatrix}0&1&0\\0&0&-1\\-1&0&0\end{pmatrix},\begin{pmatrix}0&0&-1\\1&0&0\\0&-1&0\end{pmatrix},\right.$$
$$\left.\begin{pmatrix}0&0&1\\-1&0&0\\0&-1&0\end{pmatrix},\begin{pmatrix}0&0&-1\\-1&0&0\\0&1&0\end{pmatrix},\begin{pmatrix}0&0&1\\1&0&0\\0&1&0\end{pmatrix},\begin{pmatrix}1&0&0\\0&-1&0\\0&0&-1\end{pmatrix},\begin{pmatrix}-1&0&0\\0&1&0\\0&0&-1\end{pmatrix},\begin{pmatrix}-1&0&0\\0&-1&0\\0&0&1\end{pmatrix}\right\}$$

We have three detailed projections to do, one for each diagonal element. Turn the **ϕ** symbols into concrete functions **x**, **y**, **z**; project them, then turn the projections back into **ϕ** symbols :

```
ϕFnsOfxyz = mockOrbitalsT /. ϕToCartesian;
prjsRaw = ProjectED[ϕFnsOfxyz, Trep, 1, "T"];
prjs = prjsRaw /. CartesianToϕ
```

```
{ϕ[1] - ϕ[2] + ϕ[3] - ϕ[4], ϕ[1] - ϕ[2] + ϕ[3] - ϕ[4],
 -ϕ[1] + ϕ[2] - ϕ[3] + ϕ[4], ϕ[1] - ϕ[2] + ϕ[3] - ϕ[4]}
```

The four projected functions look like sign variations on only one function. So use **MOtoStandardForm** to put them all in the same format, then throw a **Union** around them to get rid of redundancies :

```
mockTx =
  (Map[MOtoStandardForm[#, ϕ] &, prjs] // Union)[[1]]
```

```
ϕ[1] - ϕ[2] + ϕ[3] - ϕ[4]
```

Indeed, they were all the same function. To do the others the same way, make this procedure into a little **Module** named **ϕProjection**:

```
ϕProjection[xpr_, n_] := Module[{xpr1, xpr2, xpr3},
   xpr1 = ProjectED[xpr, Trep, n, "T"];
```

```
xpr2 = xpr1 /. CartesianToϕ;
xpr3 = Map[MOtoStandardForm[#, ϕ] &, xpr2];
xpr3 // Union ]
```

Remember, when using a 3×3 representation (like **Trep**) to implement the detailed projection, you can do it with any one of the three diagonal elements, using **n = 1**, **2**, or **3**. So to see everything the **ϕProjection** can do to **ϕFnsOfxyz**, we **Map** as follows:

```
{mockTx, mockTy, mockTz} =
  Map[ϕProjection[ϕFnsOfxyz, #] &, {1, 2, 3}]
```

```
{{ϕ[1] - ϕ[2] + ϕ[3] - ϕ[4]},
 {ϕ[1] - ϕ[2] - ϕ[3] + ϕ[4]}, {ϕ[1] + ϕ[2] - ϕ[3] - ϕ[4]}}
```

There. It gave us three different functions spanning the four functions in the list **ϕFnsOfxyz**. Put the trace projection of species **A1g** with them (we called it **mockA1g**), and **//Sort//Reverse** them :

```
mockAOs =
  {mockA1g, mockTx, mockTy, mockTz} // Flatten //
    Sort // Reverse
```

```
{ϕ[1] + ϕ[2] + ϕ[3] + ϕ[4], ϕ[1] - ϕ[2] - ϕ[3] + ϕ[4],
 ϕ[1] - ϕ[2] + ϕ[3] - ϕ[4], ϕ[1] + ϕ[2] - ϕ[3] - ϕ[4]}
```

That is the complete projection of the **mockHybrids** under group **T**. As you see, the nature of the **ϕ** functions does not matter, as long as their symmetry is tetrahedral. We used the ridiculous elliptical functions we cooked up, but they could have been the most elaborately accurate functions known to quantum chemistry. Whichever, the hybrid linear combinations are the same.

41.3.5 Look at the `mockAOs`

This is a purely academic exercise; don't take it too seriously. We wonder, out of idle curiosity, what do the mock atomic orbitals look like? If you perform **mockAOs/.ϕToCartesian** you have them algebraically, so it is not much trouble to do the graphics:

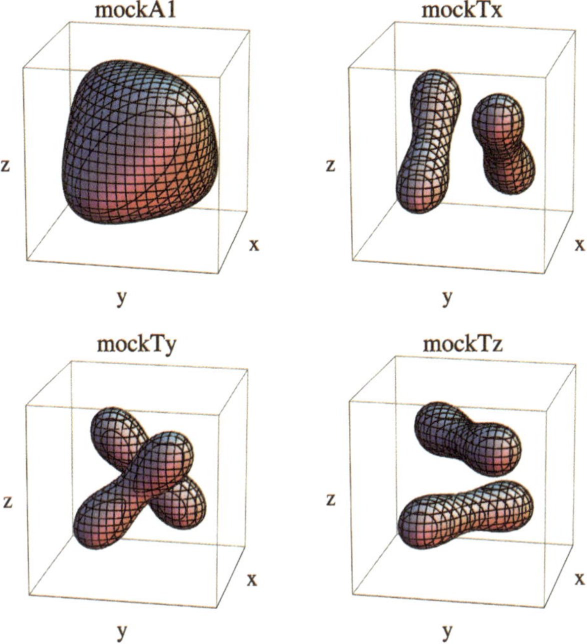

Fig. 41.4 These mock atomic orbitals, when hybridized, give the tetrahedrally directed mock hybrids of Fig. 41.3. Interactive readers can rotate these figures around with the mouse.

They are weird, unchemical shapes. They are not even centered on the carbon atom. But the point is, each belongs to the proper species of the tetrahedral group. And it is nice to see that they are related squarely to the axes (like the functions **2 p$_\mathbf{x}$**, etc.). When linearly combined, they produce four tetrahedrally pointed ellipsoids (also weird and unchemical). So when we make the same linear combinations of Schrödinger's carbon orbitals of the same species, we will get chemically meaningful hybrid orbitals pointing in tetrahedral geometry.

41.4 The SALC shortcut

There is another way to get the **mockOrbitals** in just a few lines. Of course, they are pretty sophisticated lines, but we already have the operators for them. Just watch this, and then we will talk:

```
Clear[f, ϕ];
allSALCstuff = MakeSALCs["methane", "H", σ, 1]
```

```
The group is Td

Using T2 basis {x, y, z}
```

$$\{\text{methane}, \text{H}, \{\{\text{A1}, \{\sigma[1]+\sigma[2]+\sigma[3]+\sigma[4]\}\},$$
$$\{\text{T2}, \{\sigma[1]-\sigma[2]-\sigma[3]+\sigma[4],$$
$$\sigma[1]-\sigma[2]+\sigma[3]-\sigma[4], \sigma[1]+\sigma[2]-\sigma[3]-\sigma[4]\}\}\},$$
$$\{\sigma[1]\to f\left[\frac{118579107}{100000000}-\frac{6287\,x}{5000}+x^2-\frac{6287\,y}{5000}+y^2-\frac{6287\,z}{5000}+z^2\right],$$
$$\sigma[2]\to f\left[\frac{118579107}{100000000}+\frac{6287\,x}{5000}+x^2+\frac{6287\,y}{5000}+y^2-\frac{6287\,z}{5000}+z^2\right],$$
$$\sigma[3]\to f\left[\frac{118579107}{100000000}+\frac{6287\,x}{5000}+x^2-\frac{6287\,y}{5000}+y^2+\frac{6287\,z}{5000}+z^2\right],$$
$$\sigma[4]\to f\left[\frac{118579107}{100000000}-\frac{6287\,x}{5000}+x^2+\frac{6287\,y}{5000}+y^2+\frac{6287\,z}{5000}+z^2\right]\}\}$$

```
methaneSALCs = allSALCstuff[[3]]
```

$$\{\{\text{A1}, \{\sigma[1]+\sigma[2]+\sigma[3]+\sigma[4]\}\},$$
$$\{\text{T2}, \{\sigma[1]-\sigma[2]-\sigma[3]+\sigma[4],$$
$$\sigma[1]-\sigma[2]+\sigma[3]-\sigma[4], \sigma[1]+\sigma[2]-\sigma[3]-\sigma[4]\}\}\}$$

Strip out the species information, and **//Sort//Reverse** them :

```
bareSALCs =
  Transpose[methaneSALCs][[2]] // Flatten // Sort //
   Reverse
```

$$\{\sigma[1]+\sigma[2]+\sigma[3]+\sigma[4], \sigma[1]-\sigma[2]-\sigma[3]+\sigma[4],$$
$$\sigma[1]-\sigma[2]+\sigma[3]-\sigma[4], \sigma[1]+\sigma[2]-\sigma[3]-\sigma[4]\}$$

Compare these with the **mockOrbitals** at the end of Section 41.3, above

```
mockAOs == bareSALCs
```

$$\{\phi[1]+\phi[2]+\phi[3]+\phi[4], \phi[1]-\phi[2]-\phi[3]+\phi[4],$$
$$\phi[1]-\phi[2]+\phi[3]-\phi[4], \phi[1]+\phi[2]-\phi[3]-\phi[4]\} ==$$
$$\{\sigma[1]+\sigma[2]+\sigma[3]+\sigma[4], \sigma[1]-\sigma[2]-\sigma[3]+\sigma[4],$$
$$\sigma[1]-\sigma[2]+\sigma[3]-\sigma[4], \sigma[1]+\sigma[2]-\sigma[3]-\sigma[4]\}$$

They are exactly the same. Furthermore, **MakeSALCs** gave us the symmetry information that we need. The group is $\mathbf{T_d}$, the first SALC belongs to the totally symmetric species $\mathbf{A_1}$ and the second belongs to $\mathbf{T_2}$, projected using basis **x**, **y**, **z**.

The **mockAOs** were analyzed in group **T**, and here **MakeSALCs** used T_d , the full symmetry of methane. But don't worry; **T** is a subgroup of T_d, and the species names are in one-to- one correspondence.

Why are the **bareSALCs** identical to the **mockAOs**? Is it always so easy? Well, yes, it is. As we said when we made up those ellipsoidal mock-hybrids, their shape does not matter; only their symmetry. We can consider the four H-based SALCs for the methane molecule to be carbon hybrids made of weird, unphysical atomic orbitals. (So weird that they are not even centered on carbon, but even that does not matter. Symmetry is all.)

All you have to know now is that the **2s** carbon orbital belongs to species A_1 in T_d , and the three orbitals $\{2\,p_x, 2\,p_y, 2\,p_z\}$ belong to species T_2 (because their angular factors {**x,y,z**} belong). So extract the matrix of coefficients **bMat** from the SALCs

$$\begin{pmatrix} \phi[1]+\phi[2]+\phi[3]+\phi[4] \\ \phi[1]-\phi[2]-\phi[3]+\phi[4] \\ \phi[1]-\phi[2]+\phi[3]-\phi[4] \\ \phi[1]+\phi[2]-\phi[3]-\phi[4] \end{pmatrix} == \begin{pmatrix} b_{11} & b_{12} & b_{13} & b_{14} \\ b_{21} & b_{22} & b_{23} & b_{24} \\ b_{31} & b_{32} & b_{33} & b_{34} \\ b_{41} & b_{42} & b_{43} & b_{44} \end{pmatrix} . \begin{pmatrix} \phi[1] \\ \phi[2] \\ \phi[3] \\ \phi[4] \end{pmatrix} \tag{41.3}$$

mock C orbitals ↑ mock hybrids ↑

Invert **bMat**, and left multiply by the inverse. Replace the **mockCorbitals** with **realCorbitals**, and there you have the real $\mathbf{sp}^3$ hybrids :

$$\begin{pmatrix} b_{11} & b_{12} & b_{13} & b_{14} \\ b_{21} & b_{22} & b_{23} & b_{24} \\ b_{31} & b_{32} & b_{33} & b_{34} \\ b_{41} & b_{42} & b_{43} & b_{44} \end{pmatrix}^{-1} . \begin{pmatrix} \phi[2\,s] \\ \phi[2\,px] \\ \phi[2\,py] \\ \phi[2\,pz] \end{pmatrix} == \begin{pmatrix} \psi[A] \\ \psi[B] \\ \psi[C] \\ \psi[D] \end{pmatrix} \tag{41.4}$$

real C orbitals ↑ real $\mathbf{sp}^3$ hybrids ↑

It is all quite easy:

```
vars = Map[Variables, mockAOs] // Flatten // Union
```

```
{ϕ[1], ϕ[2], ϕ[3], ϕ[4]}
```

```
bMat = MatrixOfCoefficients[mockAOs, vars];
% // GridForm
```

1	1	1	1
1	-1	-1	1
1	-1	1	-1
1	1	-1	-1

```
bInv = 4 Inverse[bMat];
% // GridForm
```

1	1	1	1
1	-1	-1	1
1	-1	1	-1
1	1	-1	-1

Now make the hybrids symbolically, and put them in standard form:

```
sp3hybridsRaw =
  bInv.{φ[2 s], φ[2 px], φ[2 py], φ[2 pz]};
sp3hybrids = Map[MOtoStandardForm[#, φ] &,
  sp3hybridsRaw]
```

```
{φ[2 px] + φ[2 py] + φ[2 pz] + φ[2 s],
 φ[2 px] + φ[2 py] - φ[2 pz] - φ[2 s],
 φ[2 px] - φ[2 py] + φ[2 pz] - φ[2 s],
 φ[2 px] - φ[2 py] - φ[2 pz] + φ[2 s]}
```

Wait just a minute. How did we know what order to use for the real orbitals ϕ? What if we used a different order? Make some alternate hybrids, using a different order for the carbon orbitals; then put them in standard form also :

```
hybridsAlt = bInv.{φ[2 py], φ[2 pz], φ[2 s], φ[2 px]};
sp3hybridsAlt =
 Map[MOtoStandardForm[#, φ] &, hybridsAlt]
```

```
{φ[2 px] + φ[2 py] + φ[2 pz] + φ[2 s],
 φ[2 px] + φ[2 py] - φ[2 pz] - φ[2 s],
 φ[2 px] - φ[2 py] + φ[2 pz] - φ[2 s],
 φ[2 px] - φ[2 py] - φ[2 pz] + φ[2 s]}
```

Are they the same?

```
sp3hybrids == sp3hybridsAlt
```

```
True
```

Symmetry takes care of everything. The order in which you list the basis functions of a space can't really matter for functions that use them symmetrically.

41.5 Electron density in sp^3 orbitals

Finally, we put the shell 2 hydrogen orbitals into Eq. 41.4 , producing the four famous **sp**^{3} hybrids already shown in the introduction. Click to them if you want to see them again. Here are the analytical forms :

```
{fn2s, fn2px, fn2py, fn2pz} =
  e^(-1/2 Sqrt[x^2+y^2+z^2]) {(2 - Sqrt[x^2 + y^2 + z^2]), x, y, z}
```

$$\left\{e^{-\frac{1}{2}\sqrt{x^2+y^2+z^2}}\left(2-\sqrt{x^2+y^2+z^2}\right),\; e^{-\frac{1}{2}\sqrt{x^2+y^2+z^2}}\,x,\; e^{-\frac{1}{2}\sqrt{x^2+y^2+z^2}}\,y,\; e^{-\frac{1}{2}\sqrt{x^2+y^2+z^2}}\,z\right\}$$

Now make the **sp**^{3} hybrids :

```
{sp3A, sp3B, sp3C, sp3D} =
  bInv.{fn2s, fn2px, fn2py, fn2pz};
% // Size[7]
```

$$\left\{e^{-\frac{1}{2}\sqrt{x^2+y^2+z^2}}\,x+e^{-\frac{1}{2}\sqrt{x^2+y^2+z^2}}\,y+e^{-\frac{1}{2}\sqrt{x^2+y^2+z^2}}\,z+e^{-\frac{1}{2}\sqrt{x^2+y^2+z^2}}\left(2-\sqrt{x^2+y^2+z^2}\right),\right.$$
$$-e^{-\frac{1}{2}\sqrt{x^2+y^2+z^2}}\,x-e^{-\frac{1}{2}\sqrt{x^2+y^2+z^2}}\,y+e^{-\frac{1}{2}\sqrt{x^2+y^2+z^2}}\,z+e^{-\frac{1}{2}\sqrt{x^2+y^2+z^2}}\left(2-\sqrt{x^2+y^2+z^2}\right),$$
$$-e^{-\frac{1}{2}\sqrt{x^2+y^2+z^2}}\,x+e^{-\frac{1}{2}\sqrt{x^2+y^2+z^2}}\,y-e^{-\frac{1}{2}\sqrt{x^2+y^2+z^2}}\,z+e^{-\frac{1}{2}\sqrt{x^2+y^2+z^2}}\left(2-\sqrt{x^2+y^2+z^2}\right),$$
$$\left.e^{-\frac{1}{2}\sqrt{x^2+y^2+z^2}}\,x-e^{-\frac{1}{2}\sqrt{x^2+y^2+z^2}}\,y-e^{-\frac{1}{2}\sqrt{x^2+y^2+z^2}}\,z+e^{-\frac{1}{2}\sqrt{x^2+y^2+z^2}}\left(2-\sqrt{x^2+y^2+z^2}\right)\right\}$$

Three-D contour plots of these four functions are given in Fig. 41.2. Plot the amplitude squared of the **sp3A** hybrid orbital along a line **x = y = z** passing right down through its center :

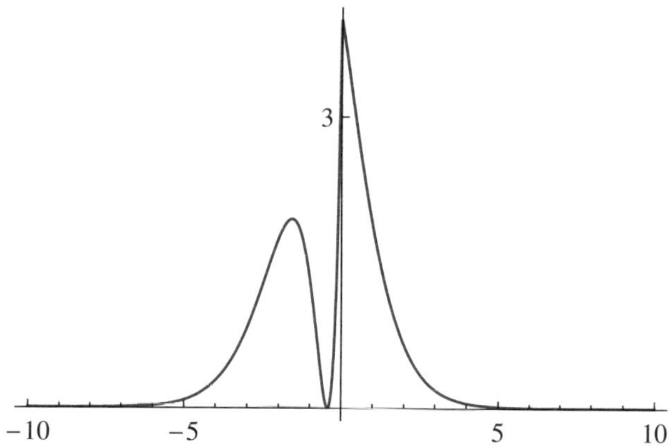

Fig. 41.5 Electron density on the center line of an sp^3 hybrid. It bonds to the right, where the mushroom cap is.

41.6 End Notes

41.6.1 Reference

F. A. Cotton, 3rd ed. Section 8.5, pp. 222- 226

42. Vibration analysis

Preliminaries

42.1. Problem and strategy

Every molecule possess a small, finite number of normal modes of vibration, in which every atom moves harmonically around its rest position with the same frequency. The general vibrational motion, as excited by thermal collisions, is simply a sum of the normal mode motions, with random amplitudes and phases. But normal modes are easy to excite as pure motions. When an electromagnetic wave of the right frequency washes over a molecule, every atom feels a push-pull force, and if that force has the frequency of a normal mode, the whole molecule begins to move in that pure normal mode. This is the classical view of infrared absorption spectroscopy.

In a molecule with **n** atoms, there are **3n** atomic coordinates. Generally 3 are devoted to center of mass motion and three are devoted to rotation. That leaves **3n-6** internal configuration coordinates. The proper linear combination of these internal coordinates are the normal modes. If the forces between atoms are Hooke's Law forces (as they are for all small amplitude motions), it turns out that the problem is exactly soluble, and if the molecule has any symmetry at all, the normal modes have a beautiful geometric symmetry.

A question now arises quite naturally: Of the **3n-6** modes of vibration, in a molecule of given group, how many belong to each symmetry species of the group? Are there species have no mode? How many modes are allowed for infrared absorption, and how many for Raman scattering, and how many are inactive for both? This chapter shows you how to answer these questions without a full-blown mode calculation.

For example, when you read a reference to "the A_{2u} modes of benzene" you know something about them immediately. Pull out the **BoxedCharacterTa\ ble** for group D_{6h} and have a look. Since species A_{2u} transforms like **z**, you know immediately that (1) the A_{2u} modes are allowed in infrared spectra, and (2) they involve out-of-plane motion. This chapter shows you how to count how many modes will belong to each species, even *before* you do the mode calculation.

The method is to make a representation of the group based on the **3-n** atomic displacement coordinates of an **n**-atom molecule. This is a large representation, size **3n 3n**, and it is highly reducible. Its analysis tells you everything you can know about the symmetry of the molecule's modes. It won't give you frequen-

W.M. McClain, *Symmetry Theory in Molecular Physics with Mathematica*,
DOI 10.1007/b13137_42, © Springer Science+Business Media, LLC 2009

cies or exact mode geometries, but it will reveal everything about the mode symmetries.

We will follow the recipe implicit in the master representation equation from Chapter 27 (MakeReps). The symmetrically transforming objects $\mathbf{f_k}$ are the **3n** atomic displacement coordinates used to describe the modes :

The master representation equation

$$\mathcal{T}_R[f_k] = \sum_{h=1}^{N[f]} f_h \, T_{h,k}^{(R)} \qquad \text{for } k = \{1,\ldots,N[f]\} \tag{42.1}$$

where

$\mathcal{T}_R$ is the transform for element **R**

f_h is basis object number **h**

`N[f]` is the number of basis objects

$T_{h,k}^{(R)}$ (or $\mathbb{T}^{(R)}$) is the desired rep matrix for element **R**

Here is the calculational strategy: The first job is to make an explicit list of the displacements $\mathbf{f_k}$ and calculate the left side, $\mathcal{T}_R[f_k]$, the result of applying all the Cartesian transforms of the group to the displacements. The second job is to find the matrix $T_{h,k}^{(R)}$ on right side. This will be done as usual by **MatrixOfCo\
efficients**.

42.2. A step-by-step example

42.2.1 The methane example

We will develop our formulas using methane as the example. The formulas lead to an algorithm (**VibrationAnalysis[molecule,group]**, Section 42.4) that works for all molecules with one hit. But we develop it, as always by working out a concrete example. Here is the methane molecule, in standard position for its group, group $\mathbf{T_d}$.

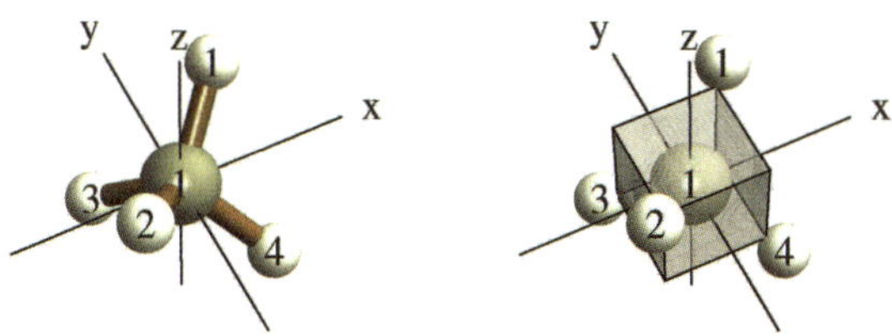

Fig. 42.1 Two views of the methane molecule in standard position. On the left, you can see the central carbon atom and the tetrahedrally arrayed C-H bonds. The right view emphasizes that the H atoms are at alternate corners of a cube.

Here is its character table:

```
BoxedCharacterTable["Td"]
```

Td	1	8	3	6	6	← Class populations
	E	C3	C2	S4	σd	↓ Basis functions
A1	1	1	1	1	1	$\{(x^2+y^2+z^2)\}$
A2	1	1	1	-1	-1	$\{(-x^4y^2+x^2y^4+x^4z^2-y^4z^2-x^2z^4+y^2z^4)\}$
E	2	-1	2	0	0	$\left\{\begin{pmatrix}-x^2-y^2+2z^2\\ \sqrt{3}\,x^2-\sqrt{3}\,y^2\end{pmatrix}\right\}$
T1	3	0	-1	1	-1	$\left\{\begin{pmatrix}xy^2-xz^2\\ -x^2y+yz^2\\ x^2z-y^2z\end{pmatrix},\begin{pmatrix}Ix\\ Iy\\ Iz\end{pmatrix}\right\}$
T2	3	0	-1	-1	1	$\left\{\begin{pmatrix}x\\ y\\ z\end{pmatrix},\begin{pmatrix}xy\\ yz\\ xz\end{pmatrix}\right\}$

It will turn out that methane has no vibrations of species A_2 or T_1. But you cannot see this ahead of time, without the calculation.

42.2.2 Make a displacement representation, step by step

Every atom needs a displacement vector $\{\delta_x, \delta_y, \delta_z\}$, which, when added to its rest position, tells where the atom is at every moment. But different molecules need different numbers of displacement vectors. A convenient way to construct the general set of displacement vectors is

```
mol = "methane";
gp = "Td";
nDim = AtomCount[mol];
δxs = Array[x, nDim];
δys = Array[y, nDim];
δzs = Array[z, nDim];
dspColumns = {δxs, δys, δzs};
% // GridForm // Size[9]
```

x[1]	x[2]	x[3]	x[4]	x[5]
y[1]	y[2]	y[3]	y[4]	y[5]
z[1]	z[2]	z[3]	z[4]	z[5]

One of the Cartesian transforms of this group is **C**$_z$[π], a two-fold rotation about **z**. In the "before" picture below we show **H** atoms **1** and **2**, and the top of the cube that encloses the methane molecule. There is a small displacement {δ_x, δ_y} on atom **1** and a zero displacement on atom 2. In the "after" picture, the molecule has rotated by **C**$_z$[π], carrying all its displacements with it.

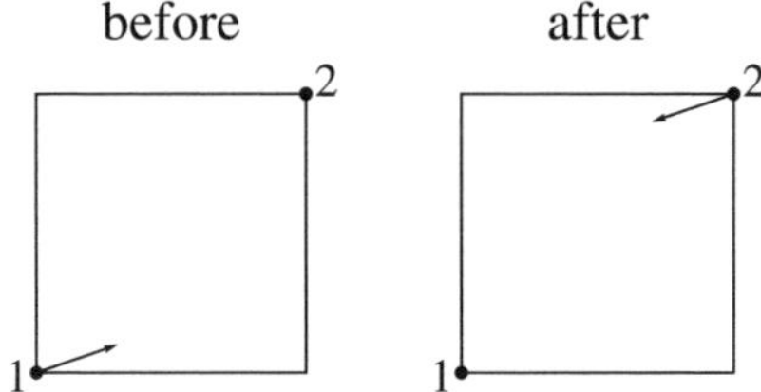

Fig. 42.2 A small vibration displacement is subjected to C$_z$[π]

Before the rotation, the 2-atom displacement list is $\begin{pmatrix} \delta_x & 0 \\ \delta_y & 0 \end{pmatrix}$. After, it is $\begin{pmatrix} 0 & -\delta_x \\ 0 & -\delta_y \end{pmatrix}$. Two things have happened: First, the displacement vector $\begin{pmatrix} \delta_x \\ \delta_y \end{pmatrix}$ rotated by half a turn about **z**, becoming $\begin{pmatrix} -\delta_x \\ -\delta_y \end{pmatrix}$. Second, the rotation of the whole molecule carried it from atom **1** to atom **2**. Therefore, it moved from the first column to the second. This two-step transform has to be carried out on all the displacements in the molecule. The rotation matrix is

```
Tmat = Cz[π]

{{-1, 0, 0}, {0, -1, 0}, {0, 0, 1}}
```

Apply it to the displacement columns:

```
dspTrf10 = Tmat.dspColumns;
% // GridForm // Size[9]
```

-x[1]	-x[2]	-x[3]	-x[4]	-x[5]
-y[1]	-y[2]	-y[3]	-y[4]	-y[5]
z[1]	z[2]	z[3]	z[4]	z[5]

Sure enough, every displacement vector was rotated by half a turn about **z**. Now we need something that will permute the columns appropriately. In fact, we already have it. When we were working with atomic orbital representations, we needed the permutation matrices that describe how identical atoms shift around in the molecule under each transform. Here they are again :

```
permuMatsMol =
  MakeRepAtomic[mol, gp][[1]];
```

This is too big to print completely. Lift out the 10th matrix (representing permutation **{2,1,4,3,5}**) :

```
permuMatsMol[[10]] // MatrixForm
```

$$\begin{pmatrix} 0 & 1 & 0 & 0 & 0 \\ 1 & 0 & 0 & 0 & 0 \\ 0 & 0 & 0 & 1 & 0 \\ 0 & 0 & 1 & 0 & 0 \\ 0 & 0 & 0 & 0 & 1 \end{pmatrix}$$

It is a **5×5** matrix. We want to permute the columns of the 10th transformed displacement matrix, which is a **3 x 5** matrix. The only way these matrices can multiply is in the order **(3 x 5).(5 x 5)**. But first, check out directly that this gives us what we want. We **Dot** the permutation matrix with an alphabetical **(5 x 5)** from both sides :

$$\left\{\begin{pmatrix} a & b & c & d & e \\ f & g & h & i & j \\ k & l & m & n & o \\ p & q & r & s & t \\ u & v & w & x & y \end{pmatrix}.\begin{pmatrix} 0 & 1 & 0 & 0 & 0 \\ 1 & 0 & 0 & 0 & 0 \\ 0 & 0 & 0 & 1 & 0 \\ 0 & 0 & 1 & 0 & 0 \\ 0 & 0 & 0 & 0 & 1 \end{pmatrix}, \begin{pmatrix} 0 & 1 & 0 & 0 & 0 \\ 1 & 0 & 0 & 0 & 0 \\ 0 & 0 & 0 & 1 & 0 \\ 0 & 0 & 1 & 0 & 0 \\ 0 & 0 & 0 & 0 & 1 \end{pmatrix}.\begin{pmatrix} a & b & c & d & e \\ f & g & h & i & j \\ k & l & m & n & o \\ p & q & r & s & t \\ u & v & w & x & y \end{pmatrix}\right\} \text{ // MatrixList}$$

$$\left\{\begin{pmatrix} b & a & d & c & e \\ g & f & i & h & j \\ l & k & n & m & o \\ q & p & s & r & t \\ v & u & x & w & y \end{pmatrix}, \begin{pmatrix} f & g & h & i & j \\ a & b & c & d & e \\ p & q & r & s & t \\ k & l & m & n & o \\ u & v & w & x & y \end{pmatrix}\right\}$$

The first product, **mat.perm**, permutes the columns, as we desire. The other way, **perm.mat**, permutes the rows. So the calculation is

```
dspTrf10.permuMatsMol[[10]];
% // MatrixForm
```

```
-x[2]  -x[1]  -x[4]  -x[3]  -x[5]
-y[2]  -y[1]  -y[4]  -y[3]  -y[5]
z[2]   z[1]   z[4]   z[3]   z[5]
```

Perfect. The **H** atoms **1** and **2** have switched places, as have the other two, **3** and **4**, while atom **5**, the **C** atom, stayed put. So let's expand our horizons and do it all for all **24** transforms at once:

```
nGp = GroupOrder[gp];
dspTrfPrm = Table[Inverse[CartRep[gp][[i]]].
    dspColumns.permuMatsMol[[i]], {i, 1, nGp}];
dspTrfPrm // MatrixList // Short[#, 5] & // Size[7]
Dimensions[dspTrfPrm]
```

```
{ x[1] x[2] x[3] x[4] x[5]     -y[2] -y[3] -y[1] -y[4] -y[5]
  y[1] y[2] y[3] y[4] y[5]  ,  z[2]  z[3]  z[1]  z[4]  z[5]   , <<20>>,
  z[1] z[2] z[3] z[4] z[5]     -x[2] -x[3] -x[1] -x[4] -x[5]

  -y[2] -y[1] -y[3] -y[4] -y[5]     x[4]  x[2]  x[3]  x[1]  x[5]
  -x[2] -x[1] -x[3] -x[4] -x[5]  ,  -z[4] -z[2] -z[3] -z[1] -z[5]  }
  z[2]  z[1]  z[3]  z[4]  z[5]      -y[4] -y[2] -y[3] -y[1] -y[5]
```

```
{24, 3, 5}
```

Each one of these 3×5 matrices needs to be made into a 15-element vector, reading down the first column, then down the second column, etc.

```
dspTrfPrm2415 =
  Map[Flatten, Map[Transpose, dspTrfPrm]];
% // Dimensions
```

```
{24, 15}
```

We need the untransformed list explicitly, to use as a list of the original **displacementVariables**, or **dspVars**, for short. It is the first one above:

```
dspVars = dspTrfPrm2415[[1]]
```

```
{x[1], y[1], z[1], x[2], y[2], z[2], x[3],
 y[3], z[3], x[4], y[4], z[4], x[5], y[5], z[5]}
```

This also checks that we put them in the desired order when we **Transposed** and **Flattened**. Now we are ready to extract the representation matrices. We will get one from each row of the **12 x 15** result above. We do not forget that **MatrixOfCoefficients** gives us the transpose of what we really want, so we wrap its output in a **Transpose** operator :

```
repMatsMethaneTd = Map[Transpose,
   Map[MatrixOfCoefficients[#, dspVars] &,
    dspTrfPrm2415]];
%[[{10}]] // GridList // Size[6]
```

{

0	0	0	-1	0	0	0	0	0	0	0	0	0	0	0
0	0	0	0	-1	0	0	0	0	0	0	0	0	0	0
0	0	0	0	0	1	0	0	0	0	0	0	0	0	0
-1	0	0	0	0	0	0	0	0	0	0	0	0	0	0
0	-1	0	0	0	0	0	0	0	0	0	0	0	0	0
0	0	1	0	0	0	0	0	0	0	0	0	0	0	0
0	0	0	0	0	0	0	0	0	-1	0	0	0	0	0
0	0	0	0	0	0	0	0	0	0	-1	0	0	0	0
0	0	0	0	0	0	0	0	0	0	0	1	0	0	0
0	0	0	0	0	0	-1	0	0	0	0	0	0	0	0
0	0	0	0	0	0	0	-1	0	0	0	0	0	0	0
0	0	0	0	0	0	0	0	1	0	0	0	0	0	0
0	0	0	0	0	0	0	0	0	0	0	0	-1	0	0
0	0	0	0	0	0	0	0	0	0	0	0	0	-1	0
0	0	0	0	0	0	0	0	0	0	0	0	0	0	1

}

This is matrix **10**, of size **15 × 15**, from the list of **24** matrices. Interactive readers can see the whole list if they want to. Is the list a rep of group $\mathbf{T_d}$?

```
MorphTest[repMatsMethaneTd, gp]
```

```
Faithful, or Isomorphic
```

Indeed it is. So the old recipe worked again.

42.3. Automate the displacement rep

42.3.1 Displacement representation automated

We pull together the steps used above in 42.2.2, to make the displacement rep for methane. We replace **methane** by **mol**, and **Td** by **gp** , and generalize any variable names specific to the methane example.

```
MakeRepDisplacement[mol_, gp_] := Module[
  {nGp, nDim, δxs, x, δys, y, δzs, z, dspCols,
   permuMats, dspTrfPrm, dspTrfPrmMat, dspVars},
  nGp = GroupOrder[gp];
  nDim = AtomCount[mol];
  δxs = Array[x, nDim];
  δys = Array[y, nDim];
  δzs = Array[z, nDim];
  dspCols = {δxs, δys, δzs};
  permuMats = MakeRepAtomic[mol, gp]⟦1⟧;
  dspTrfPrm = Table[Inverse[CartRep[gp]⟦i⟧] .
      dspCols.permuMats⟦i⟧, {i, 1, nGp}];
  dspTrfPrmMat = Map[Flatten,
     Map[Transpose, dspTrfPrm]];
```

```
dspVars = dspTrfPrmMat[[1]];
Map[Transpose,
   Map[MatrixOfCoefficients[#, dspVars] &,
    dspTrfPrmMat]]  ]
```

We test it in the next section.

42.3.2 One-click tests of MakeRepDisplacement

methane, group T_d

It had better work on the methane example:

```
repMethaneTd =
  MakeRepDisplacement["methane", "Td"];
Dimensions[repMethaneTd]
```

```
{24, 15, 15}
```

```
MorphTest[repMethaneTd, "Td"]
```

```
Faithful, or Isomorphic
```

It does. Here are a few molecules it has not seen before:

ammonia, group C_{3v}

```
repAmmoniaC3v =
  MakeRepDisplacement["ammonia", "C3v"];
Dimensions[repAmmoniaC3v]
```

```
{6, 12, 12}
```

```
MorphTest[repAmmoniaC3v, "C3v"]
```

```
Faithful, or Isomorphic
```

ethylene, group D_{2h}

```
repEthyleneD2h =
  MakeRepDisplacement["ethylene", "D2h"];
Dimensions[repEthyleneD2h]
```

```
{8, 18, 18}
```

```
MorphTest[repEthyleneD2h, "D2h"]
```

```
Faithful, or Isomorphic
```

an octahedral complex OX_6, group O

```
repOctaComplexO =
  MakeRepDisplacement["octaComplex", "O"];
Dimensions[repEthyleneD2h]
```

```
{8, 18, 18}
```

```
Dimensions[repOctaComplexO]
```

```
{24, 21, 21}
```

```
MorphTest[repOctaComplexO, "O"]
```

```
Faithful, or Isomorphic
```

uranium nonafluoride, group D_{3h}

```
repUF9D3h = MakeRepDisplacement["UF9", "D3h"];
Dimensions[repUF9D3h]
```

```
{12, 30, 30}
```

```
MorphTest[repUF9D3h, "D3h"]
```

```
Faithful, or Isomorphic
```

42.4. Analyze the vibrations

42.4.1 Make the character vector for the entire rep

We continue with the methane example worked out in 42.2.2 . The final result was **repMatsMethaneTd**:. Make sure that *Mathematica* knows it :

```
repMethaneTd = MakeRepDisplacement[mol, "Td"];
Dimensions[repMethaneTd]
```

```
{24, 15, 15}
```

If the return above is anything other than **{24,15,15}**, go back and run 42.2.2 top to bottom with one click.

First find the character of each matrix in the representation:

```
extendedChs = Map[Spur, repMatsMethaneTd]
```

```
{15, 0, 0, 0, 0, 0, 0, 0, 0, -1, -1,
 -1, -1, -1, -1, -1, -1, -1, 3, 3, 3, 3, 3, 3}
```

```
shortChs = OnePerClass[extendedChs, "Td"]
```

```
{15, 0, -1, -1, 3}
```

```
analysisRaw = Analyze[shortChs, "Td"]
```

```
{{A1, 1}, {A2, 0}, {E, 1}, {T1, 1}, {T2, 3}}
```

So the displacement representation of methane, if block diagonalized, would contain the irreducible representations $\mathbf{A_1}$, $\mathbf{E}$, and $\mathbf{T_1}$, and three occurrences of $\mathbf{T_2}$. You are guaranteed that no $\mathbf{A_2}$ will be present. More complicated $\mathbf{T_d}$ molecules could have $\mathbf{A_2}$ modes, but methane does not have enough atoms to move with that symmetry. This is the kind of thing that symmetry analysis can tell you.

How many modes is that? $\mathbf{A_1}$ is nondegenerate, $\mathbf{E}$ is doubly degenerate, while $\mathbf{T_1}$ and $\mathbf{T_2}$ are triply degenerate. So the number of raw modes is $(1\times 1) + (2\times 1) + (3\times 1) + (3\times 3) = 15$ modes. Good, that was the total number of displacement coordinates that we started with, so nothing has been lost. But not all these modes are vibrations. Three of them are translations and three are rotations, and we must remove them.

42.4.2 Remove translation modes and rotation modes

Put the raw characters into a **List** without their species names :

```
{nmsRaw, chsRaw} = Transpose[analysisRaw]
```

```
{{A1, A2, E, T1, T2}, {1, 0, 1, 1, 3}}
```

First, make the subtractions due to **x**, **y**, and **z** translations. Fortunately, we have an operator that looks for them and brings them forth:

```
translAnalysis = TranslationSpecies[gp]
```

```
{{A1, 0}, {A2, 0}, {E, 0}, {T1, 0}, {T2, {x, y, z}}}
```

```
{nmsT, translBases} = translAnalysis // Transpose
```

```
{{A1, A2, E, T1, T2}, {0, 0, 0, 0, {x, y, z}}}
```

```
translLens = Map[Length, translBases]
```

```
{0, 0, 0, 0, 3}
```

```
translRemovals = translLens / SpeciesDims[gp]
```

```
{0, 0, 0, 0, 1}
```

Now make the subtractions due to **Ix**, **Iy**, and **Iz** rotations. We have an operator that fetches them:

```
rotAnalysis = RotationSpecies[gp]
```

```
{{A1, 0}, {A2, 0}, {E, 0}, {T1, {Ix, Iy, Iz}}, {T2, 0}}
```

```
{nmsR, rotBases} = rotAnalysis // Transpose
```

```
{{A1, A2, E, T1, T2}, {0, 0, 0, {Ix, Iy, Iz}, 0}}
```

```
rotLens = Map[Length, rotBases]
```

```
{0, 0, 0, 3, 0}
```

```
rotRemovals = rotLens / SpeciesDims[gp]
```

```
{0, 0, 0, 1, 0}
```

```
{SpeciesNames[gp], chsRaw} = Transpose[analysisRaw]
```

```
{{A1, A2, E, T1, T2}, {1, 0, 1, 1, 3}}
```

And finally, here is the big subtraction:

```
vibAnalysis = Transpose[
  {nmsRaw, chsRaw - rotRemovals - translRemovals}]
```

```
{{A1, 1}, {A2, 0}, {E, 1}, {T1, 0}, {T2, 2}}
```

These are the vibrational modes of the molecule. The next step is to get it to run all in one cell; then, in the next section we make the cell into a module.

42.4.3 Automate it

The entire vibrational analysis can be automated:

```
VibrationAnalysis[mol_, gp_] := Module[

  {repMol, extendedChs, shortChs, analysisRaw,
   nmsRaw, chsRaw, transAnalysis, nmsT, translBases,
   translLens, translRemovals, rotAnalysis,
   nmsR, rotBases, rotLens, rotRemovals, spMS},

repMol = MakeRepDisplacement[mol, gp];
extendedChs = Map[Spur, repMol];
```

```
shortChs = OnePerClass[extendedChs, gp];
analysisRaw = Analyze[shortChs, gp];
{nmsRaw, chsRaw} = Transpose[analysisRaw];
translAnalysis = TranslationSpecies[gp];
{nmsT, translBases} =
   translAnalysis // Transpose;
translLens = Map[Length, translBases];
translRemovals = translLens / SpeciesDims[gp];
rotAnalysis = RotationSpecies[gp];
{nmsR, rotBases} = rotAnalysis // Transpose;
rotLens = Map[Length, rotBases];
rotRemovals = rotLens / SpeciesDims[gp];
{SpeciesNames[gp], chsRaw} =
   Transpose[analysisRaw];
vibAnalysis = Transpose[{nmsRaw,
     chsRaw - rotRemovals - translRemovals}]  ]
```

The tests of this module are in the next Section.

42.5. One-click examples

We have tested the **VibrationAnalysis** algorithm on a number of molecules, and have not seen any failures so far. Here is a miscellaneous collection of some of the molecules we have tried it on:

1. Ethylene: **3Ag + Au + 2B1g + B1u + B2g + 2B2u + 2B3u**
(pictures in Herzberg, IR & Raman, p. 107)

```
VibrationAnalysis["ethylene", "D2h"]
```

```
{{Ag, 3}, {B1g, 2}, {B2g, 1}, {B3g, 0},
 {Au, 1}, {B1u, 1}, {B2u, 2}, {B3u, 2}}
```

Perfect agreement for ethylene.

2. Carbonate: $\mathbf{A_1' + A_2'' + 2\,E'}$
(pictures in Herzberg, IR & Raman, p. 179)

```
VibrationAnalysis["carbonate anion", "D3h"]
```

$\{\{A_1', 1\}, \{A_2', 0\}, \{E', 2\}, \{A_1'', 0\}, \{A_2'', 1\}, \{E'', 0\}\}$

Exactly as listed by Herzberg.

3. Ammonia: **2 A1 + 2 E**
(pictures in Herzberg, IR & Raman, p. 110)

```
VibrationAnalysis["ammonia", "C3v"]
```

```
{{A1, 2}, {A2, 0}, {E, 2}}
```

Again.

4. Equilateral ozone: $\mathbf{A_1' + E'}$
(pictures in Herzberg, IR & Raman, p. 84, leftmost column)

```
VibrationAnalysis["equilateral ozone", "D3h"]
```

$\{\{A_1', 1\}, \{A_2', 0\}, \{E', 1\}, \{A_1'', 0\}, \{A_2'', 0\}, \{E'', 0\}\}$

And again.

43. Multiple symmetries

Preliminaries

43.1 Inspiration

In quantum mechanics there are many occasions where quantum states are written as products of two or more functions, each with its own set of variables. For instance, spin-dependent wave functions $\Psi[x_i, y_i, z_i, s_i]$ are written as $\psi[x_i, y_i, z_i]\,\phi[s_i]$, or linear combinations thereof. In the Born-Oppenheimer approximation the electronic and vibrational wave functions are multiplied together to make the "vibronic" wave function :

$$\Psi[\text{electronic, nuclear}] = \psi[\text{electronic}]\,\phi[\text{nuclear}]$$

In all such cases the energy is a sum of two parts. In this chapter we address the question of how to make symmetry group representations using basis functions that are the product of two parts.

43.2 Reps made from products of basis functions

Click back to the basic representation matrix formula, Eq. 25.1, and reread it and review the meanings of its symbols. We copy it into this chapter with a small modification: we add an $\mathbf{f}$ to the superscript on matrix $\mathbb{T}^{(R)}_{h,k}$ to indicate that the matrix was constructed on the basis of vectors named $\mathbf{f}$.

$$\mathcal{T}_R[f_i] = \sum_{h=1}^{nf} f_h\,\mathbb{T}^{(R,f)}_{h,i}$$

Then another matrix $\mathbb{T}^{(R,g)}_{J,K}$, representing element $\mathbf{R}$ on the basis of some other functions $\mathbf{g}$ would be calculated by solving for $\mathbb{T}^{(R,g)}_{J,K}$ in the equation

$$\mathcal{T}_R[g_K] = \sum_{J=1}^{ng} g_J\,\mathbb{T}^{(R,g)}_{J,K}$$

W.M. McClain, *Symmetry Theory in Molecular Physics with Mathematica*,
DOI 10.1007/b13137_43, © Springer Science+Business Media, LLC 2009

What would happen if we tried to make a representation on the basis of the products $\mathbf{f}_i\, \mathbf{g}_K$? The transform of the products will be written as $\mathcal{T}_R[\mathbf{f}_i\, \mathbf{g}_K]$, where we assume that $\mathcal{T}_R$ can work on both the coordinates of the $\mathbf{f}$'s and the coordinates of the $\mathbf{g}$'s. By linearity of the transform

$$\mathcal{T}_R[\mathbf{f}_i\, \mathbf{g}_K] = \mathcal{T}_R[\mathbf{f}_i]\, \mathcal{T}_R[\mathbf{g}_K] = \left(\sum_{h=1}^{nf} \mathbf{f}_h\, \mathbb{T}_{h,i}^{(R,f)}\right)\left(\sum_{J=1}^{NG} \mathbf{g}_J\, \mathbb{T}_{J,K}^{(R,g)}\right)$$

Rearranging the sum

$$\mathcal{T}_R[\mathbf{f}_i\, \mathbf{g}_K] = \sum_{h=1}^{nf}\sum_{J=1}^{NG} (\mathbf{f}_h\, \mathbf{g}_J)\left(\mathbb{T}_{h,i}^{(R,f)}\, \mathbb{T}_{J,K}^{(R,g)}\right)$$

Let $\mathbf{f}_u\, \mathbf{g}_V = (\mathbf{fg})_{u,V}$ and rewrite as

$$\mathcal{T}_R\left[(\mathbf{fg})_{i,K}\right] = \sum_{h=1}^{nf}\sum_{J=1}^{NG} (\mathbf{fg})_{h,J}\left(\mathbb{T}_{h,i}^{(R,f)}\, \mathbb{T}_{J,K}^{(R,g)}\right)$$

As its four indices run over all possible values, $\left(\mathbb{T}_{h,i}^{(R,f)}\, \mathbb{T}_{J,K}^{(R,g)}\right)$ runs over all possible products of pairs of elements of the two 2-index matrices $\mathbb{T}_{h,i}^{(R,f)}$ and $\mathbb{T}_{J,K}^{(R,g)}$. But to set things up so that they will fall into the form of a matrix multiplication, we must let

Direct product of two representations

$$\mathbb{T}_{(h,J),(i,K)}^{(R,fg)} = \mathbb{T}_{h,i}^{(R,f)}\, \mathbb{T}_{J,K}^{(R,g)}$$

Note the re-pairing of indices on the left. Matrices $\mathbb{T}_{h,i}^{(R,f)}$ and $\mathbb{T}_{J,K}^{(R,g)}$ are each square, but of different sizes. So $\mathbb{T}_{(h,i),(J,K)}^{(R,fg)}$ would generally be rectangular, of size $\mathbf{nf}^2 \times \mathbf{NG}^2$. But $\{\mathbf{h},\mathbf{J}\}$ runs over $\mathbf{nf}*\mathbf{NG}$ values, as does $\{\mathbf{i},\mathbf{K}\}$, so the new re-paired matrix $\mathbb{T}_{(h,J),(i,K)}^{(R,fg)}$ is square. In this notation

$$\mathcal{T}_R\left[(\mathbf{fg})_{i,K}\right] = \sum_{h=1}^{nf}\sum_{J=1}^{NG} (\mathbf{fg})_{h,J}\, \mathbb{T}_{(h,J),(i,K)}^{(R,fg)}$$

Renaming $\{\mathbf{i},\mathbf{K}\}$ as β and $\{\mathbf{h},\mathbf{J}\}$ as α,

$$\mathcal{T}_{\mathbf{R}}[(\mathbf{fg})_\beta] = \sum_{\alpha=1}^{\mathtt{nf*NG}} (\mathbf{fg})_\alpha \, \mathbb{T}^{(\mathtt{R},\mathtt{fg})}_{\alpha,\beta}$$

We have recovered the defining equation for a representation matrix. It says that matrix $\mathbb{T}^{(\mathtt{R},\mathtt{fg})}_{\alpha,\beta}$ represents transform **R** in **fg** basis. Indices α and β run from **1** to **nf*NG**.

In the EndNotes we automate this process with **DirectProduct[A,B]**.

43.3 Lemma: Character of a direct product

The character of a direct product is a true one-liner :

$$\chi^{(\mathtt{R},\mathtt{fg})} = \mathbb{T}^{(\mathtt{R},\mathtt{fg})}_{\alpha,\alpha} = \mathbb{T}^{(\mathtt{R},\mathtt{fg})}_{(\mathtt{h},\mathtt{J}),(\mathtt{h},\mathtt{J})} = \mathbb{T}^{(\mathtt{R},\mathtt{f})}_{\mathtt{h},\mathtt{h}} \, \mathbb{T}^{(\mathtt{R},\mathtt{g})}_{\mathtt{J},\mathtt{J}} = \chi^{(\mathtt{R},\mathtt{f})} \, \chi^{(\mathtt{R},\mathtt{g})}$$

There it is. The character of the direct product of $\mathbb{A}$ and $\mathbb{B}$ is the character of $\mathbb{A}$ times the character of $\mathbb{B}$. We enshrine it in a box:

$$\chi^{(\mathtt{R},\mathtt{f*g})} = \chi^{(\mathtt{R},\mathtt{f})} * \chi^{(\mathtt{R},\mathtt{g})}$$

The character of a product function is the product of the characters of its factors.

Remember this; it's profound and very useful.

43.4 Vibronic examples

43.4.1 A one-dimensional example

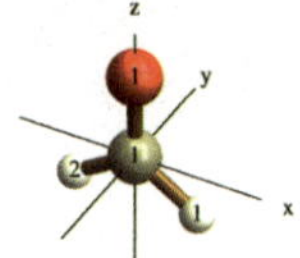

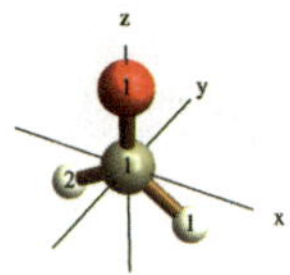

Fig. 43.1 StereoView of formaldehyde. Note the z,x reflecting plane.

The group is $\mathbf{C_{2\,v}}$. The character table is

```
$DefaultGroup = "C2v";
BoxedCharacterTable[]
```

C2v	1	1	1	1	← Class populations
	E	C2	σvzx	σvyz	↓ Basis functions
A1	1	1	1	1	{(z), (x^2), (y^2), (z^2)}
A2	1	1	-1	-1	{(xy), (Iz)}
B1	1	-1	1	-1	{(x), (xz), (Iy)}
B2	1	-1	-1	1	{(y), (yz), (Ix)}

C2v	1	1	1	1	← Class populations
	E	C2	σvzx	σvyz	↓ Basis functions
A1	1	1	1	1	{(z), (x^2), (y^2), (z^2)}
A2	1	1	-1	-1	{(xy), (Iz)}
B1	1	-1	1	-1	{(x), (xz), (Iy)}
B2	1	-1	-1	1	{(y), (yz), (Ix)}

Suppose your molecule is in a $\mathbf{B_1}$ electronic state and a $\mathbf{B_2}$ vibrational state. Then its vibronic wave function is

$\Psi_?$ **[electronic, nuclear]** = ψ_{B1} **[electronic coords]** ϕ_{B2} **[nuclear coords]**.

What is the symmetry species of $\Psi_?$?

We answer by constructing a representation using $\psi_{B1}\,\phi_{B2}$ as the basis function. Function ψ_{B1} is species $\mathbf{B_1}$ because it transforms like **x**; similarly, ϕ_{B2} is species $\mathbf{B_2}$ because it transforms like **y**. So the product $\psi_{B1}\,\phi_{B2}$ transforms like **x*y**. But looking at the character table, the function **x*y** is already there; it is the basis for species $\mathbf{A_2}$. So that is the answer; the vibronic wave function is Ψ_{A2} **[electronic, nuclear]**.

Do it again, just a little more abstractly (this moves us toward examples that are not so easy). The Lemma was written for just one transform, but of course we can do it for the first transform of each class and get a whole character vector. This time we do it without thinking about the basis functions, but only at the character vectors. Taking the **Times** product of these two character vectors, we get

```
ChVec["B1"] * ChVec["B2"]
```

```
{1, 1, -1, -1}
```

Remember, our theorem said "The character of the product is the product of the characters", and here we see this borne out :

```
ChVec["A2"]
```

```
{1, 1, -1, -1}
```

43.4.2 A two-dimensional example

Say you are working with a molecule of group C_{4v}. The character table is

```
$DefaultGroup = "C4v";
BoxedCharacterTable[]
```

C4v	1	1	2	2	2	← Class populations
	E	C2	C4	σv	σd	↓ Basis functions
A1	1	1	1	1	1	{(z), (x^2+y^2), (z^2)}
A2	1	1	1	-1	-1	{(Iz)}
B1	1	1	-1	1	-1	{(x^2-y^2)}
B2	1	1	-1	-1	1	{(x y)}
E	2	-2	0	0	0	$\left\{\begin{pmatrix}x\\y\end{pmatrix}, \begin{pmatrix}x\,z\\y\,z\end{pmatrix}, \begin{pmatrix}Ix\\Iy\end{pmatrix}\right\}$

C4v	1	1	2	2	2	← Class populations
	E	C2	C4	σv	σd	↓ Basis functions
A1	1	1	1	1	1	{(z), (x^2+y^2), (z^2)}
A2	1	1	1	-1	-1	{(Iz)}
B1	1	1	-1	1	-1	{(x^2-y^2)}
B2	1	1	-1	-1	1	{(x y)}
E	2	-2	0	0	0	$\left\{\begin{pmatrix}x\\y\end{pmatrix}, \begin{pmatrix}x\,z\\y\,z\end{pmatrix}, \begin{pmatrix}Ix\\Iy\end{pmatrix}\right\}$

Suppose your molecule is in an **E** electronic state and an $\mathbf{A_1}$ vibrational state. This means the electronic excitation was provoked by a polarization vector in the **x,y** plane, while the vibration was provoked by a **z**-polarized IR photon. Its vibronic wave function is

$$\Psi_?[\text{electronic, nuclear}] = \psi_E[\text{electronic}]\ \phi_{A1}[\text{nuclear}].$$

What is the symmetry of $\Psi_?$? If the electronic state is **E**, there must be two linearly independent wave functions at the same energy, and when they are used as the bases of a representation its character vector is **{2,-2,0,0,0}**. The vibration is not degenerate, and it supports a representation with character vector **{1,1,1,1,1}**. The dimension of the product representation is **1*2 = 2**. The product of the characters is

```
ChVec["E"] * ChVec["A1"] == ChVec["E"]
```

```
True
```

By our Lemma this is the character of the product; namely, just species E again. So the vibronic wave function **Ψ[electronic, nuclear]** is of overall species E.

43.4.3 A four-dimensional example

Now we will begin to see the power of the little lemma. Stick with the C_{4v} molecule but consider an excited state that is **E** in the electronic excitation and also **E** in the vibrational excitation. Now the dimension of the product representation must be **2*2 = 4**. This obviously cannot be a pure simple symmetry because the irreducible reps are all of dimension **1** or **2**. Nevertheless, the Lemma still works :

```
ChVec["E"] * ChVec["E"]
```

```
{4, 4, 0, 0, 0}
```

It is really beautiful to be able to compute the product state character vector without going to the trouble of computing its representation. To extract the information in this result, we determine what irreducible representations occur on the block diagonals when the (unknown) rep is reduced. Fortunately, this is also possible without knowing the actual 4-D rep :

```
Analyze[{4, 4, 0, 0, 0}]
```

```
{{A1, 1}, {A2, 1}, {B1, 1}, {B2, 1}, {E, 0}}
```

So in group C_{4v} the **E*E** excitation is a linear combination of the four simple symmetries, one of each species except **E**. So the vibronic wave function is not necessarily of any pure symmetry type.

43.4.4 Example using simple functions

We can easily show a concrete example of this calculation. We will not use actual molecular wave functions, but simple functions that have the same symmetry. You can see from the character table that two simple functions that support the **E** rep are **{x,y}**, and we will use these as stand-ins for the electronic degenerate electronic states. We need a different pair that transforms according to **E** to use as stand-ins for the vibrational state. It is easy to concoct such a pair :

```
ProjectET[{x³, y³}]
```

```
{{A1, {0, 0}}, {A2, {0, 0}},
 {B1, {0, 0}}, {B2, {0, 0}}, {E, {x³, y³}}}
```

Now we make an "electronic state" as a linear combination of the two **E** functions above, and we make the "vibrational state" as a linear combination of **x** and **y** (using GothicCapitals as arbitrary coefficients) :

$$\psi_{\text{vibronic}} = \left(\mathfrak{A}\,x^3 + \mathfrak{B}\,y^3\right)\;(\mathfrak{C}\,x + \mathfrak{D}\,y)\;//\;\text{Expand}$$

$$x^4\,\mathfrak{A}\,\mathfrak{C} + x\,y^3\,\mathfrak{B}\,\mathfrak{C} + x^3\,y\,\mathfrak{A}\,\mathfrak{D} + y^4\,\mathfrak{B}\,\mathfrak{D}$$

Our character calculation above said that this "vibronic state" could have components of any species but **E**. So we project it and see if that is true :

```
vibrProj = ProjectET[ψvibronic];
Column[%]
```

$$\left\{A1,\ \tfrac{1}{2}x^4\,\mathfrak{A}\,\mathfrak{C} + \tfrac{1}{2}y^4\,\mathfrak{A}\,\mathfrak{C} + \tfrac{1}{2}x^4\,\mathfrak{B}\,\mathfrak{D} + \tfrac{1}{2}y^4\,\mathfrak{B}\,\mathfrak{D}\right\}$$
$$\left\{A2,\ -\tfrac{1}{2}x^3\,y\,\mathfrak{B}\,\mathfrak{C} + \tfrac{1}{2}x\,y^3\,\mathfrak{B}\,\mathfrak{C} + \tfrac{1}{2}x^3\,y\,\mathfrak{A}\,\mathfrak{D} - \tfrac{1}{2}x\,y^3\,\mathfrak{A}\,\mathfrak{D}\right\}$$
$$\left\{B1,\ \tfrac{1}{2}x^4\,\mathfrak{A}\,\mathfrak{C} - \tfrac{1}{2}y^4\,\mathfrak{A}\,\mathfrak{C} - \tfrac{1}{2}x^4\,\mathfrak{B}\,\mathfrak{D} + \tfrac{1}{2}y^4\,\mathfrak{B}\,\mathfrak{D}\right\}$$
$$\left\{B2,\ \tfrac{1}{2}x^3\,y\,\mathfrak{B}\,\mathfrak{C} + \tfrac{1}{2}x\,y^3\,\mathfrak{B}\,\mathfrak{C} + \tfrac{1}{2}x^3\,y\,\mathfrak{A}\,\mathfrak{D} + \tfrac{1}{2}x\,y^3\,\mathfrak{A}\,\mathfrak{D}\right\}$$
$$\{E,\ 0\}$$

Egad. It is true. Maybe all this malarkey means something, after all. Do they add up to the original? Pull out all the projections, and add them up:

```
((vibrProj // Transpose)[[2]]) /. List → Plus //
 Simplify
```

$$\left(x^3\,\mathfrak{A} + y^3\,\mathfrak{B}\right)\;(x\,\mathfrak{C} + y\,\mathfrak{D})$$

This is exactly what we started with. Look at a numerical example to understand this a little better. If we let **{𝔄→0, 𝔅→2, ℭ→1, 𝔇→1}** then the projection is

```
ProjectET[ 2 y³ (x + y) ]
```

```
{{A1, x⁴ + y⁴}, {A2, -x³ y + x y³},
 {B1, -x⁴ + y⁴}, {B2, x³ y + x y³}, {E, 0}}
```

and here are the plots:

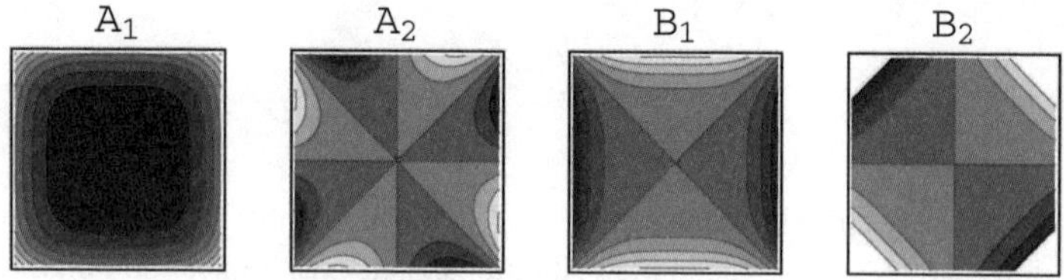

Fig. 43.2 In C_{4v}, the product of two functions of species E is a function that contains species A_1, A_2, B_1, and B_2.

43.5 End Notes

43.5.1 If $\mathcal{H}$ is a sum and Ψ is a product, the energy is a sum

Assume that the Hamiltonian is a sum of two parts $\mathcal{H} = \mathcal{H}_1 + \mathcal{H}_2$ where $\mathcal{H}_1$ and $\mathcal{H}_2$ work on different coordinates. Assume also that the wave function is a product $\Psi = \psi_1\,\psi_2$, where $\mathcal{H}_1[\psi_1] = \mathcal{E}_1\,\psi_1$ and $\mathcal{H}_2[\psi_2] = \mathcal{E}_2\,\psi_2$. Then Schrödinger's equation $\mathcal{H}[\Psi] = \mathcal{E}_{tot}\,\Psi$ becomes

$$\mathcal{H}[\Psi] = (\mathcal{H}_1 + \mathcal{H}_2)[\psi_1\,\psi_2] = \mathcal{H}_1[\psi_1\,\psi_2] + \mathcal{H}_2[\psi_1\,\psi_2]$$

But $\mathcal{H}_1$ and $\mathcal{H}_2$ work on different coordinates, so to $\mathcal{H}_1$ the function ψ_2 is just a constant, and vice-versa, so it may be factored out of the operand. So the calculation continues with

$$\psi_2\,\mathcal{H}_1[\psi_1] + \psi_1\,\mathcal{H}_2[\psi_2] = \psi_2\,\mathcal{E}_1\,\psi_1 + \psi_1\,\mathcal{E}_2\,\psi_2$$

and this becomes

$$(\mathcal{E}_1 + \mathcal{E}_2)\,\psi_1\,\psi_2 = \mathcal{E}_{tot}\,\Psi$$

Under these assumptions, $\mathcal{E}_{tot}$ is the sum of two parts $\mathcal{E}_1 + \mathcal{E}_2$.

43.5.2 Example: (general 3×3) ⊗ (general 2×2)

IMPORTANT NOTE
In the standard preliminaries we **Symbolize** anything with a subscript. However, in this chapter we block this symbolization, because now we are going to perform transformations on matrix subscripts. Test whether subscript symbolization is active:

```
Head[x₀]
```

```
Subscript
```

The return above should be `Subscript`. If it is `Symbol`, you must quit *Mathematica* and start this chapter again in a fresh session. Otherwise, everything below will be garbage.

Construct the direct product matrix

You will rarely need to construct a full direct product matrix, but it is very clarifying to carry out one simple example. Let $\mathbb{T}_{h,i}^{(R,f)}$ be

$$\text{MatA} = \begin{pmatrix} a_{1,1} & a_{1,2} & a_{1,3} \\ a_{2,1} & a_{2,2} & a_{2,3} \\ a_{3,1} & a_{3,2} & a_{3,3} \end{pmatrix};$$

and let $\mathbb{T}_{J,K}^{(R,g)}$ be

$$\text{MatB} = \begin{pmatrix} b_{1,1} & b_{1,2} \\ b_{2,1} & b_{2,2} \end{pmatrix};$$

We construct $\left(\mathbb{T}_{h,i}^{(R,f)}\ \mathbb{T}_{J,K}^{(R,g)}\right)$ using the **Outer** operator :

```
outerAB = Outer[Times, MatA, MatB]
```

$$\{\{\{\{a_{1,1}\, b_{1,1},\ a_{1,1}\, b_{1,2}\},\ \{a_{1,1}\, b_{2,1},\ a_{1,1}\, b_{2,2}\}\},$$
$$\{\{a_{1,2}\, b_{1,1},\ a_{1,2}\, b_{1,2}\},\ \{a_{1,2}\, b_{2,1},\ a_{1,2}\, b_{2,2}\}\},$$
$$\{\{a_{1,3}\, b_{1,1},\ a_{1,3}\, b_{1,2}\},\ \{a_{1,3}\, b_{2,1},\ a_{1,3}\, b_{2,2}\}\}\},$$
$$\{\{\{a_{2,1}\, b_{1,1},\ a_{2,1}\, b_{1,2}\},\ \{a_{2,1}\, b_{2,1},\ a_{2,1}\, b_{2,2}\}\},$$
$$\{\{a_{2,2}\, b_{1,1},\ a_{2,2}\, b_{1,2}\},\ \{a_{2,2}\, b_{2,1},\ a_{2,2}\, b_{2,2}\}\},$$
$$\{\{a_{2,3}\, b_{1,1},\ a_{2,3}\, b_{1,2}\},\ \{a_{2,3}\, b_{2,1},\ a_{2,3}\, b_{2,2}\}\}\},$$
$$\{\{\{a_{3,1}\, b_{1,1},\ a_{3,1}\, b_{1,2}\},\ \{a_{3,1}\, b_{2,1},\ a_{3,1}\, b_{2,2}\}\},$$
$$\{\{a_{3,2}\, b_{1,1},\ a_{3,2}\, b_{1,2}\},\ \{a_{3,2}\, b_{2,1},\ a_{3,2}\, b_{2,2}\}\},$$
$$\{\{a_{3,3}\, b_{1,1},\ a_{3,3}\, b_{1,2}\},\ \{a_{3,3}\, b_{2,1},\ a_{3,3}\, b_{2,2}\}\}\}\}$$

What happened here? The **Outer** product came out as a 3×3 matrix of 3×3 matrices that might be written as

$$\begin{pmatrix} a_{1,1}\,\text{MatB} & a_{1,2}\,\text{MatB} & a_{1,3}\,\text{MatB} \\ a_{2,1}\,\text{MatB} & a_{2,2}\,\text{MatB} & a_{2,3}\,\text{MatB} \\ a_{3,1}\,\text{MatB} & a_{3,2}\,\text{MatB} & a_{3,3}\,\text{MatB} \end{pmatrix} == \text{outerAB}$$

```
True
```

On the left above it is clear that all possible products of the elements of the two matrices were taken. This is the essence of **Outer**-ness. However, it is not very useful as a raw outer product. First, get rid of the complicated bracketing:

```
outerAB2 = Flatten[outerAB]
```

$$\{a_{1,1}\,b_{1,1},\ a_{1,1}\,b_{1,2},\ a_{1,1}\,b_{2,1},\ a_{1,1}\,b_{2,2},\ a_{1,2}\,b_{1,1},\ a_{1,2}\,b_{1,2},$$
$$a_{1,2}\,b_{2,1},\ a_{1,2}\,b_{2,2},\ a_{1,3}\,b_{1,1},\ a_{1,3}\,b_{1,2},\ a_{1,3}\,b_{2,1},\ a_{1,3}\,b_{2,2},$$
$$a_{2,1}\,b_{1,1},\ a_{2,1}\,b_{1,2},\ a_{2,1}\,b_{2,1},\ a_{2,1}\,b_{2,2},\ a_{2,2}\,b_{1,1},\ a_{2,2}\,b_{1,2},$$
$$a_{2,2}\,b_{2,1},\ a_{2,2}\,b_{2,2},\ a_{2,3}\,b_{1,1},\ a_{2,3}\,b_{1,2},\ a_{2,3}\,b_{2,1},\ a_{2,3}\,b_{2,2},$$
$$a_{3,1}\,b_{1,1},\ a_{3,1}\,b_{1,2},\ a_{3,1}\,b_{2,1},\ a_{3,1}\,b_{2,2},\ a_{3,2}\,b_{1,1},\ a_{3,2}\,b_{1,2},$$
$$a_{3,2}\,b_{2,1},\ a_{3,2}\,b_{2,2},\ a_{3,3}\,b_{1,1},\ a_{3,3}\,b_{1,2},\ a_{3,3}\,b_{2,1},\ a_{3,3}\,b_{2,2}\}$$

Now re-pair the indices, implementing $\mathbb{T}^{(\mathrm{R},\mathrm{f})}_{\mathrm{h},\mathrm{i}}\ \mathbb{T}^{(\mathrm{R},\mathrm{g})}_{\mathrm{J},\mathrm{K}} \to \mathbb{T}^{(\mathrm{R},\mathrm{fg})}_{(\mathrm{h},\mathrm{J}),(\mathrm{i},\mathrm{K})}$

$$\mathbf{outerAB3 = outerAB2\ /.\ \{a_{h\_,i\_}\ b_{J\_,K\_} \to ab_{\{\{h,J\},\{i,K\}\}}\}}$$

$$\{ab_{\{\{1,1\},\{1,1\}\}},\ ab_{\{\{1,1\},\{1,2\}\}},\ ab_{\{\{1,2\},\{1,1\}\}},\ ab_{\{\{1,2\},\{1,2\}\}},$$
$$ab_{\{\{1,1\},\{2,1\}\}},\ ab_{\{\{1,1\},\{2,2\}\}},\ ab_{\{\{1,2\},\{2,1\}\}},\ ab_{\{\{1,2\},\{2,2\}\}},$$
$$ab_{\{\{1,1\},\{3,1\}\}},\ ab_{\{\{1,1\},\{3,2\}\}},\ ab_{\{\{1,2\},\{3,1\}\}},\ ab_{\{\{1,2\},\{3,2\}\}},$$
$$ab_{\{\{2,1\},\{1,1\}\}},\ ab_{\{\{2,1\},\{1,2\}\}},\ ab_{\{\{2,2\},\{1,1\}\}},\ ab_{\{\{2,2\},\{1,2\}\}},$$
$$ab_{\{\{2,1\},\{2,1\}\}},\ ab_{\{\{2,1\},\{2,2\}\}},\ ab_{\{\{2,2\},\{2,1\}\}},\ ab_{\{\{2,2\},\{2,2\}\}},$$
$$ab_{\{\{2,1\},\{3,1\}\}},\ ab_{\{\{2,1\},\{3,2\}\}},\ ab_{\{\{2,2\},\{3,1\}\}},\ ab_{\{\{2,2\},\{3,2\}\}},$$
$$ab_{\{\{3,1\},\{1,1\}\}},\ ab_{\{\{3,1\},\{1,2\}\}},\ ab_{\{\{3,2\},\{1,1\}\}},\ ab_{\{\{3,2\},\{1,2\}\}},$$
$$ab_{\{\{3,1\},\{2,1\}\}},\ ab_{\{\{3,1\},\{2,2\}\}},\ ab_{\{\{3,2\},\{2,1\}\}},\ ab_{\{\{3,2\},\{2,2\}\}},$$
$$ab_{\{\{3,1\},\{3,1\}\}},\ ab_{\{\{3,1\},\{3,2\}\}},\ ab_{\{\{3,2\},\{3,1\}\}},\ ab_{\{\{3,2\},\{3,2\}\}}\}$$

They are re-paired, but they are not in **Sort**ed order. This step is important but cannot easily be discussed in general notation, so we don't have a formula from the derivation to show you. Nevertheless, we sort them :

```
sortedAB = outerAB3 // Sort
```

$$\{ab_{\{\{1,1\},\{1,1\}\}},\ ab_{\{\{1,1\},\{1,2\}\}},\ ab_{\{\{1,1\},\{2,1\}\}},\ ab_{\{\{1,1\},\{2,2\}\}},$$
$$ab_{\{\{1,1\},\{3,1\}\}},\ ab_{\{\{1,1\},\{3,2\}\}},\ ab_{\{\{1,2\},\{1,1\}\}},\ ab_{\{\{1,2\},\{1,2\}\}},$$
$$ab_{\{\{1,2\},\{2,1\}\}},\ ab_{\{\{1,2\},\{2,2\}\}},\ ab_{\{\{1,2\},\{3,1\}\}},\ ab_{\{\{1,2\},\{3,2\}\}},$$
$$ab_{\{\{2,1\},\{1,1\}\}},\ ab_{\{\{2,1\},\{1,2\}\}},\ ab_{\{\{2,1\},\{2,1\}\}},\ ab_{\{\{2,1\},\{2,2\}\}},$$
$$ab_{\{\{2,1\},\{3,1\}\}},\ ab_{\{\{2,1\},\{3,2\}\}},\ ab_{\{\{2,2\},\{1,1\}\}},\ ab_{\{\{2,2\},\{1,2\}\}},$$
$$ab_{\{\{2,2\},\{2,1\}\}},\ ab_{\{\{2,2\},\{2,2\}\}},\ ab_{\{\{2,2\},\{3,1\}\}},\ ab_{\{\{2,2\},\{3,2\}\}},$$
$$ab_{\{\{3,1\},\{1,1\}\}},\ ab_{\{\{3,1\},\{1,2\}\}},\ ab_{\{\{3,1\},\{2,1\}\}},\ ab_{\{\{3,1\},\{2,2\}\}},$$
$$ab_{\{\{3,1\},\{3,1\}\}},\ ab_{\{\{3,1\},\{3,2\}\}},\ ab_{\{\{3,2\},\{1,1\}\}},\ ab_{\{\{3,2\},\{1,2\}\}},$$
$$ab_{\{\{3,2\},\{2,1\}\}},\ ab_{\{\{3,2\},\{2,2\}\}},\ ab_{\{\{3,2\},\{3,1\}\}},\ ab_{\{\{3,2\},\{3,2\}\}}\}$$

The **Sort** operator does a nice job here, but generally we will need this same reordering in cases where there are no explicit indices for **Sort** to work on. To reorder a product matrix without subscripts, we would need the permutation that turns **outerAB3** into **sortedAB** :

```
perm36 =
  Map[Position[outerAB3, #] &, sortedAB] // Flatten
```

```
{1, 2, 5, 6, 9, 10, 3, 4, 7, 8, 11, 12,
 13, 14, 17, 18, 21, 22, 15, 16, 19, 20, 23, 24,
 25, 26, 29, 30, 33, 34, 27, 28, 31, 32, 35, 36}
```

This permutation would be used as shown on the left side below :

```
outerAB3[[perm36]] == sortedAB
```

```
True
```

Now we are ready to carry out the transform from double indices to single indices ($\mathbb{T}^{(R,fg)}_{(h,J),(i,K)} \to \mathbb{T}^{(R,fg)}_{\alpha,\beta}$). In the MatrixReview, it is shown that the desired rule is

```
{r_,c_} → cMax (r-1)+c
```

so we make a rule that uses it twice :

```
Clear[h, i, J, K];
toTwoIndices =
  {ab{{h_,J_},{i_,K_}} → ab{2 (h-1)+J,2 (i-1)+K}};
```

Convert from a four-index array (**sortedAB**) to a two-index array (**sortedAB2**):

```
sortedAB2 = sortedAB /. toTwoIndices
```

$\{ab_{\{1,1\}}, ab_{\{1,2\}}, ab_{\{1,3\}}, ab_{\{1,4\}}, ab_{\{1,5\}}, ab_{\{1,6\}},$
$ab_{\{2,1\}}, ab_{\{2,2\}}, ab_{\{2,3\}}, ab_{\{2,4\}}, ab_{\{2,5\}}, ab_{\{2,6\}},$
$ab_{\{3,1\}}, ab_{\{3,2\}}, ab_{\{3,3\}}, ab_{\{3,4\}}, ab_{\{3,5\}}, ab_{\{3,6\}},$
$ab_{\{4,1\}}, ab_{\{4,2\}}, ab_{\{4,3\}}, ab_{\{4,4\}}, ab_{\{4,5\}}, ab_{\{4,6\}},$
$ab_{\{5,1\}}, ab_{\{5,2\}}, ab_{\{5,3\}}, ab_{\{5,4\}}, ab_{\{5,5\}}, ab_{\{5,6\}},$
$ab_{\{6,1\}}, ab_{\{6,2\}}, ab_{\{6,3\}}, ab_{\{6,4\}}, ab_{\{6,5\}}, ab_{\{6,6\}}\}$

And finally

```
sortedAB3 = Partition[sortedAB2, 6];
% // GridForm
```

$ab_{\{1,1\}}$	$ab_{\{1,2\}}$	$ab_{\{1,3\}}$	$ab_{\{1,4\}}$	$ab_{\{1,5\}}$	$ab_{\{1,6\}}$
$ab_{\{2,1\}}$	$ab_{\{2,2\}}$	$ab_{\{2,3\}}$	$ab_{\{2,4\}}$	$ab_{\{2,5\}}$	$ab_{\{2,6\}}$
$ab_{\{3,1\}}$	$ab_{\{3,2\}}$	$ab_{\{3,3\}}$	$ab_{\{3,4\}}$	$ab_{\{3,5\}}$	$ab_{\{3,6\}}$
$ab_{\{4,1\}}$	$ab_{\{4,2\}}$	$ab_{\{4,3\}}$	$ab_{\{4,4\}}$	$ab_{\{4,5\}}$	$ab_{\{4,6\}}$
$ab_{\{5,1\}}$	$ab_{\{5,2\}}$	$ab_{\{5,3\}}$	$ab_{\{5,4\}}$	$ab_{\{5,5\}}$	$ab_{\{5,6\}}$
$ab_{\{6,1\}}$	$ab_{\{6,2\}}$	$ab_{\{6,3\}}$	$ab_{\{6,4\}}$	$ab_{\{6,5\}}$	$ab_{\{6,6\}}$

The final direct product matrix is square, and 6×6. Reading it like a page, the a factors go $a_{1,1}$, .., $a_{3,3}$ three times, while the b factors go $b_{1,1}$ nine times, then $b_{1,2}$ nine times, then b_{21}; then b_{22}.

Verification of the character lemma

```
dpSpur1 = Spur[sortedAB3]
```

$ab_{\{1,1\}} + ab_{\{2,2\}} + ab_{\{3,3\}} + ab_{\{4,4\}} + ab_{\{5,5\}} + ab_{\{6,6\}}$

The theorem says this should be the product of the character of **MatA** times the character of **MatB** :

```
dpSpur1 == Expand[Spur[MatA] Spur[MatB]]
```

$ab_{\{1,1\}} + ab_{\{2,2\}} + ab_{\{3,3\}} + ab_{\{4,4\}} + ab_{\{5,5\}} + ab_{\{6,6\}} ==$
$a_{1,1}\, b_{1,1} + a_{2,2}\, b_{1,1} + a_{3,3}\, b_{1,1} + a_{1,1}\, b_{2,2} + a_{2,2}\, b_{2,2} + a_{3,3}\, b_{2,2}$

Let's see if we can transform it step by step to give this. In **dpSpur1** we make both single indices into double indices. (Here is where it is essential to have symbolization of subscript expressions turned off.)

```
dpSpur2 = dpSpur1 /. ab{i1_,i2_} →
    ab{MakeDoubleIndex[i1,{2,3}],MakeDoubleIndex[i2,{2,3}]}
```

$ab_{\{\{1,1\},\{1,1\}\}} + ab_{\{\{1,2\},\{1,2\}\}} + ab_{\{\{1,3\},\{1,3\}\}} +$
$ab_{\{\{2,1\},\{2,1\}\}} + ab_{\{\{2,2\},\{2,2\}\}} + ab_{\{\{2,3\},\{2,3\}\}}$

Resort them:

```
dpSpur3 = dpSpur2 /. ab{{i_,K_},{j_,L_}} → a{K,L} b{i,j}
```

$a_{\{1,1\}}\, b_{\{1,1\}} + a_{\{2,2\}}\, b_{\{1,1\}} + a_{\{3,3\}}\, b_{\{1,1\}} +$
$a_{\{1,1\}}\, b_{\{2,2\}} + a_{\{2,2\}}\, b_{\{2,2\}} + a_{\{3,3\}}\, b_{\{2,2\}}$

and factor:

```
dpSpur4 = dpSpur3 // Factor
```

$(a_{\{1,1\}} + a_{\{2,2\}} + a_{\{3,3\}})\ (b_{\{1,1\}} + b_{\{2,2\}})$

The theorem is verified.

- If the return above is anything other than
  $(a_{\{1,1\}} + a_{\{2,2\}} + a_{\{3,3\}})\ (b_{\{1,1\}} + b_{\{2,2\}})$
  you need to click back and read the note at the top of Subsection 43.5.2.

43.5.3 Automation

Module **MakeDirectProductPerm**

The example showed us that we need a permutation to reorder the elements of the raw **Outer** product. It will, of course, depend on the dimensions of the example you are working with. The following operator carries out a fully indexed direct product calculation; then it uses the **Sort** operator to create the final ordering. Comparing the initial and final ordering, it returns the needed permutation.

```
MakeDirectProductPerm[szA_, szB_] := Module[
  {MatA, i, j, MatB, outerAB, directAB4,
   to2Indcs, directAB2, sortedAB2},

  MatA = Table[a_{i,j}, {i, 1, szA}, {j, 1, szA}];
  MatB = Table[b_{i,j}, {i, 1, szB}, {j, 1, szB}];
  outerAB = Outer[Times, MatA, MatB] // Flatten;
  directAB4 = outerAB /. a_{i_,j_} b_{m_,n_} → ab_{{i,m},{j,n}};
  to2Indcs =
   {ab_{{i_,j_},{m_,n_}} → ab_{szB (i-1)+j,szB (m-1)+n}};
  directAB2 = directAB4 /. to2Indcs;
  sortedAB2 = directAB2 // Sort;
  Map[Position[directAB2, #] &, sortedAB2] //
   Flatten (*endModule*)]
```

Test it :

```
newPerm36 = MakeDirectProductPerm[3, 2]
```

```
{1, 2, 5, 6, 9, 10, 3, 4, 7, 8, 11, 12,
 13, 14, 17, 18, 21, 22, 15, 16, 19, 20, 23, 24,
 25, 26, 29, 30, 33, 34, 27, 28, 31, 32, 35, 36}
```

The perm we made by direct sorting was called **perm36**. Is this result the same?

```
newPerm36 == perm36
```

```
True
```

Module DirectProduct

Once the required permutation is known, construction of the direct product is straightforward :

```
Clear[DirectProduct];
DirectProduct[matA_, matB_] := Module[
  {dimA, dimB, permAB, outerAB, directAB},
  dimA = Dimensions[matA] [[1]];
  dimB = Dimensions[matB] [[1]];
  permAB = MakeDirectProductPerm[dimA, dimB];
  outerAB = Outer[Times, matA, matB] // Flatten;
  directAB = outerAB[[permAB]] ;
  Partition[directAB, dimA * dimB]  ]
```

DirectProduct differs from **Outer** because of that **Permute** operation (**outerAB[[permAB]]**) that put the elements in sorted index order.

Automated examples

To a human this looks pretty much like the primary example above. But to a computer it is quite different; it has no explicit indices inside the matrices, so **Sort** cannot put the product elements in sorted index order. We must to rely on **MakeDirectProductPerm** to reorder the elements.

$$\text{MatC} = \begin{pmatrix} \mathbb{a} & \mathbb{b} & \mathbb{c} \\ \mathbb{d} & \mathbb{e} & \mathbb{f} \\ \mathbb{g} & \mathbb{h} & \mathbb{i} \end{pmatrix}; \quad \text{MatD} = \begin{pmatrix} \mathbb{A} & \mathbb{B} \\ \mathbb{C} & \mathbb{D} \end{pmatrix};$$

```
outerCD = Outer[Times, MatC, MatD] // Flatten
```

```
{a A, a B, a C, a D, A b, b B, b C, b D, A c, B c, c C, c D,
 A d, B d, C d, d D, A e, B e, C e, D e, A f, B f, C f, D f,
 A g, B g, C g, D g, A h, B h, C h, D h, A i, B i, C i, D i}
```

```
permCD = MakeDirectProductPerm[3, 2]
```

```
{1, 2, 5, 6, 9, 10, 3, 4, 7, 8, 11, 12,
 13, 14, 17, 18, 21, 22, 15, 16, 19, 20, 23, 24,
 25, 26, 29, 30, 33, 34, 27, 28, 31, 32, 35, 36}
```

```
directCD = Partition[outerCD[[permCD]], 6];
% // GridForm
```

a A	a B	A b	b B	A c	B c
a C	a D	b C	b D	c C	c D

𝔸 𝕕	𝔹 𝕕	𝔸 𝕖	𝔹 𝕖	𝔸 𝕗	𝔹 𝕗
ℂ 𝕕	𝕕 𝔻	ℂ 𝕖	𝔻 𝕖	ℂ 𝕗	𝔻 𝕗
𝔸 𝕘	𝔹 𝕘	𝔸 𝕙	𝔹 𝕙	𝔸 𝕚	𝔹 𝕚
ℂ 𝕘	𝔻 𝕘	ℂ 𝕙	𝔻 𝕙	ℂ 𝕚	𝔻 𝕚

The **DirectProduct** operator constructed above should give the same answer as all these maneuvers :

```
directCD == DirectProduct[MatC, MatD]
```

```
True
```

It does.

44. One-photon selection rules

Preliminaries

44.1 The main idea

In spectroscopy the quantum amplitude of a process is given by a matrix element of the form $\langle \psi_o \mid \mathcal{P} \mid \psi_f \rangle$, where ψ_{original} is the wave function for the starting state of the transition, ψ_{final} is the final state, and $\mathcal{P}$ is the operator for a particular process, such as electric dipole moment, magnetic dipole moment, electric quadrupole moment, etc. In absorption spectroscopy, ψ_o is nearly always the ground state, and ψ_f is the excited state created by the absorption. In emission spectroscopy, vice-versa. It is frequently found that this matrix element vanishes, in which case the imagined process does not occur, and is called "forbidden". What is not forbidden is "allowed", and group theory makes it very easy to tell which is which for any given transition in any kind of spectroscopy (for any given spectroscopic operator $\mathcal{P}$).

44.2 Polarization vectors and light amplitude

We eschew the full development of the operators $\mathcal{P}$. They are described in every quantum mechanics book, and we will just use the results. But we do want to make a little refinement that helps tremendously with the planning of experiments; we work with explicit, normalized polarization vectors.

First, we assume that the light has a pure polarization of some kind (linear, circular, or elliptical). Truly "unpolarized" light is rarely found in reality, and is handled by a further development of the theory. (Most light is "partly polarized", a statistical concept that we do not need here.) For one-photon electric dipole processes (absorption, fluorescence, phosphorescence) the theory says that the operator in the matrix element must be

$$\mathcal{P} = \mathbf{Ea}\ \underset{\rightharpoondown}{\lambda}\ .\{\mathbf{x},\mathbf{y},\mathbf{z}\}$$

where real number **Ea** is the amplitude of the electric field of the light, $\underset{\rightharpoondown}{\lambda}$ is the polarization vector of the light, and **{x,y,z}** are the Cartesian coordinates of the "particle" (electron or nucleus) that is interacting with the light. If there are many electrons or nuclei you simply use a sum over their coordinates,

W.M. McClain, *Symmetry Theory in Molecular Physics with Mathematica*,
DOI 10.1007/b13137_44, © Springer Science+Business Media, LLC 2009

$\sum_{i=1}^{n} \{x_i, y_i z_i\}$, so nothing important is lost if we simplify to one particle.

The polarization vector is always normalized by $\underset{\rightarrow}{\lambda}.\underset{\rightarrow}{\lambda}^* = 1$, and it tells how the electric field oscillates. If $\underset{\rightarrow}{\lambda}$ is real, $\underset{\rightarrow}{\mathbf{E}}$ oscillates in a in a straight line; if $\underset{\rightarrow}{\lambda}$ is complex it oscillates in an ellipse or a circle. Click to an End Note if you need to brush up on this.

The amplitude of the spectroscopic process thus becomes

$$\mathbf{A} = \left\langle \psi_o \,\middle|\, \{x, y, z\}.\underset{\rightarrow}{\lambda}\,\mathbf{Ea} \,\middle|\, \psi_f \right\rangle = \mathbf{Ea}\,\underset{\rightarrow}{\lambda}.\langle \psi_o \mid \{x, y, z\} \mid \psi_f \rangle$$

The spectroscopic signal $\mathbf{S}$ (events per second) is proportional to $|\mathbf{A}|^2$, so the signal is

$$\mathbf{S} \propto |\mathbf{A}|^2 = \left| \mathbf{Ea}\ \underset{\rightarrow}{\lambda}.\langle \psi_o \mid \{x, y, z\} \mid \psi_f \rangle \right|^2$$

We mention this here because the electric field $\mathbf{A}$ is often complex. This is perfectly alright. The absolute square of $\mathbf{A}$ is the only thing that relates to the signal, and it is always real.

Switching to a more systematic notation, we let $\{x, y, z\} = \{\xi_1, \xi_2, \xi_3\}$ and $\{\lambda_x, \lambda_y, \lambda_z\} = \{\lambda_1, \lambda_2, \lambda_3\}$, and $\mathbf{S}$ becomes

$$\mathbf{S} \propto \mathbf{Ea}^2 \left| \lambda_i \langle \psi_o \mid \xi_i \mid \psi_f \rangle \right|^2$$

where the sum over i is implied by its double occurrence. So the allowed or forbidden nature of the transition comes down to whether or not we get zero for the matrix elements $\langle \psi_o \mid \xi_i \mid \psi_f \rangle$ for $i = 1,2,3$.

This question is easy to decide using direct product theory. We bring together two previous results :

1. (Chapter 43) The character of a product of functions is the product of the characters of its factor functions.

The function is the whole integrand of the matrix element :

$$\psi_o^*\ \xi_i\ \psi_f$$

We want the character of this product. We know the character of each of the factors: ψ_o and ψ_f are eigenfunctions of a molecular Hamiltonian that belongs to some symmetry group $\mathcal{G}$, and they must have some pure symmetry (i.e., they must belong to some species of group $\mathcal{G}$). The function ξ_i also has a pure symmetry (it is just one of the Cartesian coordinates), and you can look up its species in the character table for group $\mathcal{G}$. So we can easily take the product of the characters and reduce it to find the species present in the integrand.

2. (Chapter 39) If a function has no totally symmetric component, its integral vanishes.

The product of the characters generally comes out as the character of a reducible representation. If so, use the **Analyze** operator to split it into its irreducible components. If the totally symmetric species is not present, the integral vanishes. Let's do an example immediately.

We will work with the benzene orbitals, so click back and pull them to the side so you can see them while you read. Remember, these are not full benzene wave functions; they are just one-electron orbitals for six attractive centers. Also, remember that these model orbitals have no **z** dependence. So this example is not quite like doing the real molecule, but it serves. The group is $\mathbf{D_{6\,h}}$:

```
BoxedCharacterTable["D6h", 6]
```

D6h	1	2	2	1	3	3	1	2	2	1	3	3	← Class populations
	E	C6	C3	C_2	C'_2	C''_2	inv	S3	S6	σh	σd	σv	↓ Basis functions
A1g	1	1	1	1	1	1	1	1	1	1	1	1	$\{(x^2+y^2), (z^2)\}$
A2g	1	1	1	1	-1	-1	1	1	1	1	-1	-1	$\{(-3x^5y+10x^3y^3-3xy^5), (Iz)\}$
B1g	1	-1	1	-1	1	-1	1	-1	1	-1	1	-1	$\{(x^3z-3xy^2z)\}$
B2g	1	-1	1	-1	-1	1	1	-1	1	-1	-1	1	$\{(-3x^2yz+y^3z)\}$
E1g	2	1	-1	-2	0	0	2	1	-1	-2	0	0	$\left\{\begin{pmatrix} xz \\ yz \end{pmatrix}, \begin{pmatrix} Ix \\ Iy \end{pmatrix}\right\}$
E2g	2	-1	-1	2	0	0	2	-1	-1	2	0	0	$\left\{\begin{pmatrix} \frac{x^2}{2}-\frac{y^2}{2} \\ xy \end{pmatrix}\right\}$
A1u	1	1	1	1	1	1	-1	-1	-1	-1	-1	-1	$\{(-3x^5yz+10x^3y^3z-3xy^5z)\}$
A2u	1	1	1	1	-1	-1	-1	-1	-1	-1	1	1	$\{(z)\}$
B1u	1	-1	1	-1	1	-1	-1	1	-1	1	-1	1	$\{(3x^2y-y^3)\}$
B2u	1	-1	1	-1	-1	1	-1	1	-1	1	1	-1	$\{(-x^3+3xy^2)\}$
E1u	2	1	-1	-2	0	0	-2	-1	1	2	0	0	$\left\{\begin{pmatrix} x \\ y \end{pmatrix}\right\}$
E2u	2	-1	-1	2	0	0	-2	1	1	-2	0	0	$\left\{\begin{pmatrix} \frac{x^2z}{2}-\frac{y^2z}{2} \\ xyz \end{pmatrix}\right\}$

The totally symmetric species is named $\mathbf{A_{1\,g}}$ in this group. What are the species of the Cartesian coordinates? The table says **x** and **y** transform together under species $\mathbf{E_{1\,u}}$, while **z** transforms under species $\mathbf{A_{2\,u}}$.

Suppose we have an electron in the lowest orbital $\mathbf{A_{1\,g}}$ and we want to promote it to the $\mathbf{E_{1\,u}}$level. Is there a photon that will do this? We need the character of a product function. So we take the character of $(\psi_o{}^*)(\xi_i)(\psi_f)$:

```
$DefaultGroup = "D6h";
productChVec =
  ChVec["A1g"] * ChVec["E1u"] * ChVec["E1u"]
```

```
{4, 1, 1, 4, 0, 0, 4, 1, 1, 4, 0, 0}
```

From the 4 in the first position we see that this is the character vector of a 4×4 representation; definitely reducible. How does it block diagonalize?

```
Analyze[productChVec]
```

```
{{A1g, 1}, {A2g, 1}, {B1g, 0}, {B2g, 0}, {E1g, 0}, {E2g, 1},
 {A1u, 0}, {A2u, 0}, {B1u, 0}, {B2u, 0}, {E1u, 0}, {E2u, 0}}
```

Yes! The total integrand has an $\mathbf{A_{1\,g}}$ component, so the transition $\mathbf{A_{1\,g}} \to \mathbf{E_{1\,u}}$ is *allowed* for linear photons polarized in some direction that lies in the **x,y** plane. But if it is **z** polarized (species **A2u**) :

```
(ChVec["A1g"] * ChVec["A2u"] * ChVec["E1u"]) //
 Analyze
```

```
{{A1g, 0}, {A2g, 0}, {B1g, 0}, {B2g, 0}, {E1g, 1}, {E2g, 0},
 {A1u, 0}, {A2u, 0}, {B1u, 0}, {B2u, 0}, {E1u, 0}, {E2u, 0}}
```

No! The $\mathbf{A_{1\,g}}$ component is zero, so the transition $\mathbf{A_{1\,g}} \to \mathbf{E_{1\,u}}$ is *forbidden* for out-of-plane (or **z**) polarization.

44.3 Projection of the one-photon dipole operator

Here is a nice way to look at this problem if you have automated projection operators. We return to the spectroscopic amplitude and pull out a part of it :

$$\underset{\rightarrow}{\lambda}\cdot\langle\psi_o \mid \{x, y, z\} \mid \psi_f\rangle$$

We project the dipole moment operator under group $\mathbf{D_{6\,h}}$, selecting only non-zero results :

```
ProjectET[{x, y, z}] //
 Select[#, FreeQ[#, {0, 0, 0}] &] &
```

```
{{A2u, {0, 0, z}}, {E1u, {x, y, 0}}}
```

It has split into two parts: **{0,0,z}** (species $\mathbf{A_{2\,u}}$) and **{x,y,0}** (species $\mathbf{E_{1\,u}}$). So suppose you are thinking about an $\mathbf{A_{1\,g}} \to \mathbf{E_{1\,u}}$ transition. You will want to do the matrix element

$$\mathrm{ME} = \langle\psi_{A1g} \mid \{x, y, 0\} \mid \psi_{E1u}\rangle$$

But of course ψ_{E1u} is degenerate; it belongs to a family. You have to pick a particular family member to do the integral. Family members are given by

$$\psi_{E1u}[\theta] = \mathrm{Cos}[\theta]\,\psi_{E1ux} + \mathrm{Sin}[\theta]\,\psi_{E1uy}$$

and the matrix element becomes

$$\mathrm{ME} = \langle\psi_{A1g} \mid \{x, y, 0\} \mid \mathrm{Cos}[\theta]\,\psi_{E1ux} + \mathrm{Sin}[\theta]\,\psi_{E1uy}\rangle$$

or

$$\mathbf{ME} = \mathrm{Cos}[\theta]\ \langle \psi_{A1g} \mid \{x, y, 0\} \mid \psi_{E1ux}\rangle + \mathrm{Sin}[\theta]\ \langle \psi_{A1g} \mid \{x, y, 0\} \mid \psi_{E1uy}\rangle$$

or

$$\mathbf{ME} = \mathrm{Cos}[\theta]\ \{\langle \psi_{A1g} \mid x \mid \psi_{E1ux}\rangle, \langle \psi_{A1g} \mid y \mid \psi_{E1ux}\rangle, 0\} + \mathrm{Sin}[\theta]\ \{\langle \psi_{A1g} \mid x \mid \psi_{E1uy}\rangle, \langle \psi_{A1g} \mid y \mid \psi_{E1uy}\rangle, 0\}$$

Function $\psi_{A1\,g}\,\mathbf{x}$ is of is of species $\mathbf{E}_{1\,ux}$, like $\mathbf{x}$ itself, so we call it ϕ_{E1ux}, and similarly for ϕ_{E1uy}. Then the expression above becomes

$$\mathbf{ME} = \mathrm{Cos}[\theta]\ \{\langle \phi_{E1u,x} \mid \psi_{E1u,x}\rangle, \langle \phi_{E1u,y} \mid \psi_{E1u,x}\rangle, 0\} + \mathrm{Sin}[\theta]\ \{\langle \phi_{E1u,x} \mid \psi_{E1u,y}\rangle, \langle \phi_{E1ux,y} \mid \psi_{E1u,y}\rangle, 0\}$$

Now $\langle \phi_{E1u,y} \mid \psi_{E1u,x}\rangle$ and $\langle \phi_{E1u,x} \mid \psi_{E1u,y}\rangle$ have different species in them and by the Bedrock Theorem they must vanish, leaving

$$\mathbf{ME} = \mathrm{Cos}[\theta]\ \{\langle \phi_{E1u,x} \mid \psi_{E1u,x}\rangle, 0, 0\} + \mathrm{Sin}[\theta]\ \left\{0, \left\langle \phi_{E1ux,y} \middle| \psi_{E1u,y}\right\rangle, 0\right\}$$

But by the Bedrock Lemma, $\langle \phi_{E1u,x} \mid \psi_{E1u,x}\rangle$ and $\langle \phi_{E1ux,y} \mid \psi_{E1u,y}\rangle$ must have the same value (call it μ). So the matrix element becomes

$$\mathbf{ME} = \mathrm{Cos}[\theta]\ \{\mu, 0, 0\} + \mathrm{Sin}[\theta]\ \{0, \mu, 0\}$$

or

The matrix element comes down to a wonderfully simple formula :

$$\langle \psi_{A1g} \mid \{x, y, 0\} \mid \psi_{E1u}[\theta]\rangle = \mu\ \{\mathrm{Cos}[\theta], \mathrm{Sin}[\theta], 0\}$$

where θ is the family parameter of the excited state.

This must be dotted with the polarization vector $\underset{\rightarrow}{\lambda}$. Assuming that $\underset{\rightarrow}{\lambda}$ is linear and $\mathbf{x}$-polarized (**{1,0,0}**), we have

```
A == Ea μ {1, 0, 0}.{Cos[θ], Sin[θ], 0}
```

```
A == Ea μ Cos[θ]
```

and the differential signal due to family members within $d\theta$ of θ is

$$dS \propto Ea^2\ \mu^2\ \mathrm{Cos}[\theta]^2\ d\theta$$

Conveniently, $\mathbf{Ea}^2$ is proportional to light intensity (photons/second). Remember, the family formula was

$$\psi_{E1u}[\theta] = \text{Cos}[\theta]\ \psi_{E1ux} + \text{Sin}[\theta]\ \psi_{E1uy}$$

Because the light is **x**-polarized, the signal is maximum for $\theta = 0°$, decreasing to **0** for $\theta = 90°$. The state produced at $\theta = 0°$ is pure ψ_{E1ux}, mixing with increasing amounts of ψ_{E1uy} as θ increases and signal decreases. Finally, just as the signal vanishes, the **x** polarized photon produces infinitesimal amounts of nearly pure ψ_{E1uy}. Contour plots of the benzene basis orbitals are on file in the **Symmetry`** package. We pull these two out again for inspection :

```
xAndy = Show[GraphicsRow[{cpBzE1ga, cpBzE1gb}],
  ImageSize → 150]
```

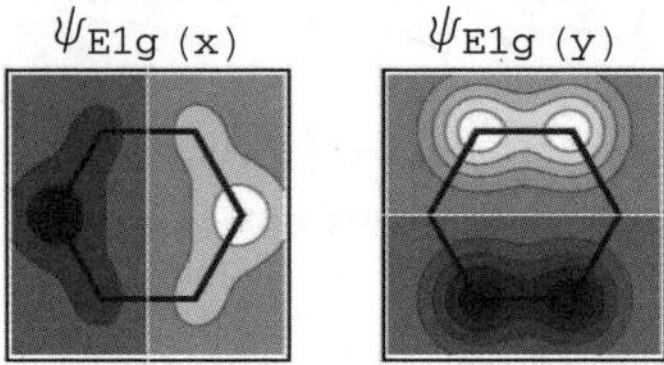

Fig. 44.1 Excited states produced by x- and y- polarized photons

It is quite clear from these diagrams that **x**-polarized light maximally provokes oscillations in the **x** direction, but cannot provoke the orthogonal partner. The other family members, the states that lie between pure **x** and pure **y** were all shown long ago in the $\mathbf{E_{1u}}$ movie in Chapter 23, SymmetryAndQuantumMechanics.

44.4 Transition oscillations

Why do we talk about oscillations in eigenstates, which are also called stationary states ? Yes, it's true, the eigenstate probability density clouds do not oscillate. But there is a time dependence that goes with every eigenstate (usually ignored). It is

$$\psi_s[x, y, z]\ \text{Exp}[i\ (En_s / \hbar)\ t]$$

where **En** means eigenenergy. When you construct the probability density functions by multiplying the wave function by its own conjugate, the time dependence consumes itself and you get a stationary probability density :

$$\psi_o[x, y, z]\ \text{Exp}\left[i\ \frac{En_o}{\hbar}\ t\right] \left(\psi_o[x, y, z]\ \text{Exp}\left[i\ \frac{En_o}{\hbar}\ t\right]\right)^*$$

$\psi_o[x, y, z]^2$

However, apply this recipe to a state that is moving from ψ_o to ψ_f. When it is halfway through the quantum transition, it is a linear combination of old and new

```
Clear[t]
```

$$\psi_{halfway} = \frac{1}{\sqrt{2}}\left(\psi_o \,\mathbf{Exp}\left[\mathrm{i}\,\frac{\mathbf{En_o}}{\hbar}\,\mathbf{t}\right] + \psi_f \,\mathbf{Exp}\left[\mathrm{i}\,\frac{\mathbf{En_f}}{\hbar}\,\mathbf{t}\right]\right);$$

$$\psi^*_{halfway} = \frac{1}{\sqrt{2}}\left(\psi_o \,\mathbf{Exp}\left[-\mathrm{i}\,\frac{\mathbf{En_o}}{\hbar}\,\mathbf{t}\right] + \psi_f \,\mathbf{Exp}\left[-\mathrm{i}\,\frac{\mathbf{En_f}}{\hbar}\,\mathbf{t}\right]\right);$$

Construct the probability density for $\psi_{halfway}$ by the usual recipe :

$$\rho = \psi_{halfway}\,\psi^*_{halfway} \;//\; \mathbf{FullSimplify}$$

$$\frac{1}{2}\left(\psi_f^2 + \psi_o^2 + 2\,\psi_f\,\psi_o\,\mathrm{Cos}\left[\frac{(\mathrm{En_f} - \mathrm{En_o})\,t}{\hbar}\right]\right)$$

The time dependence survived. Its angular frequency is $\omega == \frac{(\mathbf{En_f}-\mathbf{En_o})}{\hbar}$.

But this is exactly the Einstein's Law $\Delta\mathbf{E} = \hbar\,\omega$. Something is going right here!

For plotting purposes we would rather have $2\pi\mathbf{t}$ in the **Cos** function; we will see one whole cycle as **t** runs from **0** to **1**. We make this change at the same time that we put in the orbitals for the $\mathbf{A_{1\,g}} \to \mathbf{E_{1\,u\,x}}$ transition in benzene :

$$\psi\mathbf{sq}[\mathbf{t\_}] := \mathbf{Evaluate}\Big[\rho\;/.\;\left\{\psi_o \to \psi\mathbf{BzA2u},\; \psi_f \to \psi\mathbf{BzE1ga},\; \frac{(\mathbf{En_f} - \mathbf{En_o})}{\hbar} \to 2\,\pi\right\}\Big];$$

The probability density ψ**sq** is a monster, as you might suspect. You can look at it on your own if you want to. But *Mathematica* loves to evaluate monster functions, and does so very quickly. So without further ado, interactive readers can see a movie of it; hard copy readers will have to make do with one frame.

```
trTable =
  Table[ContourPlot[ψsq[t], {x, -3, 3}, {y, -3, 3},
    Contours → 4, ImageSize → 100, FrameTicks → None,
    ColorFunction → BlueShades], {t, 0, .95, .05}];
ListAnimate[trTable]
```

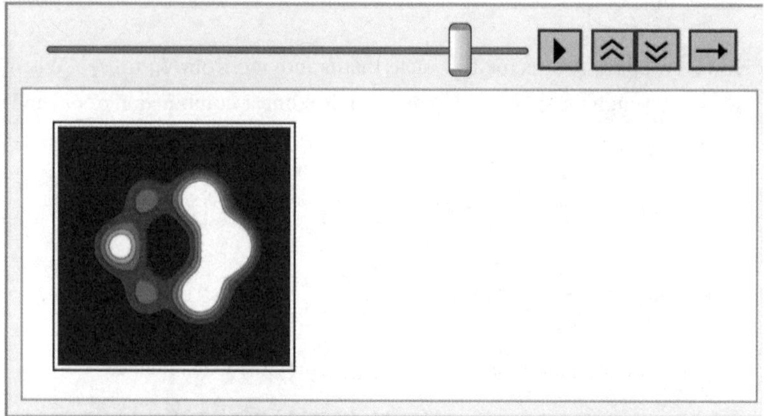

Fig. 44.2 The transition oscillations in the $A_{2u} \to E_{1g}$ halfway state

Run the movie and you will see exactly what is meant by transition oscillations. These oscillations are exactly in synch with the **x**- polarized oscillations of the electric field. James Clerk Maxwell deduced that accelerating charge is the one and only source or sink of radiation. As you see, quantum mechanics supports this very nicely.

On your own
Make a movie of the transition oscillations from A_{1g} to another member of the E_{1u} family. The direction will not be exactly parallel to **x**, but there is a component in that direction. That is why they are less probably excited than **ψBzE1ux**.

44.5 End Notes

44.5.1 Polarization vectors

Light is described most fundamentally by its vector potential, $\underset{\rightharpoondown}{\mathbf{A}}[\xi,\eta,\varsigma,\mathbf{t}]$, where $\underset{\rightharpoondown}{\rho} = \{\xi, \eta, \varsigma\}$ are the Cartesian coordinates. From this both the electric and magnetic fields can be extracted by simple formulas. The electric field formula is

$$\underset{\rightharpoondown}{\mathbf{E}}[\xi, \eta, \varsigma, \mathbf{t}] == \mathbf{K}\,\mathbf{Re}\left[\partial_t \underset{\rightharpoondown}{\mathbf{A}}[\xi, \eta, \varsigma, \mathbf{t}]\right]$$

where **K** is a combination of fundamental constants. If the vector potential is a plane wave

```
A[ξ, η, ζ, t] == A⁰ λ Exp[ⅈ (κ.ρ - ω t + ϕ)]
```

The explicit phase factor ϕ allows us to take the amplitude $\mathbf{A}^0$ pure real. Without loss of generality we may take κ to be `{0,0,κ}`. But λ has to be perpendicular to κ, so then `λ = {λξ, λη, 0}`. In 2D graphic work below, we will use only the transverse part `{λξ, λη}`. The formula for the electric field is

```
Efield = K (A⁰ Re[∂t (λ Exp[ⅈ (κ ζ - ω t + ϕ)])] /.
      {(κ ζ - ω t + ϕ) → χ} // Factor)

A⁰ K Im[ⅇ^(ⅈ χ) ω λ]
```

After taking the derivative above we replaced the wave argument `k ζ - ω t + ϕ` by χ. As Lord Rayleigh famously observed, the factor of ω generated by taking the derivative is what makes the sky blue. The harmonic factor is

```
Clear[HrmFctr];
HrmFctr[λvec_] :=
  Cos[χ] Im[λvec] + Re[λvec] Sin[χ];
```

Now you see why the field oscillates linearly in the direction of λ if λ is pure real; `HrmFctr` becomes just `λ Sin[χ]`,

If λ is complex, the answer is much more interesting. First make a normalized transverse complex vector :

```
rndC := RandomComplex[{-1 - ⅈ, +1 + ⅈ}];
vct = {rndC, rndC};
λ = vct / Sqrt[vct.vct*]

{-0.299276 - 0.946134 ⅈ, -0.014462 + 0.122699 ⅈ}
```

Look at the `HrmFctr` that this produces :

```
HrmFctr[λ]

{-0.946134 Cos[χ] - 0.299276 Sin[χ],
 0.122699 Cos[χ] - 0.014462 Sin[χ]}
```

This is exactly the format needed by `ParametricPlot`. Make a fresh λ vector and plot as χ runs over one cycle:

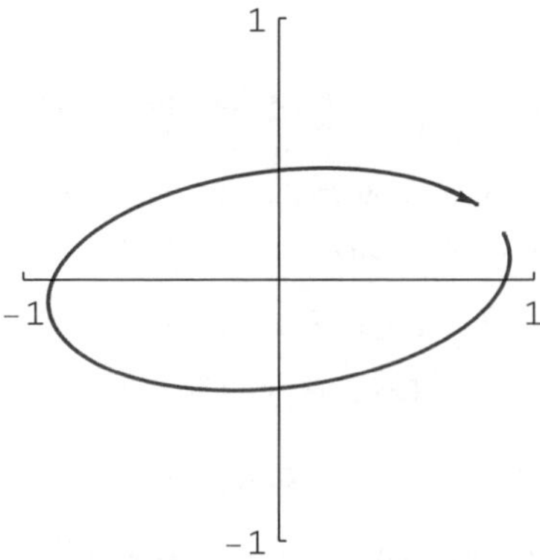

Fig. 44.3 Elliptical polarization. Motion of the tip of the electric vector, looking into the light source.

Select the little closed cell above this figure and evaluate it a few times. It uses **rndC** to make a new random ellipse every time it runs. The arrow shows the sense of rotation; sometimes it goes clockwise, sometimes counterclockwise.

The transverse right- and left- handed circular polarization vectors are

```
{Eright, Eleft} =
{HrmFctr[{1, -i}/√2], HrmFctr[{1, i}/√2]}
```

$$\left\{\left\{\frac{\text{Sin}[\chi]}{\sqrt{2}}, -\frac{\text{Cos}[\chi]}{\sqrt{2}}\right\}, \left\{\frac{\text{Sin}[\chi]}{\sqrt{2}}, \frac{\text{Cos}[\chi]}{\sqrt{2}}\right\}\right\}$$

Lest there be any misunderstanding, we plot these functions from a start on the **x**-axis through about 350°, ending in a blue arrowhead.

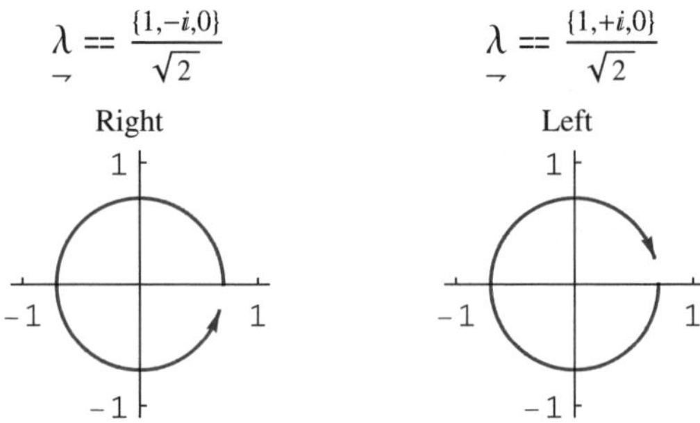

Fig. 44.4 Circular polarization. Motion of the tip of the electric vector, looking into the light source.

Here is where the names "right" and "left" come from: Point at your nose with the thumb of your right hand, fingers curled. Your thumb represents the propagation vector, so you are looking into the light source. The electric vector turns in the same direction as your fingers (counterclockwise) , so the figure labelled "Right" is indeed right-handed light. Use your left hand for the other one.

45. Two-photon tensor projections

Preliminaries

45.1. The operator for two-photon processes

45.1.1 The matrix element

We depend on other sources (Sakurai, Bunker, McClain and Harris) for the derivation of the two-photon operator, which applies to both two-photon absorption (TPA; two in, none out) and Raman and Rayleigh scattering (RRS; one in, one out). In this chapter we show how the two-photon operator poses a symmetry problem, and how the tools we have developed can be used to solve it.

The theory was first developed by Maria Göppert-Mayer in 1931. She built on the theory of one-photon events (first order perturbation by oscillating electric field), taking it to second order to obtain the two-photon result. In TPA, the molecule is located at the intersection of two laser beams with frequencies ω_a and ω_b, and polarization vectors $\underline{\lambda}^a$ and $\underline{\lambda}^b$. The amplitude $\mathcal{A}_{\gamma,\phi}$ for a two-photon transition (from ground state ψ_γ to final state ψ_ϕ) is given by

$$\mathcal{A}_{\gamma,\phi} = \sum_{\iota} \left(\frac{\langle \psi_\gamma | \underline{\lambda}^a . \underline{r} | \psi_\iota \rangle \langle \psi_\iota | \underline{r} . \underline{\lambda}^b | \psi_\phi \rangle}{\omega_{\gamma,\iota} - \omega_a} + \frac{\langle \psi_\gamma | \underline{\lambda}^b . \underline{r} | \psi_\iota \rangle \langle \psi_\iota | \underline{r} . \underline{\lambda}^a | \psi_\phi \rangle}{\omega_{\gamma,\iota} - \omega_b} \right) \tag{45.1}$$

where γ, ι, and ϕ are mnemonics for *ground*, *intermediate*, and *final*. The sum runs over all molecular eigenstates ψ_ι; $\omega_{\gamma,\iota}$ is the frequency of the transition $\gamma \to \iota$; $\underline{r}$ is the Cartesian coordinate vector **{x, y, z}**.

This state sum is symmetric for exchange of photons; if **a→b** and **b→a**, the first and second summands change places, but the sum is the same. We will work exclusively with the amplitude $\mathcal{A}_{\gamma,\phi}$; but never forget that the observable is the rate constant for the two-photon event, proportional to $\left|\mathcal{A}_{\gamma,\phi}\right|^2$, or

W.M. McClain, *Symmetry Theory in Molecular Physics with Mathematica*, DOI 10.1007/b13137_45, © Springer Science+Business Media, LLC 2009

when the final state is an $\mathbf{E}$ state with basis functions ϕ^{Ea} and ϕ^{Eb} the formula becomes $\left|\mathcal{A}_{\gamma,\phi^{Ea}}\right|^2 + \left|\mathcal{A}_{\gamma,\phi^{Eb}}\right|^2$, and so forth for higher degeneracies.

We now factor the polarization vectors out of the matrix elements :

$$\mathcal{A}_{\gamma,\phi} = \sum_L \left(\underset{\rightharpoondown}{\lambda}^{a} \cdot \left(\frac{\langle \psi_\gamma \mid \underset{\rightharpoondown}{r} \mid \psi_L \rangle \langle \psi_L \mid \underset{\rightharpoondown}{r} \mid \psi_\phi \rangle}{\omega_{\gamma,L} - \omega_a} \right) \cdot \underset{\rightharpoondown}{\lambda}^{b} + \underset{\rightharpoondown}{\lambda}^{b} \cdot \left(\frac{\langle \psi_\gamma \mid \underset{\rightharpoondown}{r} \mid \psi_L \rangle \langle \psi_L \mid \underset{\rightharpoondown}{r} \mid \psi_\phi \rangle}{\omega_{\gamma,L} - \omega_b} \right) \cdot \underset{\rightharpoondown}{\lambda}^{a} \right)$$

or in index notation

$$\mathcal{A}_{\gamma,\phi} = \lambda^a_j \lambda^b_k \sum_L \left(\left(\frac{\langle \psi_\gamma \mid r_j \mid \psi_L \rangle \langle \psi_L \mid r_k \mid \psi_\phi \rangle}{\omega_{\gamma,L} - \omega_a} \right) + \left(\frac{\langle \psi_\gamma \mid r_k \mid \psi_L \rangle \langle \psi_L \mid r_j \mid \psi_\phi \rangle}{\omega_{\gamma,L} - \omega_b} \right) \right)$$

where the sum over $\mathbf{j,k}$ is understood. It also helps to move the bra $\langle \psi_\gamma |$ and the ket $| \psi_\phi \rangle$ as far outward as possible, leaving a single big operator between them

$$\mathcal{A}_{\gamma,\phi} = \sum\nolimits_{j,k} \lambda^a_j \lambda^b_k \left\langle \psi_\gamma \middle| \sum\nolimits_L \left(\frac{r_j \,|\psi_L\rangle \langle \psi_L|\, r_k}{\omega_{\gamma,L} - \omega_a} + \frac{r_k \,|\psi_L\rangle \,\text{"}\langle \psi_L|\, r_j}{\omega_{\gamma,L} - \omega_b} \right) \middle| \psi_\phi \right\rangle \tag{45.2}$$

We need to distinguish between the operator and the matrix element. The operator is a double-struck $\mathbb{S}$

$$\mathbb{S}^{a,b}_{j,k} = \sum\nolimits_L \left(\frac{r_j \,|\psi_L\rangle \;\langle \psi_L|\, r_k}{\omega_{\gamma,L} - \omega_a} + \frac{r_k \,|\psi_L\rangle \;\langle \psi_L|\, r_j}{\omega_{\gamma,L} - \omega_b} \right) \tag{45.3}$$

whereas the tensor matrix element (a 3×3 array of numbers) is a Gothic $\mathfrak{S}$:

$$\mathfrak{S}^{a,b}_{j,k}[\gamma,\ \phi] = \left\langle \psi_\gamma \middle| \mathbb{S}^{a,b}_{j,k} \middle| \psi_\phi \right\rangle \tag{45.4}$$

The explicit γ,ϕ dependence is usually redundant, so except when we need specific state names we write just $\mathfrak{S}^{a,b}_{j,k}$. Finally it is clear that the two-photon amplitude is just a matrix dotted from each side by a polarization vector. In index notation it is $\sum_{j,k} \lambda^a_j\ \mathfrak{S}^{a,b}_{j,k} \lambda^b_k$, or in vector notation it is

$$\mathcal{A}_{TPA} = \underline{\lambda}^{a} . \mathfrak{S}^{a,b} . \underline{\lambda}^{b}$$

This is a good place to note that the formulae for TPA and RRS are just a little different, because the basic formalism of physics requires that annihilated photons and created photons be treated differently. In Raman, the polarization vector of the created photon **b** must be conjugated:

$$\mathcal{A}_{Raman} = \underline{\lambda}^{a} . \mathfrak{S}^{a,b} . \underline{\lambda}^{b*}$$

The Raman emission is of course an incoherent mixture of polarizations. What we really mean above is that the detector is looking through a polarization filter that passes only the pure polarization $\underline{\lambda}^{b}$.

Define the two-photon matrix as a *Mathematica* function :

$$\mathfrak{S}^{a\_,b\_}_{j\_,k\_} := \sum_{L} \left(\left(\frac{\langle \psi_\gamma \mid r_j \mid \psi_L \rangle \langle \psi_L \mid r_k \mid \psi_\phi \rangle}{\omega_{\gamma,L} - \omega_a} \right) + \left(\frac{\langle \psi_\gamma \mid r_k \mid \psi_L \rangle \langle \psi_L \mid r_j \mid \psi_\phi \rangle}{\omega_{\gamma,L} - \omega_b} \right) \right) /. \{r_1 \to x,\ r_2 \to y,\ r_3 \to z\}$$

The matrix itself is not symmetric for exchange of `j` and `k` alone; its symmetry is a little more complicated:

$$\mathfrak{S}^{a,b}_{j,k} == \mathfrak{S}^{b,a}_{k,j}$$

```
True
```

45.1.2 Proof that the matrix element is a tensor

Make a harmless maneuver: insert a unit matrix at two places in the **Dot** product of Eq. 45.4, but write it as a rotation matrix $\mathbb{R}$ times its inverse:

$$\mathcal{A} = \underline{\lambda}^{a} . \left(\mathbb{R}^{-1} . \mathbb{R}\right) . \mathfrak{S}^{a,b} . \left(\mathbb{R}^{-1} . \mathbb{R}\right) . \underline{\lambda}^{b}$$

Regrouping we see something interesting:

$$\mathcal{A} = \left(\underline{\lambda}^{a} . \mathbb{R}^{-1}\right) . \left(\mathbb{R} . \mathfrak{S}^{a,b} . \mathbb{R}^{-1}\right) . \left(\mathbb{R} . \underline{\lambda}^{b}\right)$$

It gets better if we use the transpose identity on the leftmost factor :

$$\mathcal{A} = \left(\left(\mathbb{R}^{-1}\right)^{T} . \left(\underline{\lambda}^{a}\right)^{T}\right) . \left(\mathbb{R} . \mathfrak{S}^{a,b} . \mathbb{R}^{-1}\right) . \left(\mathbb{R} . \underline{\lambda}^{b}\right)$$

But $\mathbb{R}$ is a unitary matrix, so $\left(\mathbb{R}^{-1}\right)^{T} = \mathbb{R}$. And as the writers of *Mathematica* have so usefully pointed out, there is no need to keep track of whether vectors are row vectors or column vectors. They are just simple lists that take on row or column properties according to their position in the **Dot** product (left of the dot ⇒ row ; right of the dot ⇒ column). So

$$\mathcal{A} = \left(\mathbb{R}.\underrightarrow{\lambda^{a}}\right).\left(\mathbb{R}.\mathfrak{S}^{a,b}.\mathbb{R}^{-1}\right).\left(\mathbb{R}.\underrightarrow{\lambda^{b}}\right) \tag{45.5}$$

This expression is a scalar amplitude, invariant under symmetry transforms. But $\mathbb{R}.\underrightarrow{\lambda^{a}}$ is well known as the rotational transform of vector $\underrightarrow{\lambda^{a}}$, and similarly for $\underrightarrow{\lambda^{b}}$. Therefore the object in the middle must be the formula for rotating $\mathfrak{S}^{a,b}$ by $\mathbb{R}$. These objects give the same scalar $\mathcal{A}$ when they all transform together to a new absolute orientation in space. The rotation formula $\mathbb{R}.\mathfrak{S}.\mathbb{R}^{-1}$ is in fact the defining property of a Cartesian tensor. Therefore, the very complicated quantum sum $\mathfrak{S}$ is a Cartesian tensor.

45.2. Operators for tensor projection

45.2.1. The projection step by step

We illustrate the projection in detail, using the **x,y** element of the general tensor as defined above , except that we take the state sum over ι as understood. We must avoid setting **T[x,y]** to anything, so we call it **Txy** instead:

Txy = $\mathfrak{S}^{a,b}_{1,2}$[[2]]

$$\frac{\langle\psi_{\Upsilon} \mid x \mid \psi_{\iota}\rangle \langle\psi_{\iota} \mid y \mid \psi_{\phi}\rangle}{-\omega_{a} + \omega_{\Upsilon,\iota}} + \frac{\langle\psi_{\Upsilon} \mid y \mid \psi_{\iota}\rangle \langle\psi_{\iota} \mid x \mid \psi_{\phi}\rangle}{-\omega_{b} + \omega_{\Upsilon,\iota}}$$

The **1,2** subscript has brought forth the **x,y** element. Remember, this is only one of nine tensor elements that we could have chosen to use as the example.

Below, we calculate three results labelled **Txy1** to **Txy3**, all identical to each other. First, do a trace projection of **Txy**. The whole result is longer than we need to see, so we display only part [[1,2]], the species **A** projection :

```
Txy1 = ProjectET[Txy, "C3"][[1, 2]]
```

$$\frac{\left\langle \psi_{\Upsilon} \left| -\frac{x}{2}+\frac{\sqrt{3}\,y}{2} \right| \psi_{L} \right\rangle \left\langle \psi_{L} \left| -\frac{\sqrt{3}\,x}{2}-\frac{y}{2} \right| \psi_{\Phi} \right\rangle}{3\,(-\omega_{a}+\omega_{\Upsilon,L})}+$$

$$\frac{\left\langle \psi_{\Upsilon} \left| -\frac{x}{2}-\frac{\sqrt{3}\,y}{2} \right| \psi_{L} \right\rangle \left\langle \psi_{L} \left| \frac{\sqrt{3}\,x}{2}-\frac{y}{2} \right| \psi_{\Phi} \right\rangle}{3\,(-\omega_{a}+\omega_{\Upsilon,L})}+$$

$$\frac{\langle \psi_{\Upsilon} \mid x \mid \psi_{L} \rangle \langle \psi_{L} \mid y \mid \psi_{\Phi} \rangle}{3\,(-\omega_{a}+\omega_{\Upsilon,L})}+\frac{\langle \psi_{\Upsilon} \mid y \mid \psi_{L} \rangle \langle \psi_{L} \mid x \mid \psi_{\Phi} \rangle}{3\,(-\omega_{b}+\omega_{\Upsilon,L})}+$$

$$\frac{\left\langle \psi_{\Upsilon} \left| \frac{\sqrt{3}\,x}{2}-\frac{y}{2} \right| \psi_{L} \right\rangle \left\langle \psi_{L} \left| -\frac{x}{2}-\frac{\sqrt{3}\,y}{2} \right| \psi_{\Phi} \right\rangle}{3\,(-\omega_{b}+\omega_{\Upsilon,L})}+$$

$$\frac{\left\langle \psi_{\Upsilon} \left| -\frac{\sqrt{3}\,x}{2}-\frac{y}{2} \right| \psi_{L} \right\rangle \left\langle \psi_{L} \left| -\frac{x}{2}+\frac{\sqrt{3}\,y}{2} \right| \psi_{\Phi} \right\rangle}{3\,(-\omega_{b}+\omega_{\Upsilon,L})}$$

As expected, it is a linear combination of expressions in which **x** and **y** have been transformed by the operators of group **C₃**. We simplify this carefully by hand. First, several integrands are a sum of two expressions, so we rewrite them as the sum of two integrals :

```
Txy2 = Txy1 /.
      〈g_ | r_ + s_ | f_〉 → 〈g | r | f〉 + 〈g | s | f〉 // Expand;
```

Interactive readers can remove the semicolon and see the rather long intermediate expression. Now all the operands are either **x** or **y** times some coefficient. We move the coefficients out in front of the integrals, after which *Mathematica* does a lot of cancellation :

```
Txy3 = Txy2 /. {
      〈g_ | c_ x | f_〉 → c 〈g | x | f〉,
      〈g_ | c_ y | f_〉 → c 〈g | y | f〉} // Expand
```

$$-\frac{\langle \psi_{\Upsilon} \mid y \mid \psi_{L} \rangle \langle \psi_{L} \mid x \mid \psi_{\Phi} \rangle}{2\,(-\omega_{a}+\omega_{\Upsilon,L})}+\frac{\langle \psi_{\Upsilon} \mid x \mid \psi_{L} \rangle \langle \psi_{L} \mid y \mid \psi_{\Phi} \rangle}{2\,(-\omega_{a}+\omega_{\Upsilon,L})}+$$
$$\frac{\langle \psi_{\Upsilon} \mid y \mid \psi_{L} \rangle \langle \psi_{L} \mid x \mid \psi_{\Phi} \rangle}{2\,(-\omega_{b}+\omega_{\Upsilon,L})}-\frac{\langle \psi_{\Upsilon} \mid x \mid \psi_{L} \rangle \langle \psi_{L} \mid y \mid \psi_{\Phi} \rangle}{2\,(-\omega_{b}+\omega_{\Upsilon,L})}$$

It is done. The trace projection of `T[x,y]` under group **C₃** has only four terms.

45.3 Self-purifying tensor elements

Now think what we did. We defined

$$\texttt{T[x, y]} == \frac{\langle \psi_\gamma | \mathbf{x} | \psi_\iota \rangle \langle \psi_\iota | \mathbf{y} | \psi_\phi \rangle}{-\omega_{\mathbf{a}} + \omega_{\gamma,\iota}} + \tag{45.6}$$

$$\frac{\langle \psi_\gamma | \mathbf{y} | \psi_\iota \rangle \langle \psi_\iota | \mathbf{x} | \psi_\phi \rangle}{-\omega_{\mathbf{b}} + \omega_{\gamma,\iota}}$$

(with sum over ι understood) and projected it, giving an answer with four terms like $\frac{\langle \psi_\gamma | \Box | \psi_\iota \rangle \langle \psi_\iota | \Box | \psi_\phi \rangle}{-\omega_\Box + \omega_{\gamma,\iota}}$. The quantum sum **T[x,y]** is incredibly difficult to compute, but it has one simple feature: it is linear in the Cartesian operators. In the preliminaries we make special rules for the function **T[u,v]** (where **u** and **v** are Cartesian variables **x,y,z**) that makes it automatically linear in its Cartesian variables. It uses a test that identifies the Cartesian variables:

```
CartesianQ[u_]:=(u===x||u===y||u===z)
```

Here is a copy of the definitions that give **T** its linear properties :

```
T[a_ + b_, c_] := T[a, c] + T[b, c];
T[a_, c_ + d_] := T[a, c] + T[a, d];
T[a_ *u_?CartesianQ, v_] := a*T[u, v];
T[u_, b_*v_?CartesianQ] := b*T[u, v];
```

These rules are active if you ran the preliminaries, so observe how **T** automatically simplifies itself :

```
T[a x + b y, c x + d y]
```

```
a c T[x, x] + a d T[x, y] + b c T[y, x] + b d T[y, y]
```

If we project **T[i,j]** as a surrogate for $\mathfrak{S}^{\mathbf{a},\mathbf{b}}_{\mathbf{i},\mathbf{j}}$, each summand will automatically expand itself to a linear combination of the 9 possible simple objects. Then *Mathematica* will automatically perform cancellations, producing an object that cannot be longer than 9 terms. Here is an example:

```
ProjectET[T[x, y], "C3"]
```

$$\left\{\left\{\mathrm{A},\ \frac{1}{2}\,\mathrm{T[x,\ y]} - \frac{1}{2}\,\mathrm{T[y,\ x]}\right\},\right.$$

```
{Ea, 1/4 ⅈ T[x, x] + 1/4 T[x, y] + 1/4 T[y, x] - 1/4 ⅈ T[y, y]},
{Eb, -1/4 ⅈ T[x, x] + 1/4 T[x, y] + 1/4 T[y, x] + 1/4 ⅈ T[y, y]}}
```

All the work of linear expansion has been done automatically.

45.4. Operators for tensor projection

45.4.1. Projection of the whole tensor

This means we can easily do the whole tensor of **T**-functions at once. In the preliminaries we set up a tensor of nine **T** functions:

```
Tij = ( T[x, x]  T[x, y]  T[x, z]
        T[y, x]  T[y, y]  T[y, z]
        T[z, x]  T[z, y]  T[z, z] );
```

so we just project the whole **Tij** tensor. We make it a little slicker by defining a specialized projection operator called **TraceProjectTij**. It will be fully explained below. It consists of perfectly ordinary trace projection, plus a filter that gets rid of any all-zero projections.

45.4.2. The simplest example

As our first demo we do a very simple case, group **C2**, with only two species. It has only one transform, rotation about **z** by half a turn. But it gives a non-trivial trace projection:

```
trPrjC2 = TraceProjectTij["C2"]
```

```
{C2, {{A, {{T[x, x], T[x, y], 0},
     {T[y, x], T[y, y], 0}, {0, 0, T[z, z]}}},
   {B, {{0, 0, T[x, z]}, {0, 0, T[y, z]},
     {T[z, x], T[z, y], 0}}}}}
```

This is easier to read in a **Grid** :

```
{A, GridForm[trPrjC2[[2, 1, 2]]]}
{B, GridForm[trPrjC2[[2, 2, 2]]]}
```

{A,

T[x, x]	T[x, y]	0
T[y, x]	T[y, y]	0
0	0	T[z, z]

}

{B,				}
	0	0	T[x, z]	
	0	0	T[y, z]	
	T[z, x]	T[z, y]	0	

Now you can see at a glance that the two projected parts add up to the original. Those elements that are unchanged by the two-fold rotation about **z** are in species A; those that change sign are in species B. All groups work out pretty much like this, but with many more complications, of course.

45.4.3. Trace projection, TraceProjectTij

Do this again with $\mathbf{D_{3\,h}}$, which has two true 2D species. As always, the 2D species name is **E**, or something anchored on an **E**:

```
BoxedCharacterTable["D3h"]
```

D3h	1	2	3	1	2	3	← Class populations
	E	C3	C2	σh	S3	σv	↓ Basis functions
A_1'	1	1	1	1	1	1	$\{(x^2+y^2),\ (z^2)\}$
A_2'	1	1	-1	1	1	-1	$\{(y\,(3x^2-y^2)),\ (Iz)\}$
E'	2	-1	0	2	-1	0	$\left\{\begin{pmatrix}x\\y\end{pmatrix},\ \begin{pmatrix}\frac{1}{2}(x^2-y^2)\\xy\end{pmatrix}\right\}$
A_1''	1	1	1	-1	-1	-1	$\{(x\,(x^2-3y^2)\,z)\}$
A_2''	1	1	-1	-1	-1	1	$\{(z)\}$
E''	2	-1	0	-2	1	0	$\left\{\begin{pmatrix}xz\\yz\end{pmatrix},\ \begin{pmatrix}Ix\\Iy\end{pmatrix}\right\}$

First do a trace projection:

```
tracePrjD3h = TraceProjectTij["D3h"]
```

```
{D3h, {{A'1, {{1/2 (T[x, x] + T[y, y]), 0, 0},
    {0, 1/2 (T[x, x] + T[y, y]), 0}, {0, 0, T[z, z]}}},
  {A'2, {{0, 1/2 (T[x, y] - T[y, x]), 0},
    {1/2 (-T[x, y] + T[y, x]), 0, 0}, {0, 0, 0}}},
  {E', {{1/2 (T[x, x] - T[y, y]), 1/2 (T[x, y] + T[y, x]), 0},
    {1/2 (T[x, y] + T[y, x]), 1/2 (-T[x, x] + T[y, y]), 0},
    {0, 0, 0}}}, {E'', {{0, 0, T[x, z]},
    {0, 0, T[y, z]}, {T[z, x], T[z, y], 0}}}}}
```

You must look carefully to see that several elements above are the same, or closely related. To help with this we need a routine that scans the whole tensor, picking out the first linear combination it comes to (that would be $\frac{1}{2}$ **T[x, x]** + $\frac{1}{2}$ **T[y, y]**), and then searches out all other instances of this same expression and replaces them all by some single letter. Then it continues assigning letters to expressions until all the tensor elements have been replaced by some single letter. This will let us see the "pattern" of the tensor at a glance.

Such an operator is in the Preliminaries under the name **SymbolizeTij**. It takes its symbols from two lists prepared in the Preliminaries: **LatUC** (Latin upper case, for expressions with no **i** in them) and **GrkLC** (Greek lower case for expressions that do contain **i**). When it finds a complex expression, it divides it into a real part **a** and imaginary part **b**, and makes up rules that include

$$\mathbf{a} + i\,\mathbf{b} \to \alpha \quad \text{and} \quad \mathbf{a} - i\,\mathbf{b} \to \alpha^*$$

When used with a symbolizer that automatically detects and reports proportionalities, this gives the complex notation that humans expect. Watch it work on the raw projections displayed above :

```
traceMatsD3h = Transpose[tracePrjD3h[[2]]][[2]];
SymbolizeTij[traceMatsD3h, LatUC, GrkLC] //
  GridList // Size[8]
```

{

A	0	0
0	A	0
0	0	B

,

0	C	0
-C	0	0
0	0	0

,

D	F	0
F	-D	0
0	0	0

,

0	0	G
0	0	H
J	K	0

}

These are the $\mathfrak{S}$ tensors, with all interior matrix elements $\langle \psi_\gamma \mid \mathbf{x} \mid \psi_\iota \rangle$ and $\langle \psi_\iota \mid \mathbf{y} \mid \psi_\phi \rangle$ etc. done symbolically, and the whole sum over all intermediate states performed symbolically. The numbers A, B, ... may or may not be computable by quantum theorists, but they can certainly be measured by spectroscopists. So these tensor "patterns" are the meeting point of theory and experiment.

We see only Latin capitals, so the trace projections have no explicit i in them. Compare to the raw tensors above to see that symbolization worked correctly. After you symbolize, you may want to know what the operators are. Just call up a quantity named **Symbolizations** that was created while **SymbolizeTij** was operating. Here they are :

```
Symbolizations // Column
```

$$\left\{\frac{1}{2}\,\mathrm{T}[\mathrm{x},\,\mathrm{x}] + \frac{1}{2}\,\mathrm{T}[\mathrm{y},\,\mathrm{y}] \to \mathrm{A}\right\}$$
$$\{\mathrm{T}[\mathrm{z},\,\mathrm{z}] \to \mathrm{B}\}$$
$$\left\{\frac{1}{2}\,\mathrm{T}[\mathrm{x},\,\mathrm{y}] - \frac{1}{2}\,\mathrm{T}[\mathrm{y},\,\mathrm{x}] \to \mathrm{C}\right\}$$
$$\left\{\frac{1}{2}\,\mathrm{T}[\mathrm{x},\,\mathrm{x}] - \frac{1}{2}\,\mathrm{T}[\mathrm{y},\,\mathrm{y}] \to \mathrm{D}\right\}$$
$$\left\{\frac{1}{2}\,\mathrm{T}[\mathrm{x},\,\mathrm{y}] + \frac{1}{2}\,\mathrm{T}[\mathrm{y},\,\mathrm{x}] \to \mathrm{F}\right\}$$
$$\{\mathrm{T}[\mathrm{x},\,\mathrm{z}] \to \mathrm{G}\}$$
$$\{\mathrm{T}[\mathrm{y},\,\mathrm{z}] \to \mathrm{H}\}$$
$$\{\mathrm{T}[\mathrm{z},\,\mathrm{x}] \to \mathrm{J}\}$$
$$\{\mathrm{T}[\mathrm{z},\,\mathrm{y}] \to \mathrm{K}\}$$

Symbolizations is overwritten each time **SymbolizeTij** runs, so it must be preserved immediately if you want to use it.

45.4.4. Detailed projection, `DetailedProjectTij`

When a group has an **E** species, the trace projection gives only the sum of the two detailed projections. To see them individually, you must do a detailed projection. The operator for this has a lot to do. First it has to make a full representation of the group, and the basis it uses for this is an important point. It always uses the first basis listed in the character table, which in the case of species **E**′of $\mathbf{D}_{3\,\mathbf{h}}$ is **{x,y}**. But the algorithm automatically transforms basis **{a,b}** into
{a+ⅈb,a-ⅈb}. So in this case it uses **{x+ⅈy,x-ⅈy}**. The **rep** is quickly found by **MakeRepPoly**, and it is used by

ProjectED[expr,rep,component,"D3h"]

with **component**s **1** and **2**. It always appends letters **a** and **b** to the **E**-type species name to form the names of the detailed projections that are found. Apply it and display only the **E**′species from the result:

```
detailedPrjD3h = DetailedProjectTij[tracePrjD3h];
detailedPrjD3h[[2, 3]]
```

```
{{E'a, {{1/4 (T[x, x] - i (T[x, y] + T[y, x]) - T[y, y]),
    1/4 (i T[x, x] + T[x, y] + T[y, x] - i T[y, y]), 0},
   {1/4 (i T[x, x] + T[x, y] + T[y, x] - i T[y, y]),
    1/4 (-T[x, x] + i (T[x, y] + T[y, x]) + T[y, y]),
    0}, {0, 0, 0}}},
  {E'b, {{1/4 (T[x, x] + i (T[x, y] + T[y, x] + i T[y, y])),
    1/4 (-i T[x, x] + T[x, y] + T[y, x] + i T[y, y]), 0},
   {1/4 (-i T[x, x] + T[x, y] + T[y, x] + i T[y, y]),
    1/4 (-T[x, x] - i (T[x, y] + T[y, x]) + T[y, y]),
    0}, {0, 0, 0}}}}
```

The **E′a** projection is from component **1**; **E′b** is from **2**. That is a lot of output to wade through. Symbolization is now quite essential.

45.4.5. Symbolization by SymbolizeProjectionTij

How SymbolizeProjectionTij works

Our projections contain a group name and species names that must be protected against symbolization, so we have written a useful operator that separates names from tensors, send the tensors to **SymbolizeTij**, makes a labelled copy (not overwritable) of the **Symbolizations**, then finally restores the names and returns the symbolized projection. Click into the Preliminaries read it; or just watch it work :

```
entryD3h = SymbolizeProjectionTij[detailedPrjD3h]
```

```
{D3h, {{A'1, {{A, 0, 0}, {0, A, 0}, {0, 0, B}}},
  {A'2, {{0, C, 0}, {-C, 0, 0}, {0, 0, 0}}},
  {{E'a, {{α, i α, 0}, {i α, -α, 0}, {0, 0, 0}}},
   {E'b, {{α*, -i α*, 0}, {-i α*, -α*, 0}, {0, 0, 0}}}},
  {{E''a, {{0, 0, β}, {0, 0, -i β}, {γ, -i γ, 0}}},
   {E''b, {{0, 0, β*}, {0, 0, i β*}, {γ*, i γ*, 0}}}}}}
```

The **E′a** and **E′b** results collapsed to a single line each. It preserved **SymbolizationS** under the name **SymbolRules["D3h"]** :

```
SymbolRules["D3h"] // Column
```

$\{A \to \frac{1}{2} T[x, x] + \frac{1}{2} T[y, y]\}$

$\{B \to T[z, z]\}$

$\{C \to \frac{1}{2} T[x, y] - \frac{1}{2} T[y, x]\}$

$\{\alpha \to \frac{1}{4} T[x, x] - \frac{1}{4} i T[x, y] - \frac{1}{4} i T[y, x] - \frac{1}{4} T[y, y]\}$

$\{\beta \to \frac{1}{2} T[x, z] + \frac{1}{2} i T[y, z]\}$

$\{\gamma \to \frac{1}{2} T[z, x] + \frac{1}{2} i T[z, y]\}$

Cyclic axis permutation symmetry

The tetrahedron has a three-fold symmetry rotation about the **{1,1,1}** axis. We bring it out as a symmetry rule:

```
Thread[{x, y, z} →
  AxialRotation[{1, 1, 1}, -2 π / 3].{x, y, z}]
```

```
{x → y, y → z, z → x}
```

So this rotation is just a cyclic permutation of axes. In group **T**, this is rule 2:

```
SymmetryRules["T"][[2]]
```

```
{x → y, y → z, z → x}
```

How many other groups also possess this rule?

```
Select[GroupCatalog,
 Not[FreeQ[SymmetryRules[#], {x → y, y → z, z → x}]] &]
```

```
{I, O, Oh, T, Td, Th}
```

All the Platonic groups have it.

As we argued in detail in Chapter 39 (Matrix elements) integrals are invariant under a unitary transformation of the entire integrand. So if the matrix element

$$T[x, x] = \frac{\langle \psi_\Upsilon \mid x \mid \psi^x_\iota \rangle \langle \psi^x_\iota \mid x \mid \psi_\phi \rangle}{-\omega_a + \omega_{\Upsilon,\iota}} + \frac{\langle \psi_\Upsilon \mid x \mid \psi^x_\iota \rangle \langle \psi^x_\iota \mid x \mid \psi_\phi \rangle}{-\omega_b + \omega_{\Upsilon,\iota}}$$

(where ψ_Υ and ψ_ϕ are totally symmetric) is subjected to axis permutation (including axis permutation in the intermediate wave functions), its value does not change, but its formula becomes

$$\frac{\langle\psi_\gamma \mid y \mid \psi_L^y\rangle \langle\psi_L^y \mid y \mid \psi_\phi\rangle}{-\omega_a + \omega_{\gamma,L}} + \frac{\langle\psi_\gamma \mid y \mid \psi_L^y\rangle \langle\psi_L^y \mid y \mid \psi_\phi\rangle}{-\omega_b + \omega_{\gamma,L}}$$

But this is exactly **T[y,y]**. This means that **T[x,x]** = **T[y,y]**, and similarly, **T[y,y]** = **T[z,z]** in the Platonic groups. A similar thing happens in the off-diagonal elements. The formula for the **x,y** element is

$$\frac{\langle\psi_\gamma \mid x \mid \psi_L^x\rangle \langle\psi_L^x \mid y \mid \psi_\phi^{x,y}\rangle}{-\omega_a + \omega_{\gamma,L}} + \frac{\langle\psi_\gamma \mid y \mid \psi_L^y\rangle \langle\psi_L^y \mid x \mid \psi_\phi^{x,y}\rangle}{-\omega_b + \omega_{\gamma,L}}$$

Subject this to axis permutation:

$$\frac{\langle\psi_\gamma \mid y \mid \psi_L^y\rangle \langle\psi_L^y \mid z \mid \psi_\phi^{y,z}\rangle}{-\omega_a + \omega_{\gamma,L}} + \frac{\langle\psi_\gamma \mid z \mid \psi_L^z\rangle \langle\psi_L^z \mid y \mid \psi_\phi^{y,z}\rangle}{-\omega_b + \omega_{\gamma,L}}$$

where the final states $\psi_\phi^{x,y}$ and $\psi_\phi^{y,z}$ are degenerate. But the operator here is exactly the operator of **T[y,z]**, so after integration **T[x,y]=T[y,z]**; and similarly **T[y,z]=T[z,x]**. The same argument shows **T[y,x]=T[z,y] = T[x,z]**. However, because the denominators are generally different, it is not true that **T[x,y]** ≡ **T[y,x]**, etc. Here is a rule that embodies these results:

```
xyzPermSymmTij = {
    T[x, x] → PP, T[y, y] → PP, T[z, z] → PP,
    T[x, y] → QQ, T[y, z] → QQ, T[z, x] → QQ,
    T[y, x] → RR, T[z, y] → RR, T[x, z] → RR};
```

SymbolizeProjectionTij applies this to all appropriate groups. The names **PP**, etc., are just dummies and are replaced by single letters when submitted to the final symbolization routine.

Another axis permutation symmetry

Group $\mathbf{D_{2\,d}}$ has a dihedral reflection between the **x** and **y** axes, so this reflection swaps **x** and **y** while leaving **z** alone: **{x→y,y→x,z→z}**. Do any other groups have this symmetry?

```
Select[GroupCatalog,
  Not[FreeQ[SymmetryRules[#], {x → y, y → x, z → z}]] &]
```

```
{C4v, D2d, D4h, D6d, Oh, Td}
```

Note that **O** and $\mathbf{O_h}$ have both symmetries. Therefore we must divide the axis permutation groups into three types: nonsquare Platonic (**I**, **T**, $\mathbf{T_d}$, $\mathbf{T_h}$), square Platonic (**O**, $\mathbf{O_h}$) , and square nonPlatonic ($\mathbf{C_{4\,v}}$, $\mathbf{D_{2\,d}}$, $\mathbf{D_{4\,h}}$, $\mathbf{D_{6\,d}}$). Each of these has its own set of axis permutation symmetries. The nonsquare Platonic rules have already been given. The square NonPlatonic rules are

```
xyPermSymmTij = {
   T[x, x] → PP, T[y, y] → PP,
   T[x, y] → QQ, T[y, x] → QQ,
   T[z, x] → RR, T[z, y] → RR,
   T[x, z] → SS, T[y, z] → SS};
```

and the rules for the square Platonics are

```
xyAndxyzPermSymmTij = {
   T[x, x] → PP, T[y, y] → PP, T[z, z] → PP,
   T[x, y] → QQ, T[y, z] → QQ, T[z, x] → QQ,
   T[y, x] → QQ, T[z, y] → QQ, T[x, z] → QQ};
```

SymbolizeProjectionTij applies these rules to all appropriate groups.

By our orientation rules, no other axis permutation symmetries will exist.

45.4.6. The combined operator MakeRawEntryTij

The three operators above (**TraceProjectTij**, **DetailedProjectTij**, and **SymbolizeProjectionTij**) are conveniently combined by a little driver called **MakeRawEntryTij[groupName]**. Click into the Preliminaries to read it, then watch it work:

```
entryC3 = MakeRawEntryTij["C3"]
```

```
{C3, {{A, {{A, B, 0}, {-B, A, 0}, {0, 0, C}}},
  {Ea, {{α, ⅈ α, β}, {ⅈ α, -α, -ⅈ β}, {γ, -ⅈ γ, 0}}},
  {Eb, {{α*, -ⅈ α*, β*}, {-ⅈ α*, -α*, ⅈ β*}, {γ*, ⅈ γ*, 0}}}}}
```

While **SymbolizeProjectionTij** was operating, it made the **SymbolRules** for this group:

```
SymbolRules["C3"]
```

$$\left\{\left\{A\to\frac{1}{2}T[x,x]+\frac{1}{2}T[y,y]\right\},\right.$$
$$\left\{B\to\frac{1}{2}T[x,y]-\frac{1}{2}T[y,x]\right\},\ \{C\to T[z,z]\},$$
$$\left\{\alpha\to\frac{1}{4}T[x,x]-\frac{1}{4}\,i\,T[x,y]-\frac{1}{4}\,i\,T[y,x]-\frac{1}{4}T[y,y]\right\},$$
$$\left.\left\{\beta\to\frac{1}{2}T[x,z]+\frac{1}{2}\,i\,T[y,z]\right\},\ \left\{\gamma\to\frac{1}{2}T[z,x]+\frac{1}{2}\,i\,T[z,y]\right\}\right\}$$

45.4.7. The operator `TPTdisplay`

The output of **`SymbolizeProjectionTij`** is still a little hard to read for non-*Mathematica* people. So we have written an operator **`TPTdisplay`** that cleans them up like a typesetter might. Its input is a list of raw table entries and a style specification. Only fanatics will open the preliminaries to see how it works. Here is an example:

```
MakeTijdisplay[{entryD3h},
  {FontFamily → "Arial", 7}]
```

D3h $\{A'_1, A'_2, \{E'a, E'b\}, \{E''a, E''b\}\}$

$$\begin{array}{|c|c|c|}\hline A & 0 & 0 \\ \hline 0 & A & 0 \\ \hline 0 & 0 & B \\ \hline \end{array} \quad \begin{array}{|c|c|c|}\hline 0 & C & 0 \\ \hline -C & 0 & 0 \\ \hline 0 & 0 & 0 \\ \hline \end{array} \quad \left\{ \begin{array}{|c|c|c|}\hline \alpha & i\alpha & 0 \\ \hline i\alpha & -\alpha & 0 \\ \hline 0 & 0 & 0 \\ \hline \end{array}\, , \begin{array}{|c|c|c|}\hline \alpha^{*} & -\mathbb{i}\ \alpha^{*} & 0 \\ \hline -\mathbb{i}\ \alpha^{*} & -\alpha^{*} & 0 \\ \hline 0 & 0 & 0 \\ \hline \end{array} \right\}$$

$$\left\{ \begin{array}{|c|c|c|}\hline 0 & 0 & \beta \\ \hline 0 & 0 & -i\beta \\ \hline \gamma & -i\gamma & 0 \\ \hline \end{array}\, , \begin{array}{|c|c|c|}\hline 0 & 0 & \beta^{*} \\ \hline 0 & 0 & \mathbb{i}\ \beta^{*} \\ \hline \gamma^{*} & \mathbb{i}\ \gamma^{*} & 0 \\ \hline \end{array} \right\}$$

The name of the group is shown in **boldface**, followed by a list of species that have nonzero tensor projections; then on the next line are the respective patterns, in grids that help the eye. Degenerate patterns are bracketed together. All-zero patterns are omitted.

The patterns of many groups are identical. So the display generator also searches the raw entries input list for cases that have identical patterns, and groups them together, printing the patterns only once. This shortens the final output considerably.

45.5 Two-photon tensor projection table

45.5.1. Make all non-zero table entries (no styling)

We will now do all our tabulated groups in one fell swoop.

```
GroupCatalog
```

```
{C1, C2, C2h, C2v, C3, C3h, C3v, C4, C4h, C4v, C5, C6, C6h,
 C6v, Ch, Ci, D2, D2d, D2h, D3, D3d, D3h, D4, D4d, D4h, D5,
 D5d, D5h, D6, D6d, D6h, I, O, Oh, S4, S6, S8, T, Td, Th}
```

Apply **MakeRawEntryTij** to this long list of group names. You will see a lot of flashing as it works. It is trying to display the name of the group it is working on, but most flash by too quickly to read. Only the larger groups take a perceptible time; in particular **I** is quite slow (be patient; it will finish). Be aware that all this is done in infinite precision, even the groups with five-fold axes.

```
allRaw = MakeRawEntryTij[GroupCatalog];
```

Look at just one of the 40 group results :

```
allRaw[[25]]
```

```
{D4h, {{A1g, {{A, 0, 0}, {0, A, 0}, {0, 0, B}}},
  {A2g, {{0, 0, 0}, {0, 0, 0}, {0, 0, 0}}},
  {B1g, {{0, 0, 0}, {0, 0, 0}, {0, 0, 0}}},
  {B2g, {{0, C, 0}, {C, 0, 0}, {0, 0, 0}}},
  {{Ega, {{0, 0, α}, {0, 0, α*}, {β, β*, 0}}},
   {Egb, {{0, 0, α*}, {0, 0, α}, {β*, β, 0}}}}}}
```

45.5.2. Table of two-photon tensor patterns

All that is left is the display operator **MakeTijdisplay**. It has already been described, so we just use it :

```
MakeTijdisplay[allRaw, {FontFamily → "Arial", 8}];
```

C1 {A}

Ci {Ag}

A	B	C
D	F	G
H	J	K

C2 {A, B}

C2h {Ag, Bg}

Ch {A, B}

A	B	0
C	D	0
0	0	F

0	0	G
0	0	H
J	K	0

C2v {A1 , A2, B1, B2}

D2 {A, B1, B2, B3}

D2h {Ag, B1g, B2g, B3g}

A	0	0
0	B	0
0	0	C

0	D	0
F	0	0
0	0	0

0	0	G
0	0	0
H	0	0

0	0	0
0	0	J
0	K	0

C3 {A, Ea, Eb}

S6 {Ag, Ega, Egb}

A	B	0
−B	A	0
0	0	C

α	$i\alpha$	β
$i\alpha$	$-\alpha$	$-i\beta$
γ	$-i\gamma$	0

α^*	$-i\,\alpha^*$	β^*
$-i\,\alpha^*$	$-\alpha^*$	$i\,\beta^*$
γ^*	$i\,\gamma^*$	0

C3h {A′, E'_a, E'_b, E''_a, E''_b}

A	B	0
–B	A	0
0	0	C

α	$i\alpha$	0
$i\alpha$	$-\alpha$	0
0	0	0

α^*	$-i\,\alpha^*$	0
$-i\,\alpha^*$	$-\alpha^*$	0
0	0	0

0	0	β
0	0	$-i\beta$
γ	$-i\gamma$	0

0	0	β^*
0	0	$i\,\beta^*$
γ^*	$i\,\gamma^*$	0

C3v {A1, A2, {Ea, Eb}}

D3 {A1, A2, {Ea, Eb}}

D3d {A1g, A2g, {Ega, Egb}}

A	0	0
0	A	0
0	0	B

0	C	0
–C	0	0
0	0	0

{

α	$i\alpha$	β
$i\alpha$	$-\alpha$	$-i\beta$
γ	$-i\gamma$	0

,

α^*	$-i\,\alpha^*$	β^*
$-i\,\alpha^*$	$-\alpha^*$	$i\,\beta^*$
γ^*	$i\,\gamma^*$	0

}

C4 {A, B, Ea, Eb}

C4h {Ag, Bg, Ega, Egb}

S4 {A, B, Ea, Eb}

A	B	0
–B	A	0
0	0	C

D	F	0
F	–D	0
0	0	0

0	0	α
0	0	$i\alpha$
β	$i\beta$	0

0	0	α^*
0	0	$-i\,\alpha^*$
β^*	$-i\,\beta^*$	0

C4v {A1, A2, B1, B2, {Ea, Eb}}

D2d {A_1, A_2, B_1, B_2, {Ea, Eb}}

D4h {A1g, A2g, B1g, B2g, {Ega, Egb}}

A	0	0
0	A	0
0	0	B

0	0	0
0	0	0
0	0	0

0	0	0
0	0	0
0	0	0

0	C	0
C	0	0
0	0	0

{

0	0	α
0	0	α^*
β	β^*	0

,

0	0	α^*
0	0	α
β^*	β	0

}

C5 {A, E1a, E1b, E2a, E2b}

C6 {A, E1a, E1b, E2a, E2b}

C6h {Ag, E1ga, E1gb, E2ga, E2gb}

A	B	0
−B	A	0
0	0	C

0	0	α
0	0	$i\alpha$
β	$i\beta$	0

0	0	α^*
0	0	$-\mathbb{i}\,\alpha^*$
β^*	$-\mathbb{i}\,\beta^*$	0

γ	$i\gamma$	0
$i\gamma$	$-\gamma$	0
0	0	0

γ^*	$-\mathbb{i}\,\gamma^*$	0
$-\mathbb{i}\,\gamma^*$	$-\gamma^*$	0
0	0	0

C6v {A1, A2, {E1a, E1b}, {E2a, E2b}}

D5 {A1, A2, {E1a, E1b}, {E2a, E2b}}

D5d $\{A_{1g}, A_{2g}, \{E_{1g}a, E_{1g}b\}, \{E_{2g}a, E_{2g}b\}\}$

D6 {A1, A2, {E1a, E1b}, {E2a, E2b}}

D6h {A1g, A2g, {E1ga, E1gb}, {E2ga, E2gb}}

A	0	0
0	A	0
0	0	B

0	C	0
−C	0	0
0	0	0

{

0	0	α
0	0	$-i\alpha$
β	$-i\beta$	0

,

0	0	α^*
0	0	$\mathbb{i}\,\alpha^*$
β^*	$\mathbb{i}\,\beta^*$	0

}

{

γ	$-i\gamma$	0
$-i\gamma$	$-\gamma$	0
0	0	0

,

γ^*	$\mathbb{i}\,\gamma^*$	0
$\mathbb{i}\,\gamma^*$	$-\gamma^*$	0
0	0	0

}

D3h $\{A_1', A_2', \{E'a, E'b\}, \{E''a, E''b\}\}$

A	0	0
0	A	0
0	0	B

0	C	0
−C	0	0
0	0	0

{

α	$i\alpha$	0
$i\alpha$	$-\alpha$	0
0	0	0

,

α^*	$-\mathbb{i}\,\alpha^*$	0
$-\mathbb{i}\,\alpha^*$	$-\alpha^*$	0
0	0	0

}

{

0	0	β
0	0	$-i\beta$
γ	$-i\gamma$	0

,

0	0	β^*
0	0	$\mathbb{i}\,\beta^*$
γ^*	$\mathbb{i}\,\gamma^*$	0

}

D4 **{A1, A2, B1, B2, {Ea, Eb}}**

$$\begin{array}{|c|c|c|}\hline A & 0 & 0 \\ \hline 0 & A & 0 \\ \hline 0 & 0 & B \\ \hline\end{array} \quad \begin{array}{|c|c|c|}\hline 0 & C & 0 \\ \hline -C & 0 & 0 \\ \hline 0 & 0 & 0 \\ \hline\end{array} \quad \begin{array}{|c|c|c|}\hline D & 0 & 0 \\ \hline 0 & -D & 0 \\ \hline 0 & 0 & 0 \\ \hline\end{array}$$

$$\begin{array}{|c|c|c|}\hline 0 & F & 0 \\ \hline F & 0 & 0 \\ \hline 0 & 0 & 0 \\ \hline\end{array} \quad \left\{ \begin{array}{|c|c|c|}\hline 0 & 0 & \alpha \\ \hline 0 & 0 & -i\alpha \\ \hline \beta & -i\beta & 0 \\ \hline\end{array} , \begin{array}{|c|c|c|}\hline 0 & 0 & \alpha^* \\ \hline 0 & 0 & \mathbb{i}\,\alpha^* \\ \hline \beta^* & \mathbb{i}\,\beta^* & 0 \\ \hline\end{array} \right\}$$

D4d **{A1, A2, {E2a, E2b}, {E3a, E3b}}**

D5h **$\{A'_1, A'_2, \{E'_2a, E'_2b\}, \{E''_1a, E''_1b\}\}$**

$$\begin{array}{|c|c|c|}\hline A & 0 & 0 \\ \hline 0 & A & 0 \\ \hline 0 & 0 & B \\ \hline\end{array} \quad \begin{array}{|c|c|c|}\hline 0 & C & 0 \\ \hline -C & 0 & 0 \\ \hline 0 & 0 & 0 \\ \hline\end{array} \quad \left\{ \begin{array}{|c|c|c|}\hline \alpha & -i\alpha & 0 \\ \hline -i\alpha & -\alpha & 0 \\ \hline 0 & 0 & 0 \\ \hline\end{array} , \begin{array}{|c|c|c|}\hline \alpha^* & \mathbb{i}\,\alpha^* & 0 \\ \hline \mathbb{i}\,\alpha^* & -\alpha^* & 0 \\ \hline 0 & 0 & 0 \\ \hline\end{array} \right\}$$

$$\left\{ \begin{array}{|c|c|c|}\hline 0 & 0 & \beta \\ \hline 0 & 0 & -i\beta \\ \hline \gamma & -i\gamma & 0 \\ \hline\end{array} , \begin{array}{|c|c|c|}\hline 0 & 0 & \beta^* \\ \hline 0 & 0 & \mathbb{i}\,\beta^* \\ \hline \gamma^* & \mathbb{i}\,\gamma^* & 0 \\ \hline\end{array} \right\}$$

D6d **{A1, A2, {E2a, E2b}, {E5a, E5b}}**

$$\begin{array}{|c|c|c|}\hline A & 0 & 0 \\ \hline 0 & A & 0 \\ \hline 0 & 0 & B \\ \hline\end{array} \quad \begin{array}{|c|c|c|}\hline 0 & 0 & 0 \\ \hline 0 & 0 & 0 \\ \hline 0 & 0 & 0 \\ \hline\end{array} \quad \left\{ \begin{array}{|c|c|c|}\hline \alpha & C & 0 \\ \hline C & -\alpha & 0 \\ \hline 0 & 0 & 0 \\ \hline\end{array} , \begin{array}{|c|c|c|}\hline -\alpha & C & 0 \\ \hline C & \alpha & 0 \\ \hline 0 & 0 & 0 \\ \hline\end{array} \right\}$$

$$\left\{ \begin{array}{|c|c|c|}\hline 0 & 0 & \beta \\ \hline 0 & 0 & \beta^* \\ \hline \gamma & \gamma^* & 0 \\ \hline\end{array} , \begin{array}{|c|c|c|}\hline 0 & 0 & \beta^* \\ \hline 0 & 0 & \beta \\ \hline \gamma^* & \gamma & 0 \\ \hline\end{array} \right\}$$

I **{A, {T1a, T1b, T1c}, {Ha, Hb, Hc, Hd, He}}**

$$\begin{array}{|c|c|c|}\hline A & 0 & 0 \\ \hline 0 & A & 0 \\ \hline 0 & 0 & A \\ \hline\end{array} \quad \left\{ \begin{array}{|c|c|c|}\hline 0 & 0 & 0 \\ \hline 0 & 0 & B \\ \hline 0 & -B & 0 \\ \hline\end{array} , \begin{array}{|c|c|c|}\hline 0 & 0 & -B \\ \hline 0 & 0 & 0 \\ \hline B & 0 & 0 \\ \hline\end{array} , \begin{array}{|c|c|c|}\hline 0 & B & 0 \\ \hline -B & 0 & 0 \\ \hline 0 & 0 & 0 \\ \hline\end{array} \right\}$$

$$\left\{ \begin{array}{|c|c|c|}\hline 0 & 0 & 0 \\ \hline 0 & 0 & 0 \\ \hline 0 & 0 & 0 \\ \hline\end{array} , \begin{array}{|c|c|c|}\hline 0 & 0 & C \\ \hline 0 & 0 & 0 \\ \hline C & 0 & 0 \\ \hline\end{array} , \begin{array}{|c|c|c|}\hline 0 & 0 & 0 \\ \hline 0 & 0 & C \\ \hline 0 & C & 0 \\ \hline\end{array} , \begin{array}{|c|c|c|}\hline 0 & C & 0 \\ \hline C & 0 & 0 \\ \hline 0 & 0 & 0 \\ \hline\end{array} , \begin{array}{|c|c|c|}\hline 0 & 0 & 0 \\ \hline 0 & 0 & 0 \\ \hline 0 & 0 & 0 \\ \hline\end{array} \right\}$$

O {A1, {Ea, Eb}, {T1a, T1b, T1c}, {T2a, T2b, T2c}}

A	0	0
0	A	0
0	0	A

{

0	0	0
0	0	0
0	0	0

,

0	0	0
0	0	0
0	0	0

}

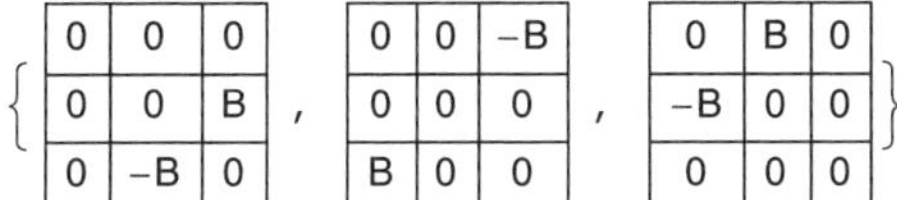

{

0	0	0
0	0	B
0	−B	0

,

0	0	−B
0	0	0
B	0	0

,

0	B	0
−B	0	0
0	0	0

}

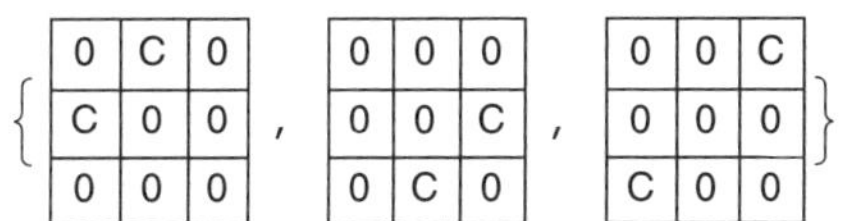

{

0	C	0
C	0	0
0	0	0

,

0	0	0
0	0	C
0	C	0

,

0	0	C
0	0	0
C	0	0

}

Oh {A1g, {Ega, Egb}, {T1ga, T1gb, T1gc}, {T2ga, T2gb, T2gc}}

Td {A1, {Ea, Eb}, {T1a, T1b, T1c}, {T2a, T2b, T2c}}

A	0	0
0	A	0
0	0	A

{

0	0	0
0	0	0
0	0	0

,

0	0	0
0	0	0
0	0	0

}

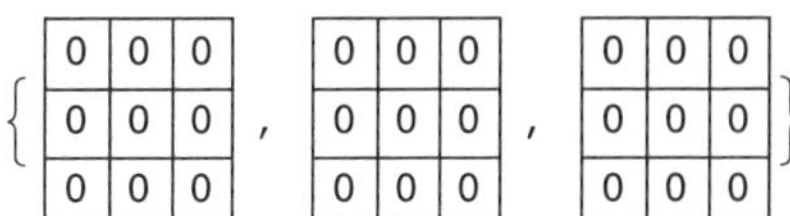

{

0	0	0
0	0	0
0	0	0

,

0	0	0
0	0	0
0	0	0

,

0	0	0
0	0	0
0	0	0

}

{

0	B	0
B	0	0
0	0	0

,

0	0	0
0	0	B
0	B	0

,

0	0	B
0	0	0
B	0	0

}

S8 {A, E1a, E1b, E2a, E2b, E3a, E3b}

A	B	0
−B	A	0
0	0	C

0	0	α
0	0	β
γ	δ	0

0	0	$-\alpha$
0	0	$-\beta$
$-\gamma$	$-\delta$	0

ε	$-i\varepsilon$	0
$-i\varepsilon$	$-\varepsilon$	0
0	0	0

ε^*	$i\,\varepsilon^*$	0
$i\,\varepsilon^*$	$-\varepsilon^*$	0
0	0	0

0	0	D
0	0	F
G	H	0

0	0	D
0	0	F
G	H	0

T {A1, Ea, Eb, {Ta, Tb, Tc}}

Th {Ag, Ega, Egb, {Tga, Tgb, Tgc}}

A	0	0
0	A	0
0	0	A

0	0	0
0	0	0
0	0	0

0	0	0
0	0	0
0	0	0

{

0	0	0
0	0	B
0	C	0

,

0	0	C
0	0	0
B	0	0

,

0	B	0
C	0	0
0	0	0

}

After each group was symbolized, its **Symbolizations** were given the name **SymbolRules[gp]**, before the next **SymbolizeProjectionTij** overwrote it. You can look at any of them you want:

```
SymbolRules["D2h"]
```

```
{{A → T[x, x]}, {B → T[y, y]}, {C → T[z, z]},
 {D → T[x, y]}, {F → T[y, x]}, {G → T[x, z]},
 {H → T[z, x]}, {J → T[y, z]}, {K → T[z, y]}}
```

45.6 Axis permutation forbiddenness

Some of the results for the Platonic groups are quite surprising, and seem to be a new discovery. Here we review one of them step by step. Look at the detailed projection of species **Ea** of group **T**:

```
tracePrj = TraceProjectTij["T"];
detPrjTEa = DetailedProjectTij[tracePrj][[2, 2]]
```

$$\Big\{\text{Ea},\ \Big\{\Big\{\frac{1}{3}\,\text{T[x, x]} - \frac{1}{6}\,\text{T[y, y]} - \frac{\text{i}\,\text{T[y, y]}}{2\sqrt{3}} - \frac{1}{6}\,\text{T[z, z]} + \frac{\text{i}\,\text{T[z, z]}}{2\sqrt{3}},\ 0,\ 0\Big\},\ \Big\{0,\ -\frac{1}{6}\,\text{T[x, x]} + \frac{\text{i}\,\text{T[x, x]}}{2\sqrt{3}} + \frac{1}{3}\,\text{T[y, y]} - \frac{1}{6}\,\text{T[z, z]} - \frac{\text{i}\,\text{T[z, z]}}{2\sqrt{3}},\ 0\Big\},\ \Big\{0,\ 0,\ -\frac{1}{6}\,\text{T[x, x]} - \frac{\text{i}\,\text{T[x, x]}}{2\sqrt{3}} - \frac{1}{6}\,\text{T[y, y]} + \frac{\text{i}\,\text{T[y, y]}}{2\sqrt{3}} + \frac{1}{3}\,\text{T[z, z]}\Big\}\Big\}\Big\}$$

This is a nice non-zero result, and was not filtered out by the all-zero filter in **TraceProjectTij**. Now put in the axis permutation symmetry:

```
detPrjTEa /. xyzPermSymmTij
```

```
{Ea, {{0, 0, 0}, {0, 0, 0}, {0, 0, 0}}}
```

This is a big surprise. It appears that in group **T** the transition **A→Ea** is forbidden. All the published projections of two-photon or Raman tensors (including our own), done laboriously by hand forty or more years ago, failed to pick this up.

On your own
Do the same thing for species **Eb**. Then do **Ega** and **Egb** in groups $\mathbf{T_h}$, and do the **E** species in **O**, $\mathbf{O_h}$ and $\mathbf{T_d}$, and the **H** species of group **I**. You will find surprising vanishings in all of them. Species **H** of **I** is particularly weird, as only two of its five components vanish.

45.7 Alternative forms of the tensors

The tensor patterns that you get by the projection process depend on the representation that you use in detailed projection. For instance in group D3, species E, one person might say that the degenerate patterns are

$$\left\{\begin{array}{|c|c|c|}\hline D & 0 & 0\\\hline 0 & -D & F\\\hline 0 & G & 0\\\hline\end{array}\ ,\ \begin{array}{|c|c|c|}\hline 0 & H & J\\\hline H & 0 & 0\\\hline K & 0 & 0\\\hline\end{array}\right\}$$

while another insists that this is wrong and the true answer is

$$\left\{\begin{array}{|c|c|c|}\hline \alpha & i\,\alpha & \beta\\\hline i\,\alpha & -\alpha & -i\,\beta\\\hline \gamma & -i\,\gamma & 0\\\hline\end{array}\ ,\ \begin{array}{|c|c|c|}\hline \alpha^* & -i\,\alpha^* & \beta^*\\\hline -i\,\alpha^* & -\alpha^* & i\,\beta^*\\\hline \gamma^* & i\,\gamma^* & 0\\\hline\end{array}\right\}$$

But both would be correct. Remember, these are just amplitudes. It is just that the first person used a rep based on **{x,y}**, while the second used a rep based on **{x+iy,x-iy}**. When they calculate an intensity, each using his own tensor amplitude pattern, they get physically correct answers.

This is a basic property of quantum mechanics. Only intensities are observable; amplitudes are not observable and many different amplitudes can give the same intensity.

46. Three-photon tensor projections

Preliminaries

46.1 The operator for three-photon processes

46.1.1 The matrix element

This chapter is structured exactly like Chapter 45 (TwoPhotonProjections), so we will close some sections that are very similar the the corresponding section in Chapter 45. They may be read from CD if you want to see them explicitly.

One-photon processes were elucidated by first order perturbation theory; two-photon processes, by second order; now three-photon processes, by third order perturbation theory. These processes include simultaneous three-photon absorption and two-in, one-out up-conversion processes, like frequency summing, with or without conservation of photon energy. The amplitude $\mathcal{A}$ for a two-photon transition (from ground state ψ_γ to final state ψ_ϕ) is given by an analog of Eq. 45.1 :

$$\mathcal{A}_{\gamma,\phi} = \sum_{\alpha,\beta} \frac{\left\langle \psi_\gamma \middle| \underset{\rightharpoondown}{\lambda^{a}} . \underset{\rightharpoondown}{r} \middle| \psi_\alpha \right\rangle \left\langle \psi_\alpha \middle| \underset{\rightharpoondown}{r} . \underset{\rightharpoondown}{\lambda^{b}} \middle| \psi_\beta \right\rangle \left\langle \psi_\beta \middle| \underset{\rightharpoondown}{r} . \underset{\rightharpoondown}{\lambda^{c}} \middle| \psi_\phi \right\rangle}{\left(\omega_{\gamma,\alpha} - \omega_a\right) \quad \left(\omega_{\gamma,\beta} - \omega_b\right)} + \frac{\left\langle \psi_\gamma \middle| \underset{\rightharpoondown}{\lambda^{b}} . \underset{\rightharpoondown}{r} \middle| \psi_\alpha \right\rangle \left\langle \psi_\alpha \middle| \underset{\rightharpoondown}{r} . \underset{\rightharpoondown}{\lambda^{c}} \middle| \psi_\beta \right\rangle \left\langle \psi_\beta \middle| \underset{\rightharpoondown}{r} . \underset{\rightharpoondown}{\lambda^{a}} \middle| \psi_\phi \right\rangle}{\left(\omega_{\gamma,\alpha} - \omega_b\right) \quad \left(\omega_{\gamma,\beta} - \omega_c\right)} + \frac{\left\langle \psi_\gamma \middle| \underset{\rightharpoondown}{\lambda^{c}} . \underset{\rightharpoondown}{r} \middle| \psi_\alpha \right\rangle \left\langle \psi_\alpha \middle| \underset{\rightharpoondown}{r} . \underset{\rightharpoondown}{\lambda^{a}} \middle| \psi_\beta \right\rangle \left\langle \psi_\beta \middle| \underset{\rightharpoondown}{r} . \underset{\rightharpoondown}{\lambda^{b}} \middle| \psi_\phi \right\rangle}{\left(\omega_{\gamma,\alpha} - \omega_c\right) \quad \left(\omega_{\gamma,\beta} - \omega_a\right)} \tag{46.1}$$

where γ, α, β, and ϕ index states of the molecule. This matrix element is written for the simultaneous absorption of three photons. Usually γ is the ground state; ϕ is the final state, and α and β must run over all states of the molecule. In photon summing processes, γ and ϕ both stand for the ground state, so that the subject molecule makes only a transient excursion out of the ground state while the frequency sum generation process is going on.

The tensor appears when the polarization vectors $\underset{\rightharpoondown}{\lambda^{a}}$, $\underset{\rightharpoondown}{\lambda^{b}}$, and $\underset{\rightharpoondown}{\lambda^{c}}$ are factored out of the scalar expression above. Converting the $\underset{\rightharpoondown}{\lambda^{n}} . \underset{\rightharpoondown}{r}$ expressions to explicit

DOI 10.1007/b13137_46,

index notation, the formula becomes

$$\mathcal{A}_{\gamma,\phi} = \sum_{\alpha,\beta} \sum_{i,j,k} \lambda_i^a \lambda_j^b \lambda_k^g \; \frac{\langle\psi_\gamma \mid r_i \mid \psi_\alpha\rangle \langle\psi_\alpha \mid r_j \mid \psi_\beta\rangle \langle\psi_\beta \mid r_k \mid \psi_\phi\rangle}{(\omega_{\gamma,\alpha} - \omega_a)\;(\omega_{\gamma,\beta} - \omega_b)} + \frac{\langle\psi_\gamma \mid r_j \mid \psi_\alpha\rangle \langle\psi_\alpha \mid r_k \mid \psi_\beta\rangle \langle\psi_\beta \mid r_i \mid \psi_\phi\rangle}{(\omega_{\gamma,\alpha} - \omega_b)\;(\omega_{\gamma,\beta} - \omega_c)} + \frac{\langle\psi_\gamma \mid r_k \mid \psi_\alpha\rangle \langle\psi_\alpha \mid r_i \mid \psi_\beta\rangle \langle\psi_\beta \mid r_j \mid \psi_\phi\rangle}{(\omega_{\gamma,\alpha} - \omega_c)\;(\omega_{\gamma,\beta} - \omega_a)} \qquad (46.2)$$

where the tensor operator is the analog of Eq. 45.2 :

$$\mathbb{S}_{i,j,k}^{a,b,c} = \sum_{\alpha,\beta} \frac{\langle\psi_\gamma \mid r_i \mid \psi_\alpha\rangle \langle\psi_\alpha \mid r_j \mid \psi_\beta\rangle \langle\psi_\beta \mid r_k \mid \psi_\phi\rangle}{(\omega_{\gamma,\alpha} - \omega_a)\;(\omega_{\gamma,\beta} - \omega_b)} + \frac{\langle\psi_\gamma \mid r_j \mid \psi_\alpha\rangle \langle\psi_\alpha \mid r_k \mid \psi_\beta\rangle \langle\psi_\beta \mid r_i \mid \psi_\phi\rangle}{(\omega_{\gamma,\alpha} - \omega_b)\;(\omega_{\gamma,\beta} - \omega_c)} + \frac{\langle\psi_\gamma \mid r_k \mid \psi_\alpha\rangle \langle\psi_\alpha \mid r_i \mid \psi_\beta\rangle \langle\psi_\beta \mid r_j \mid \psi_\phi\rangle}{(\omega_{\gamma,\alpha} - \omega_c)\;(\omega_{\gamma,\beta} - \omega_a)} \qquad (46.2)$$

It is traditional to use a capital S (meaning **S**tate**S**um) for this quantity. As before, the double-struck $\mathbb{S}$ above is the operator and the Gothic $\mathfrak{S}$ below is the transition amplitude, the matrix element between ground (γ) and final (ϕ) states :

$$\mathfrak{S}_{i,j,k}^{a,b,c}[\gamma,\ \phi] = \left\langle \psi_\gamma \,\middle|\, \mathbb{S}_{i,j,k}^{a,b,c} \,\middle|\, \psi_\phi \right\rangle \qquad (46.3)$$

In the preliminaries, we declare a **Notation** that prevents the superscripts on $\mathfrak{S}$ from being interpreted as a strange kind of power. Therefore, we can use them as function arguments:

```
Clear[𝔖];
```

$$\mathfrak{S}_{i\_,j\_,k\_}^{a\_,b\_,c\_} := \left(\frac{\langle\psi_\gamma \mid r_i \mid \psi_\alpha\rangle \langle\psi_\alpha \mid r_j \mid \psi_\beta\rangle \langle\psi_\beta \mid r_k \mid \psi_\phi\rangle}{(\omega_{\gamma,\alpha} - \omega_a)\;(\omega_{\gamma,\beta} - \omega_b)} + \right.$$

$$\frac{\langle\psi_\gamma \mid r_j \mid \psi_\alpha\rangle\,\langle\psi_\alpha \mid r_k \mid \psi_\beta\rangle\,\langle\psi_\beta \mid r_i \mid \psi_\phi\rangle}{(\omega_{\gamma,\alpha}-\omega_b)\quad(\omega_{\gamma,\beta}-\omega_c)} + \left.\frac{\langle\psi_\gamma \mid r_k \mid \psi_\alpha\rangle\,\langle\psi_\alpha \mid r_i \mid \psi_\beta\rangle\,\langle\psi_\beta \mid r_j \mid \psi_\phi\rangle}{(\omega_{\gamma,\alpha}-\omega_c)\quad(\omega_{\gamma,\beta}-\omega_a)}\right) /.$$

$\{r_1 \to x,\ r_2 \to y,\ r_3 \to z\}$

Try the function out. Use numerical indices (**1,2,3**) for input; the output automatically contains the corresponding Cartesian variables (x, y, z):

$\mathfrak{S}^{a,b,c}_{1,1,2}$

$$\frac{\langle\psi_\alpha \mid x \mid \psi_\beta\rangle\,\langle\psi_\beta \mid x \mid \psi_\phi\rangle\,\langle\psi_\gamma \mid y \mid \psi_\alpha\rangle}{(-\omega_c+\omega_{\gamma,\alpha})\ (-\omega_a+\omega_{\gamma,\beta})} + \frac{\langle\psi_\alpha \mid x \mid \psi_\beta\rangle\,\langle\psi_\beta \mid y \mid \psi_\phi\rangle\,\langle\psi_\gamma \mid x \mid \psi_\alpha\rangle}{(-\omega_a+\omega_{\gamma,\alpha})\ (-\omega_b+\omega_{\gamma,\beta})} + \frac{\langle\psi_\alpha \mid y \mid \psi_\beta\rangle\,\langle\psi_\beta \mid x \mid \psi_\phi\rangle\,\langle\psi_\gamma \mid x \mid \psi_\alpha\rangle}{(-\omega_b+\omega_{\gamma,\alpha})\ (-\omega_c+\omega_{\gamma,\beta})}$$

It works as desired. We verify that index permutation is a symmetry of this formula, but only if the Cartesian indices **i,j,k** and the photon indices **a,b,c** are permuted together:

$\mathfrak{S}^{a,b,c}_{i,j,k} == \mathfrak{S}^{b,c,a}_{j,k,i} == \mathfrak{S}^{c,a,b}_{k,i,j}$

True

46.1.2 Proof that the matrix element is a tensor

Closed; please read from CD

46.2. The projection

Because of the similarity to 45.2.1 we give only a sketch. A concrete example of the object to be projected is

$\mathfrak{S}^{a,b,c}_{1,2,3}$

$$\frac{\langle\psi_\alpha \mid x \mid \psi_\beta\rangle\,\langle\psi_\beta \mid y \mid \psi_\phi\rangle\,\langle\psi_\gamma \mid z \mid \psi_\alpha\rangle}{(-\omega_c+\omega_{\gamma,\alpha})\ (-\omega_a+\omega_{\gamma,\beta})} +$$

$$\frac{\langle\psi_\alpha \mid y \mid \psi_\beta\rangle \langle\psi_\beta \mid z \mid \psi_\phi\rangle \langle\psi_\gamma \mid x \mid \psi_\alpha\rangle}{(-\omega_a + \omega_{\gamma,\alpha})\ (-\omega_b + \omega_{\gamma,\beta})} + \frac{\langle\psi_\alpha \mid z \mid \psi_\beta\rangle \langle\psi_\beta \mid x \mid \psi_\phi\rangle \langle\psi_\gamma \mid y \mid \psi_\alpha\rangle}{(-\omega_b + \omega_{\gamma,\alpha})\ (-\omega_c + \omega_{\gamma,\beta})}$$

Our operator **ProjectET** works on any expression that has **x**, **y**, and **z** in it, so it will work on this one. We project and look at just a fragment of the output:

```
test = ProjectET[𝔖^{a,b,c}_{1,2,3}, "C3"];
ByteCount[test]
```

73 416

It' s huge. Look at just a small fragment of it:

```
test[[2, 2, {1, 8}]]
```

$$\frac{\langle\psi_\alpha \mid y \mid \psi_\beta\rangle \langle\psi_\beta \mid z \mid \psi_\phi\rangle \langle\psi_\gamma \mid x \mid \psi_\alpha\rangle}{3\ (\omega_a\,\omega_b - \omega_b\,\omega_{\gamma,\alpha} - \omega_a\,\omega_{\gamma,\beta} + \omega_{\gamma,\alpha}\,\omega_{\gamma,\beta})} - \frac{\langle\psi_\alpha \mid z \mid \psi_\beta\rangle \left\langle\psi_\beta \left| -\frac{x}{2} - \frac{\sqrt{3}\,y}{2} \right| \psi_\phi\right\rangle \left\langle\psi_\gamma \left| \frac{\sqrt{3}\,x}{2} - \frac{y}{2} \right| \psi_\alpha\right\rangle}{6\ (\omega_b\,\omega_c - \omega_c\,\omega_{\gamma,\alpha} - \omega_b\,\omega_{\gamma,\beta} + \omega_{\gamma,\alpha}\,\omega_{\gamma,\beta})}$$

As expected, many of the Cartesian variables have been replaced by linear combinations of Cartesian variables, and the resulting expressions need to be expanded. We omit the step-by step expansion that we showed in the two-photon chapter. In the end, you have some linear combination of some or all of the 27 simple $\mathfrak{S}$ elements, $\mathfrak{S}^{a,b,c}_{1,1,1}$ to $\mathfrak{S}^{a,b,c}_{3,3,3}$.

46.3 Self-purifying tensor elements

Closed; please read from CD

46.4. Operators for tensor projection

46.4.1. Projection of the whole tensor

Now with this linearity definition active, observe how the projection of **T[x,x,x]** under group C_{3h} automatically simplifies itself :

```
ProjectET[T[x, y, z], "C3"];
% // Column
```

$\{A, \frac{1}{2}$ T[x, y, z] $- \frac{1}{2}$ T[y, x, z]$\}$

$\{$Ea,

$\frac{1}{4}$ i T[x, x, z] + $\frac{1}{4}$ T[x, y, z] + $\frac{1}{4}$ T[y, x, z] $- \frac{1}{4}$ i T[y, y, z]$\}$

$\{$Eb,

$-\frac{1}{4}$ i T[x, x, z] + $\frac{1}{4}$ T[x, y, z] + $\frac{1}{4}$ T[y, x, z] + $\frac{1}{4}$ i T[y, y, z]$\}$

When we did this with the authentic $\mathfrak{S}^{a,b,c}_{1,2,3}$ we got an expression 73,000 bytes long. The surrogate is doing its job.

This means we can easily do the whole tensor of **T**-functions at once. In the preliminaries we set up a tensor **Tijk** of the twenty-seven simple **T** functions:

```
Tijk
```

```
{{{T[x, x, x], T[x, x, y], T[x, x, z]},
  {T[x, y, x], T[x, y, y], T[x, y, z]},
  {T[x, z, x], T[x, z, y], T[x, z, z]}},
 {{T[y, x, x], T[y, x, y], T[y, x, z]},
  {T[y, y, x], T[y, y, y], T[y, y, z]},
  {T[y, z, x], T[y, z, y], T[y, z, z]}},
 {{T[z, x, x], T[z, x, y], T[z, x, z]},
  {T[z, y, x], T[z, y, y], T[z, y, z]},
  {T[z, z, x], T[z, z, y], T[z, z, z]}}}
```

so we just project the whole **Tijk** tensor.

We make it a little slicker by defining a specialized projection operator called **TraceProjectTijk**, consisting of a perfectly ordinary projection, plus a filter that gets rid of any all-zero projections.

46.4.2. The simplest example

As our first demo we do a very simple case, group **C2**, with only two species. It has only one transform, rotation about **z** by half a turn. But it gives a non-trivial trace projection:

```
trPrjC2 = TraceProjectTijk["C2"]
```

```
{C2, {{A, {{{0, 0, T[x, x, z]}, {0, 0, T[x, y, z]},
      {T[x, z, x], T[x, z, y], 0}}, {{0, 0, T[y, x, z]},
      {0, 0, T[y, y, z]}, {T[y, z, x], T[y, z, y], 0}},
     {{T[z, x, x], T[z, x, y], 0},
```

```
    {T[z, y, x], T[z, y, y], 0}, {0, 0, T[z, z, z]}}}},
{B, {{{T[x, x, x], T[x, x, y], 0},
    {T[x, y, x], T[x, y, y], 0}, {0, 0, T[x, z, z]}},
   {{T[y, x, x], T[y, x, y], 0}, {T[y, y, x], T[y, y, y], 0},
    {0, 0, T[y, z, z]}}, {{0, 0, T[z, x, z]},
    {0, 0, T[z, y, z]}, {T[z, z, x], T[z, z, y], 0}}}}}
```

This is easier to read with a little display help (explained below) :

```
C2display = MakeTijkDisplay[{trPrjC2},
   {FontFamily → "Arial", 8}] // Size[5]
```

C2 {A, B}

0	0	T[x, x, z]
0	0	T[x, y, z]
T[x, z, x]	T[x, z, y]	0

0	0	T[y, x, z]
0	0	T[y, y, z]
T[y, z, x]	T[y, z, y]	0

T[z, x, x]	T[z, x, y]	0
T[z, y, x]	T[z, y, y]	0
0	0	T[z, z, z]

T[x, x, x]	T[x, x, y]	0
T[x, y, x]	T[x, y, y]	0
0	0	T[x, z, z]

T[y, x, x]	T[y, x, y]	0
T[y, y, x]	T[y, y, y]	0
0	0	T[y, z, z]

0	0	T[z, x, z]
0	0	T[z, y, z]
T[z, z, x]	T[z, z, y]	0

Null

The species **A** projection is the top box; species **B**, the bottom. Now you can see at a glance that the two projections add up to the original. In a two-fold rotation about **z** both **x** and **y** change sign; **z** remains unchanged. Read carefully and you will see that those elements that are unchanged by the rotation are in species **A**; those that change sign are in species **B**. This is the basic idea of symmetry projection.

46.4.3. Trace projection, `TraceProjectTijk`

Do this again with $\mathbf{C_{3\,v}}$, which has a true 2D species. As always, the 2D species name is **E**, or something anchored on an **E**:

```
BoxedCharacterTable["C3v"]
```

C3v	1	2	3	← Class populations
	E	C3	σv	↓ Basis functions
A1	1	1	1	$\{(z),\ (x^2+y^2),\ (z^2)\}$
A2	1	1	-1	$\{(Iz)\}$
E	2	-1	0	$\left\{\begin{pmatrix}x\\y\end{pmatrix},\ \begin{pmatrix}\frac{1}{2}(x^2-y^2)\\xy\end{pmatrix},\ \begin{pmatrix}xz\\yz\end{pmatrix},\ \begin{pmatrix}Ix\\Iy\end{pmatrix}\right\}$

First do a trace projection. The display is so long that we suppress it, but interactive readers may remove the semicolon and see the whole thing.

```
tracePrjTijkC3v = TraceProjectTijk["C3v"];
% // ByteCount
```

```
17232
```

From its size alone, and more especially if you look at it, you will appreciate the utter impossibility of grasping its structure without symbolization. So, of course, an appropriate symbolizer has been written It works just like the one in the two-photon chapter:

```
entryC3v = SymbolizeProjectionTijk[tracePrjTijkC3v]
```

```
{C3v, {{A1, {{{A, 0, B}, {0, -A, 0}, {C, 0, 0}},
      {{0, -A, 0}, {-A, 0, B}, {0, C, 0}},
      {{D, 0, 0}, {0, D, 0}, {0, 0, F}}}},
   {A2, {{{0, G, 0}, {G, 0, H}, {0, J, 0}},
      {{G, 0, -H}, {0, -G, 0}, {-J, 0, 0}},
      {{0, K, 0}, {-K, 0, 0}, {0, 0, 0}}}},
   {E, {{{L, M, N}, {P, Q, R}, {S, U, V}},
      {{W, X, R}, {Y, Z, -N}, {U, -S, ZA}},
      {{ZB, ZC, ZD}, {ZC, -ZB, ZE}, {ZF, ZG, 0}}}}}}
```

We see only Latin capitals, so the trace projections have no explicit i in them.

We also have a 2D display for this kind of output. We show it off without discussing its innards:

```
MakeTijkDisplay[{entryC3v},
  {FontFamily → "Arial", 8}]
```

C3v {A1, A2, E}

A	0	B
0	–A	0
C	0	0

0	–A	0
–A	0	B
0	C	0

D	0	0
0	D	0
0	0	F

0	G	0
G	0	H
0	J	0

G	0	–H
0	–G	0
–J	0	0

0	K	0
–K	0	0
0	0	0

L	M	N
P	Q	R
S	U	V

W	X	R
Y	Z	–N
U	–S	ZA

ZB	ZC	ZD
ZC	–ZB	ZE
ZF	ZG	0

Each 27-element tensor is enclosed in its own box, with the group name and the species names in an introductory line.

If you want to know what these symbols stand for, you may call up a quantity named **Symbolizations** that was created while **SymbolizeTijk** was operating. Here it is in **Short** view:

```
Symbolizations // Short[#, 8] &
```

$$\left\{\left\{\frac{1}{4}\,T[x,x,x]-\frac{1}{4}\,T[x,y,y]-\frac{1}{4}\,T[y,x,y]-\frac{1}{4}\,T[y,y,x]\to A\right\},\ \left\{\frac{1}{2}\,T[x,x,z]+\frac{1}{2}\,T[y,y,z]\to B\right\},\ \left\{\frac{1}{2}\,T[x,z,x]+\frac{1}{2}\,T[y,z,y]\to C\right\},\ \ll 23\gg,\ \{T[z,y,z]\to ZE\},\ \{T[z,z,x]\to ZF\},\ \{T[z,z,y]\to ZG\}\right\}$$

Symbolizations is overwritten each time **SymbolizeTijk** runs, so it must be preserved immediately if you want to use it.

46.4.4. Detailed projection, `DetailedProjectTijk`

When a group has an **E** species, the trace projection gives only the sum of the two detailed projections. To see them individually, you must do a detailed projection. The operator for this has a lot to do. First it has to make a full representation of the group, and the basis it uses for this is an important point. It always uses the first basis listed in the character table, which in the case of species **E′** of $\mathbf{D_{3h}}$ is **{x,y}**. But complications are minimized if it transforms basis **{a,b}** into **{a+ⅈb,a-ⅈb}**. So in this case it uses **{x+ⅈy,x-ⅈy}**. The **rep** is quickly found by **MakeRepPoly**, and it is used by

ProjectED[expr,rep,component,"C3"]

with **component**s **1** and **2**. It always appends an **a** and a **b** to the **E**-type species name to form the names of the detailed projections that are found. Watch it work:

```
detailedPrjC3v =
  DetailedProjectTijk[tracePrjTijkC3v];
```

The display is too long; we suppress it. There is an **Ea** projection from component **1**, and an **Eb** from **2**. We want to look at the **Ea** projection by itself. So where is the **Ea** label? We find it by systematic winnowing:

```
detailedPrjC3v[[2, 3, 1, 1]]
```

```
Ea
```

So the three-index tensor itself should be part **[[2,3,1,2]]**.

```
EaRaw = detailedPrjC3v[[2, 3, 1, 2]];
% // Dimensions
```

```
{3, 3, 3}
```

Correct; it is a three-index Cartesian tensor. So how big is it?

```
ByteCount[EaRaw]
```

```
18 856
```

Interactive readers can look at it if they want to, but no human can see patterns in that much output. We have gone through all this just to show that symbolization is now quite essential.

46.4.5. Symbolization by SymbolizeProjectionTijk

Closed; please read from CD

46.4.6. The combined operator MakeRawEntryTijk

Closed; please read from CD

46.4.7. The operator TPTdisplay

The output of **SymbolizeProjectionTijk** is still a little hard to read for non-*Mathematica* people. So we have written an operator **TPTdisplay** that cleans them up like a typesetter might. Its input is a list of raw table entries and a style specification. Only fanatics will open the preliminaries to see how it works. Here is an example:

```
C3display = MakeTijkDisplay[
  {entryC3}, {FontFamily → "Arial", 8}]
```

C3 **{A, Ea, Eb}**

A	B	C	B	−A	−D	H	J	0
B	−A	D	−A	−B	C	−J	H	0
F	G	0	−G	F	0	0	0	K

α	β	γ	$-i\varepsilon$	$i\delta$	$i\gamma$	θ	$i\theta$	ι
δ	ε	$i\gamma$	$i\beta$	$-i\alpha$	$-\gamma$	$i\theta$	$-\theta$	$-i\iota$
ζ	$i\zeta$	η	$i\zeta$	$-\zeta$	$-i\eta$	κ	$-i\kappa$	0

α^*	β^*	γ^*	$\mathbb{i}\,\varepsilon^*$	$-\mathbb{i}\,\delta^*$	$-\mathbb{i}\,\gamma^*$	θ^*	$-\mathbb{i}\,\theta^*$	ι^*
δ^*	ε^*	$-\mathbb{i}\,\gamma^*$	$-\mathbb{i}\,\beta^*$	$\mathbb{i}\,\alpha^*$	$-\gamma^*$	$-\mathbb{i}\,\theta^*$	$-\theta^*$	$\mathbb{i}\,\iota^*$
ζ^*	$-\mathbb{i}\,\zeta^*$	η^*	$-\mathbb{i}\,\zeta^*$	$-\zeta^*$	$\mathbb{i}\,\eta^*$	κ^*	$\mathbb{i}\,\kappa^*$	0

```
Column[Null]
```

The name of the group is shown in **boldface**, followed by a list of species that have nonzero tensor projections; then on the next line are the respective patterns, in grids that help the eye. Degenerate patterns are bracketed together. All-zero patterns are omitted.

The patterns of several groups are identical. So the display generator also searches the raw entries input list for cases that have identical patterns, and groups them together, printing the patterns only once.

46.5 Three-photon tensor projection table

46.5.1. Make all non-zero table entries (no styling)

We could do all our tabulated groups in one fell swoop, but group **I** takes so long that we separate it out.

```
GroupCatalogNoI = Complement[GroupCatalog, {"I"}]
```

```
{C1, C2, C2h, C2v, C3, C3h, C3v, C4, C4h, C4v, C5, C6, C6h,
 C6v, Ch, Ci, D2, D2d, D2h, D3, D3d, D3h, D4, D4d, D4h,
 D5, D5d, D5h, D6, D6d, D6h, O, Oh, S4, S6, S8, T, Td, Th}
```

Apply **MakeRawEntryTijk** to this long list of group names. In the beginning you will see a lot of flashing. It is trying to display the name of the group it is working on, but the small, simple groups flash by too quickly to read. Only the larger groups take a perceptible time; in particular, groups with five-fold axes are quite slow (be patient; they will finish). All this is done in infinite precision.

```
allRaw = MakeRawEntryTijk[GroupCatalogNoI];
```

Look at just one of the results :

```
prjD4h = allRaw[[25]]
```

```
{D4h, {{A1u, {{{0, 0, 0}, {0, 0, 0}, {0, 0, 0}},
     {{0, 0, 0}, {0, 0, 0}, {0, 0, 0}},
     {{0, 0, 0}, {0, 0, 0}, {0, 0, 0}}}},
   {A2u, {{{0, 0, A}, {0, 0, 0}, {B, 0, 0}},
     {{0, 0, 0}, {0, 0, A}, {0, B, 0}},
     {{C, 0, 0}, {0, C, 0}, {0, 0, D}}}},
   {B1u, {{{0, 0, 0}, {0, 0, F}, {0, G, 0}},
     {{0, 0, F}, {0, 0, 0}, {G, 0, 0}},
     {{0, H, 0}, {H, 0, 0}, {0, 0, 0}}}},
   {B2u, {{{0, 0, 0}, {0, 0, 0}, {0, 0, 0}},
     {{0, 0, 0}, {0, 0, 0}, {0, 0, 0}},
     {{0, 0, 0}, {0, 0, 0}, {0, 0, 0}}}},
   {{Eua, {{{α, β, 0}, {γ, δ, 0}, {0, 0, ε}},
      {{δ*, γ*, 0}, {β*, α*, 0}, {0, 0, ε*}},
```

```
      {{0, 0, ζ}, {0, 0, ζ*}, {η, η*, 0}}}},
    {Eub, {{{α*, β*, 0}, {γ*, δ*, 0}, {0, 0, ε*}},
      {{δ, γ, 0}, {β, α, 0}, {0, 0, ε}},
      {{0, 0, ζ*}, {0, 0, ζ}, {η*, η, 0}}}}}}}
```

This is the information that will be tabulated for $\mathbf{D_{4\,h}}$, but in a more readable, typeset format, provided by the operator **MakeTijkDisplay**.

MakeTijkDisplay[{prjD4h}, {FontFamily → "Arial", 8}]

D4h **{A1u, A2u, B1u, B2u, {Eua, Eub}}**

$$\begin{pmatrix}0&0&0\\0&0&0\\0&0&0\end{pmatrix}\begin{pmatrix}0&0&0\\0&0&0\\0&0&0\end{pmatrix}\begin{pmatrix}0&0&0\\0&0&0\\0&0&0\end{pmatrix}\qquad\begin{pmatrix}0&0&A\\0&0&0\\B&0&0\end{pmatrix}\begin{pmatrix}0&0&0\\0&0&A\\0&B&0\end{pmatrix}\begin{pmatrix}C&0&0\\0&C&0\\0&0&D\end{pmatrix}$$

$$\begin{pmatrix}0&0&0\\0&0&F\\0&G&0\end{pmatrix}\begin{pmatrix}0&0&F\\0&0&0\\G&0&0\end{pmatrix}\begin{pmatrix}0&H&0\\H&0&0\\0&0&0\end{pmatrix}\qquad\begin{pmatrix}0&0&0\\0&0&0\\0&0&0\end{pmatrix}\begin{pmatrix}0&0&0\\0&0&0\\0&0&0\end{pmatrix}\begin{pmatrix}0&0&0\\0&0&0\\0&0&0\end{pmatrix}$$

$$\left\{\begin{pmatrix}\alpha&\beta&0\\\gamma&\delta&0\\0&0&\varepsilon\end{pmatrix}\begin{pmatrix}\delta^*&\gamma^*&0\\\beta^*&\alpha^*&0\\0&0&\varepsilon^*\end{pmatrix}\begin{pmatrix}0&0&\zeta\\0&0&\zeta^*\\\eta&\eta^*&0\end{pmatrix},\right.$$

$$\left.\begin{pmatrix}\alpha^*&\beta^*&0\\\gamma^*&\delta^*&0\\0&0&\varepsilon^*\end{pmatrix}\begin{pmatrix}\delta&\gamma&0\\\beta&\alpha&0\\0&0&\varepsilon\end{pmatrix}\begin{pmatrix}0&0&\zeta^*\\0&0&\zeta\\\eta^*&\eta&0\end{pmatrix}\right\}$$

46.5.2. Table of three-photon tensor patterns

All that is left is the display operator **MakeTijkDisplay**. This where the axial permutation rules are applied, so if you see an all-zero tensor, axial symmetry caused it. More normal all zero projections have been filtered out.

```
MakeTijkDisplay[allRaw, {FontFamily → "Arial", 8}];
```

C1 {A}

Ci {Au}

A	B	C	L	M	N	W	X	Y
D	F	G	P	Q	R	Z	ZA	ZB
H	J	K	S	U	V	ZC	ZD	ZE

C2 {A, B}

C2h {Au, Bu}

0	0	A	0	0	F	K	L	0
0	0	B	0	0	G	M	N	0
C	D	0	H	J	0	0	0	P

Q	R	0	W	X	0	0	0	ZB
S	U	0	Y	Z	0	0	0	ZC
0	0	V	0	0	ZA	ZD	ZE	0

C2v {A1 , A2, B1, B2}

0	0	A	0	0	0	F	0	0
0	0	0	0	0	C	0	G	0
B	0	0	0	D	0	0	0	H

0	0	0	0	0	L	0	N	0
0	0	J	0	0	0	P	0	0
0	K	0	M	0	0	0	0	0

Q	0	0	0	U	0	0	0	W
0	R	0	V	0	0	0	0	0
0	0	S	0	0	0	X	0	0

0	Y	0	ZA	0	0	0	0	0
Z	0	0	0	ZB	0	0	0	ZD
0	0	0	0	0	ZC	0	ZE	0

C3 {A, Ea, Eb}

S6 {Au, Eua, Eub}

A	B	C	B	–A	–D	H	J	0
B	–A	D	–A	–B	C	–J	H	0
F	G	0	–G	F	0	0	0	K

α	β	γ	$-i\varepsilon$	$i\delta$	$i\gamma$	θ	$i\theta$	ι
δ	ε	$i\gamma$	$i\beta$	$-i\alpha$	$-\gamma$	$i\theta$	$-\theta$	$-i\iota$
ζ	$i\zeta$	η	$i\zeta$	$-\zeta$	$-i\eta$	κ	$-i\kappa$	0

α^*	β^*	γ^*	$\mathbb{i}\ \varepsilon^*$	$-\mathbb{i}\ \delta^*$	$-\mathbb{i}\ \gamma^*$	θ^*	$-\mathbb{i}\ \theta^*$	ι^*
δ^*	ε^*	$-\mathbb{i}\ \gamma^*$	$-\mathbb{i}\ \beta^*$	$\mathbb{i}\ \alpha^*$	$-\gamma^*$	$-\mathbb{i}\ \theta^*$	$-\theta^*$	$\mathbb{i}\ \iota^*$
ζ^*	$-\mathbb{i}\ \zeta^*$	η^*	$-\mathbb{i}\ \zeta^*$	$-\zeta^*$	$\mathbb{i}\ \eta^*$	κ^*	$\mathbb{i}\ \kappa^*$	0

C3h $\{A', E'_a, E'_b, A'', E''_a, E''_b\}$

A	B	0	B	−A	0	0	0	0
B	−A	0	−A	−B	0	0	0	0
0	0	0	0	0	0	0	0	0

α	β	0	$-i\delta$	$i\gamma$	0	0	0	ζ
γ	δ	0	$i\beta$	$-i\alpha$	0	0	0	$-i\zeta$
0	0	ε	0	0	$-i\varepsilon$	η	$-i\eta$	0

α^*	β^*	0	$\mathbb{i}\,\delta^*$	$-\mathbb{i}\,\gamma^*$	0	0	0	ζ^*
γ^*	δ^*	0	$-\mathbb{i}\,\beta^*$	$\mathbb{i}\,\alpha^*$	0	0	0	$\mathbb{i}\,\zeta^*$
0	0	ε^*	0	0	$\mathbb{i}\,\varepsilon^*$	η^*	$\mathbb{i}\,\eta^*$	0

0	0	C	0	0	−D	H	J	0
0	0	D	0	0	C	−J	H	0
F	G	0	−G	F	0	0	0	K

0	0	θ	0	0	$i\theta$	κ	$i\kappa$	0
0	0	$i\theta$	0	0	$-\theta$	$i\kappa$	$-\kappa$	0
ι	$i\iota$	0	$i\iota$	$-\iota$	0	0	0	0

0	0	θ^*	0	0	$-\mathbb{i}\,\theta^*$	κ^*	$-\mathbb{i}\,\kappa^*$	0
0	0	$-\mathbb{i}\,\theta^*$	0	0	$-\theta^*$	$-\mathbb{i}\,\kappa^*$	$-\kappa^*$	0
ι^*	$-\mathbb{i}\,\iota^*$	0	$-\mathbb{i}\,\iota^*$	$-\iota^*$	0	0	0	0

C3v {A1, A2, {Ea, Eb}}

A	0	B	0	–A	0	D	0	0
0	–A	0	–A	0	B	0	D	0
C	0	0	0	C	0	0	0	F

0	G	0	G	0	–H	0	K	0
G	0	H	0	–G	0	–K	0	0
0	J	0	–J	0	0	0	0	0

{

α	β	γ	$-i\varepsilon$	$i\delta$	$i\gamma$	θ	$i\theta$	ι
δ	ε	$i\gamma$	$i\beta$	$-i\alpha$	$-\gamma$	$i\theta$	$-\theta$	$-i\iota$
ζ	$i\zeta$	η	$i\zeta$	$-\zeta$	$-i\eta$	κ	$-i\kappa$	0

,

α^*	β^*	γ^*	$\mathbb{i}\,\varepsilon^*$	$-\mathbb{i}\,\delta^*$	$-\mathbb{i}\,\gamma^*$	θ^*	$-\mathbb{i}\,\theta^*$	ι^*
δ^*	ε^*	$-\mathbb{i}\,\gamma^*$	$-\mathbb{i}\,\beta^*$	$\mathbb{i}\,\alpha^*$	$-\gamma^*$	$-\mathbb{i}\,\theta^*$	$-\theta^*$	$\mathbb{i}\,\iota^*$
ζ^*	$-\mathbb{i}\,\zeta^*$	η^*	$-\mathbb{i}\,\zeta^*$	$-\zeta^*$	$\mathbb{i}\,\eta^*$	κ^*	$\mathbb{i}\,\kappa^*$	0

}

C4 **{A, B, Ea, Eb}**

C4h **{Au, Bu, Eua, Eub}**

0	0	A	0	0	–B	F	G	0
0	0	B	0	0	A	–G	F	0
C	D	0	–D	C	0	0	0	H

0	0	J	0	0	K	N	P	0
0	0	K	0	0	–J	P	–N	0
L	M	0	M	–L	0	0	0	0

α	β	0	$i\delta$	$-i\gamma$	0	0	0	ζ
γ	δ	0	$-i\beta$	$i\alpha$	0	0	0	$i\zeta$
0	0	ε	0	0	$i\varepsilon$	η	$i\eta$	0

α^*	β^*	0	$-\mathbb{i}\ \delta^*$	$\mathbb{i}\ \gamma^*$	0	0	0	ζ^*
γ^*	δ^*	0	$\mathbb{i}\ \beta^*$	$-\mathbb{i}\ \alpha^*$	0	0	0	$-\mathbb{i}\ \zeta^*$
0	0	ε^*	0	0	$-\mathbb{i}\ \varepsilon^*$	η^*	$-\mathbb{i}\ \eta^*$	0

C4v {A1, A2, B1, B2, {Ea, Eb}}

$$\left[\begin{array}{ccc}0&0&A\\0&0&0\\B&0&0\end{array}\right]\left[\begin{array}{ccc}0&0&0\\0&0&A\\0&B&0\end{array}\right]\left[\begin{array}{ccc}C&0&0\\0&C&0\\0&0&D\end{array}\right]\quad\left[\begin{array}{ccc}0&0&0\\0&0&0\\0&0&0\end{array}\right]\left[\begin{array}{ccc}0&0&0\\0&0&0\\0&0&0\end{array}\right]\left[\begin{array}{ccc}0&0&0\\0&0&0\\0&0&0\end{array}\right]$$

$$\left[\begin{array}{ccc}0&0&0\\0&0&0\\0&0&0\end{array}\right]\left[\begin{array}{ccc}0&0&0\\0&0&0\\0&0&0\end{array}\right]\left[\begin{array}{ccc}0&0&0\\0&0&0\\0&0&0\end{array}\right]\quad\left[\begin{array}{ccc}0&0&0\\0&0&F\\0&G&0\end{array}\right]\left[\begin{array}{ccc}0&0&F\\0&0&0\\G&0&0\end{array}\right]\left[\begin{array}{ccc}0&H&0\\H&0&0\\0&0&0\end{array}\right]$$

$$\left\{\left[\begin{array}{ccc}\alpha&\beta&0\\\gamma&\delta&0\\0&0&\varepsilon\end{array}\right]\left[\begin{array}{ccc}\delta^*&\gamma^*&0\\\beta^*&\alpha^*&0\\0&0&\varepsilon^*\end{array}\right]\left[\begin{array}{ccc}0&0&\zeta\\0&0&\zeta^*\\\eta&\eta^*&0\end{array}\right],\right.$$

$$\left.\left[\begin{array}{ccc}\alpha^*&\beta^*&0\\\gamma^*&\delta^*&0\\0&0&\varepsilon^*\end{array}\right]\left[\begin{array}{ccc}\delta&\gamma&0\\\beta&\alpha&0\\0&0&\varepsilon\end{array}\right]\left[\begin{array}{ccc}0&0&\zeta^*\\0&0&\zeta\\\eta^*&\eta&0\end{array}\right]\right\}$$

C5 {A, E1a, E1b, E2a, E2b}

0	0	A	0	0	−B	F	G	0
0	0	B	0	0	A	−G	F	0
C	D	0	−D	C	0	0	0	H

α	β	0	$i\delta$	$-i\gamma$	0	0	0	ζ
γ	δ	0	$-i\beta$	$i\alpha$	0	0	0	$i\zeta$
0	0	ε	0	0	$i\varepsilon$	η	$i\eta$	0

α^*	β^*	0	$-\mathbb{i}\ \delta^*$	$\mathbb{i}\ \gamma^*$	0	0	0	ζ^*
γ^*	δ^*	0	$\mathbb{i}\ \beta^*$	$-\mathbb{i}\ \alpha^*$	0	0	0	$-\mathbb{i}\ \zeta^*$
0	0	ε^*	0	0	$-\mathbb{i}\ \varepsilon^*$	η^*	$-\mathbb{i}\ \eta^*$	0

θ	$-i\theta$	ι	$-i\theta$	$-\theta$	$i\iota$	λ	$i\lambda$	0
$-i\theta$	$-\theta$	$i\iota$	$-\theta$	$i\theta$	$-\iota$	$i\lambda$	$-\lambda$	0
κ	$i\kappa$	0	$i\kappa$	$-\kappa$	0	0	0	0

θ^*	$\mathbb{i}\ \theta^*$	ι^*	$\mathbb{i}\ \theta^*$	$-\theta^*$	$-\mathbb{i}\ \iota^*$	λ^*	$-\mathbb{i}\ \lambda^*$	0
$\mathbb{i}\ \theta^*$	$-\theta^*$	$-\mathbb{i}\ \iota^*$	$-\theta^*$	$-\mathbb{i}\ \theta^*$	$-\iota^*$	$-\mathbb{i}\ \lambda^*$	$-\lambda^*$	0
κ^*	$-\mathbb{i}\ \kappa^*$	0	$-\mathbb{i}\ \kappa^*$	$-\kappa^*$	0	0	0	0

C6 **{A, B, E1a, E1b, E2a, E2b}**

C6h **{Au, Bu, E1ua, E1ub, E2ua, E2ub}**

0	0	A		0	0	−B		F	G	0
0	0	B		0	0	A		−G	F	0
C	D	0		−D	C	0		0	0	H

J	K	0		K	−J	0		0	0	0
K	−J	0		−J	−K	0		0	0	0
0	0	0		0	0	0		0	0	0

α	β	0		$i\delta$	$-i\gamma$	0		0	0	ζ
γ	δ	0		$-i\beta$	$i\alpha$	0		0	0	$i\zeta$
0	0	ε		0	0	$i\varepsilon$		η	$i\eta$	0

α^*	β^*	0		$-\mathbb{i}\,\delta^*$	$\mathbb{i}\,\gamma^*$	0		0	0	ζ^*
γ^*	δ^*	0		$\mathbb{i}\,\beta^*$	$-\mathbb{i}\,\alpha^*$	0		0	0	$-\mathbb{i}\,\zeta^*$
0	0	ε^*		0	0	$-\mathbb{i}\,\varepsilon^*$		η^*	$-\mathbb{i}\,\eta^*$	0

0	0	θ		0	0	$i\theta$		κ	$i\kappa$	0
0	0	$i\theta$		0	0	$-\theta$		$i\kappa$	$-\kappa$	0
ι	$i\iota$	0		$i\iota$	$-\iota$	0		0	0	0

0	0	θ^*		0	0	$-\mathbb{i}\,\theta^*$		κ^*	$-\mathbb{i}\,\kappa^*$	0
0	0	$-\mathbb{i}\,\theta^*$		0	0	$-\theta^*$		$-\mathbb{i}\,\kappa^*$	$-\kappa^*$	0
ι^*	$-\mathbb{i}\,\iota^*$	0		$-\mathbb{i}\,\iota^*$	$-\iota^*$	0		0	0	0

C6v **{A1, A2, B1, B2, {E1a, E1b}, {E2a, E2b}}**

0	0	A		0	0	0		C	0	0
0	0	0		0	0	A		0	C	0
B	0	0		0	B	0		0	0	D

0	0	0		0	0	−F		0	H	0
0	0	F		0	0	0		−H	0	0
0	G	0		−G	0	0		0	0	0

0	J	0		J	0	0		0	0	0
J	0	0		0	−J	0		0	0	0
0	0	0		0	0	0		0	0	0

K	0	0		0	−K	0		0	0	0
0	−K	0		−K	0	0		0	0	0
0	0	0		0	0	0		0	0	0

$\{$

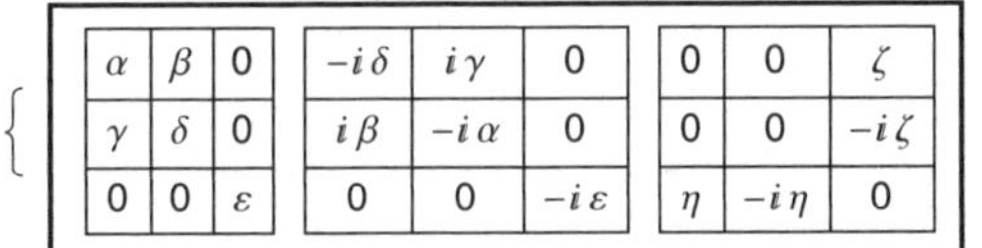

α	β	0		$-i\delta$	$i\gamma$	0		0	0	ζ
γ	δ	0		$i\beta$	$-i\alpha$	0		0	0	$-i\zeta$
0	0	ε		0	0	$-i\varepsilon$		η	$-i\eta$	0

,

α^*	β^*	0		$\mathbb{i}\,\delta^*$	$-\mathbb{i}\,\gamma^*$	0		0	0	ζ^*
γ^*	δ^*	0		$-\mathbb{i}\,\beta^*$	$\mathbb{i}\,\alpha^*$	0		0	0	$\mathbb{i}\,\zeta^*$
0	0	ε^*		0	0	$\mathbb{i}\,\varepsilon^*$		η^*	$\mathbb{i}\,\eta^*$	0

$\}$

$\{$

0	0	θ		0	0	$-i\theta$		κ	$-i\kappa$	0
0	0	$-i\theta$		0	0	$-\theta$		$-i\kappa$	$-\kappa$	0
ι	$-i\iota$	0		$-i\iota$	$-\iota$	0		0	0	0

,

0	0	θ^*		0	0	$\mathbb{i}\,\theta^*$		κ^*	$\mathbb{i}\,\kappa^*$	0
0	0	$\mathbb{i}\,\theta^*$		0	0	$-\theta^*$		$\mathbb{i}\,\kappa^*$	$-\kappa^*$	0
ι^*	$\mathbb{i}\,\iota^*$	0		$\mathbb{i}\,\iota^*$	$-\iota^*$	0		0	0	0

$\}$

Ch {A, B}

A	B	0		G	H	0		0	0	M
C	D	0		J	K	0		0	0	N
0	0	F		0	0	L		P	Q	0

0	0	R		0	0	W		ZA	ZB	0
0	0	S		0	0	X		ZC	ZD	0
U	V	0		Y	Z	0		0	0	ZE

D2 {A, B1, B2, B3}

D2h {Au, B1u, B2u, B3u}

0	0	0		0	0	C		0	F	0
0	0	A		0	0	0		G	0	0
0	B	0		D	0	0		0	0	0

0	0	H		0	0	0		M	0	0
0	0	0		0	0	K		0	N	0
J	0	0		0	L	0		0	0	P

0	Q	0		S	0	0		0	0	0
R	0	0		0	U	0		0	0	W
0	0	0		0	0	V		0	X	0

Y	0	0		0	ZB	0		0	0	ZD
0	Z	0		ZC	0	0		0	0	0
0	0	ZA		0	0	0		ZE	0	0

D2d {A_1, A_2, B_1, B_2, {Ea, Eb}}

0	0	0	0	0	A	0	C	0
0	0	A	0	0	0	C	0	0
0	B	0	B	0	0	0	0	0

0	0	0	0	0	0	0	0	0
0	0	0	0	0	0	0	0	0
0	0	0	0	0	0	0	0	0

0	0	0	0	0	0	0	0	0
0	0	0	0	0	0	0	0	0
0	0	0	0	0	0	0	0	0

0	0	D	0	0	0	G	0	0
0	0	0	0	0	D	0	G	0
F	0	0	0	F	0	0	0	H

{

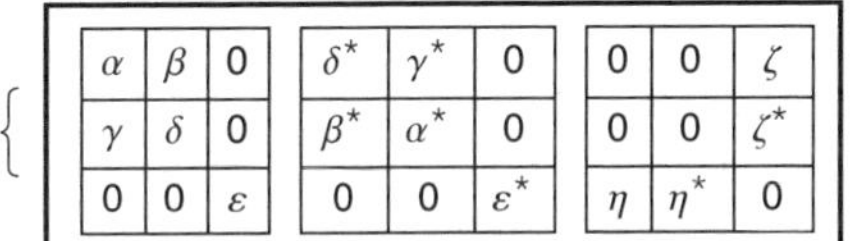

α	β	0	δ^*	γ^*	0	0	0	ζ
γ	δ	0	β^*	α^*	0	0	0	ζ^*
0	0	ε	0	0	ε^*	η	η^*	0

,

α^*	β^*	0	δ	γ	0	0	0	ζ^*
γ^*	δ^*	0	β	α	0	0	0	ζ
0	0	ε^*	0	0	ε	η^*	η	0

}

D3 {A1, A2, {Ea, Eb}}

D3d {A1u, A2u, {Eua, Eub}}

A	0	0	0	–A	–B	0	D	0
0	–A	B	–A	0	0	–D	0	0
0	C	0	–C	0	0	0	0	0

0	F	G	F	0	0	J	0	0
F	0	0	0	–F	G	0	J	0
H	0	0	0	H	0	0	0	K

{

α	β	γ	$-i\varepsilon$	$i\delta$	$i\gamma$	θ	$i\theta$	ι
δ	ε	$i\gamma$	$i\beta$	$-i\alpha$	$-\gamma$	$i\theta$	$-\theta$	$-i\iota$
ζ	$i\zeta$	η	$i\zeta$	$-\zeta$	$-i\eta$	κ	$-i\kappa$	0

,

α^*	β^*	γ^*	$\mathbb{i}\,\varepsilon^*$	$-\mathbb{i}\,\delta^*$	$-\mathbb{i}\,\gamma^*$	θ^*	$-\mathbb{i}\,\theta^*$	ι^*
δ^*	ε^*	$-\mathbb{i}\,\gamma^*$	$-\mathbb{i}\,\beta^*$	$\mathbb{i}\,\alpha^*$	$-\gamma^*$	$-\mathbb{i}\,\theta^*$	$-\theta^*$	$\mathbb{i}\,\iota^*$
ζ^*	$-\mathbb{i}\,\zeta^*$	η^*	$-\mathbb{i}\,\zeta^*$	$-\zeta^*$	$\mathbb{i}\,\eta^*$	κ^*	$\mathbb{i}\,\kappa^*$	0

}

D3h {A_1', A_2', {E′a, E′b}, A_1'', A_2'', {E″a, E″b}}

A	0	0	0	–A	0	0	0	0
0	–A	0	–A	0	0	0	0	0
0	0	0	0	0	0	0	0	0

0	B	0	B	0	0	0	0	0
B	0	0	0	–B	0	0	0	0
0	0	0	0	0	0	0	0	0

{

α	β	0	$-i\delta$	$i\gamma$	0	0	0	ζ
γ	δ	0	$i\beta$	$-i\alpha$	0	0	0	$-i\zeta$
0	0	ε	0	0	$-i\varepsilon$	η	$-i\eta$	0

,

$$\left(\begin{array}{ccc}\alpha^* & \beta^* & 0\\ \gamma^* & \delta^* & 0\\ 0 & 0 & \varepsilon^*\end{array}\right)\quad\left(\begin{array}{ccc}\mathbb{i}\,\delta^* & -\mathbb{i}\,\gamma^* & 0\\ -\mathbb{i}\,\beta^* & \mathbb{i}\,\alpha^* & 0\\ 0 & 0 & \mathbb{i}\,\varepsilon^*\end{array}\right)\quad\left(\begin{array}{ccc}0 & 0 & \zeta^*\\ 0 & 0 & \mathbb{i}\,\zeta^*\\ \eta^* & \mathbb{i}\,\eta^* & 0\end{array}\right)\Bigg\}$$

$$\left(\begin{array}{ccc}0 & 0 & 0\\ 0 & 0 & C\\ 0 & D & 0\end{array}\right)\quad\left(\begin{array}{ccc}0 & 0 & -C\\ 0 & 0 & 0\\ -D & 0 & 0\end{array}\right)\quad\left(\begin{array}{ccc}0 & F & 0\\ -F & 0 & 0\\ 0 & 0 & 0\end{array}\right)$$

$$\left(\begin{array}{ccc}0 & 0 & G\\ 0 & 0 & 0\\ H & 0 & 0\end{array}\right)\quad\left(\begin{array}{ccc}0 & 0 & 0\\ 0 & 0 & G\\ 0 & H & 0\end{array}\right)\quad\left(\begin{array}{ccc}J & 0 & 0\\ 0 & J & 0\\ 0 & 0 & K\end{array}\right)$$

$$\Bigg\{\left(\begin{array}{ccc}0 & 0 & \theta\\ 0 & 0 & i\theta\\ \iota & i\iota & 0\end{array}\right)\quad\left(\begin{array}{ccc}0 & 0 & i\theta\\ 0 & 0 & -\theta\\ i\iota & -\iota & 0\end{array}\right)\quad\left(\begin{array}{ccc}\kappa & i\kappa & 0\\ i\kappa & -\kappa & 0\\ 0 & 0 & 0\end{array}\right),$$

$$\left(\begin{array}{ccc}0 & 0 & \theta^*\\ 0 & 0 & -\mathbb{i}\,\theta^*\\ \iota^* & -\mathbb{i}\,\iota^* & 0\end{array}\right)\quad\left(\begin{array}{ccc}0 & 0 & -\mathbb{i}\,\theta^*\\ 0 & 0 & -\theta^*\\ -\mathbb{i}\,\iota^* & -\iota^* & 0\end{array}\right)\quad\left(\begin{array}{ccc}\kappa^* & -\mathbb{i}\,\kappa^* & 0\\ -\mathbb{i}\,\kappa^* & -\kappa^* & 0\\ 0 & 0 & 0\end{array}\right)\Bigg\}$$

D4 **{A1, A2, B1, B2, {Ea, Eb}}**

0	0	0		0	0	–A		0	C	0
0	0	A		0	0	0		–C	0	0
0	B	0		–B	0	0		0	0	0

0	0	D		0	0	0		G	0	0
0	0	0		0	0	D		0	G	0
F	0	0		0	F	0		0	0	H

0	0	0		0	0	J		0	L	0
0	0	J		0	0	0		L	0	0
0	K	0		K	0	0		0	0	0

0	0	M		0	0	0		P	0	0
0	0	0		0	0	–M		0	–P	0
N	0	0		0	–N	0		0	0	0

{

α	β	0		$-i\delta$	$i\gamma$	0		0	0	ζ
γ	δ	0		$i\beta$	$-i\alpha$	0		0	0	$-i\zeta$
0	0	ε		0	0	$-i\varepsilon$		η	$-i\eta$	0

,

α^*	β^*	0		$i\,\delta^*$	$-i\,\gamma^*$	0		0	0	ζ^*
γ^*	δ^*	0		$-i\,\beta^*$	$i\,\alpha^*$	0		0	0	$i\,\zeta^*$
0	0	ε^*		0	0	$i\,\varepsilon^*$		η^*	$i\,\eta^*$	0

}

""

D4d **{B1, B2, {E1a, E1b}, {E2a, E2b}, {E3a, E3b}}**

0	0	0		0	0	–A		0	C	0
0	0	A		0	0	0		–C	0	0
0	B	0		–B	0	0		0	0	0

0	0	D		0	0	0		G	0	0
0	0	0		0	0	D		0	G	0
F	0	0		0	F	0		0	0	H

$$\left\{ \boxed{\begin{array}{|c|c|c|}\hline \alpha & \beta & 0 \\ \hline \gamma & \delta & 0 \\ \hline 0 & 0 & \varepsilon \\ \hline\end{array} \quad \begin{array}{|c|c|c|}\hline -i\delta & i\gamma & 0 \\ \hline i\beta & -i\alpha & 0 \\ \hline 0 & 0 & -i\varepsilon \\ \hline\end{array} \quad \begin{array}{|c|c|c|}\hline 0 & 0 & \zeta \\ \hline 0 & 0 & -i\zeta \\ \hline \eta & -i\eta & 0 \\ \hline\end{array}} \right. ,$$

$$\left. \boxed{\begin{array}{|c|c|c|}\hline \alpha^{\star} & \beta^{\star} & 0 \\ \hline \gamma^{\star} & \delta^{\star} & 0 \\ \hline 0 & 0 & \varepsilon^{\star} \\ \hline\end{array} \quad \begin{array}{|c|c|c|}\hline \mathbb{i}\ \delta^{*} & -\mathbb{i}\ \gamma^{*} & 0 \\ \hline -\mathbb{i}\ \beta^{*} & \mathbb{i}\ \alpha^{*} & 0 \\ \hline 0 & 0 & \mathbb{i}\ \varepsilon^{*} \\ \hline\end{array} \quad \begin{array}{|c|c|c|}\hline 0 & 0 & \zeta^{\star} \\ \hline 0 & 0 & \mathbb{i}\ \zeta^{*} \\ \hline \eta^{\star} & \mathbb{i}\ \eta^{*} & 0 \\ \hline\end{array}} \right\}$$

$$\left\{ \boxed{\begin{array}{|c|c|c|}\hline 0 & 0 & \theta \\ \hline 0 & 0 & i\theta \\ \hline \iota & i\iota & 0 \\ \hline\end{array} \quad \begin{array}{|c|c|c|}\hline 0 & 0 & i\theta \\ \hline 0 & 0 & -\theta \\ \hline i\iota & -\iota & 0 \\ \hline\end{array} \quad \begin{array}{|c|c|c|}\hline \kappa & i\kappa & 0 \\ \hline i\kappa & -\kappa & 0 \\ \hline 0 & 0 & 0 \\ \hline\end{array}} \right. ,$$

$$\left. \boxed{\begin{array}{|c|c|c|}\hline 0 & 0 & \theta^{\star} \\ \hline 0 & 0 & -\mathbb{i}\ \theta^{*} \\ \hline \iota^{\star} & -\mathbb{i}\ \iota^{*} & 0 \\ \hline\end{array} \quad \begin{array}{|c|c|c|}\hline 0 & 0 & -\mathbb{i}\ \theta^{*} \\ \hline 0 & 0 & -\theta^{*} \\ \hline -\mathbb{i}\ \iota^{*} & -\iota^{*} & 0 \\ \hline\end{array} \quad \begin{array}{|c|c|c|}\hline \kappa^{\star} & -\mathbb{i}\ \kappa^{*} & 0 \\ \hline -\mathbb{i}\ \kappa^{*} & -\kappa^{*} & 0 \\ \hline 0 & 0 & 0 \\ \hline\end{array}} \right\}$$

$$\left\{ \boxed{\begin{array}{|c|c|c|}\hline \lambda & i\lambda & 0 \\ \hline i\lambda & -\lambda & 0 \\ \hline 0 & 0 & 0 \\ \hline\end{array} \quad \begin{array}{|c|c|c|}\hline i\lambda & -\lambda & 0 \\ \hline -\lambda & -i\lambda & 0 \\ \hline 0 & 0 & 0 \\ \hline\end{array} \quad \begin{array}{|c|c|c|}\hline 0 & 0 & 0 \\ \hline 0 & 0 & 0 \\ \hline 0 & 0 & 0 \\ \hline\end{array}} \right. ,$$

$$\left. \boxed{\begin{array}{|c|c|c|}\hline \lambda^{\star} & -\mathbb{i}\ \lambda^{*} & 0 \\ \hline -\mathbb{i}\ \lambda^{*} & -\lambda^{*} & 0 \\ \hline 0 & 0 & 0 \\ \hline\end{array} \quad \begin{array}{|c|c|c|}\hline -\mathbb{i}\ \lambda^{*} & -\lambda^{*} & 0 \\ \hline -\lambda^{*} & \mathbb{i}\ \lambda^{*} & 0 \\ \hline 0 & 0 & 0 \\ \hline\end{array} \quad \begin{array}{|c|c|c|}\hline 0 & 0 & 0 \\ \hline 0 & 0 & 0 \\ \hline 0 & 0 & 0 \\ \hline\end{array}} \right\}$$

D4h **{A1u, A2u, B1u, B2u, {Eua, Eub}}**

$$\begin{array}{|c|c|c|}\hline 0&0&0\\\hline 0&0&0\\\hline 0&0&0\\\hline\end{array}\quad\begin{array}{|c|c|c|}\hline 0&0&0\\\hline 0&0&0\\\hline 0&0&0\\\hline\end{array}\quad\begin{array}{|c|c|c|}\hline 0&0&0\\\hline 0&0&0\\\hline 0&0&0\\\hline\end{array}$$

$$\begin{array}{|c|c|c|}\hline 0&0&\mathrm{A}\\\hline 0&0&0\\\hline \mathrm{B}&0&0\\\hline\end{array}\quad\begin{array}{|c|c|c|}\hline 0&0&0\\\hline 0&0&\mathrm{A}\\\hline 0&\mathrm{B}&0\\\hline\end{array}\quad\begin{array}{|c|c|c|}\hline \mathrm{C}&0&0\\\hline 0&\mathrm{C}&0\\\hline 0&0&\mathrm{D}\\\hline\end{array}$$

$$\begin{array}{|c|c|c|}\hline 0&0&0\\\hline 0&0&\mathrm{F}\\\hline 0&\mathrm{G}&0\\\hline\end{array}\quad\begin{array}{|c|c|c|}\hline 0&0&\mathrm{F}\\\hline 0&0&0\\\hline \mathrm{G}&0&0\\\hline\end{array}\quad\begin{array}{|c|c|c|}\hline 0&\mathrm{H}&0\\\hline \mathrm{H}&0&0\\\hline 0&0&0\\\hline\end{array}$$

$$\begin{array}{|c|c|c|}\hline 0&0&0\\\hline 0&0&0\\\hline 0&0&0\\\hline\end{array}\quad\begin{array}{|c|c|c|}\hline 0&0&0\\\hline 0&0&0\\\hline 0&0&0\\\hline\end{array}\quad\begin{array}{|c|c|c|}\hline 0&0&0\\\hline 0&0&0\\\hline 0&0&0\\\hline\end{array}$$

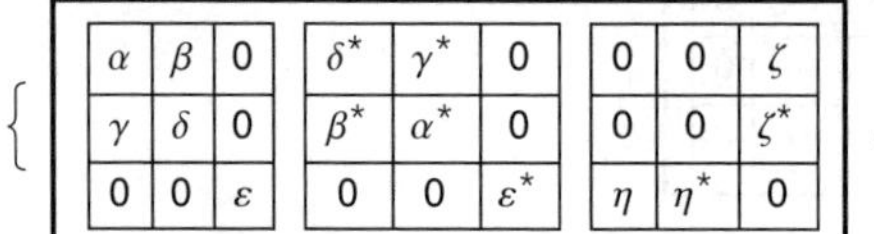

$$\Bigg\{\begin{array}{|c|c|c|}\hline \alpha&\beta&0\\\hline \gamma&\delta&0\\\hline 0&0&\varepsilon\\\hline\end{array}\quad\begin{array}{|c|c|c|}\hline \delta^*&\gamma^*&0\\\hline \beta^*&\alpha^*&0\\\hline 0&0&\varepsilon^*\\\hline\end{array}\quad\begin{array}{|c|c|c|}\hline 0&0&\zeta\\\hline 0&0&\zeta^*\\\hline \eta&\eta^*&0\\\hline\end{array}\;,$$

$$\begin{array}{|c|c|c|}\hline \alpha^*&\beta^*&0\\\hline \gamma^*&\delta^*&0\\\hline 0&0&\varepsilon^*\\\hline\end{array}\quad\begin{array}{|c|c|c|}\hline \delta&\gamma&0\\\hline \beta&\alpha&0\\\hline 0&0&\varepsilon\\\hline\end{array}\quad\begin{array}{|c|c|c|}\hline 0&0&\zeta^*\\\hline 0&0&\zeta\\\hline \eta^*&\eta&0\\\hline\end{array}\Bigg\}$$

D5 {A1, A2, {E1a, E1b}, {E2a, E2b}}

D5d $\{A_{1u}, A_{2u}, \{E_{1u}a, E_{1u}b\}, \{E_{2u}a, E_{2u}b\}\}$

0	0	0	0	0	−A	0	C	0
0	0	A	0	0	0	−C	0	0
0	B	0	−B	0	0	0	0	0

0	0	D	0	0	0	G	0	0
0	0	0	0	0	D	0	G	0
F	0	0	0	F	0	0	0	H

{

α	β	0	$-i\delta$	$i\gamma$	0	0	0	ζ
γ	δ	0	$i\beta$	$-i\alpha$	0	0	0	$-i\zeta$
0	0	ε	0	0	$-i\varepsilon$	η	$-i\eta$	0

,

α^*	β^*	0	$𝕚\ \delta^*$	$-𝕚\ \gamma^*$	0	0	0	ζ^*
γ^*	δ^*	0	$-𝕚\ \beta^*$	$𝕚\ \alpha^*$	0	0	0	$𝕚\ \zeta^*$
0	0	ε^*	0	0	$𝕚\ \varepsilon^*$	η^*	$𝕚\ \eta^*$	0

}

{

θ	$i\theta$	ι	$i\theta$	$-\theta$	$-i\iota$	λ	$-i\lambda$	0
$i\theta$	$-\theta$	$-i\iota$	$-\theta$	$-i\theta$	$-\iota$	$-i\lambda$	$-\lambda$	0
κ	$-i\kappa$	0	$-i\kappa$	$-\kappa$	0	0	0	0

,

θ^*	$-𝕚\ \theta^*$	ι^*	$-𝕚\ \theta^*$	$-\theta^*$	$𝕚\ \iota^*$	λ^*	$𝕚\ \lambda^*$	0
$-𝕚\ \theta^*$	$-\theta^*$	$𝕚\ \iota^*$	$-\theta^*$	$𝕚\ \theta^*$	$-\iota^*$	$𝕚\ \lambda^*$	$-\lambda^*$	0
κ^*	$𝕚\ \kappa^*$	0	$𝕚\ \kappa^*$	$-\kappa^*$	0	0	0	0

}

D5h $\{\{E_1'a, E_1'b\}, \{E_2'a, E_2'b\}, A_1'', A_2'', \{E_2''a, E_2''b\}\}$

{

α	β	0	$-i\delta$	$i\gamma$	0	0	0	ζ
γ	δ	0	$i\beta$	$-i\alpha$	0	0	0	$-i\zeta$
0	0	ε	0	0	$-i\varepsilon$	η	$-i\eta$	0

,

$$\left.\begin{pmatrix} \alpha^* & \beta^* & 0 \\ \gamma^* & \delta^* & 0 \\ 0 & 0 & \varepsilon^* \end{pmatrix} \begin{pmatrix} \mathbb{i}\,\delta^* & -\mathbb{i}\,\gamma^* & 0 \\ -\mathbb{i}\,\beta^* & \mathbb{i}\,\alpha^* & 0 \\ 0 & 0 & \mathbb{i}\,\varepsilon^* \end{pmatrix} \begin{pmatrix} 0 & 0 & \zeta^* \\ 0 & 0 & \mathbb{i}\,\zeta^* \\ \eta^* & \mathbb{i}\,\eta^* & 0 \end{pmatrix}\right\}$$

$$\left\{\begin{pmatrix} \theta & i\theta & 0 \\ i\theta & -\theta & 0 \\ 0 & 0 & 0 \end{pmatrix} \begin{pmatrix} i\theta & -\theta & 0 \\ -\theta & -i\theta & 0 \\ 0 & 0 & 0 \end{pmatrix} \begin{pmatrix} 0 & 0 & 0 \\ 0 & 0 & 0 \\ 0 & 0 & 0 \end{pmatrix}\right.,$$

$$\left.\begin{pmatrix} \theta^* & -\mathbb{i}\,\theta^* & 0 \\ -\mathbb{i}\,\theta^* & -\theta^* & 0 \\ 0 & 0 & 0 \end{pmatrix} \begin{pmatrix} -\mathbb{i}\,\theta^* & -\theta^* & 0 \\ -\theta^* & \mathbb{i}\,\theta^* & 0 \\ 0 & 0 & 0 \end{pmatrix} \begin{pmatrix} 0 & 0 & 0 \\ 0 & 0 & 0 \\ 0 & 0 & 0 \end{pmatrix}\right\}$$

$$\begin{pmatrix} 0 & 0 & 0 \\ 0 & 0 & A \\ 0 & B & 0 \end{pmatrix} \begin{pmatrix} 0 & 0 & -A \\ 0 & 0 & 0 \\ -B & 0 & 0 \end{pmatrix} \begin{pmatrix} 0 & C & 0 \\ -C & 0 & 0 \\ 0 & 0 & 0 \end{pmatrix}$$

$$\begin{pmatrix} 0 & 0 & D \\ 0 & 0 & 0 \\ F & 0 & 0 \end{pmatrix} \begin{pmatrix} 0 & 0 & 0 \\ 0 & 0 & D \\ 0 & F & 0 \end{pmatrix} \begin{pmatrix} G & 0 & 0 \\ 0 & G & 0 \\ 0 & 0 & H \end{pmatrix}$$

$$\left\{\begin{pmatrix} 0 & 0 & \iota \\ 0 & 0 & -i\iota \\ \kappa & -i\kappa & 0 \end{pmatrix} \begin{pmatrix} 0 & 0 & -i\iota \\ 0 & 0 & -\iota \\ -i\kappa & -\kappa & 0 \end{pmatrix} \begin{pmatrix} \lambda & -i\lambda & 0 \\ -i\lambda & -\lambda & 0 \\ 0 & 0 & 0 \end{pmatrix}\right.,$$

$$\left.\begin{pmatrix} 0 & 0 & \iota^* \\ 0 & 0 & \mathbb{i}\,\iota^* \\ \kappa^* & \mathbb{i}\,\kappa^* & 0 \end{pmatrix} \begin{pmatrix} 0 & 0 & \mathbb{i}\,\iota^* \\ 0 & 0 & -\iota^* \\ \mathbb{i}\,\kappa^* & -\kappa^* & 0 \end{pmatrix} \begin{pmatrix} \lambda^* & \mathbb{i}\,\lambda^* & 0 \\ \mathbb{i}\,\lambda^* & -\lambda^* & 0 \\ 0 & 0 & 0 \end{pmatrix}\right\}$$

D6 {A1, A2, B1, B2, {E1a, E1b}, {E2a, E2b}}

0	0	0	0	0	−A	0	C	0
0	0	A	0	0	0	−C	0	0
0	B	0	−B	0	0	0	0	0

0	0	D	0	0	0	G	0	0
0	0	0	0	0	D	0	G	0
F	0	0	0	F	0	0	0	H

J	0	0	0	−J	0	0	0	0
0	−J	0	−J	0	0	0	0	0
0	0	0	0	0	0	0	0	0

0	K	0	K	0	0	0	0	0
K	0	0	0	−K	0	0	0	0
0	0	0	0	0	0	0	0	0

{

α	β	0	$-i\delta$	$i\gamma$	0	0	0	ζ
γ	δ	0	$i\beta$	$-i\alpha$	0	0	0	$-i\zeta$
0	0	ε	0	0	$-i\varepsilon$	η	$-i\eta$	0

,

α^*	β^*	0	$i\delta^*$	$-i\gamma^*$	0	0	0	ζ^*
γ^*	δ^*	0	$-i\beta^*$	$i\alpha^*$	0	0	0	$i\zeta^*$
0	0	ε^*	0	0	$i\varepsilon^*$	η^*	$i\eta^*$	0

}

{

0	0	θ	0	0	$-i\theta$	κ	$-i\kappa$	0
0	0	$-i\theta$	0	0	$-\theta$	$-i\kappa$	$-\kappa$	0
ι	$-i\iota$	0	$-i\iota$	$-\iota$	0	0	0	0

,

0	0	θ^*	0	0	$i\theta^*$	κ^*	$i\kappa^*$	0
0	0	$i\theta^*$	0	0	$-\theta^*$	$i\kappa^*$	$-\kappa^*$	0
ι^*	$i\iota^*$	0	$i\iota^*$	$-\iota^*$	0	0	0	0

}

D6d **{B1, B2, {E1a, E1b}, {E3a, E3b}, {E4a, E4b}}**

$$\left[\begin{array}{ccc}0&0&0\\0&0&0\\0&0&0\end{array}\right]\left[\begin{array}{ccc}0&0&0\\0&0&0\\0&0&0\end{array}\right]\left[\begin{array}{ccc}0&0&0\\0&0&0\\0&0&0\end{array}\right] \quad \left[\begin{array}{ccc}0&0&A\\0&0&0\\B&0&0\end{array}\right]\left[\begin{array}{ccc}0&0&0\\0&0&A\\0&B&0\end{array}\right]\left[\begin{array}{ccc}C&0&0\\0&C&0\\0&0&D\end{array}\right]$$

$$\left\{\left[\begin{array}{ccc}\alpha&\beta&0\\\gamma&\delta&0\\0&0&\varepsilon\end{array}\right]\left[\begin{array}{ccc}\delta^*&\gamma^*&0\\\beta^*&\alpha^*&0\\0&0&\varepsilon^*\end{array}\right]\left[\begin{array}{ccc}0&0&\zeta\\0&0&\zeta^*\\\eta&\eta^*&0\end{array}\right],\right.$$

$$\left.\left[\begin{array}{ccc}\alpha^*&\beta^*&0\\\gamma^*&\delta^*&0\\0&0&\varepsilon^*\end{array}\right]\left[\begin{array}{ccc}\delta&\gamma&0\\\beta&\alpha&0\\0&0&\varepsilon\end{array}\right]\left[\begin{array}{ccc}0&0&\zeta^*\\0&0&\zeta\\\eta^*&\eta&0\end{array}\right]\right\}$$

$$\left\{\left[\begin{array}{ccc}\theta&-\theta^*&0\\-\theta^*&-\theta&0\\0&0&0\end{array}\right]\left[\begin{array}{ccc}-\theta^*&-\theta&0\\-\theta&\theta^*&0\\0&0&0\end{array}\right]\left[\begin{array}{ccc}0&0&0\\0&0&0\\0&0&0\end{array}\right],\right.$$

$$\left.\left[\begin{array}{ccc}\theta^*&-\theta&0\\-\theta&-\theta^*&0\\0&0&0\end{array}\right]\left[\begin{array}{ccc}-\theta&-\theta^*&0\\-\theta^*&\theta&0\\0&0&0\end{array}\right]\left[\begin{array}{ccc}0&0&0\\0&0&0\\0&0&0\end{array}\right]\right\}$$

$$\left\{\left[\begin{array}{ccc}0&0&\iota\\0&0&-i\iota\\\kappa&-i\kappa&0\end{array}\right]\left[\begin{array}{ccc}0&0&-i\iota\\0&0&-\iota\\-i\kappa&-\kappa&0\end{array}\right]\left[\begin{array}{ccc}\lambda&-i\lambda&0\\-i\lambda&-\lambda&0\\0&0&0\end{array}\right],\right.$$

$$\left.\left[\begin{array}{ccc}0&0&-\iota\\0&0&-i\iota\\-\kappa&-i\kappa&0\end{array}\right]\left[\begin{array}{ccc}0&0&-i\iota\\0&0&\iota\\-i\kappa&\kappa&0\end{array}\right]\left[\begin{array}{ccc}-\lambda&-i\lambda&0\\-i\lambda&\lambda&0\\0&0&0\end{array}\right]\right\}$$

D6h {A1u, A2u, B1u, B2u, {E1ua, E1ub}, {E2ua, E2ub}}

0	0	0	0	0	−A	0	C	0
0	0	A	0	0	0	−C	0	0
0	B	0	−B	0	0	0	0	0

0	0	D	0	0	0	G	0	0
0	0	0	0	0	D	0	G	0
F	0	0	0	F	0	0	0	H

0	J	0	J	0	0	0	0	0
J	0	0	0	−J	0	0	0	0
0	0	0	0	0	0	0	0	0

K	0	0	0	−K	0	0	0	0
0	−K	0	−K	0	0	0	0	0
0	0	0	0	0	0	0	0	0

{

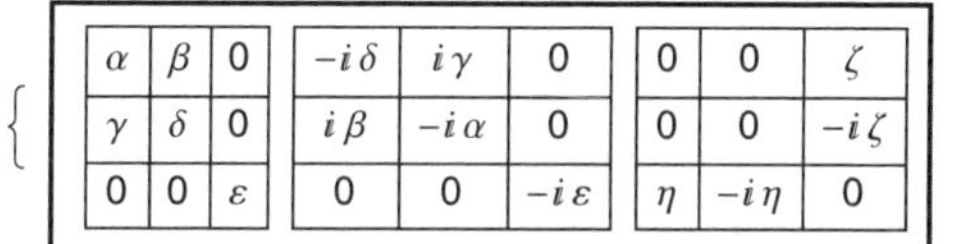

α	β	0	$-i\delta$	$i\gamma$	0	0	0	ζ
γ	δ	0	$i\beta$	$-i\alpha$	0	0	0	$-i\zeta$
0	0	ε	0	0	$-i\varepsilon$	η	$-i\eta$	0

,

α^*	β^*	0	$i\,\delta^*$	$-i\,\gamma^*$	0	0	0	ζ^*
γ^*	δ^*	0	$-i\,\beta^*$	$i\,\alpha^*$	0	0	0	$i\,\zeta^*$
0	0	ε^*	0	0	$i\,\varepsilon^*$	η^*	$i\,\eta^*$	0

}

{

0	0	θ	0	0	$-i\theta$	κ	$-i\kappa$	0
0	0	$-i\theta$	0	0	$-\theta$	$-i\kappa$	$-\kappa$	0
ι	$-i\iota$	0	$-i\iota$	$-\iota$	0	0	0	0

,

0	0	θ^*	0	0	$i\,\theta^*$	κ^*	$i\,\kappa^*$	0
0	0	$i\,\theta^*$	0	0	$-\theta^*$	$i\,\kappa^*$	$-\kappa^*$	0
ι^*	$i\,\iota^*$	0	$i\,\iota^*$	$-\iota^*$	0	0	0	0

}

O **{A1, A2, {Ea, Eb}, {T1a, T1b, T1c}, {T2a, T2b, T2c}}**

Oh **{A1u, A2u, {Eua, Eub}, {T1ua, T1ub, T1uc}, {T2ua, T2ub, T2uc}}**

$$\begin{bmatrix}0&0&0\\0&0&A\\0&-A&0\end{bmatrix}\begin{bmatrix}0&0&-A\\0&0&0\\A&0&0\end{bmatrix}\begin{bmatrix}0&A&0\\-A&0&0\\0&0&0\end{bmatrix}$$

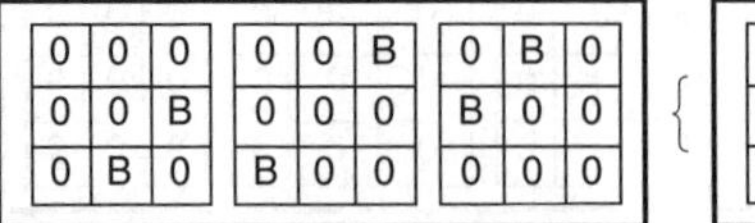

$$\begin{bmatrix}0&0&0\\0&0&B\\0&B&0\end{bmatrix}\begin{bmatrix}0&0&B\\0&0&0\\B&0&0\end{bmatrix}\begin{bmatrix}0&B&0\\B&0&0\\0&0&0\end{bmatrix}$$

$$\left\{\begin{bmatrix}0&0&0\\0&0&\alpha\\0&\beta&0\end{bmatrix}\begin{bmatrix}0&0&\gamma\\0&0&0\\\delta&0&0\end{bmatrix}\begin{bmatrix}0&\varepsilon&0\\\zeta&0&0\\0&0&0\end{bmatrix},\right.$$

$$\left.\begin{bmatrix}0&0&0\\0&0&\alpha^*\\0&\beta^*&0\end{bmatrix}\begin{bmatrix}0&0&\gamma^*\\0&0&0\\\delta^*&0&0\end{bmatrix}\begin{bmatrix}0&\varepsilon^*&0\\\zeta^*&0&0\\0&0&0\end{bmatrix}\right\}$$

$$\left\{\begin{bmatrix}C&0&0\\0&D&0\\0&0&D\end{bmatrix}\begin{bmatrix}0&F&0\\G&0&0\\0&0&0\end{bmatrix}\begin{bmatrix}0&0&F\\0&0&0\\G&0&0\end{bmatrix},\begin{bmatrix}0&H&0\\J&0&0\\0&0&0\end{bmatrix}\begin{bmatrix}K&0&0\\0&L&0\\0&0&K\end{bmatrix}\begin{bmatrix}0&0&0\\0&0&J\\0&H&0\end{bmatrix},\right.$$

$$\left.\begin{bmatrix}0&0&M\\0&0&0\\N&0&0\end{bmatrix}\begin{bmatrix}0&0&0\\0&0&M\\0&N&0\end{bmatrix}\begin{bmatrix}P&0&0\\0&P&0\\0&0&Q\end{bmatrix}\right\}$$

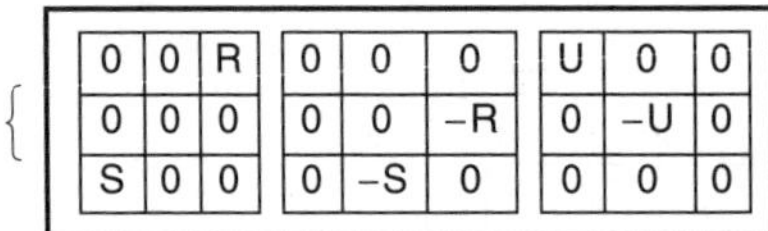

$$\left\{\begin{bmatrix}0&0&R\\0&0&0\\S&0&0\end{bmatrix}\begin{bmatrix}0&0&0\\0&0&-R\\0&-S&0\end{bmatrix}\begin{bmatrix}U&0&0\\0&-U&0\\0&0&0\end{bmatrix},\right.$$

$$\begin{bmatrix}0&0&0\\0&V&0\\0&0&-V\end{bmatrix}\begin{bmatrix}0&W&0\\X&0&0\\0&0&0\end{bmatrix}\begin{bmatrix}0&0&-W\\0&0&0\\-X&0&0\end{bmatrix},$$

$$\left.\begin{bmatrix}0&Y&0\\Z&0&0\\0&0&0\end{bmatrix}\begin{bmatrix}ZA&0&0\\0&0&0\\0&0&-ZA\end{bmatrix}\begin{bmatrix}0&0&0\\0&0&-Z\\0&-Y&0\end{bmatrix}\right\}$$

S4 **{A, B, Ea, Eb}**

0	0	A	0	0	B	F	G	0
0	0	B	0	0	–A	G	–F	0
C	D	0	D	–C	0	0	0	0

0	0	H	0	0	–J	M	N	0
0	0	J	0	0	H	–N	M	0
K	L	0	–L	K	0	0	0	P

α	β	0	$-i\delta$	$i\gamma$	0	0	0	ζ
γ	δ	0	$i\beta$	$-i\alpha$	0	0	0	$-i\zeta$
0	0	ε	0	0	$-i\varepsilon$	η	$-i\eta$	0

α^*	β^*	0	$\mathbb{i}\ \delta^*$	$-\mathbb{i}\ \gamma^*$	0	0	0	ζ^*
γ^*	δ^*	0	$-\mathbb{i}\ \beta^*$	$\mathbb{i}\ \alpha^*$	0	0	0	$\mathbb{i}\ \zeta^*$
0	0	ε^*	0	0	$\mathbb{i}\ \varepsilon^*$	η^*	$\mathbb{i}\ \eta^*$	0

S8 {B, E1a, E1b, E2a, E2b, E3a, E3b}

0	0	A		0	0	–B		F	G	0
0	0	B		0	0	A		–G	F	0
C	D	0		–D	C	0		0	0	H

α	β	0		ζ	η	0		0	0	λ
γ	δ	0		θ	ι	0		0	0	μ
0	0	ε		0	0	κ		ν	ξ	0

α^*	β^*	0		ζ^*	η^*	0		0	0	λ
γ^*	δ^*	0		θ^*	ι^*	0		0	0	μ
0	0	ε		0	0	κ		ν	ξ	0

0	0	ρ		0	0	$i\rho$		ς	$i\varsigma$	0
0	0	$i\rho$		0	0	$-\rho$		$i\varsigma$	$-\varsigma$	0
σ	$i\sigma$	0		$i\sigma$	$-\sigma$	0		0	0	0

0	0	ρ^*		0	0	$-\mathbb{i}\,\rho^*$		ς^*	$-\mathbb{i}\,\varsigma^*$	0
0	0	$-\mathbb{i}\,\rho^*$		0	0	$-\rho^*$		$-\mathbb{i}\,\varsigma^*$	$-\varsigma^*$	0
σ^*	$-\mathbb{i}\,\sigma^*$	0		$-\mathbb{i}\,\sigma^*$	$-\sigma^*$	0		0	0	0

$-\mathbb{i}\,\iota^*$	$\mathbb{i}\,\theta^*$	0		$\mathbb{i}\,\delta^*$	$-\mathbb{i}\,\gamma^*$	0		0	0	$-i\mu$
$\mathbb{i}\,\eta^*$	$-\mathbb{i}\,\zeta^*$	0		$-\mathbb{i}\,\beta^*$	$\mathbb{i}\,\alpha^*$	0		0	0	$i\lambda$
0	0	$-i\kappa$		0	0	$i\varepsilon$		$-i\xi$	$i\nu$	0

$i\iota$	$-i\theta$	0		$-i\delta$	$i\gamma$	0		0	0	$i\mu$
$-i\eta$	$i\zeta$	0		$i\beta$	$-i\alpha$	0		0	0	$-i\lambda$
0	0	$i\kappa$		0	0	$-i\varepsilon$		$i\xi$	$-i\nu$	0

T **{A1, Ea, Eb, {Ta, Tb, Tc}}**

Th **{Au, Eua, Eub, {Tua, Tub, Tuc}}**

$$\begin{bmatrix} 0 & 0 & 0 \\ 0 & 0 & \text{A} \\ 0 & \text{B} & 0 \end{bmatrix} \begin{bmatrix} 0 & 0 & \text{B} \\ 0 & 0 & 0 \\ \text{A} & 0 & 0 \end{bmatrix} \begin{bmatrix} 0 & \text{A} & 0 \\ \text{B} & 0 & 0 \\ 0 & 0 & 0 \end{bmatrix} \qquad \begin{bmatrix} 0 & 0 & 0 \\ 0 & 0 & \alpha \\ 0 & \beta & 0 \end{bmatrix} \begin{bmatrix} 0 & 0 & \gamma \\ 0 & 0 & 0 \\ \delta & 0 & 0 \end{bmatrix} \begin{bmatrix} 0 & \varepsilon & 0 \\ \zeta & 0 & 0 \\ 0 & 0 & 0 \end{bmatrix}$$

$$\begin{bmatrix} 0 & 0 & 0 \\ 0 & 0 & \alpha^* \\ 0 & \beta^* & 0 \end{bmatrix} \begin{bmatrix} 0 & 0 & \gamma^* \\ 0 & 0 & 0 \\ \delta^* & 0 & 0 \end{bmatrix} \begin{bmatrix} 0 & \varepsilon^* & 0 \\ \zeta^* & 0 & 0 \\ 0 & 0 & 0 \end{bmatrix}$$

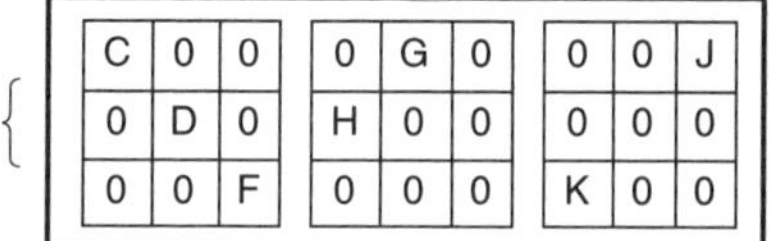

$$\left\{ \begin{bmatrix} \text{C} & 0 & 0 \\ 0 & \text{D} & 0 \\ 0 & 0 & \text{F} \end{bmatrix} \begin{bmatrix} 0 & \text{G} & 0 \\ \text{H} & 0 & 0 \\ 0 & 0 & 0 \end{bmatrix} \begin{bmatrix} 0 & 0 & \text{J} \\ 0 & 0 & 0 \\ \text{K} & 0 & 0 \end{bmatrix}, \right.$$

$$\begin{bmatrix} 0 & \text{L} & 0 \\ \text{M} & 0 & 0 \\ 0 & 0 & 0 \end{bmatrix} \begin{bmatrix} \text{N} & 0 & 0 \\ 0 & \text{P} & 0 \\ 0 & 0 & \text{Q} \end{bmatrix} \begin{bmatrix} 0 & 0 & 0 \\ 0 & 0 & \text{R} \\ 0 & \text{S} & 0 \end{bmatrix},$$

$$\left. \begin{bmatrix} 0 & 0 & \text{U} \\ 0 & 0 & 0 \\ \text{V} & 0 & 0 \end{bmatrix} \begin{bmatrix} 0 & 0 & 0 \\ 0 & 0 & \text{W} \\ 0 & \text{X} & 0 \end{bmatrix} \begin{bmatrix} \text{Y} & 0 & 0 \\ 0 & \text{Z} & 0 \\ 0 & 0 & \text{ZA} \end{bmatrix} \right\}$$

Td {A1, A2, {Ea, Eb}, {T1a, T1b, T1c}, {T2a, T2b, T2c}}

$$\left[\begin{bmatrix}0&0&0\\0&0&A\\0&A&0\end{bmatrix}\begin{bmatrix}0&0&A\\0&0&0\\A&0&0\end{bmatrix}\begin{bmatrix}0&A&0\\A&0&0\\0&0&0\end{bmatrix}\right]\quad\left[\begin{bmatrix}0&0&0\\0&0&B\\0&-B&0\end{bmatrix}\begin{bmatrix}0&0&-B\\0&0&0\\B&0&0\end{bmatrix}\begin{bmatrix}0&B&0\\-B&0&0\\0&0&0\end{bmatrix}\right]$$

$$\left\{\left[\begin{bmatrix}0&0&0\\0&0&\alpha\\0&\beta&0\end{bmatrix}\begin{bmatrix}0&0&\gamma\\0&0&0\\\delta&0&0\end{bmatrix}\begin{bmatrix}0&\varepsilon&0\\\zeta&0&0\\0&0&0\end{bmatrix}\right],\left[\begin{bmatrix}0&0&0\\0&0&\alpha^*\\0&\beta^*&0\end{bmatrix}\begin{bmatrix}0&0&\gamma^*\\0&0&0\\\delta^*&0&0\end{bmatrix}\begin{bmatrix}0&\varepsilon^*&0\\\zeta^*&0&0\\0&0&0\end{bmatrix}\right]\right\}$$

$$\left\{\left[\begin{bmatrix}0&0&0\\0&C&0\\0&0&-C\end{bmatrix}\begin{bmatrix}0&D&0\\F&0&0\\0&0&0\end{bmatrix}\begin{bmatrix}0&0&-D\\0&0&0\\-F&0&0\end{bmatrix}\right],\right.$$

$$\left[\begin{bmatrix}0&G&0\\H&0&0\\0&0&0\end{bmatrix}\begin{bmatrix}J&0&0\\0&0&0\\0&0&-J\end{bmatrix}\begin{bmatrix}0&0&0\\0&0&-H\\0&-G&0\end{bmatrix}\right],$$

$$\left.\left[\begin{bmatrix}0&0&K\\0&0&0\\L&0&0\end{bmatrix}\begin{bmatrix}0&0&0\\0&0&-K\\0&-L&0\end{bmatrix}\begin{bmatrix}M&0&0\\0&-M&0\\0&0&0\end{bmatrix}\right]\right\}\left\{\left[\begin{bmatrix}0&0&N\\0&0&0\\P&0&0\end{bmatrix}\begin{bmatrix}0&0&0\\0&0&N\\0&P&0\end{bmatrix}\begin{bmatrix}Q&0&0\\0&Q&0\\0&0&R\end{bmatrix}\right],\right.$$

$$\left.\left[\begin{bmatrix}S&0&0\\0&U&0\\0&0&U\end{bmatrix}\begin{bmatrix}0&V&0\\W&0&0\\0&0&0\end{bmatrix}\begin{bmatrix}0&0&V\\0&0&0\\W&0&0\end{bmatrix}\right],\left[\begin{bmatrix}0&X&0\\Y&0&0\\0&0&0\end{bmatrix}\begin{bmatrix}Z&0&0\\0&ZA&0\\0&0&Z\end{bmatrix}\begin{bmatrix}0&0&0\\0&0&Y\\0&X&0\end{bmatrix}\right]\right\}$$

Group `I` is too slow to include in the automatic constructions above. However, it does succeed if you wait long enough. Its projections are

I {A, {T1a, T1b, T1c}, {F2a, F2b, F2c}, {Ga, Gb, Gc, Gd}, {Ha, Hb, Hc, Hd, He}}

$$\boxed{\begin{bmatrix}0&0&0\\0&0&A\\0&-A&0\end{bmatrix}\begin{bmatrix}0&0&-A\\0&0&0\\A&0&0\end{bmatrix}\begin{bmatrix}0&A&0\\-A&0&0\\0&0&0\end{bmatrix}}$$

$$\{\boxed{\begin{bmatrix}B&0&0\\0&C&0\\0&0&C\end{bmatrix}\begin{bmatrix}0&D&0\\F&0&0\\0&0&0\end{bmatrix}\begin{bmatrix}0&0&D\\0&0&0\\F&0&0\end{bmatrix}},\ \boxed{\begin{bmatrix}0&F&0\\D&0&0\\0&0&0\end{bmatrix}\begin{bmatrix}C&0&0\\0&B&0\\0&0&C\end{bmatrix}\begin{bmatrix}0&0&0\\0&0&D\\0&F&0\end{bmatrix}},$$

$$\boxed{\begin{bmatrix}0&0&F\\0&0&0\\D&0&0\end{bmatrix}\begin{bmatrix}0&0&0\\0&0&F\\0&D&0\end{bmatrix}\begin{bmatrix}C&0&0\\0&C&0\\0&0&B\end{bmatrix}}\}\{\boxed{\begin{bmatrix}0&0&G\\0&0&0\\G&0&0\end{bmatrix}\begin{bmatrix}0&0&0\\0&0&H\\0&H&0\end{bmatrix}\begin{bmatrix}G&0&0\\0&H&0\\0&0&J\end{bmatrix}},$$

$$\boxed{\begin{bmatrix}J&0&0\\0&G&0\\0&0&H\end{bmatrix}\begin{bmatrix}0&G&0\\G&0&0\\0&0&0\end{bmatrix}\begin{bmatrix}0&0&H\\0&0&0\\H&0&0\end{bmatrix}},\ \boxed{\begin{bmatrix}0&H&0\\H&0&0\\0&0&0\end{bmatrix}\begin{bmatrix}H&0&0\\0&J&0\\0&0&G\end{bmatrix}\begin{bmatrix}0&0&0\\0&0&G\\0&G&0\end{bmatrix}}\}$$

$$\{\boxed{\begin{bmatrix}K&0&0\\0&L&0\\0&0&M\end{bmatrix}\begin{bmatrix}0&L&0\\L&0&0\\0&0&0\end{bmatrix}\begin{bmatrix}0&0&M\\0&0&0\\M&0&0\end{bmatrix}},\ \boxed{\begin{bmatrix}0&0&L\\0&0&0\\L&0&0\end{bmatrix}\begin{bmatrix}0&0&0\\0&0&M\\0&M&0\end{bmatrix}\begin{bmatrix}L&0&0\\0&M&0\\0&0&K\end{bmatrix}},$$

$$\boxed{\begin{bmatrix}0&M&0\\M&0&0\\0&0&0\end{bmatrix}\begin{bmatrix}M&0&0\\0&K&0\\0&0&L\end{bmatrix}\begin{bmatrix}0&0&0\\0&0&L\\0&L&0\end{bmatrix}},\ \boxed{\begin{bmatrix}0&0&0\\0&0&N\\0&N&0\end{bmatrix}\begin{bmatrix}0&0&N\\0&0&0\\N&0&0\end{bmatrix}\begin{bmatrix}0&N&0\\N&0&0\\0&0&0\end{bmatrix}}\}$$

$$\{\boxed{\begin{bmatrix}0&0&0\\0&0&0\\0&0&0\end{bmatrix}\begin{bmatrix}0&0&0\\0&0&0\\0&0&0\end{bmatrix}\begin{bmatrix}0&0&0\\0&0&0\\0&0&0\end{bmatrix}},\ \boxed{\begin{bmatrix}0&P&0\\Q&0&0\\0&0&0\end{bmatrix}\begin{bmatrix}R&0&0\\0&0&0\\0&0&-R\end{bmatrix}\begin{bmatrix}0&0&0\\0&0&-Q\\0&-P&0\end{bmatrix}},$$

$$\boxed{\begin{bmatrix}0&0&0\\0&-R&0\\0&0&R\end{bmatrix}\begin{bmatrix}0&-Q&0\\-P&0&0\\0&0&0\end{bmatrix}\begin{bmatrix}0&0&Q\\0&0&0\\P&0&0\end{bmatrix}},$$

$$\boxed{\begin{bmatrix}0&0&-P\\0&0&0\\-Q&0&0\end{bmatrix}\begin{bmatrix}0&0&0\\0&0&P\\0&Q&0\end{bmatrix}\begin{bmatrix}-R&0&0\\0&R&0\\0&0&0\end{bmatrix}},\ \boxed{\begin{bmatrix}0&0&0\\0&0&0\\0&0&0\end{bmatrix}\begin{bmatrix}0&0&0\\0&0&0\\0&0&0\end{bmatrix}\begin{bmatrix}0&0&0\\0&0&0\\0&0&0\end{bmatrix}}\}$$

After each group was symbolized, its **Symbolizations** were given the name **SymbolRules[gp]**, before the next **SymbolizeProjectionTijk** overwrote it. For instance, look at parts **1** through **3**:

```
SymbolRules["D3d"][[1 ;; 3]]
```

$$\left\{\left\{A\to \frac{1}{4}T[x,x,x]-\frac{1}{4}T[x,y,y]-\frac{1}{4}T[y,x,y]-\frac{1}{4}T[y,y,x]\right\},\left\{B\to \frac{1}{2}T[x,y,z]-\frac{1}{2}T[y,x,z]\right\},\left\{C\to \frac{1}{2}T[x,z,y]-\frac{1}{2}T[y,z,x]\right\}\right\}$$

If you see things like

```
SymbolRules["D2d"][[1 ;; 3]]
```

```
{{A → v06}, {B → v08}, {C → v11}}
```

it means that axial permutation rules have been invoked, and you will have to work a little harder to answer your question. The answer is somewhere before **MakeTijkDisplay** was invoked.

46.6 Axis permutation forbiddenness

Some of the results for the Platonic groups are quite surprising, and seem to be a new discovery. Here we review one of them step by step. Look at the detailed projection of species **Ea** of group **T**:

```
tracePrj = TraceProjectTijk["T"];
detPrjTEa = DetailedProjectTijk[tracePrj][[2, 2]];
Short[%, 9]
```

```
{Ea, {{{0, 0, 0}, {0, 0, 1/3 T[x, y, z] - 1/6 T[y, z, x] -
      (i T[y, z, x])/(2 Sqrt[3]) - 1/6 T[z, x, y] + (i T[z, x, y])/(2 Sqrt[3])},
    {0, 1/3 T[x, z, y] - 1/6 T[y, x, z] - (i T[y, x, z])/(2 Sqrt[3]) -
      1/6 T[z, y, x] + (i T[z, y, x])/(2 Sqrt[3]), 0}}, {<<1>>},
  {{0, <<6>> + 1/3 T[<<1>>], 0}, {<<1>>}, {0, 0, 0}}}}
```

This is a non-zero result, and was not filtered out by the all-zero filter in **Trace`ProjectTijk**. Now put in the axis permutation symmetry:

```
detPrjTEa /. xyzCyclicRules
```

```
{Ea, {{{0, 0, 0}, {0, 0, 0}, {0, 0, 0}},
  {{0, 0, 0}, {0, 0, 0}, {0, 0, 0}},
  {{0, 0, 0}, {0, 0, 0}, {0, 0, 0}}}}
```

As in the two-photon tensors, it appears that in group **T** the transition **A→Ea** is forbidden.

> **On your own**
> Do the same thing for species **Eb**. Then do **Ega** and **Egb** in groups $\mathbf{T_h}$, and do the **E** species in **O**, $\mathbf{O_h}$ and $\mathbf{T_d}$, and the **H** species of group **I**. You will find surprising vanishings in all of them. Species **H** of **I** is particularly weird, as only two of its five components vanish.

46.7 End Notes

46.7.1. Rotation formula for three-index tensors

Closed; please read from CD

46.7.2. Derivation of axis permutation rules

Closed; please read from CD

47. Class sums and their products

Preliminaries

47.1 Motivation

This is the first of two chapters that teach the automated construction of character tables. In the analysis of flexible molecules you may need a character table that is not tabulated in this book or any book. Such molecules can be analyzed using permutation groups, but there are so many permutation groups that no one tabulates their character tables. When you need the character table of a permutation group, you usually have to make it yourself. This is a treacherous and uncertain job by hand, but the *Mathematica* automation offered here will do any group without requiring human thought at all. All you need to know about the group is its multiplication table.

This topic requires an idea not treated by most applied group theory books; namely, the idea of a group algebra. In this chapter, we will define **group algebra** and use it to define a **class sum**, then prove some surprising and beautiful little theorems about class sums and their products. We end this chapter with the automated construction of the **class sum multiplication table** (the CSMT).

In the next chapter we will show how to construct the **character table** of the group, using only the CSMT.

47.2 Declaration of a "group algebra"

By postulate, group elements possess a pairwise group operator conventionally called "multiplication", but nothing in the group postulates assures us that we can take linear combinations of group elements. But they do not prevent it, either.

Group algebra
So now we announce as a separate postulate that we are going to allow linear combinations of group elements.

Any set of elements on which multiplication and linear combination are permitted is called "an algebra". Linear combinations of group elements are not

W.M. McClain, *Symmetry Theory in Molecular Physics with Mathematica*, DOI 10.1007/b13137_47, © Springer Science+Business Media, LLC 2009

members of the group (they do not appear in the closed multiplication table), but they will quickly prove themselves interesting and useful.

You may think of the group elements as the basis vectors of a space, and linear combinations of group elements as points in the space. We call this space the "group space".

Multiplication of two points of the group space is defined as ordinary scalar multiplication of numerical coefficients, with group theoretical multiplication of group elements (with preservation of the given order). Therefore it needs a special symbol that is neither * nor ⊗. We choose the **Diamond**, ⋄ , which in *Mathematica* is undefined.

Multiplication of two points in group space

Using Greek letters for numerical coefficients and Latin capitals for group elements

$$(\ldots + \alpha \mathbf{A} + \ldots) \diamond (\ldots + \beta \mathbf{B} + \ldots) = (\ldots + (\alpha * \beta)\ (\mathbf{A} \otimes \mathbf{B}) + \ldots)$$

On the right side we intend for $\alpha * \beta$ and $\mathbf{A} \otimes \mathbf{B}$ to be actually carried out, each according to its own rules.

47.3 Class sums

One set of points in the group space will be of great interest to us. They are the class sums, defined as the sum of all the group elements in a class. The notation we will use is

Class sum notation

All class sums will be denoted by a DoubleStruck capital ℂ, with a subscript indicating the class. If class **A** is $\{A_1, .., A_m\}$, its class sum will be
$\mathbb{C}_A = A_1 + \ldots + A_m$. Often the class names are just $\mathbb{C}_1$, $\mathbb{C}_2$, etc.

Class sum multiplication is just a special case of group point multiplication, defined above, but it is so important that we write it explicitly :

Class sum multiplication

If $\mathbb{C}_A = A_1 + \ldots + A_m$ is one class sum and $\mathbb{C}_B = B_1 + \ldots + B_n$ is another, the class sum product is

$$\mathbb{C}_A \diamond \mathbb{C}_B = (A_1 + \ldots + A_m) \diamond (B_1 + \ldots + B_n) == A_1 \otimes B_1 + \ldots + A_m \otimes B_n$$

This is just a normal algebraic expansion, but with preservation of factor order.

47.4 ClassSum theorems

Class sums have some remarkable properties that we will need. Here they are.

1. Every class sum $\mathbb{C}_A$ commutes with every element M of the group.

This is written as $M \diamond \mathbb{C}_A = \mathbb{C}_A \diamond M$. Multiply on the right by M^{-1}: $M \diamond \mathbb{C}_A \diamond M^{-1} = \mathbb{C}_A$, and expand the $\mathbb{C}$'s to give

$$M A_1 M^{-1} + \ldots + M A_n M^{-1} = A_1 + \ldots + A_n$$

Look back at Theorem 4, Chapter 20, to justify that equality. See if you can argue it on from here. If you get confused, you can always click here.

2. All the class sums in a group commute with each other.

This is easy in light of the theorem above. As usual, start by writing the theorem in symbols : $\mathbb{C}_A \diamond \mathbb{C}_B = \mathbb{C}_B \diamond \mathbb{C}_A$. Expand just one of the sums:

$$(A_1 + \ldots + A_n) \diamond \mathbb{C}_B = \mathbb{C}_B \ (A_1 + \ldots + A_n)$$

$$A_1 \mathbb{C}_B + \ldots + A_n \mathbb{C}_B = \mathbb{C}_B A_1 + \ldots + \mathbb{C}_B A_n$$

But by Theorem 1, the first term on the left is the same as the first term on the right; etc. So q.e.d.

3. The product of two class sums is unchanged if similarity transformed by any member G_k of the group.

Since $G_k \diamond G_k^{-1}$ is the unit element, we can write the first equality below: the second and third equalities are true because G_k commutes with both $\mathbb{C}_A$ and $\mathbb{C}_B$ (Theorem 1).

$$\mathbb{C}_A \diamond \mathbb{C}_B = \mathbb{C}_A \diamond \mathbb{C}_B \diamond \left(G_k \diamond G_k^{-1}\right) = \mathbb{C}_A \diamond G_k \diamond \mathbb{C}_B \diamond G_k^{-1} = G_k \diamond (\mathbb{C}_A \diamond \mathbb{C}_B) \diamond G_k^{-1}$$

4. The product of two class sums is a linear combination of class sums, with integer coefficients. In symbols,

$$\mathbb{C}_A \diamond \mathbb{C}_B = \sum_{M=1}^{nC} \beta_M \, \mathbb{C}_M$$

where nC is the number of classes and the β_M are integers. The proof is a bit long. It follows below.

Proof of Theorem 4:

Let $\mathbb{C}_A = A_1 + \ldots + A_m$ and $\mathbb{C}_B = B_1 + \ldots + B_n$, and expand the class sums:

$$\mathbb{C}_A \diamond \mathbb{C}_B = (A_1 + \ldots + A_m) \diamond (B_1 + \ldots + B_n) = A_1 \diamond B_1 + \ldots + A_m \diamond B_n$$

But each element product on the right is a single element of the group. Collect them into class sums as far as possible

$$\mathbb{C}_A \diamond \mathbb{C}_B = \sum_{m=1}^{nC} \alpha_m \, \mathbb{C}_m + \beta_r \, G_r + \ldots$$

where α_m is the number of times class $\mathbb{C}_m$ occurs in the product, and G_r is one of several leftovers after all possible class sums have been collected together. It is left over β_r times, etc. But by Theorem 3, above, this is unchanged if we similarity transform it by some element G_p.

$$\mathbb{C}_A \diamond \mathbb{C}_B = G_p \left(\sum_{m=1}^{nC} \alpha_m \, \mathbb{C}_m \right) G_p^{-1} + \beta_r \, G_p \, G_r \, G_p^{-1} + \ldots$$

But by Theorem 4 of Chapter 20, classes are invariant under similarity transforms. Therefore, so are class sums, and we can write

$$\mathbb{C}_A \diamond \mathbb{C}_B = \sum_{m=1}^{nC} \alpha_m\, \mathbb{C}_m + \beta_r \left(G_p\, G_r\, G_p^{-1}\right) + \ldots$$

Let G_s be a member of the class of G_r that is *not* among the leftovers. But there must be some G_p in the group such that $G_p\ G_r\ G_p^{-1} = G_s$ so for this G_p we have

$$\mathbb{C}_A \diamond \mathbb{C}_B = \sum_{m=1}^{nC} \alpha_m\, \mathbb{C}_m + \beta_r\, G_s + \ldots$$

But this is a contradiction, since G_s was specifically *not* among the leftovers. So there can be no leftovers. Also the α_m are integers (class counts), and we have

$$\mathbb{C}_A \diamond \mathbb{C}_B = \sum_{m=1}^{nC} \alpha_m\, \mathbb{C}_m$$

which was to be proved.

47.5 Class sum multiplication for a familiar group

We want to make a class sum multiplication table for a familiar group. Pick one, and begin by constructing its group multiplication table:

```
gp = "T";
MT = MultiplicationTable[Dot,
   CartRep[gp], Range[GroupOrder[gp]]];
% // GridForm
```

1	2	3	4	5	6	7	8
2	9	7	6	8	10	12	11
3	6	8	9	7	12	10	1
4	8	6	7	9	11	1	10
5	7	9	8	6	1	11	12
6	11	10	12	1	5	2	4
7	10	11	1	12	3	4	2
8	12	1	11	10	2	5	3
9	1	12	10	11	4	3	5
10	5	4	3	2	8	9	6
11	3	2	5	4	9	8	7
12	4	5	2	3	7	6	9

Here **1** stands for the first group element in the **CartRep**; **2**, for the second; etc. Now construct the class list, using the same integer name code:

```
toIntegerNames =
  Thread[ElementNames[gp] → Range[GroupOrder[gp]]];
CL = Classes[gp] /. toIntegerNames
```

```
{{1}, {2, 3, 4, 5}, {6, 7, 8, 9}, {10, 11, 12}}
```

We relied above on the operators **CartRep** and **Classes**, which produced respectively, the **MT** in standard order and the **CL** in standard order. This will produce, in the end, a character table with its columns (but not its rows) in standard order, making it fairly easy to compare our construction with the standard character table. But on a desert island you would not have this help, so in the last section of the next chapter we will construct the character table of an untabulated group so you can see (1) that it can be done, and (2) the issues that arise when there is no standard order for anything.

47.6 The ClassSumProduct operator

47.6.1 Step by step construction

We want to make a function with the calling expression

ClassSumProduct[n1,n2,CL,MT]

where **n1** is the class number of one class; **n2**, the other; **CL** is the class list; and **MT** is the group multiplication table.

To be concrete, we take the class sum product of class number **2** with class number **3**, step by step. The inputs are

```
n1 = 2;
n2 = 3;
```

We will need the lengths of the two classes:

```
{len1, len2} = Map[Length, {CL[[n1]], CL[[n2]]}]
```

```
{4, 4}
```

Now multiply each element in class number **n1** by each element of class number **n2**. We use the old integer name trick, pulling the product name directly from the integer-named multiplication table:

```
expr1 = Table[MT[[CL[[n1, i]], CL[[n2, j]]]],
   {i, 1, len1}, {j, 1, len2}] // Flatten
```

```
{10, 12, 11, 1, 12, 10, 1, 11, 11, 1, 10, 12, 1, 11, 12, 10}
```

These are the summands of the class sum product. The definition now says to add them up, but they are just integers and we do not want them to add in the arithmetic sense. So protect them by wrapping them with an undefined head **Q** :

```
expr2 = expr1 /. n_Integer → Q[n]
```

```
{Q[10], Q[12], Q[11], Q[1], Q[12], Q[10], Q[1], Q[11],
 Q[11], Q[1], Q[10], Q[12], Q[1], Q[11], Q[12], Q[10]}
```

Now we can safely add them :

```
expr3 = expr2 /. List → Plus
```

```
4 Q[1] + 4 Q[10] + 4 Q[11] + 4 Q[12]
```

All this has been collected into a single operator named **ClassSumProduct**, defined in the preliminaries We demonstrate it:

```
ClassSumProduct[n1, n2, CL, MT]
```

```
4 Q[1] + 4 Q[10] + 4 Q[11] + 4 Q[12]
```

This is the same as **expr3**, above. By Theorem 4, it must be a linear combination of class sums, with integer coefficients. So we need an operator that will recognize classes automatically. Fortunately, with *Mathematica*'s pattern recognition abilities this is not too hard. We do it in the next section.

47.6.2 From sum of elements to the symbolic class sum

In the preliminaries we defined an operator MakeTagRules that generates the appropriate rules :

```
tagRules = MakeTagRules[CL]
```

```
{Q[1] → ℂ1, Q[2] → -Q[3] - Q[4] - Q[5] + ℂ2,
 Q[6] → -Q[7] - Q[8] - Q[9] + ℂ3, Q[10] → -Q[11] - Q[12] + ℂ4}
```

Look carefully. Each rule replaces the first element in each class by the class name minus the other class elements. After using this rule and expanding, all the element names cancel and you are left only with the class name, times the appropriate numerical factor.

```
expr3 /. tagRules // Expand
```

```
4 ℂ1 + 4 ℂ4
```

There is the symbolic answer we wanted. The **Expand** after application of the **tagRules** is very important; this is what causes the cancellation of the Q-wrapped element names. Take it away and rerun. You will see.

47.7 ClassSumMultiplicationTable

Now we are ready for mass production of class sum products.

```
CSMT =
  Table[ClassSumProduct[i, j, CL, MT], {i, 1, nC},
     {j, 1, nC}] /. tagRules // Expand;
% // GridForm
```

$\mathbb{C}_1$	$\mathbb{C}_2$	$\mathbb{C}_3$	$\mathbb{C}_4$
$\mathbb{C}_2$	$4\,\mathbb{C}_3$	$4\,\mathbb{C}_1 + 4\,\mathbb{C}_4$	$3\,\mathbb{C}_2$
$\mathbb{C}_3$	$4\,\mathbb{C}_1 + 4\,\mathbb{C}_4$	$4\,\mathbb{C}_2$	$3\,\mathbb{C}_3$
$\mathbb{C}_4$	$3\,\mathbb{C}_2$	$3\,\mathbb{C}_3$	$3\,\mathbb{C}_1 + 2\,\mathbb{C}_4$

In the next chapter we will verify that this is correct, by showing that it leads to the correct character table.

The construction of the **CSMT** that you just saw above has been condensed into a single operator **ClassSumMultiplicationTable[mltTbl_]**. It is in the preliminaries. Test it:

```
CSMT == ClassSumMultiplicationTable[MT]
```

```
True
```

In the next chapter, you will learn how to turn the class multiplication table into the essential square part of the character table. In other words, we will determine the traces of all the irreducible representation matrices without ever knowing the matrices themselves.

47.8 End Notes

47.8.1 Theorem 1, finished

We left off with

$$\mathbf{M}\,\mathbf{A}_1\,\mathbf{M}^{-1} + \mathbf{M}\,\mathbf{A}_2\,\mathbf{M}^{-1} + \ldots + \mathbf{M}\,\mathbf{a}_n\,\mathbf{M}^{-1} = \mathbf{A}_1 + \mathbf{A}_2 + \ldots + \mathbf{A}_n$$

where no two of the elements $\mathbf{A_i}$ are the same. There are **n** summands on both sides. But by the referenced theorem, every summand on the left belongs to the class, and therefore is the same as some summand on the right. The only way the equation could fail is if two terms on the left were the same.

So assume that two terms on the left are the same: $\mathbf{M}\,\mathbf{A}_i\,\mathbf{M}^{-1} = \mathbf{M}\,\mathbf{A}_k\,\mathbf{M}^{-1}$. Then left multiply by $\mathbf{M}^{-1}$ and right multiply by $\mathbf{M}$, and you are left with $\mathbf{A}_i = \mathbf{A}_k$. But by hypothesis all the $\mathbf{A}_i$ are different. Contradiction.

Therefore, the left side is just a reordering of the right side, and every class sum commutes with every element of the group, as was to be shown.

48. Make a character table

Preliminaries

48.1 Introduction

This is the second of two chapters that teach the automated construction of character tables. In the last chapter we showed how to construct the class sum multiplication table (CSMT), using only information in the group multiplication table (GMT). In this chapter we will construct the character table, using only information in the CSMT.

48.2 Make the CSMT again

Partly as review, and partly to get ourselves started again, we work quickly down to where we left off in the last chapter, at the **ClassSumMultiplicationTable** (**CSMT**). You can use any tabulated group, but we continue on with group **T**, as in the last chapter.

```
gp = "T";
```

At the end, you will be encouraged to run other groups. Most groups use **Dot** as the group multiplication operator, but a few groups need special multiplication operators. We choose them automatically for the groups that need them:

```
dotOp = Which[
  gp == "I", IcosaDot,
  gp == "C5", PentaDot,
  gp == "D5", PentaDot,
  gp == "D5h", PentaDot,
  gp == "D5d", PentaDot,
  True, Dot]
```

```
Dot
```

Remember, **IcosaDot** and **PentaDot** work just like **Dot**, except that they also resolve some format ambiguities in the output. Construct the multiplication table :

DOI 10.1007/b13137_48, © Springer Science+Business Media, LLC 2009

```
MT = MultiplicationTable[dotOp,
   CartRep[gp], Range[GroupOrder[gp]]];
% // GridForm
```

1	2	3	4	5	6	7	8
2	9	7	6	8	10	12	11
3	6	8	9	7	12	10	1
4	8	6	7	9	11	1	10
5	7	9	8	6	1	11	12
6	11	10	12	1	5	2	4
7	10	11	1	12	3	4	2
8	12	1	11	10	2	5	3
9	1	12	10	11	4	3	5
10	5	4	3	2	8	9	6
11	3	2	5	4	9	8	7
12	4	5	2	3	7	6	9

```
CSMT = ClassSumMultiplicationTable[MT];
% // GridForm
```

$\mathbb{C}_1$	$\mathbb{C}_2$	$\mathbb{C}_3$	$\mathbb{C}_4$
$\mathbb{C}_2$	$4\,\mathbb{C}_3$	$4\,\mathbb{C}_1 + 4\,\mathbb{C}_4$	$3\,\mathbb{C}_2$
$\mathbb{C}_3$	$4\,\mathbb{C}_1 + 4\,\mathbb{C}_4$	$4\,\mathbb{C}_2$	$3\,\mathbb{C}_3$
$\mathbb{C}_4$	$3\,\mathbb{C}_2$	$3\,\mathbb{C}_3$	$3\,\mathbb{C}_1 + 2\,\mathbb{C}_4$

This is where we left off in the previous chapter.

48.3 Class sum product coefficients

We showed, as Theorem 5 of Chapter 47, that $\mathbb{C}_a \diamond \mathbb{C}_b = \sum_{m=1}^{nC} \beta_m\, \mathbb{C}_m$. This is a fine equation, but computation-oriented people will notice that something is a little wrong with it. Free indices **a** and **b** on the left specify which product of class sums you mean, but they do not appear on the right. This is easily corrected. We make it read

$$\mathbb{C}_a \diamond \mathbb{C}_b = \sum_{m=1}^{nC} c_{a,b,m}\, \mathbb{C}_m \tag{48.1}$$

We let $c_{a,b,m}$ be the "class sum product coefficients" (**CSPC**). They are easily lifted out of the **CSMT**. As always, we start with something easy and build up. First, we find all the **CSPC**'s for just one row of the **CSMT** (we take row **2**). The row itself is

```
CSMTrow2 = CSMT[[2]]
```

$\{\mathbb{C}_2,\ 4\,\mathbb{C}_3,\ 4\,\mathbb{C}_1 + 4\,\mathbb{C}_4,\ 3\,\mathbb{C}_2\}$

The variables that occur in these expressions are listed in the first row (provided the first class is the unit class) :

```
CDvars = CSMT[[1]]
```

$\{\mathbb{C}_1,\ \mathbb{C}_2,\ \mathbb{C}_3,\ \mathbb{C}_4\}$

Each expression in **CSMTrow4** is linear in the **CDvars** (as assured by Theorem 5 of Chapter 47) so the matrix relating them can be found by

```
CSPCrow2 = MatrixOfCoefficients[CSMTrow2, CDvars];
% // GridForm
```

0	1	0	0
0	0	4	0
4	0	0	4
0	3	0	0

Check this:

```
CSPCrow2.CDvars == CSMTrow2
```

True

Yes, it had to work. The number of classes will be **nC**. Run it over all rows :

```
nC = Length[CSMT];
CSPC = Table[MatrixOfCoefficients[
    CSMT[[i]], CDvars], {i, 1, nC}];
% // GridList
```

{

1	0	0	0
0	1	0	0
0	0	1	0
0	0	0	1

,

0	1	0	0
0	0	4	0
4	0	0	4
0	3	0	0

,

0	0	1	0
4	0	0	4
0	4	0	0
0	0	3	0

,

0	0	0	1
0	3	0	0
0	0	3	0
3	0	0	2

}

There is the whole three-index array $\mathbf{c}_{\mathbf{a},\mathbf{b},\mathbf{m}}$. For maximum resemblance to the definition, Eq. 48.1, (click to see it again) we make a function definition for eacl array element :

```
Clear[c, a, b, m];
```

```
Do[c_{a_,b_,m_} := CSPC[[a, b, m]],
  {a, 1, nC}, {b, 1, nC}, {m, 1, nC}]
```

- If you get an error above, it is probably because subscripted quantities were Symbolized when you ran a different chapter. We want this in most chapters, but not here. Quit *Mathematica* and do this chapter in a fresh session of its own. The symbolization of subscripts is blocked in this chapter.

Make sure the indices of the **c** functions work as variables :

```
CSPC == Table[c_{r,s,t}, {r, 1, nC}, {s, 1, nC}, {t, 1, nC}]
```

```
True
```

Since class sums commute, the **c** array should be symmetric in its first two indices :

```
Table[c_{r,s,t} == c_{s,r,t}, {r, 1, nC},
   {s, 1, nC}, {t, 1, nC}] // Flatten // Union
```

```
{True}
```

In the coming calculation we will need the populations of the classes, symbolized as P_i:

```
Clear[P];
Plist = Table[P_i, {i, 1, nC}]
```

```
{P_1, P_2, P_3, P_4}
```

```
popList = Map[Length, Classes[gp]]
```

```
{1, 4, 4, 3}
```

```
toPops = Thread[Plist → popList]
```

```
{P_1 → 1, P_2 → 4, P_3 → 4, P_4 → 3}
```

48.4 A strategy for the character table

The definition of the **CSPC**, Eq. 48.1, is written in terms of class sums.

But we are more interested in class characters than class sums. All the matrices in any given class have the same character, so the trace of the class sum **i** is just the population of class **i** (P_i) times $\chi_{s,i}$, the character of class **i**:

$$\text{Trace}[\mathbb{C}_i] == P_i\, \chi_{s,i}$$

Putting this into Eq. 48.1, we have the master equation for the characters of the group (Bunker and Jensen, Eq. 5-69, p. 80)

$$\left(\mathbf{P}_{i}\,\chi_{s,i}\right)\ \left(\mathbf{P}_{j}\,\chi_{s,j}\right) == \chi_{s,1}\sum_{k=1}^{nC}\mathbf{c}_{i,j,k}\,\mathbf{P}_{k}\,\chi_{s,k} \qquad (48.2)$$

where $\mathbf{P_i}$ is the population of class **i**; $\chi_{s,c}$ is the character of class **c** in species **s**; in particular, $\chi_{s,1}$is the character of class **1** in species **s**. Class **1** is always the class of the unit element, which contains only a single matrix; namely, a unit matrix of appropriate dimension. In other words, $\chi_{s,1}$ is the dimension of species **s**.

We know how to calculate the **c** and **P** quantities from the multiplication table of the group, so the unknowns of Eq. 48.2 are all the χ quantities; i.e., the character table. If we could solve that equation, the problem would be finished. However, it is homogeneous of degree 2 in the χ's, so it will give us only relative values, not absolute values. In other words, the master equation will give us the character vector of each species, except that one character must remain symbolic. (We will let it be $\chi_{s,1}$, the dimension of the species.) So for the complete solution, we will need another relation.

Fortunately, the row and column orthogonalities of the character table give us the additional relations we need. So our strategy will have two steps: first we solve the master equation, finding the characters of each species expressed in terms of the species dimension. Then we will use the column orthogonality to get the species dimensions, and the solution will be complete and exact.

48.5 Use the master equation

Put the known values of the **c** and **P** quantities into the master equation :

$$\chi\mathbf{EqsRaw} = \mathbf{Union}\Big[\mathbf{Flatten}\Big[\mathbf{Table}\Big[\mathbf{P}_{i}\,\mathbf{P}_{j}\,\chi_{s,i}\,\chi_{s,j} == \chi_{s,1}\sum_{k=1}^{nC}\mathbf{c}_{i,j,k}\,\mathbf{P}_{k}\,\chi_{s,k},\ \{i,1,nC\},\{j,1,nC\}\Big]\ /.\ \mathbf{toPops}\Big]\Big]\ //\ \mathbf{ExpandAll}$$

$$\{\text{True},\ 16\,\chi_{s,2}^{2} == 16\,\chi_{s,1}\,\chi_{s,3},\ 16\,\chi_{s,2}\,\chi_{s,3} == 4\,\chi_{s,1}^{2} + 12\,\chi_{s,1}\,\chi_{s,4},\ 16\,\chi_{s,3}^{2} == 16\,\chi_{s,1}\,\chi_{s,2},\ 12\,\chi_{s,2}\,\chi_{s,4} == 12\,\chi_{s,1}\,\chi_{s,2},\ 12\,\chi_{s,3}\,\chi_{s,4} == 12\,\chi_{s,1}\,\chi_{s,3},\ 9\,\chi_{s,4}^{2} == 3\,\chi_{s,1}^{2} + 6\,\chi_{s,1}\,\chi_{s,4}\}$$

Get rid of the tautology :

```
χEqsRaw2 = Select[χEqsRaw, # =!= True &];
Column[χEqsRaw2]
```

$16\,\chi_{s,2}^2 == 16\,\chi_{s,1}\,\chi_{s,3}$

$16\,\chi_{s,2}\,\chi_{s,3} == 4\,\chi_{s,1}^2 + 12\,\chi_{s,1}\,\chi_{s,4}$

$16\,\chi_{s,3}^2 == 16\,\chi_{s,1}\,\chi_{s,2}$

$12\,\chi_{s,2}\,\chi_{s,4} == 12\,\chi_{s,1}\,\chi_{s,2}$

$12\,\chi_{s,3}\,\chi_{s,4} == 12\,\chi_{s,1}\,\chi_{s,3}$

$9\,\chi_{s,4}^2 == 3\,\chi_{s,1}^2 + 6\,\chi_{s,1}\,\chi_{s,4}$

Let's count equations and variables. For each species

```
{nEqs, nVars} = {Length[χEqsRaw2], nC}
```

{6, 4}

So for **nC** species, the total number of equations and variables is

```
nC * {nEqs, nVars}
```

{24, 16}

So there is redundancy; more equations than variables. This is not a problem for *Mathematica* as long as all the equations are self-consistent. The **Reduce** operator will give you the complete story:

```
reducedχEqsRaw = Reduce[χEqsRaw2] // LogicalExpand
```

$(\chi_{s,2} == 0\ \&\&\ \chi_{s,3} == 0\ \&\&\ -3\,\chi_{s,4} == \chi_{s,1})\ ||$

$(\chi_{s,1} == 0\ \&\&\ \chi_{s,2} == 0\ \&\&\ \chi_{s,3} == 0\ \&\&\ \chi_{s,4} == 0)\ ||$

$\left(\frac{\chi_{s,3}^2}{\chi_{s,4}} == \chi_{s,2}\ \&\&\ \chi_{s,4} == \chi_{s,1}\ \&\&\ \chi_{s,4} == \chi_{s,3}\ \&\&\ \chi_{s,4} \neq 0\right)\ ||$

$\left(\frac{\chi_{s,3}^2}{\chi_{s,4}} == \chi_{s,2}\ \&\&\ \chi_{s,4} == \chi_{s,1}\ \&\&\ -(-1)^{1/3}\,\chi_{s,4} == \chi_{s,3}\ \&\&\ \chi_{s,4} \neq 0\right)\ ||$

$\left(\frac{\chi_{s,3}^2}{\chi_{s,4}} == \chi_{s,2}\ \&\&\ \chi_{s,4} == \chi_{s,1}\ \&\&\ (-1)^{2/3}\,\chi_{s,4} == \chi_{s,3}\ \&\&\ \chi_{s,4} \neq 0\right)$

This is a little easier to read if we get rid of the logical **And** and **Or** operators:

```
χEqs2 = reducedχEqsRaw /. Or → List /. And → List
```

$\{\{\chi_{s,2} == 0,\ \chi_{s,3} == 0,\ -3\,\chi_{s,4} == \chi_{s,1}\},$

$\{\chi_{s,1} == 0,\ \chi_{s,2} == 0,\ \chi_{s,3} == 0,\ \chi_{s,4} == 0\},$

$$\left\{\frac{\chi_{s,3}^2}{\chi_{s,4}} == \chi_{s,2},\ \chi_{s,4} == \chi_{s,1},\ \chi_{s,4} == \chi_{s,3},\ \chi_{s,4} \neq 0\right\},$$

$$\left\{\frac{\chi_{s,3}^2}{\chi_{s,4}} == \chi_{s,2},\ \chi_{s,4} == \chi_{s,1},\ -(-1)^{1/3}\,\chi_{s,4} == \chi_{s,3},\ \chi_{s,4} \neq 0\right\},$$

$$\left\{\frac{\chi_{s,3}^2}{\chi_{s,4}} == \chi_{s,2},\ \chi_{s,4} == \chi_{s,1},\ (-1)^{2/3}\,\chi_{s,4} == \chi_{s,3},\ \chi_{s,4} \neq 0\right\}\}$$

The all-zero solution is worthless. No species has dimension $\chi_{s,1} == 0$, so use this test to discard :

```
χEqs3 = Select[χEqs2, FreeQ[#, χs,1 == 0] &]
```

$$\{\{\chi_{s,2} == 0,\ \chi_{s,3} == 0,\ -3\,\chi_{s,4} == \chi_{s,1}\},$$

$$\left\{\frac{\chi_{s,3}^2}{\chi_{s,4}} == \chi_{s,2},\ \chi_{s,4} == \chi_{s,1},\ \chi_{s,4} == \chi_{s,3},\ \chi_{s,4} \neq 0\right\},$$

$$\left\{\frac{\chi_{s,3}^2}{\chi_{s,4}} == \chi_{s,2},\ \chi_{s,4} == \chi_{s,1},\ -(-1)^{1/3}\,\chi_{s,4} == \chi_{s,3},\ \chi_{s,4} \neq 0\right\},$$

$$\left\{\frac{\chi_{s,3}^2}{\chi_{s,4}} == \chi_{s,2},\ \chi_{s,4} == \chi_{s,1},\ (-1)^{2/3}\,\chi_{s,4} == \chi_{s,3},\ \chi_{s,4} \neq 0\right\}\}$$

The condition $\chi_{s,\_} \neq 0$ does not help with the solutions, so we might as well get rid of it too. Make a selection list based on the absence on an inequality:

```
selector = Map[FreeQ[#, a_ ≠ b_] &, χEqs3, {2}]
```

```
{{True, True, True}, {True, True, True, False},
 {True, True, True, False}, {True, True, True, False}}
```

Now **Pick** will pull out only those elements that correspond to a `True` :

```
χEqs = Pick[χEqs3, selector]
```

$$\{\{\chi_{s,2} == 0,\ \chi_{s,3} == 0,\ -3\,\chi_{s,4} == \chi_{s,1}\},$$

$$\left\{\frac{\chi_{s,3}^2}{\chi_{s,4}} == \chi_{s,2},\ \chi_{s,4} == \chi_{s,1},\ \chi_{s,4} == \chi_{s,3}\right\},$$

$$\left\{\frac{\chi_{s,3}^2}{\chi_{s,4}} == \chi_{s,2},\ \chi_{s,4} == \chi_{s,1},\ -(-1)^{1/3}\,\chi_{s,4} == \chi_{s,3}\right\},$$

$$\left\{\frac{\chi_{s,3}^2}{\chi_{s,4}} == \chi_{s,2},\ \chi_{s,4} == \chi_{s,1},\ (-1)^{2/3}\,\chi_{s,4} == \chi_{s,3}\right\}\}$$

Each equation set specifies the character vector for a different species. The number of equations sets should therefore be equal to **nS**, or equivalently, to **nC**:

```
Length[χEqs] == nC
```

```
True
```

Each equation set contains **nC-1** equations in **nC** variables, so we can only solve for **nC-1** variables. In all species the one we pick to remain unsolved is $\chi_{s,1}$, the dimension of the species. Here is how we implement this decision: Make a list called **unks**, the unknowns to solve for:

```
χVars = Table[χs,i, {i, 1, nC}];
unks = Drop[χVars, 1]
```

$\{\chi_{s,2}, \chi_{s,3}, \chi_{s,4}\}$

Now just call **Solve** :

```
solnsRaw =
 Map[Solve[#, unks][[1]] &, χEqs] // FullSimplify
```

$$\left\{\left\{\chi_{s,2} \to 0,\ \chi_{s,3} \to 0,\ \chi_{s,4} \to -\frac{\chi_{s,1}}{3}\right\},\right.$$
$$\{\chi_{s,2} \to \chi_{s,1},\ \chi_{s,3} \to \chi_{s,1},\ \chi_{s,4} \to \chi_{s,1}\},$$
$$\left\{\chi_{s,2} \to (-1)^{2/3}\,\chi_{s,1},\ \chi_{s,3} \to -(-1)^{1/3}\,\chi_{s,1},\ \chi_{s,4} \to \chi_{s,1}\right\},$$
$$\left.\left\{\chi_{s,2} \to -(-1)^{1/3}\,\chi_{s,1},\ \chi_{s,3} \to (-1)^{2/3}\,\chi_{s,1},\ \chi_{s,4} \to \chi_{s,1}\right\}\right\}$$

After all this work, it is good to check for consistency of the putative solution:

```
Table[χEqs[[i]] /. solnsRaw[[i]] // Simplify,
   {i, 1, nC}] // Flatten // Union
```

```
{True}
```

If you see `{True}` consistency is verified. The character table, still with unknown species dimension, is now in the form

```
unnormalizedChVecs = χVars /. solnsRaw // Simplify;
% // GridForm
```

$\chi_{s,1}$	0	0	$-\frac{\chi_{s,1}}{3}$
$\chi_{s,1}$	$\chi_{s,1}$	$\chi_{s,1}$	$\chi_{s,1}$
$\chi_{s,1}$	$(-1)^{2/3}\,\chi_{s,1}$	$-(-1)^{1/3}\,\chi_{s,1}$	$\chi_{s,1}$
$\chi_{s,1}$	$-(-1)^{1/3}\,\chi_{s,1}$	$(-1)^{2/3}\,\chi_{s,1}$	$\chi_{s,1}$

This is as far as we can go without more equations.

48.6 Use column orthogonality

In the first row we need $\chi_{s,1} \to \chi_{1,1}$; in the second row we need $\chi_{s,1} \to \chi_{2,1}$; etc. Each row is linear in $\chi_{s,1}$, so if we multiply each row by its own variable and the set $\chi_{s,1}$ to 1, we get just what we want :

```
χs1List = Table[χs,1, {s, 1, nC}];
χs1Mat = DiagonalMatrix[χs1List];
chVecs2 = χs1Mat.unnormalizedChVecs /. χs,1 → 1;
% // GridForm
```

$\chi_{1,1}$	0	0	$-\frac{\chi_{1,1}}{3}$
$\chi_{2,1}$	$\chi_{2,1}$	$\chi_{2,1}$	$\chi_{2,1}$
$\chi_{3,1}$	$(-1)^{2/3}\,\chi_{3,1}$	$-(-1)^{1/3}\,\chi_{3,1}$	$\chi_{3,1}$
$\chi_{4,1}$	$-(-1)^{1/3}\,\chi_{4,1}$	$(-1)^{2/3}\,\chi_{4,1}$	$\chi_{4,1}$

Now use column orthogonality:

$$\sum_{s=1}^{nS}\left(\sqrt{\frac{P_C}{h}}\;\chi^{\dagger}_{s,C}\right)\left(\sqrt{\frac{P_D}{h}}\;\chi_{s,D}\right) == \delta_{C,D}$$

where $\chi^{\dagger}_{s,C}$ is $\chi^{*}_{C,s}$. Let $\sqrt{\frac{P_D}{h}}\;\chi_{s,D}$ be **Xmat**. Then the left side is $\mathbf{Xmat}^{\dagger}.\mathbf{Xmat}$.

In words: If each column of the character table is weighted by the square root of the quantity (class population P_{class} divided by group order **h**), then the columns of **Xmat** are orthonormal in the complex sense.

```
h = GroupOrder[gp];
classWtFactors = Table[Sqrt[Pi/h], {i, 1, nC}] /. toPops
```

$$\left\{\frac{1}{2\sqrt{3}},\ \frac{1}{\sqrt{3}},\ \frac{1}{\sqrt{3}},\ \frac{1}{2}\right\}$$

Those powers of **-1** in **chVecs2** are generally complex, and we need to make them explicit, in the form **a_+ ⅈ b_**:

```
chVecs3 =
```

```
chVecs2 // ComplexExpand // Collect[#, χs_,1] &;
% // GridForm
```

$\chi_{1,1}$	0	0	$-\frac{\chi_{1,1}}{3}$
$\chi_{2,1}$	$\chi_{2,1}$	$\chi_{2,1}$	$\chi_{2,1}$
$\chi_{3,1}$	$\left(-\frac{1}{2}+\frac{i\sqrt{3}}{2}\right)\chi_{3,1}$	$\left(-\frac{1}{2}-\frac{i\sqrt{3}}{2}\right)\chi_{3,1}$	$\chi_{3,1}$
$\chi_{4,1}$	$\left(-\frac{1}{2}-\frac{i\sqrt{3}}{2}\right)\chi_{4,1}$	$\left(-\frac{1}{2}+\frac{i\sqrt{3}}{2}\right)\chi_{4,1}$	$\chi_{4,1}$

We must multiply each column (each class) by its own **classWtFactor**. We can accomplish this by multiplication with a diagonal matrix. But does it go on the left or the right? Make a test:

```
M = {{a, b, c}, {d, e, f}, {g, i, j}};
D = DiagonalMatrix[{A, B, C}];
{M.D, D.M} // GridList
```

{

a A	b B	c C
A d	B e	C f
A g	B i	C j

,

a A	A b	A c
B d	B e	B f
C g	C i	C j

}

Column multiplication is given by **M.D**; the diagonal matrix goes on the right:

```
Xmat = chVecs3.DiagonalMatrix[classWtFactors] //
    Simplify;
% // GridForm
```

$\frac{\chi_{1,1}}{2\sqrt{3}}$	0	0	$-\frac{\chi_{1,1}}{6}$
$\frac{\chi_{2,1}}{2\sqrt{3}}$	$\frac{\chi_{2,1}}{\sqrt{3}}$	$\frac{\chi_{2,1}}{\sqrt{3}}$	$\frac{\chi_{2,1}}{2}$
$\frac{\chi_{3,1}}{2\sqrt{3}}$	$\frac{i\left(i+\sqrt{3}\right)\chi_{3,1}}{2\sqrt{3}}$	$-\frac{i\left(-i+\sqrt{3}\right)\chi_{3,1}}{2\sqrt{3}}$	$\frac{\chi_{3,1}}{2}$
$\frac{\chi_{4,1}}{2\sqrt{3}}$	$-\frac{i\left(-i+\sqrt{3}\right)\chi_{4,1}}{2\sqrt{3}}$	$\frac{i\left(i+\sqrt{3}\right)\chi_{4,1}}{2\sqrt{3}}$	$\frac{\chi_{4,1}}{2}$

To take the dot products of the rows "in the complex sense", we need the conjugate transpose **Xmat**†, which we dot with **Xmat**:

```
Xmat† = Transpose[
    Xmat /. Complex[a_, b_] → Complex[a, -b]];
colOrthog = dotOp[Xmat, Xmat†] // ComplexExpand //
    ToRadicals;
% // GridForm
```

$\frac{\chi_{1,1}^2}{9}$	0	0	0
0	$\chi_{2,1}^2$	0	0
0	0	$\chi_{3,1}^2$	0
0	0	0	$\chi_{4,1}^2$

If it is not diagonal, something is wrong. By the column orthogonality theorem, it must be a unit matrix. So set all the diagonal elements to **1**, and **Solve**.

```
dimEqs = Thread[Diagonal[colOrthog] == 1]
```

$$\left\{\frac{\chi_{1,1}^2}{9} == 1,\ \chi_{2,1}^2 == 1,\ \chi_{3,1}^2 == 1,\ \chi_{4,1}^2 == 1\right\}$$

```
dimSolnsRaw = Map[Solve, dimEqs] // Flatten
```

$$\{\chi_{1,1} \to -3,\ \chi_{1,1} \to 3,\ \chi_{2,1} \to -1,\ \chi_{2,1} \to 1,\ \chi_{3,1} \to -1,\ \chi_{3,1} \to 1,\ \chi_{4,1} \to -1,\ \chi_{4,1} \to 1\}$$

The answer is the dimension of the species, which must always be positive. Select out the negative solutions :

```
dimSolns = Select[dimSolnsRaw, (#[[2]] > 0) &]
```

$$\{\chi_{1,1} \to 3,\ \chi_{2,1} \to 1,\ \chi_{3,1} \to 1,\ \chi_{4,1} \to 1\}$$

At last, here is our main goal, the character table :

```
ourChVecs1 = chVecs3 /. dimSolns // ComplexExpand;
% // GridForm
```

3	0	0	-1
1	1	1	1
1	$-\frac{1}{2}+\frac{i\sqrt{3}}{2}$	$-\frac{1}{2}-\frac{i\sqrt{3}}{2}$	1
1	$-\frac{1}{2}-\frac{i\sqrt{3}}{2}$	$-\frac{1}{2}+\frac{i\sqrt{3}}{2}$	1

48.7 Final formatting

Some groups (namely, $\mathbf{D_{5h}}$ and maybe others) need a little help with the formatting of expressions that contain $\sqrt{5}$. We provide it by the rule **noRoot5InDenom**:

```
noRoot5InDenom =
  {2/(1 + √5) → -1/2 + √5/2, -2/(1 + √5) → 1/2 - √5/2};
```

If **dotOp** is not **Dot**, but rather **PentaDot** or **IcosaDot**, the character vectors above may also contain other radical expressions that need to be brought into standard form. The following ruse does it, because **PentaDot** and **Icosadot** have the desired cleanup machinery inside them :

```
ourChVecs = (dotOp[IdentityMatrix[nC], ourChVecs1] //
    ComplexExpand) /. noRoot5InDenom;
% // GridForm
```

3	0	0	-1
1	1	1	1
1	$-\frac{1}{2}+\frac{i\sqrt{3}}{2}$	$-\frac{1}{2}-\frac{i\sqrt{3}}{2}$	1
1	$-\frac{1}{2}-\frac{i\sqrt{3}}{2}$	$-\frac{1}{2}+\frac{i\sqrt{3}}{2}$	1

This is the answer, with radicals (if any) in standard format. Do these rows all occur in the standard character table, and if so, in what order ?

```
stdChVecs = ChVecs[gp] // ComplexExpand;
speciesPerm =
 Map[Position[ourChVecs, #] &, stdChVecs] // Flatten
```

```
{2, 4, 3, 1}
```

If **speciesPerm** is a permutation of the first **nC** integers, there is complete agreement. So the character table in standard order is

```
ourChVecs[[speciesPerm]];
% // GridForm
```

1	1	1	1
1	$-\frac{1}{2}-\frac{i\sqrt{3}}{2}$	$-\frac{1}{2}+\frac{i\sqrt{3}}{2}$	1
1	$-\frac{1}{2}+\frac{i\sqrt{3}}{2}$	$-\frac{1}{2}-\frac{i\sqrt{3}}{2}$	1
3	0	0	-1

On your own
Go back to the top and change **gp** to any other group in the **GroupCatalog**. It should run down to here automatically.

Groups tested by the author: I, C_5, D_5, $D_{5\,h}$, $D_{5\,d}$, $D_{6\,h}$, $C_{3\,v}$

The slowest step is often the calculation of the multiplication table.

48.8 Character table of a new group

48.8.1 Make group D_8 and its MultiplicationTable

In the character table constructions above, we drew on the standard **CartRep** for the basic group of matrices, which was arranged by classes in conventional order. This made it easy to compare the calculated result with the standard result. Here we will show what else you have to do when dealing with a newly constructed group that has nothing in any conventional order. Our example will be group D_8, a reasonably useful group not included in our tabulations. (Our tabulation of the D_n groups stops at **n = 6**.) For a standard character table of this group, see Cotton, 3rd ed., p. 432, upper left quadrant of his table for $D_{8\,h}$.

Construct group D_8:

```
rawD8 =
  MakeGroup[Dot, {Cz[2 π / 8], Cx[2 π / 2]}] // Reverse;
```

Observe that our construction method has put the unit matrix first :

```
rawD8[[1]]
```

```
{{1, 0, 0}, {0, 1, 0}, {0, 0, 1}}
```

However, nothing else is in any particular order. Make its multiplication table :

```
rawMTD8 =
  MultiplicationTable[Dot, rawD8, Range[16]];
% // GridForm // Size[7]
```

1	2	3	4	5	6	7	8	9	10	11
2	1	4	3	6	5	8	7	10	9	12
3	5	1	7	2	9	4	11	6	13	8
4	6	2	8	1	10	3	12	5	14	7
5	3	7	1	9	2	11	4	13	6	15
6	4	8	2	10	1	12	3	14	5	16
7	9	5	11	3	13	1	15	2	16	4
8	10	6	12	4	14	2	16	1	15	3
9	7	11	5	13	3	15	1	16	2	14
10	8	12	6	14	4	16	2	15	1	13
11	13	9	15	7	16	5	14	3	12	1
12	14	10	16	8	15	6	13	4	11	2
13	11	15	9	16	7	14	5	12	3	10
14	12	16	10	15	8	13	6	11	4	9
15	16	13	14	11	12	9	10	7	8	5
16	15	14	13	12	11	10	9	8	7	6

This multiplication table is the sole source of information about the group that we use in the rest of this construction.

48.8.2 Operators that use only the MT

In order to create a character table for a group known only by its multiplication table, we will need an element multiplication operator and an element inverse operator that work by reference to only the raw multiplication table.

In the special preliminaries for this chapter, we define such operators. The multiplication operator is called **TimesByTbl[name1,name2,tbl]**. What does it predict for the value of **3 ⊗ 2**, using **rawMTD8** as the lookup table?

```
TimesByTbl[3, 2, rawMTD8]
```

```
5
```

To verify this, locate **3** in the first column and **2** in the first row, and look for the intersection of their respective row and column. It has to agree, because **TimesByTbl** looked it up just like you did.

We can also find inverses using **InverseByTbl[name_, tbl_]**:

```
inv9 = InverseByTbl[9, rawMTD8]
```

```
8
```

Check it with

```
TimesByTbl[9, inv9, rawMTD8] == 1
```

```
True
```

> **On your own**
> Check all inverses by running this check over the whole group.
> Hint: **Table** is one way.

The **TimesByTbl** and **InverseByTbl** operators were created so we could write **ClassifyByTbl** to find the classes of the group without knowing anything but its multiplication table. It works just like the **Classify** operator that you are familiar with :

```
rawCLD8 = ClassifyByTbl[rawMTD8]
```

```
{{1}, {2, 7, 10, 15}, {3, 6, 11, 14},
 {4, 5}, {8, 9}, {12, 13}, {16}}
```

As you see, this group is greatly in need of some tidying up. Normally, groups are listed so that classes come together. Here they are scattered all over.

48.8.3 Tidy the group up; name the classes

Making the group tidy is not necessary, but in this didactic exercise it helps in the end when we want to compare our calculated result to the known result. And even for totally untabulated groups you probably will want to name the classes and put them in some semi-logical order. That is what we do next.

The first element in each class is

```
classFirsts = Map[First, rawCLD8]
```

```
{1, 2, 3, 4, 8, 12, 16}
```

```
Map[RecognizeMatrix[rawD8[[#]]] &, classFirsts];
% // Column
```

{identity}
{rotation, 2-fold, {{1, 0, 0}, π}}
{rotation, 2-fold, {{$\frac{2+\sqrt{2}}{\sqrt{2}}$, 1, 0}, π}}
{rotation, 8-fold, {{0, 0, -1}, $\frac{\pi}{4}$}}
{rotation, 4-fold, {{0, 0, -1}, $\frac{\pi}{2}$}}
{rotation, 2.66667-fold, {{0, 0, -1}, $\frac{3\pi}{4}$}}
{rotation, 2-fold, {{0, 0, 1}, π}}

On the basis of these matrix types, we assign class names:

1. **E** the identity

2. $\mathbf{C}_2'$ four horizontal two-fold rotations, one being the **x**-axis
3. $\mathbf{C}_2''$ another four horizontal two-folds, between the $\mathbf{C}_2'$ axes
4. $\mathbf{C}_8$ two **z**-rotations by $\mathbf{1} \times \frac{2\pi}{8}$, one in each sense
5. $\mathbf{C}_4$ two **z**-rotations by $\mathbf{2} \times \frac{2\pi}{8}$, one in each sense
6. $\mathbf{C}_8^3$ two **z**-rotations by $\mathbf{3} \times \frac{2\pi}{8}$, one in each sense
7. $\mathbf{C}_2$ one **z**-rotation by $\mathbf{4} \times \frac{2\pi}{8}$

```
rawClassNames = {E, C2′, C2″, C8, C4, C8^3, C2}
```

$\{E, C_2', C_2'', C_8, C_4, C_8^3, C_2\}$

Cotton gives the order of these classes as **{1,4,6,5,7,2,3}**. It will be easier to verify the final result if we put the classes in this order.

```
newCL = rawCLD8[[{1, 4, 6, 5, 7, 2, 3}]]
```

```
{{1}, {4, 5}, {12, 13}, {8, 9},
 {16}, {2, 7, 10, 15}, {3, 6, 11, 14}}
```

Permute the names the same way :

```
newClassNames = rawClassNames[[{1, 4, 6, 5, 7, 2, 3}]]
```

$\{E, C_8, C_8^3, C_4, C_2, C_2', C_2''\}$

It is now easy to permute the list of matrices so that classes come together:

```
newElPerm = Flatten[newCL]
```

```
{1, 4, 5, 12, 13, 8, 9, 16, 2, 7, 10, 15, 3, 6, 11, 14}
```

```
newD8 = rawD8[[newElPerm]];
```

> **On your own**
> Run **RecognizeMatrix** over **newD8** to see that the group is in fact “tidy”.

With the group in a new order, we must redo its multiplication table:

```
newMTD8 =
  MultiplicationTable[Dot, newD8, Range[16]];
% // GridForm // Size[7]
```

1	2	3	4	5	6	7	8	9	10	11
2	6	1	8	7	4	3	5	14	13	16
3	1	7	6	8	2	5	4	13	15	14
4	8	6	7	1	5	2	3	16	14	15
5	7	8	1	6	3	4	2	15	16	13
6	4	2	5	3	8	1	7	11	9	12
7	3	5	2	4	1	8	6	10	12	9
8	5	4	3	2	7	6	1	12	11	10
9	13	14	15	16	10	11	12	1	6	7
10	15	13	16	14	12	9	11	7	1	8
11	14	16	13	15	9	12	10	6	8	1
12	16	15	14	13	11	10	9	8	7	6
13	10	9	12	11	15	14	16	3	2	5
14	9	11	10	12	13	16	15	2	4	3
15	12	10	11	9	16	13	14	5	3	4
16	11	12	9	10	14	15	13	4	5	2

Verify that the elements are now arranged by classes in conventional class order:

```
newCLD8 = ClassifyByTbl[newMTD8]
```

```
{{1}, {2, 3}, {4, 5}, {6, 7},
 {8}, {9, 10, 11, 12}, {13, 14, 15, 16}}
```

Now the classes are in Cotton' s order. The order of the elements within each class is still uncontrolled, but we do not care about that.

48.8.4 Character table for group D_8

To bring things to a rapid conclusion, we now copy into a single cell all the operations from Section 48.2 to Section 48.6, eliminating everything extraneous, so that it will go through with a single hit.

```
newCSMT = ClassSumMultiplicationTable[newMTD8];
nC = Length[newCSMT];
newCDvars = newCSMT[[1]];
newCSPC =
  Table[MatrixOfCoefficients[newCSMT[[i]], newCDvars], {i, 1, nC}];
Clear[c, a, b, m];
Do[c_{a_,b_,m_} := newCSPC[[a, b, m]], {a, 1, nC}, {b, 1, nC}, {m, 1, nC}];
Clear[P];
Plist = Table[P_i, {i, 1, nC}];
newCLD8 = ClassifyByTbl[newMTD8];
newPopList = Map[Length, newCLD8];
toPops = Thread[Plist → newPopList];
χEqsRaw =
  Union[Flatten[Table[P_i P_j χ_{s,i} χ_{s,j} == χ_{s,1} Sum_{k=1}^{nC} c_{i,j,k} P_k χ_{s,k}, {i, 1, nC},
      {j, 1, nC}] /. toPops]] // ExpandAll;
χEqsRaw2 = Select[χEqsRaw, # =!= True &];
reducedχEqsRaw = Reduce[χEqsRaw2] // LogicalExpand;
χEqs2 = reducedχEqsRaw /. Or → List /. And → List;
χEqs3 = Select[χEqs2, FreeQ[#, χ_{s,1} == 0] &];
selector = Map[FreeQ[#, a_ ≠ b_] &, χEqs3, {2}];
```

```
χEqs = Pick[χEqs3, selector];
χVars = Table[χs,i, {i, 1, nC}];
unks = Drop[χVars, 1];
solnsRaw = Map[Solve[#, unks][[1]] &, χEqs] // FullSimplify;
unnormalizedChVecs = χVars /. solnsRaw // Simplify;
χs1List = Table[χs,1, {s, 1, nC}];
χs1Mat = DiagonalMatrix[χs1List];
chVecs2 = χs1Mat.unnormalizedChVecs /. χs,1 → 1;
h = Length[newMTD8];
classWtFactors = Table[Sqrt[Pi/h], {i, 1, nC}] /. toPops;
Xmat = chVecs2.DiagonalMatrix[classWtFactors] // Simplify;
Xmat† = Transpose[Xmat /. Complex[a_, b_] → Complex[a, -b]];
colOrthog = Xmat.Xmat†;
dimEqs = Thread[Diagonal[colOrthog] == 1];
dimSolnsRaw = Map[Solve, dimEqs] // Flatten;
dimSolns = Select[dimSolnsRaw, (#[[2]] > 0) &];
ourChVecs = chVecs2 /. dimSolns // Reverse;
ourChTable1 = Prepend[ourChVecs, newClassNames];
% // GridForm
```

E	C_8	C_8^3	C_4	C_2	C_2'	C_2''
1	1	1	1	1	1	1
1	-1	-1	1	1	-1	1
1	1	1	1	1	-1	-1
1	-1	-1	1	1	1	-1
2	$\sqrt{2}$	$-\sqrt{2}$	0	-2	0	0
2	$-\sqrt{2}$	$\sqrt{2}$	0	-2	0	0
2	0	0	-2	2	0	0

Comparison with Cotton shows that the rows are correct, but of course the order of the rows was determined by the solving operators. For exact agreement with Cotton, we need to permute the rows.

```
rowPerm = {1, 2, 4, 5, 3, 6, 8, 7};
ourChTable = ourChTable1[[rowPerm]];
% // GridForm
```

E	C_8	C_8^3	C_4	C_2	C_2'	C_2''
1	1	1	1	1	1	1
1	1	1	1	1	-1	-1
1	-1	-1	1	1	1	-1
1	-1	-1	1	1	-1	1
2	$\sqrt{2}$	$-\sqrt{2}$	0	-2	0	0
2	0	0	-2	2	0	0
2	$-\sqrt{2}$	$\sqrt{2}$	0	-2	0	0

Eyeball comparison against Cotton shows exact agreement. You may want to name the species as well as the classes. For this, use the Mulliken names as much as possible, as expounded in Chapter 28. In this case they should be, respectively $\mathbf{A}_1$, $\mathbf{A}_2$, $\mathbf{B}_1$, $\mathbf{B}_2$, $\mathbf{E}_1$, $\mathbf{E}_2$, $\mathbf{E}_3$. But many physicists do not bother with the Mulliken names, assigning to species the meaningless names Γ_1, Γ_2, etc.

A1. Mathematica packages

This computational chapter may be read from the CD that accompanies the book

W.M. McClain, *Symmetry Theory in Molecular Physics with Mathematica*,
DOI 10.1007/b13137, © Springer Science+Business Media, LLC 2009

A2. `SymbolizeExpressions`

This computational chapter may be read from the CD that accompanies the book

A3. Matrix review

This computational chapter may be read from the CD that accompanies the book

B1. Bibliography

Preliminaries

B1.1. Wolfram Research, Inc.

1. See the WorldWideWeb site of Wolfram Research Inc., the developers of *Mathematica,* at http://www.wolfram.com .

2. Stephen Wolfram, **The *Mathematica* Book, 3rd ed**., 1403 pages. (Wolfram Media / Cambridge University Press, 1998). This entire book is available online in *Mathematica version 3.0* and later.

3. Wolfram MathWorld, "the web's most extensive mathematics resource. Created, developed & nurtured by Eric Weisstein, with input from the world's mathematics community".

B1.2. Applied group theory books

4. Eugene Wigner, **Group Theory and its application to the quantum mechanics of atomic spectra** , Academic Press (New York, N.Y.) 1959. First edition, in German, 1931.

5. Morton Hamermesh, **Group Theory and its application to physical problems**, Addison-Wesley, (Reading, MA) 1962.

6. Volker Heine, **Group Theory in Quantum Mechanics, An introduction to its present usage** , Pergamon Press (New York, NY) 1960.

7. Leo M. Falicov, **Group Theory and Its Physical Applications** , University of Chicago Press (Chicago, IL) 1966. Notable for its concise proofs of all the important theorems. Tragically out of print for many years. Good libraries have it.

8. F. Albert Cotton, **Chemical Applications of Group Theory** , John Wiley and Sons (New York, NY) First edition 1963, Third edition1990. Library of Congress number QD461.C65 1990.

9. George H. Duffey, **Applied Group Theory for Physicists and Chemists** , Prentice-Hall (Englewood Cliffs NJ) 1991. Library of

Congress number QD174.17.S9D84 1991. Duffey explains Cayley diagrams for groups, as well as class sums. These topics are missing from most group theory applications books.

10. Philip R. Bunker and Per Jensen, **Molecular Symmetry and Spectroscopy**, Second Edition, NRC Research Press, Ottawa (1998). Library of Congress number QD461.B86 1998. Bunker and Jensen are unique in emphasizing the permutation group approach to physical applications. This permits them to treat flexible molecules (molecules that invert, or that have internal rotors) which are impossible to treat under the standard point group approach.

B1.3. Mathematica books

For a comprehensive, up-to-the-minute listing of *Mathematica* tutorial books, go to http://store.wolfram.com/catalog/books/tut-ref.html , a part of the Wolfram website. Here are my personal favorites:

11. Stan Wagon, *Mathematica* in Action, 3rd ed. (Springer Verlag, 2008). A very current book making extensive use of the new Version 6 graphics and manipulations operators. Wagon is famous for his summer tutorials "Rocky Mountain *Mathematica*", in Frisco, Colorado. About 600 pages, with CD-ROM.

12. Nancy Blachman and Colin P. Williams, ***Mathematica*: A Practical Approach**, 363 pages. (Prentice Hall, 1999). Excellent for beginners; look for a forthcoming edition 2008 or later.

13. John W. Gray, **Mastering *Mathematica*: programming methods and applications** (Academic Press Professional, 1994). John W. Gray is Professor of Mathematics at the University of Illinois, and is the father of Theo Gray, chief designer of *Mathematica*'s Front End. This book starts from the beginning with *Mathematica* and carries on to some very sophisticated mathematical materials at the end. Of special interest to Symmetry Theory is his section on Polya pattern analysis.

14. And of course the grandfather of them all,
Stephen Wolfram, **The *Mathematica* Book, 3rd ed**. The definitive reference for pre-6 versions. See Section B1.1, above.

B1.4. Chemistry and Physics books

15. E. Bright Wilson, Jr., J. C. Decius, and Paul C. Cross, **Molecular Vibrations** (McGraw-Hill, 1955; Dover Publications, Inc. 1980). 388 pages. This may have been the first Chemistry book to make systematic use of group theory.

16. L. Pauling and E. B. Wilson, *Introduction to quantum mechanics* (McGraw-Hill, New York, 1935), Atomic term symbols are on pp. 230-234.

17. A. R. Edmonds, **Angular Momentum in Quantum Mechanics** (PrincetonUniversity Press, Princeton, N.J.) Second edition, 1960, 146 pages. A concise derivation and presentation of many complex formulas related to the full rotation group, remarkably free of error.

18. G. Herzberg, **Molecular Spectra and Molecular Structure,** 3 volumes, (Prentice-Hall, 1939)

19. Condon, E. U., and Shortley, G.H., **The Theory of Atomic Spectra** (Cambridge University Press (1935)). Their phase convention for the spherical harmonics has become standard, and is linked to the Euler angle convention used in this book.

20. Landau, L. D. and Lifschitz, E. M., **Quantum Mechanics - Non-relativistic Theory**, Second Ed., , Pergamon Press, Addison-Wesley Publishing Co., Reading, Mass. (1965).

21. Greiner, W. and Müller, B., **Quantum Mechanics - Symmetries**, Second Ed., Springer Verlag (1994).

22. Peter F. Bernath, **Spectra of Atoms and Molecules**, Oxford University Press (1995)

23. J.J. Sakurai, **Modern Quantum Mechanics**, Revised Ed., San Fu Tuan, ed., Addison Wesley Publishing Co. (1994). Wigner-Eckhardt is on pp. 238 - 242.

24. J.J. Sakurai, **Advanced Quantum Mechanics**, Benjamin/Cummings Publ. Co. (1967)

25. Hans A. Bethe, **Intermediate Quantum Mechanics**, based on notes by Roman Jackiw, W. A. Benjamin (1964)

26. G. F. Koster, J. O. Dimmock, R. G. Wheeler, and H. Statz, **Properties of the 32 Crystal Point Groups**, MIT Press (1963)

B1.5. Math books

27. Margenau, Henry, and Murphy, George M.,., **The Mathematics of Physics and Chemistry**, van Nostrand & Co., New York, N.Y.. (1943, 1955). Matrices and Matrix Algebra, Ch. 10.

28. Perlis, Sam., **Theory of Matrices**,, Addison-Wesley Publishing Co., Reading, Mass. (1952). Block diagonal multiplication, pp. 15-16.

29. Isaacs, I. Martin, **Character Theory of Finite Groups**, Academic Press, New York (1976)

30. Coxeter, H. S. M., **Regular Polytopes**, Methuen, London, (1948)

31. Jeffreys, Harold, Sir, **Cartesian Tensors**, Cambridge, Eng. : The University Press (1931), 92 p. (reprinted 1961) Succinct and powerful.

32. Buck, R, Creighton , **Advanced Calculus**, McGraw-Hill (1956)

B1.6. Data collections

33. Moore, Charlotte E., **Bibliography on the analyses of optical atomic spectra;** *Sec. I, Hydrogen - Vanadium; Sec. 2, Chromium-Niobium; Sec. 3, Molybdenum - Lanthanum and Hafnium - Actinium; Sec. 4, Lanthanum - Lutecium and Actinium - Einsteinium.* U.S. National Bureau of Standards; for sale by the Supt. of Docs., U.S. Govt. Print. Ofc., Washington, 1968-69.

B1.7. Research papers

34. A. M. Schönflies,
Über Gruppen von Transformationen des Raumes in sich,
Math. Ann. **34**, 172 –203 (1889)

35. L. Pauling,
J. Am. Chem. Soc. **53**, 1367 (1931)

36. M. Göppert-Mayer,
Ann.(Leipzig) 5, **9**, 273 (1931)

37. Longuet-Higgins, H. C.
Mol. Phys. **6**, 445 (1963)

38. W. M. McClain and R. A. Harris, in
Excited States, Volume 3, E. C, Lim, Ed.,
Academic Press. (1977) pp. 2-56.

39. Spiro, T. G. and Streakas, T. C. ,
Proc. Nat. Acad. Sci. U.S.A., **69**, 2622 (1972)
http://www.pnas.org/cgi/reprint/69/9/2622.pdf

40. Robert F. Curl and John E. Kilpatrick,
Am. J. Phys. **28**, 357-365 (1960)

41. W. Heisenberg, *Ann. Phys.* **120**, 888 (1931)

Index

Printed in the United States of America